Climate Science for Serving Society

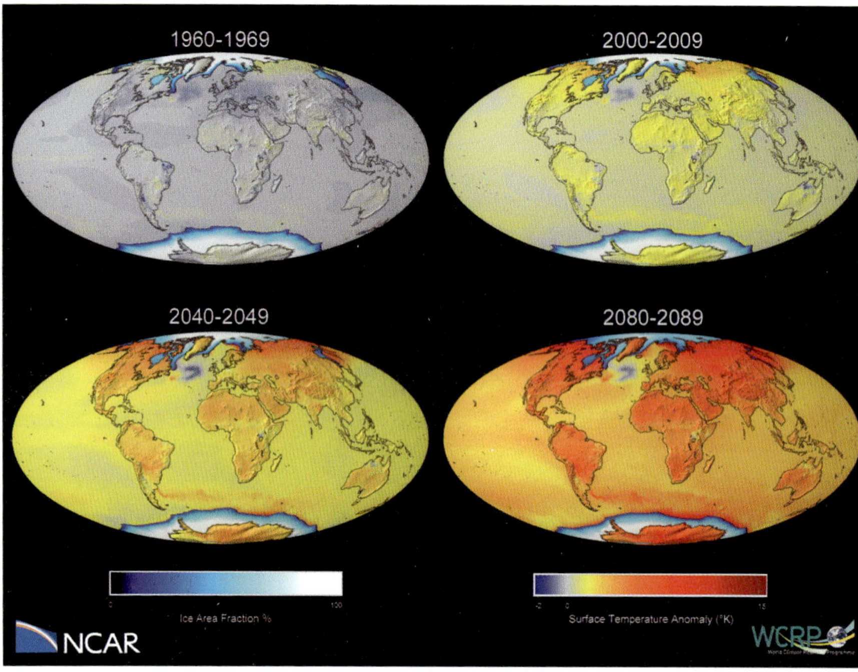

Decadal-mean, ensemble-average of near-surface air temperature (°C) and sea ice extent (%) anomalies simulated for historical and future conditions by the Community Climate System Model, version 4 (CCSM4). Future conditions are projected using the Representative Concentration Pathway 8.5 (RCP8.5) emission scenario. Anomalies are relative to 1850-1899 conditions, as simulated by six-member ensembles of CCSM4. The CCSM4 is one of more than 20 climate and Earth system models contributing to the Coupled Model Intercomparison Project Phase 5 (CMIP5) that was established by the World Climate Research Programme, through its Working Group on Coupled Modeling, as a standard experimental protocol for studying the output of coupled climate models. It provides a community-based infrastructure in support of climate model diagnosis, validation, documentation and data access, thus enabling a diverse community of scientists to analyze climate model output in a systematic fashion. Virtually, the entire international climate modeling community has participated in the CMIP project since its inception in 1995.

Ghassem R. Asrar • James W. Hurrell
Editors

Climate Science for Serving Society

Research, Modeling and Prediction Priorities

Springer
WCRP World Climate Research Programme
NCAR NATIONAL CENTER FOR ATMOSPHERIC RESEARCH

Editors
Ghassem R. Asrar
World Climate Research Programme
World Meteorological Organization
7 bis, Avenue de la Paix
1211 Geneva 2
Switzerland

James W. Hurrell
Earth System Laboratory
National Center for Atmospheric Research
3090 Center Green Drive
Boulder, CO 80301
USA

The cover image was developed for the World Climate Research Program (WCRP) Open Science Conference held on October 24 – 28, 2011 in Denver, Colorado, USA. It symbolizes the benefits of science-based climate information for decision-makers and citizens worldwide in addressing the opportunities and challenges associated with climate variability and changes.

ISBN 978-94-007-6691-4 ISBN 978-94-007-6692-1 (eBook)
DOI 10.1007/978-94-007-6692-1
Springer Dordrecht Heidelberg New York London

Library of Congress Control Number: 2013941743

Printed on acid-free paper

Springer is part of Springer Science+Business Media (www.springer.com)

Contents

List of Figures

Chapter 10

Chapter 11

Chapter 12

Chapter 14

List of Tables

Contributors

Arun Agrawal School of Natural Resources and Environment, University of Michigan, Ann Arbor, MI, USA

Domingo Alcaraz-Segura Department of Botany, University of Granada, Granada, Spain

Lisa V. Alexander Climate Change Research Centre and ARC Centre of Excellence for Climate System Science, The University of New South Wales, Sydney, NSW, Australia

M. Joan Alexander NorthWest Research Associates, Boulder, CO, USA

Rob Allan The Met Office, Hadley Centre, Exeter, Devon, UK

Myles Allen School of Geography and the Environment, University of Oxford, Oxford, UK

David Anderson The IDR Group, Lamb Chambers, the Chambers of John Cherry QC, London, UK

Richard A. Anthes University Corporation for Atmospheric Research, Boulder, CO, USA

Ghassem R. Asrar World Climate Research Program, World Meteorological Organization, Geneve 2, Switzerland

Alan Belward Institute for Environment and Sustainability, Joint Research Centre of the European Commission, Ispra, VA, Italy

Ernesto Hugo Berbery Earth System Science Interdisciplinary Center, University of Maryland, College Park, MD, USA

Eric Bertherat Pandemic and Epidemic Diseases Department, World Health Organization, Geneva, Switzerland

Eleanor M. Blyth The Centre for Ecology & Hydrology, Wallingford, Oxfordshire, UK

Sandrine Bony Laboratoire de Météorologie Dynamique/IPSL, CNRS, Pierre & Marie Curie University, Paris, France

Michael G. Bosilovich Global Modeling and Assimilation Office, Goddard Space Flight Center, NASA, Greenbelt, MD, USA

Pascale Braconnot Insitut Pierre Simon Laplace/Laboratoire des Sciences du Climat et de l'Environnement, unité mixte de recherches CEA-CNRS-UVSQ, Gif-Sur-Yvette, France

Hélène Broutin MIVEGEC, UMR 5290 CNRS- 224IRD- UM1- UM2 Centre de Recherche IRD, Montpellier, France

Andy R. Brown Met Office, Hadley Center, Exeter, UK

Otis B. Brown Cooperative Institute for Climate and Satellites – North Carolina (CICS-NC), North Carolina State University, Asheville, NC, USA

NOAA's National Climatic Data Center, Asheville, NC, USA

Gilbert Brunet Meteorological Research Division (MRD)/ASTD, Environment Canada, Dorval, QC, Canada

Antonio J. Busalacchi Earth System Science Interdisciplinary Center (ESSIC), University of Maryland, College Park, MD, USA

Nikolaos Christidis Met Office, Hadley Center, Exeter, UK

Amy Clement RSMAS/MPO, University of Miami, Miami, FL, USA

Heather Cooley Pacific Institute, Oakland, CA, USA

John S. Daniel Chemical Sciences Division, NOAA Earth System Research Laboratory, Boulder, CO, USA

Sean Davis NOAA ESRL Chemical Sciences Division, Cooperative Institute for Research in Environmental Sciences (CIRES), University of Colorado at Boulder, Boulder, CO, USA

Dick Dee European Centre for Medium Range Weather Forecasting (E.C.M.W.F.), Reading, UK

Clara Deser Climate & Global Dynamics, National Center for Atmospheric Research, Boulder, CO, USA

Randall M. Dole NOAA Earth System Research Laboratory, Boulder, CO, USA

Leo J. Donner Geophysical Fluid Dynamics Laboratory/NOAA, Princeton University, Princeton, NJ, USA

Jean-Louis Dufresne Laboratoire de Météorologie Dynamique/IPSL, CNRS, Pierre & Marie Curie University, Paris, France

Hallie Eakin Department of Geography, School of Sustainability, Arizona State University, Tempe, AZ, USA

Kerry A. Emanuel Massachusetts Institute for Technology, Cambridge, MA, USA

Nathan L. Engle School of Natural Resources and Environment, University of Michigan, Ann Arbor, MI, USA

Veronika Eyring Deutsches Zentrum für Luft- und Raumfahrt (DLR), Institut für Physik der Atmosphäre (IPA), Oberpfaffenhofen, Wessling, Germany

James S. Famiglietti Department of Earth System Science, University of California Center for Hydrologic Modeling (UCCHM), University of California, Irvine, Irvine, CA, USA

Graham Feingold Chemical Sciences Division, NOAA Earth System Research Laboratory, Boulder, CO, USA

E. Firth Pandemic and Epidemic Diseases Department, World Health Organization, Geneva, Switzerland

Pierre Friedlingstein College of Engineering, Mathematics and Physical Sciences, University of Exeter, Exeter, UK

Edwin P. Gerber Center for Atmosphere Ocean Science, Courant Institute of Mathematical Sciences, New York University, New York, NY, USA

Peter Gleckler Program for Climate Model Diagnosis and Intercomparison, Lawrence Livermore National Laboratory, Livermore, CA, USA

Peter H. Gleick Pacific Institute, Oakland, CA, USA

Hugues Goosse Earth and Life Institute, Centre de recherches sur la terre et le climat Georges Lemaître, SST/ELI/ELIC, Université catholique de Louvain, Louvain-la-Neuve, Belgium

B.N. Goswami Indian Institute of Tropical Meteorology, Pashan, India

Wojciech W. Grabowski Mesoscale and Microscale Meteorology Division, National Center for Atmospheric Research, Boulder, CO, USA

Stephen Griffies Geophysical Fluid Dynamics Laboratory, Princeton University, Princeton, NJ, USA

William Gutowski Department of Geological & Atmospheric Sciences, Department of Agronomy, Iowa State University, Ames, IA, USA

Ted Habermann National Environmental Satellite, Data, and Information Service (NESDIS) NOAA, Boulder, CO, USA

Gabriele C. Hegerl School of Geosciences, Grant Institute, University of Edinburgh, Edinburgh, UK

Isaac H. Held Geophysical Fluid Dynamics Laboratory, Princeton University, Princeton, NJ, USA

Harry Hendon Centre for Australian Weather and Climate Research, Bureau of Meteorology, Melbourne, VIC, Australia

Bruce Hewitson Department of Environmental & Geographical Science, Climate Systems Analysis Group (CSAG), University of Cape Town, Cape Town, South Africa

Martin Hoerling NOAA Earth System Research Laboratory, Boulder, CO, USA

Stephane Hugonnet Pandemic and Epidemic Diseases Department, World Health Organization, Geneva, Switzerland

Chris Huntingford The Centre for Ecology & Hydrology, Oxfordshire, UK

James W. Hurrell Earth System Laboratory, National Center for Atmospheric Research, Boulder, CO, USA

Michel Jancloes Health and Climate Foundation, Champ Courtet, Marchissy, Switzerland

Owen Johns School of Natural Resources and Environment, University of Michigan, Ann Arbor, MI, USA

Colin Jones Rossby Center, Swedish Meteorological and Hydrological Institute, Norrköping, Sweden

In-Sik Kang Atmospheric Science Program, School of Earth and Environmental Sciences, Seoul National University, Seoul, Korea

Thomas R. Karl National Environmental Satellite, Data, and Information Service (NESDIS), NOAA, Boulder, CO, USA

John Kennedy The Met Office, Hadley Centre, Exeter, Devon, UK

Ben Kirtman RSMAS/MPO, University of Miami, Miami, FL, USA

Thomas R. Knutson Geophysical Fluid Dynamics Laboratory, NOAA, Princeton University, Princeton, NJ, USA

Reto Knutti Institute for Atmospheric and Climate Science, Zürich, Switzerland

James P. Kossin NOAA's National Climatic Data Center, Asheville, NC, USA

Maria Carmen Lemos School of Natural Resources and Environment, University of Michigan, Ann Arbor, MI, USA

Dennis P. Lettenmaier Department of Civil and Environmental Engineering, University of Washington, Seattle, WA, USA

Piero Lionello Università del Salento – Dipartimento di Scienze e Tecnolog, University of Salento and CMCC, Lecce, Italy

Jose Antonio Marengo Centro de Ciencia do Sistema Terrestre (CCST), INPE, São Paulo, Brazil

Damon Matthews Department of Geography, Planning and Environment, Concordia University, Montreal, QC, Canada

Norm A. McFarlane Canadian Centre for Climate Modelling and Analysis, University of Victoria, Victoria, BC, Canada

Carlos R. Mechoso Department of Atmospheric and Oceanic Sciences, University of California, Los Angeles (UCLA), Los Angeles, CA, USA

Abere Mihretie Health, Development and Anti Malaria Association, Addis Ababa, Ethiopia

John F. Mitchell Met Office, Hadley Centre, Exeter, UK

Phillippe Naveau Laboratoire des Sciences du Climat et l'Environnement (LSCE), CNRS, Gif-sur-Yvette, France

Don R. Nelson Department of Anthropology, University of Georgia, Athens, GA, USA

Neville Nicholls School of Geography and Environmental Science, Monash University, Clayton, VIC, Australia

Slobodan Nickovic World Meteorological Organization, Geneva, Switzerland

Taikan Oki Institute of Industrial Science, The University of Tokyo, Meguro-ku, Tokyo, Japan

Alan O'Neill Department of Meteorology, National Centre for Earth Observation, The University of Reading, Reading, UK

Masami Onoda Group on Earth Observations, Geneva, Switzerland

Lawrence Berkeley National Lab, Berkeley, CA, USA

Pardeep Pall Lawrence Livermore National Laboratory, Berkeley, CA, USA

ETH Zurich, Institute for Atmospheric and Climate Science, Zurich, Switzerland

W. Perea Pandemic and Epidemic Diseases Department, World Health Organization, Geneva, Switzerland

Judith Perlwitz Physical Sciences Division, NOAA Earth System Research Laboratory, Boulder, CO, USA

University of Colorado, Boulder, CO, USA

Raymond T. Pierrehumbert Department of the Geophysical Sciences, The University of Chicago, Chicago, IL, USA

Chris C. Reason Department of Oceanography, University of Cape Town, Cape Town, South Africa

Daniel Rosenfeld Institute of Earth Sciences, The Hebrew University of Jerusalem, Jerusalem, Israel

Karen H. Rosenlof Chemical Sciences Division, NOAA Earth Systems Research Laboratory, Boulder, CO, USA

Steve Running Numerical Terradynamic Simulation Group (NTSG), The University of Montana, Missoula, MT, USA

Barbara Ryan Director, GEO Secretariat, Geneva, Switzerland

Adam A. Scaife Met Office, Hadley Center, Exeter, UK

Christoph Schär Institute for Atmospheric and Climate Science, ETH Zurich, Zurich, Switzerland

Sonia I. Seneviratne Institute for Atmospheric and Climate Science, ETH Zurich, Zurich, Switzerland

Catherine Senior Met Office, Hadley Center, Exeter, UK

Steven C. Sherwood Climate Change Research Centre (CCRC) and Australian Research Council (ARC), Centre of Excellence for Earth Systems Science, The University of New South Wales, Sydney, NSW, Australia

Doug Smith Met Office, Hadley Center, Exeter, UK

Susan Solomon Department of Earth, Atmospheric, and Planetary Sciences, Massachusetts Institute of Technology, Cambridge, MA, USA

Department of Atmospheric and Oceanic Sciences, University of Colorado, Boulder, CO, USA

S. Sow Center for Vaccine Development, CVD-Mali, Centre National D'Appui A La Lutte Contre La Maladie, Ministère de la Santé, Institut Marchoux, Bamako, Mali

Bjorn Stevens Max Planck Institute for Meteorology, Hamburg, Germany

Dáithí Stone Lawrence Livermore National Laboratory, Berkeley, CA, USA

University of Cape Town, Cape Town, South Africa

Peter A. Stott Met Office, Hadley Center, Exeter, UK

Michael Tanner National Environmental Satellite, Data, and Information Service (NESDIS), NOAA, Boulder, CO, USA

Joao Teixeira Jet Propulsion Laboratory, Pasadena, CA, USA

Laurent Terray Sciences de l'Univers au CERFACS URA1875, CERFACS/CNRS, Toulouse, France

Madeleine C. Thomson International Research Institute for Climate and Society, The Earth Institute, Columbia University, Palisades, NY, USA

Chris D. Thorncroft Department of Atmospheric and Environmental Sciences, University at Albany, SUNY, Albany, NY, USA

Kevin E. Trenberth Climate and Global Dynamics, National Center for Atmospheric Research, Boulder, CO, USA

Bart van den Hurk Royal Netherlands Meteorological Institute (KNMI), De Bilt, The Netherlands

Carolina Vera Centro de Investigaciones del Mar y la Atmosfera (CIMA/CONICET-UBA), DCAO/FCEN, UMI IFAECI/CNRS, Buenos Aires, Argentina

Charles J. Vörösmarty Department of Civil Engineering, City University of New York, New York, NY, USA

Bruce Wielicki NASA Langley Research Center, Hampton, VA, USA

Eric F. Wood Department of Civil and Environmental Engineering, Princeton University, Princeton, NJ, USA

Robert Wood Atmospheric Sciences, University of Washington, Seattle, WA, USA

Xuebin Zhang Climate Research Division, Environment Canada, Toronto, ON, Canada

Francis W. Zwiers Pacific Climate Impacts Consortium, University of Victoria, Victoria, BC, Canada

The World Climate Research Program Strategy and Priorities: Next Decade

Ghassem R. Asrar, James W. Hurrell, and Antonio J. Busalacchi

Abstract In this chapter an overview of the research planning and priorities of the World Climate Research Program (WCRP) over the next decade is provided. The research, modeling and prediction plans are significantly shaped by the major sponsors of the WCRP, as well as by its international network of scientists and stakeholders. However, major input into the planning process was also derived from sessions and discussions among the more than 1,900 scientists who attended the WCRP Open Science Conference (OSC) in October 2011. This monograph is comprised of position papers emanating from the OSC. They address many of the overall research and intellectual challenges across the WCRP spectrum of activities. A brief overview of these papers is given.

Keywords Climate research • Climate information • Actionable research • Research priorities • Climate services • Future Earth

G.R. Asrar (✉)
World Climate Research Program, World Meteorological Organization,
7 bis Avenue de la Paix, Case Postale No 2300, CH-1211 Geneve 2, Switzerland
e-mail: gasrar@wmo.int

J.W. Hurrell
Earth System Laboratory, National Center for Atmospheric Research,
PO Box 3000, Boulder, CO 80307-3000, USA
e-mail: jhurrell@ucar.edu

A.J. Busalacchi
Earth System Science Interdisciplinary Center (ESSIC),
University of Maryland, Suite 4001, M Square Office Building, #950,
5825 University Research Court, College Park, MD 20740, USA
e-mail: tonyb@essic.umd.edu

G.R. Asrar and J.W. Hurrell (eds.), *Climate Science for Serving Society: Research, Modeling and Prediction Priorities*, DOI 10.1007/978-94-007-6692-1_1,

1 Introduction

The WCRP convened a major Open Science Conference (OSC) on 24–28 October 2011, in Denver, Colorado (Asrar et al. 2012). The purpose of the OSC was to assess the current state of knowledge on climate variability and change, identify the most urgent scientific issues and research challenges, and ascertain how the WCRP can best facilitate research and develop partnerships critical for progress. The OSC also sought to facilitate dialogue and cooperation across the diverse research communities among the WCRP Projects and their network of researchers, as well as with other international research programs, including the International Geosphere-Biosphere Program (IGBP), the World Weather Research Program (WWRP) and the Earth System Science Partnership (ESSP).

The overall theme of conference was "Climate Science in Service to Society" to allow a more effective dialogue between the climate information and knowledge developers (i.e., the research community) and the decision makers faced with difficult adaptation, mitigation and risk management issues. A main goal was to identify key opportunities and challenges in observations, modeling, analysis and process research required to understand and predict Earth system variability and change. The main objectives for the WCRP since its inception have been to determine the predictability of climate and to determine the effect of human activities on climate. The OSC confirmed that these remain valid objectives today, along with the WCRP strategic priority of an enhanced focus on climate research that is of direct value and benefit to society.

More than 1,900 participants, including 541 graduate students and early career scientists from 86 nations and more than 300 scientists from developing nations, made the conference a terrific success. The conference included seven plenary sessions, 15 parallel sessions and more than 2,000 poster presentations organized around daily themes (e.g., societal needs for climate information, the state of the global climate observing system, challenges and opportunities in climate modeling and prediction, and the detection and attribution of climate extremes). The sessions were designed to allow for in-depth plenary presentations informed by a series of community-based scientific position papers, followed by parallel and poster sessions with sufficient time for discussion and one-on-one interactions among presenters and participants.

This monograph is developed around the papers prepared and presented at the Open Science Conference based on contributions by a large number of international climate scientists at the invitation of the WCRP. The conference participants and members of the broader scientific community were invited to provide their comments and feedback to the authors before they were presented and discussed during the conference. Prior to acceptance for inclusion in this book, each paper was revised based on these feedbacks, the outcome of conference deliberations, and at least three independent peer reviews. The scientific challenges and opportunities identified in the following chapters form the basis of climate research priorities for WCRP to pursue through its network of affiliated projects and scientists in the ensuing decade.

2 Evolution of WCRP Research Mandate

The WCRP was established at the conclusion of the first World Climate Conference in 1980, in Geneva, Switzerland. Under the sponsorship of the World Meteorological Organization (WMO), the Intergovernmental Oceanographic Commission (IOC) of UNESCO, and the International Council for Science (ICSU), the aim of WCRP is to facilitate the analysis and prediction of Earth's climate system variability and change for use in an increasing range of practical applications of direct relevance, benefit, and value to society (Asrar 2009). This primary objective was reaffirmed in 2005 with a new strategic framework for WCRP referred to as COPES (Coordinated Observation and Prediction of the Earth System). COPES was designed to help promote the creation of a comprehensive, reliable, end-to-end global climate observing system for the dual purpose of describing the structure and variability of the climate system, and of generating a physically consistent description of the state of the coupled climate system for numerical prediction of climate. Through this framework WCRP strives to provide the soundest possible scientific basis for the predictive capability of the total climate system for the benefit of society, including an assessment of the inherent uncertainty in probabilistic prediction of climate on various space and time scales (WCRP 2005).

In 2008, the sponsors of the World Climate Research Program (WCRP) initiated an independent review of the Program to evaluate the extent to which WCRP adds value to national efforts in climate research. The international panel of experts also identified and recommended major future research priorities for WCRP (ICSU-IGFA 2009). The Review Panel took a prospective view with the aim of maximizing the future added value of WCRP while learning from a retrospective assessment of the current program and its recent evolution. The Review focused on interactions within WCRP and also its external connections. The Review Panel considered not just science relevance, but also the policy and development relevance of WCRP. Questions considered by the review panel included: "What is the role of natural versus social science? Does the Program engage the younger generation of scientists? What is the relationship of WCRP to the Earth System Science Partnership? Is the increasing collaboration between IGBP and WCRP an impetus for even tighter working relationships? What do end-users serviced by members of the sponsoring international organizations expect from WCRP?"

Upon completion of their review and deliberations, the panel recognized the many important achievements of WCRP, and it concluded that WCRP plays a significant role in helping society meet the challenges of global climate change. The panel cited numerous examples of unique contributions by the program and pointed out that, without WCRP leadership, such contributions would not have been possible. The panel concluded, however, that WCRP lacked the focus, planning, and funding to meet the scientific challenges identified in the COPES document. A recommendation was "WCRP must focus its Projects and connect with partners and users in strategic ways, and it will need new resources to do so". The panel indicated that WCRP should continue to stay in the forefront of climate research, modeling and prediction, in order to attract international research leaders to volunteer their time and efforts to support the Program. The panel also concluded excellence in research alone is not sufficient: WCRP also needs to facilitate the use of the scientific knowledge it develops by decision makers to demonstrate the benefits to be accrued by its sponsors, stakeholders and the society at large. In this regard, the panel offered several specific recommendations aimed at building the necessary focus and connections into WCRP and its partnerships (ICSU-IGFA 2009).

In parallel with the independent review of WCRP, the Joint Scientific Committee (JSC), who has the scientific oversight for the Program, began to engage in a dialogue with the WCRP International Projects and other partner programs (e.g., IGBP, WWRP, ESSP) to develop a strategy and implementation plan for the future. The WCRP International Projects are: Global Energy and Water Experiment (GEWEX), CLImate VARiability and Change (CLIVAR), Climate and Cryosphere (CliC) and Stratospheric Processes And their Role in Climate (SPARC). The JSC and the leaders of these Projects reached agreement in short order to adopt the WCRP COPES strategic framework as a blueprint for the near-term (2008–2012) and worked to develop an accompanying implementation plan (WCRP 2009b). This approach offered sufficient time for developing a longer-term (post 2012) strategy through consultation with the international scientific community at large, and for transitioning the functions and structure of the four core Projects to support future research directions and emerging priorities. The logic behind this two-phase approach was to foster active community engagement in setting the new research directions and priorities through a "bottoms up" and participatory approach, a necessity given the voluntary nature of the Program.

3 Future Plans and Priorities of WCRP Major Sponsors

To set the stage for its deliberations, the JSC also sought guidance from the three major WCRP sponsors. This was especially important because the three sponsors had also initiated their own future planning through separate processes.

The WMO convened the World Climate Conference-3 (WCC-3) in September 2009, in Geneva, Switzerland, about 30 years after the first conference (WCC-1) that established the WCRP. More than 2,500 scientists, policymakers, Non-Governmental Organizations (NGOs) and a wide range of regional and national experts from 150 countries participated. The scientific and technical part of the conference, "the Expert Segment," identified an urgent need for establishing the Global Framework for Climate Services (GFCS) as a complement to the research, observations and assessment initiatives that resulted from the WCC-1 and WCC-2. The Expert Segment called for major strengthening and implementation of the essential elements of a global framework for climate services; however, they stated that for a GFCS to be successful it must function as an integrated and end-to-end system, with the main focus on delivery of climate information and services to the end users and stakeholders of such information. The five essential elements of the GFCS were identified as:

- The Global Climate Observing System and all its components and associated activities; and provision of free and unrestricted exchange and access to climate data;

- The World Climate Research Program, underpinned by adequate computing resources and increased interaction with other global climate relevant research initiatives;
- Climate services information systems taking advantage of enhanced existing national and international climate service arrangements in the delivery of products, including sector oriented information to support adaptation activities;
- Climate user interface mechanisms that are focused on building links and integrating information, at all levels, between the providers and users of climate services, and that are aimed at the development and efficient use of climate information products including the support of adaptation activities; and
- Efficient and enduring capacity building through education, training, and strengthened outreach and communication.

In 2011, the WMO Congress endorsed the GFCS initiative as a UN system-wide effort and requested that WMO establish a secretariat to facilitate the development of an implementation plan and a governance strategy to be presented for further deliberation and approval at an extra ordinary session of its member states in October 2012 in Geneva, Switzerland. In this process, WCRP was identified as a major contributor to the development of the Research, Modeling and Prediction pillar of the GFCS (Fig. 1), in coordination and with active engagement of other major international research programs (WMO 2011). The GFCS Implementation Plan that was recently approved by the WMO Congress includes an annex chapter that highlights research priorities and WCRP expected contributions for the GFCS during next decade (WMO 2012).

In September 2009, the ICSU General Assembly decided to develop a future strategy for ICSU environmental research. Subsequently, ICSU established an international planning panel to initiate this process through a combination of web-based consultation and several planning workshops during 2010–2011. The ICSU visioning process concluded that there is a need to expand Earth System Science research towards sustainability science that explicitly covers development and equity in relation to other global change issues. The visioning process led to the identification of five Grand Challenges for the international global change research communities (Table 1, ICSU-ISSC 2010; Reid et al. 2010). Simultaneously, the international funding agencies developed the Belmont Challenges (ICSU-ISSC 2010) with a strong regional and sectoral emphasis and user orientation.

The ICSU strategy calls for trans-disciplinary research towards the goal of global sustainability by recognizing a need for end-to-end and solution oriented approaches to Earth system research through more effective partnerships between natural, socioeconomic sciences, and engineering disciplines (ICSU-ISSC 2010). In 2011, ICSU appointed a Transition Team consisting of international scientists and experts to develop further its research strategy and a governance mechanism for its oversight. The ICSU initiative is entitled "Future Earth: Research for Global Sustainability". It is envisioned Future Earth will succeed ESSP (Leemans et al. 2009). A subset of the International Group of Funding Agencies (IGFA), called the Belmont Forum, together with some UN system organizations and ICSU, have established an Alliance to oversee the governance and implementation of the Future Earth initiative.

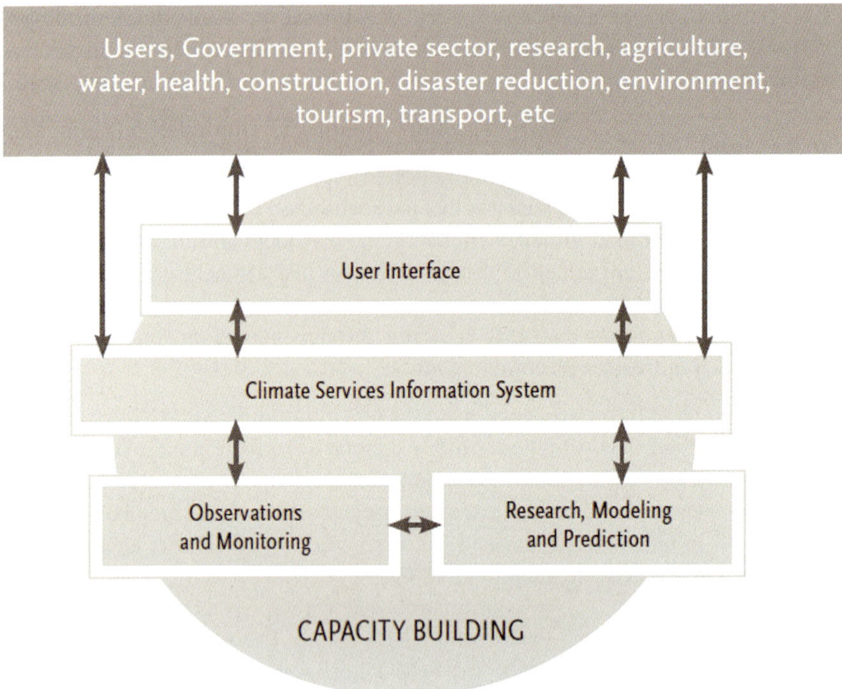

Fig. 1 A conceptual illustration of how the five major pillars of the Global Framework for Climate Services (GFCS) must function in a coordinated and integrated manner to realize the GFCS grand vision in near-, mid- and long-term (Adapted from WMO 2011)

Table 1 The five Grand Challenges according to the International Council of Science (ICSU 2010)

Challenge 1	Forecasting: improve the usefulness of forecasts of future environmental conditions and their consequences for people
Challenge 2	Observing: develop, enhance and integrate the observation systems needed to manage global and regional environmental change
Challenge 3	Confining: determine how to anticipate, avoid and manage disruptive global environmental change
Challenge 4	Responding: determine what institutional, economic and behavioral changes can enable effective steps toward global sustainability
Challenge 5	Innovating: encourage innovation (coupled with sound mechanisms for evaluation) in developing technological, policy, and social responses to achieve global sustainability

The Future Earth initiative has the strategic goal "to develop knowledge for solutions to move toward a future of integrated environmental, social, and economic well-being *[or…provide new knowledge to face risks and seize opportunities posed by global environmental change and to support transitions of societies in the world to global sustainability]*". The intent is to deliver at global and regional scales the knowledge required for sustained human development in an era of rapidly escalating global environmental risks and new opportunities. The Future Earth will include major foci in synthesis and assessment of research knowledge, communication, bridging science with policy and practice, and capacity development such as training and development of the next generation of early career scientists, especially those from developing countries. These exciting goals and objectives will require a "step change" in required funding, both flexible institutional funding and competitive research funding (ICSU-ISSC 2010).

The IOC/UNESCO, a third sponsor of WCRP, organized its second OceanObs09 Conference in 2009, in Venice, Italy to define its strategy for ocean observations and research in the ensuing decades (Hall et al. 2009). The conference objective was to take stock of progress made since the first OceanObs Conference in 1999 to identify future observations and research priorities in support of the IOC mission. More than 800 participants from 36 countries confirmed significant progress on the goals and priorities identified in the first conference. These included the establishment of global in situ observing networks in the oceans, development and operation of space-based ocean observing satellites through international cooperation, and test-beds for near real-time ocean information activities such as GODAE. The participants recognized the seminal contributions of the WCRP to the ocean observing systems through major research projects such as World Oceans Climate Experiment (WOCE), Tropical Oceans Global Atmosphere (TOGA), and the regional observing systems through the CLIVAR project. The conference identified the following priorities through a conference statement (Hall et al. 2009);

> Provision of routine and sustained global information on the marine environment sufficient to meet society's needs for describing, understanding and forecasting marine variability (including physical, biogeochemical, ecosystems and living marine resources), weather, seasonal to decadal climate variability, climate change, sustainable management of living marine resources, and assessment of longer term trends. The ocean observing system must be sustained and enhanced because; 1) systematic observation of the properties of the ocean and the information derived are changing what we know about the ocean and its implications for society; 2) the real-time flow of these observations underpin the development, production, and delivery of many ocean services and support coastal zone management; 3) global oceans information is critical to support forecasting of climate, weather and natural hazards from daily to centennial time scales; 4) the development of an increasing range of ocean assessments and climate services for planning, early warning, adaptation and mitigation, depend upon availability of accurate observations and models of the world ocean; 5) the ocean is an important sink of anthropogenic CO2, and ocean acidification potentially has significant impacts on marine ecosystems; 6) sustainable management of marine living resources depends on timely and accurate monitoring of and information on biogeochemical cycles and ecosystem function; 7) biodiversity is understood to be a key factor in ensuring sustainable ecosystem function; 8) healthy coastal environments and their interactions with the open ocean are important to society; and 9) the oceans remain seriously under-sampled, and no single nation can perform all necessary ocean observations.

4 Overview of Following Chapters

The following chapters consist of the key position papers that were prepared and presented at the OSC. Each paper addresses the overall research and intellectual challenges of the topic it covers. Together, the papers present an excellent assessment of the current state of knowledge on climate variability and change, identify the scientific challenges and most urgent research issues, and ascertain how the WCRP can best facilitate research and develop partnerships critical for progress during the next decade. Each paper benefited from comments and feedback from at least three independent reviewers. The following chapters are organized according to the major scientific themes of the OSC.

The Climate System Components and Their Interactions: The presentations in this part of the conference focused on the need for WCRP to remain focused on facilitating the discovery of key processes in each component of the Earth's climate system and the interactions among them. They also referred to the need to represent these processes in models and thereby provide the basis for improved predictions of climate variability and change. Talks emphasized, for instance, significant shortcomings in our understanding of cloud and convection-related processes, as well as uncertainties in aerosol radiative effects (Sherwood et al. and Rosenfeld et al., this volume). Gaps in our understanding of the complexity of the hydrological cycle and human influences on the character and dynamics of it were discussed, as well as the fact that these are central to our understanding of many other atmospheric, chemical, and physical processes (Gleick et al., this volume). The many research achievements, yet remaining challenges, around land-atmosphere coupling were also presented and discussed (Oki et al., this volume).

Observation and Analysis of the Climate System: Presentations in these sessions outlined significant progress over the past three decades in observations of all of the major Earth system domains (i.e., atmosphere, oceans, land and polar region). These observations are increasingly needed for planning and informed decision making related to climate services in the broadest sense. However, data gaps and other major challenges still exist, such as how best to deal with the continually changing observing system, especially from satellites, in order to provide a continuous climate record (Trenberth et al., this volume). Since most observing systems were not developed with a climate objective in mind, tremendous efforts have gone into assessing and reprocessing data records. Recent progress in reprocessing and reanalyzing observations, as well as existing challenges and next steps in these efforts, were described and discussed by OSC participants (e.g., Bosilovich et al., this volume).

Assessing and Improving Models and Predictive Capabilities: Climate and Earth System models are getting more realistic, comprehensive and capable to deliver short-, medium- and long-term predictions and projections to users. The scientific basis for prediction from weeks to decades, current capabilities and outstanding challenges were highlighted in these sessions (e.g., Kirtman et al., this volume), as were the reliability of models used for longer-term climate projections and the power of multi-model ensembles (van den Hurk et al., this volume). The presentations also emphasized, however, the need for WCRP to continue to facilitate comprehensive and coordinated model evaluation, especially in light of the long list of systematic errors that plague all models, and the importance of continued development of the "foundations" of Earth System models, namely the atmospheric, oceanic, and land components. Overall, the sessions re-confirmed the major outcomes and recommendations of the WCRP Modeling Summit (WCRP 2009a).

Climate Assessments and Future Challenges: The speakers in these sessions focused both on science-based climate and environment assessments and strengthening the policy relevance of such assessments by highlighting the need for research on anthropogenic climate change (Bony et al., this volume; Solomon et al. this volume). Priority should be given to understanding the processes and mechanisms responsible for climate variability and predictability at regional scales (Vera et al., this volume; Rosenlof et al., this volume). Promoting research on detection and attribution of extreme events and research on heat waves, tornadoes, extreme precipitation and tropical cyclones will provide a solid scientific foundation to improve their prediction (Stott et al., this volume; Zwiers et al., this volume). It was also stated that cryospheric research is rich with grand challenges and recommended that WCRP continue to promote research, modeling and analysis for addressing large uncertainties in knowledge of ice sheet mass balance for sea-level change and variability, sea-ice dynamics, and changes in solid precipitation in a changing climate, with a major regional focus.

Translating Scientific Understanding of Climate System into Climate Information for Decision Makers: The speakers in the final plenary session of the Conference focused on the development and use of climate knowledge and information for socio-economic development and societal services. They stated that, in less developed countries, there is a direct relationship between building adaptive capacity and development (Lemos et al., this volume). Many of the causes of vulnerability are connected to development deficits, which calls for a new paradigm of adaptive development and the requirement for countries to solve some of their development problems in a context of climate and environmental change. The ICSU vision for "Future Earth" initiative and associated Grand Challenges (Reid et al. 2010), for example, states that the global sustainability is a prerequisite for poverty alleviation. The concept of "planetary boundary thresholds" (Rockstrom et al. 2009) calls for innovative pathways for societal transformations to ensure global sustainability, and science-based planetary stewardship for human prosperity.

Climate services are a necessary element of this transformation, and they have the potential to bridge communities, language and value systems. The language of uncertainty employed in the climate information provision, however, casts doubts among users, and there is a merit for informing them based on the concept of probabilities and likelihood. Providing climate information that meet or exceed in time, space, and frequency the user-defined needs is powerful. The issues of responsibility, accountability, credibility, and values are largely missing from the climate services dialogue. To put them in place, producers and users of climate information need to collectively develop the language that leads to plausible, defendable and actionable messages. For example, in the area of climate and health, projects like Meningitis Environmental Risk Information Technologies (MERIT) (Thompson et al. 2006) is showing significant achievements in implementing health–climate alliance. The basis for joint action is the agreement among the stakeholders on the corresponding evidence, stemming from a strategic approach to the creation of the evidence, together with the development of a cumulative knowledge base, effective dissemination of knowledge, with development of effective means of access to knowledge, and resulting in initiatives to increase the uptake of evidence in both policy and practice.

Meeting the Climate Information Needs of Decision Makers: A clear and emerging priority from the OSC was the need for "actionable" science. The consensus that emerged at the conference was that the number one service to society provided by WCRP is the encouragement and enabling of the climate-related research that will provide the scientific basis for sound decision making over the next decades. WCRP should also help in developing an interface between climate information and its use for a particular application. The main challenge in the discussion of science support to climate services is the optimal balance of fundamental and applied research, and interface between climate research and climate information and users. Given the still existing gap between science and decision-making, the need to understand and use deliverables of climate research outcomes become even more important than in the past.

The reality is that decision-makers – including water providers, farmers, insurance companies, oil exploration companies and many more – need climate and other scientific information to guide decisions more than ever before. Future water availability in a region, for example, may guide crop selection for the ensuing year, or siting decisions for a new water treatment plant that will be operational for decades. But there is often a mismatch between the scientific data available and the information needed, so there is a need for "symbiotic" relationships between providers and users of climate information to ensure that 'actionable' (timely, accessible, and easy to understand) climate information is developed and used effectively.

The need for actionable science was also explored during an evening session with a panel of experts from the private sector. Sponsored by the University of Maryland's Earth System Science Interdisciplinary Center, the panelists were senior executives of several major companies – BP, Northrup Grumman, Zurich Financial Services, Computer Sciences Corporation and the Weather Channel. The discussion focused on the need for scientists and the private enterprise to work better together toward actionable information. They shared the perspective that conversations during scientific conferences such as the OSC were a start, in that awareness of the richness of the scientific data and information available is gained by users while scientists begin to understand what kind of data and information is required to guide business and policy decisions. Participants agreed that while gaps exist today between data and information needs and availability, those gaps are rich with opportunity. In particular, there is the chance for private companies to use the data created by the research community to deliver more detailed, relevant information to decision-makers. The participants stated the need now is to go beyond understanding natural systems alone into understanding how natural systems connect with human systems, and that requires understanding the kinds of information the users need to make decisions. A major point made repeatedly by the users of climate information throughout the conference was the importance of their early and continued engagement with scientists to define the needs from science, the type of information required, and the most effective way to convey the results to inform actionable decisions. The "symbiotic" relationship between producers and users of climate information was identified as the best model to ensure such effective partnership.

5 Summary

As a result of the week's deliberations at the OSC, several major scientific themes and priorities emerged from the conference presentations and discussions. They include: (1) the need for prediction of the Earth System bridging the physical climate system with biogeochemistry, the social sciences, and human dimensions, a problem that transcends the WCRP and one that should benefit from the proposed Earth System Science alliance; (2) the opportunity, provided by new satellite observations, to make a quantum leap in understanding of clouds and aerosols and their contributions to climate sensitivity; (3) the necessity of skilful climate information on regional scales, embodying the so-called "seamless prediction" paradigm; (4) the importance of quantifying "true" uncertainty in climate predictions; (5) the challenges and opportunities of predicting how natural modes of climate variability will modify the "forced" anthropogenic component of climate change over the coming years to decades; (6) the increasing importance of establishing the predictability of polar climate, perhaps especially with the opening of the Arctic and international negotiations regarding increased commercial traffic for shipping and extraction of natural resources; (7) the need to better understand the causes of extreme events and performing attribution studies in near real-time; (8) the challenges of improved predictions of future sea-level rise on regional scales, which will require knowledge of not only cryospheric and thermosteric contributions but also how gyre circulations, storm tracks, and tidal amplitudes will change; and (9) the requirement to train and empower the next generation of climate scientists across all corners of the globe, a priority of the future WCRP as it seeks opportunities for capacity development with its partners in the human and social sciences.

The general consensus among the OSC participants was that the WCRP and its affiliate network of international scientists and Projects must continue to provide the scientific foundation for understanding and predicting the Earth's climate system. However, they also must play a major role in providing the resulting knowledge and information in ways that yield practical solutions to the complex and interrelated challenges required to ensure a sustainable Earth for future generations. The World Climate Research Program, its leaders, and network of projects stand ready to support the research community in pursuing the challenges and opportunities identified during the conference and captured in the following chapters of this book in the spirit of pursuing Climate Science in Service to Society in the ensuing decades.

References

Asrar GR (2009) The World Climate Research Programme. Public Serv Rev Sci Technol 3:115–120

Asrar GR, Busalacchi A, Hurrell J (2012) WCRP Open Science Conference. American Geophysical Union, Eos Transactions 93(12), 20 Mar 2012

Hall J, Harrison DE, Stammer D (eds) (2009) "Conference Statement" in Proceedings of OceanObs'09: Sustained Ocean Observations and Information for Society, vol 1. Venice, 21–25 Sept 2009, Publication ESA WPP-306, doi:10.5270/OceanObs09

ICSU-IGFA (2009) Review of the World Climate Research Programme (WCRP). International Council for Science, Paris. Available at www.icsu.org, ISBN: 978-0-930357-70-2, 57 pp

ICSU-ISSC (2010) Earth system science for global sustainability: the grand challenges. International Council for Science, Paris. Available at www.icsu.org, ISBN: 978-0-930357-73-3, 20 pp

Leemans R, Asrar G, Busalacchi A, Canadell J, Ingram J, Larigauderie A, Mooney H, Nobre C, Patwardhan A, Rice M, Schmidt F, Seitzinger S, Virji H, Vorosmarty C, Young O (2009) Developing a common strategy for integrative global environmental research and outreach: the Earth System Science Partnership (ESSP). Curr Opin Env Sust 1:4–13

Reid W, Chen D, Goldfarb L, Hackmann H, Lee YT, Mokhele K, Ostrom E, Raivio K, Rockström J, Schellnhuber HJ, Whyte A (2010) Earth system science for global sustainability: grand challenges. Science 330:916–917

Rockstrom J, Steffen W, Noone K, Persson K, Chapin F III, Lambin E, Lenton M, Scheffer M, Folke C et al (2009) Planetary boundaries: exploring a safe operating space for humanity. Nature 461:472–475

Thompson M, Molesworth A, Djingarey M, Yameogo K, Belanger F, Cuevas L (2006) Potential of environmental models to predict meningitis epidemics in Africa. Trop Med Int Health 11:1–9

World Climate Research Programme (WCRP) (2005) Strategic framework 2005–2015, "Coordinated Observation and Prediction of the Earth System", WCRP-123, WMO/TD-No. 1291, 65 pp

World Climate Research Programme (2009a) World Modelling Summit for Climate Prediction, Reading, UK, 6–9 May 2008, WCRP-No. 131, WMO/TD-No. 1468, 29 pp

World Climate Research Programme (2009b) WCRP Implementation Plan 2010–2015, Geneva, WMO/TD-No. 1503, 43 pp

World Meteorological Organization (2011) Climate knowledge for action: a global framework for climate services (GFCS). The report of high level task force for GFCS, WMO-No. 1065, 240 pp

World Meteorological Organization (2012) Implementation plan for the Global Framework for Climate Services, WMO-No. (in print), 90 pp

Challenges of a Sustained Climate Observing System

Kevin E. Trenberth, Richard A. Anthes, Alan Belward, Otis B. Brown, Ted Habermann, Thomas R. Karl, Steve Running, Barbara Ryan, Michael Tanner, and Bruce Wielicki

Abstract Observations of planet Earth and especially all climate system components and forcings are increasingly needed for planning and informed decision making related to climate services in the broadest sense. Although significant progress has been made, much more remains to be done before a fully functional and dependable climate observing system exists. Observations are needed on spatial scales from local to global, and all time scales, especially to understand and document changes in extreme events. Climate change caused by human activities adds a new dimension and a vital imperative: to acquire climate observations of sufficient quality and coverage, and analyze them into products for multiple purposes to inform decisions for mitigation, adaptation, assessing vulnerability and impacts, possible geo-engineering, and predicting climate variability and change and their consequences. A major challenge

K.E. Trenberth (✉)
Climate and Global Dynamics, National Center for Atmospheric Research,
PO Box 3000, Boulder, CO 80307-3000, USA
e-mail: trenbert@ucar.edu

R.A. Anthes
University Corporation for Atmospheric Research, PO Box 3000,
Boulder, CO 80307, USA
e-mail: anthes@ucar.edu

A. Belward
Institute for Environment and Sustainability, Joint Research Centre of the European Commission, Joint Research Centre, Via E. Fermi 2749, I-21027 Ispra, VA, Italy
e-mail: alan.belward@jrc.it

O.B. Brown
Cooperative Institute for Climate and Satellites – North Carolina (CICS-NC),
North Carolina State University, Asheville, NC, USA

NOAA's National Climatic Data Center, 151 Patton Avenue,
Asheville, NC 28801, USA
e-mail: otis_brown@ncsu.edu

G.R. Asrar and J.W. Hurrell (eds.), *Climate Science for Serving Society: Research, Modeling and Prediction Priorities*, DOI 10.1007/978-94-007-6692-1_2,

is to adequately deal with the continually changing observing system, especially from satellites and other remote sensing platforms such as in the ocean, in order to provide a continuous climate record. Even with new computational tools, challenges remain to provide adequate analysis, processing, meta-data, archival, access, and management of the resulting data and the data products. As volumes of data continue to grow, so do the challenges of distilling information to allow us to understand what is happening and why, and what the implications are for the future. The case is compelling that prompt coordinated international actions are essential to provide for information-based actions and decisions related to climate variability and change.

Keywords Climate observing system • Satellite observations • Climate change • Data processing • Earth observations • Metadata • Climate data records

1 Introduction

The first rule of management is often stated to be "you can't manage what you can't measure". Indeed, Earth is observed more completely today than at any other time. Multiple observations are made from space in many different wavelengths via passive and active sensors that provide information on many geophysical and meteorological variables. However, a key question is the extent to which these observations are suitable for characterizing climate, and especially for climate monitoring and prediction.

As the climate system is continuously evolving, there is a need to measure changes globally and regionally, to understand the system, attribute the causes of the changes by linking the changes in state variables to various forcings, and to develop models that can simulate and predict the system's evolution (Trenberth et al. 2002, 2006). The observations must be processed and analyzed, often into

T. Habermann • T.R. Karl • M. Tanner
National Environmental Satellite, Data, and Information Service (NESDIS), NOAA, 325 Broadway, Boulder, CO 80305-3328, USA
e-mail: ted.habermann@noaa.gov; thomas.r.karl@noaa.gov; michael.tanner@noaa.gov

S. Running
Numerical Terradynamic Simulation Group (NTSG), CHCB room 428, The University of Montana, 32 Campus Drive, Missoula, MT 59812, USA
e-mail: swr@ntsg.umt.edu

B. Ryan
Director, GEO Secretariat, 7bis, avenue de la Paix, Case postale No. 2300, CH-1211 Geneva 2, Switzerland
e-mail: bryan@wmo.int

B. Wielicki
NASA Langley Research Center, Hampton, VA 23681-0001, USA
e-mail: b.a.wielicki@nasa.gov

globally gridded fields that can be used as an initial state for predictions using climate models. Accordingly, observations are used to document the state of the climate and how it varies and changes over time, along with documenting external influences on the system such as the sun, the Earth's radiation budget, the Earth's surface and changes in the climate system from human influences.

Moreover, because the climate is changing from natural and human influences (IPCC 2007) it is an imperative to document what is happening, understand those changes and their causes, sort out the human contribution, and make projections and predictions on various time horizons into the future (Trenberth 2008). Mitigation of the human influences, such as reducing greenhouse gas and aerosol emissions, is a major challenge yet to be adequately addressed and the effectiveness of any mitigation actions needs to be documented in order for them to continue. However, given the likelihood of large future human-induced changes, understanding and planning how to cope with the projected changes, and how well the predictions are verifying, become extremely important. Hence information related to adaptation to climate change is also vital. Process studies using special, perhaps short-term observations will help improve models and the information they can provide. Prospects of geo-engineering to offset climate change mandate diligent observations to ensure that the intended effects are in fact happening and to check for unforeseen side effects. Together, all of these activities and needs define the observation requirements for a climate information system that provides climate services to users of all kinds.

Many observations pertinent to this information system are made (Fig. 1), but most are not of sufficient quality to meet climate needs. In the atmosphere, most observations are made for weather forecasting which involves documenting the state of atmospheric weather systems such as low and high pressure systems, cold and warm fronts, tropical cyclones, rain bands, clear skies, and so forth as a first step to predicting their movement and evolution. Weather fluctuations are huge compared with climate change and so high measurement accuracy and precision have not been a priority, although this has changed as models have improved and the need to correct biases has grown. Climate change must discern relatively small changes over time, which calls for both stability and calibrated measurements of high accuracy. Knowing how the measurements of 20 or 50 years ago relate to those of today is very important.

The climate observing system challenge can be understood by considering that understanding and predicting this complex system requires many more variables than for weather prediction. The current estimate is 50 Essential Climate Variables (ECVs): 16 for atmosphere, 18 for ocean, and 16 for terrestrial (GCOS 2010). The ECV accuracy requirement is also much more stringent than for weather observations (e.g., 0.1 K vs 1 K). Space and time scales are more extreme, ranging from aerosol and cloud physics occurring at seconds and micrometers, to global decadal change at 100 years and 40,000 km: a range greater than 10^9 in time, and 10^{13} in space.

At the surface, observing instruments can be calibrated, but sites often change and the representativeness of the observations is a concern. For instance, since the

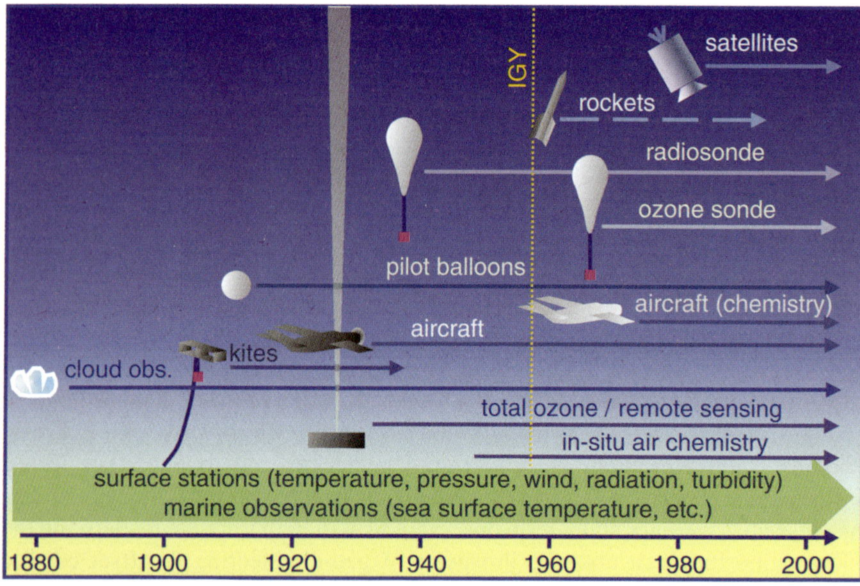

Fig. 1 Changes in the mix and increasing diversity of observations over time create challenges for a consistent climate record (Courtesy, S. Brönnimann, University of Bern. Adapted from Brönnimann et al. 2008)

1970s around 50,000 km^2 per year of natural vegetation across Africa has been converted to agricultural land or cleared (Brink and Eva 2009). Elsewhere the urban heat island effect associated with the concrete jungle of a city and its effects on runoff and heat retention plus space heating are important locally but make up less than 0.5 % of land (Schneider et al. 2009), and these changes are very small on a global basis. Radiosonde and other instrumental records suffer from biases that have changed over time.

Satellites have observed Earth for over 50 years now, and have provided a series of wonderful and enlightening imagery and measurements (NRC 2008). They help offset the otherwise uneven spatial coverage of *in situ* observations. Nonetheless, each satellite mission has a new instrument that is exposed to cosmic rays, outgassing contaminants, and a hostile environment, and the satellite orbit eventually decays and drifts in time. The instruments thus require on-board calibration and/or validation from in situ instruments. An exception is GNSS (Global Navigation Satellite System) radio occultation, which is self calibrating (Steiner et al. 2011). A mission typically lasts 5 years or so; thus determining how new measurements relate to old ones to ensure continuity of the record is a major issue (Fig. 1). Because of these issues, only a few satellite records (water vapor and microwave temperatures) were used to determine trends in the IPCC Fourth Assessment Report (AR4) (IPCC 2007).

In the following, the observing system and its suitability for climate purposes is outlined. Acronyms are given in an appendix. We describe recent improvements for cross calibrating space-based observations, for instance, and immediate prospects for the future. The needs are discussed along with the issues and challenges in meeting them. Indeed the needs are compelling and enormous, but also feasible with international cooperation and leveraging of resources.

2 The Current Climate Observing System

2.1 Status of Systematic Climate Observations

The Global Climate Observing System (GCOS) organization leads the international advisory oversight of systematic climate observations, and focuses on observations to support the United Nations Framework Convention on Climate Change (UNFCCC). Appendix A provides a brief summary of its organizational structure and charter. One of GCOS most critical roles is to produce regular assessments of the adequacy of climate observations, including suggestions for needed improvements. Recent GCOS reports provide an excellent reference point for discussing the status of climate observations.

A progress report (GCOS 2009) concluded that:

- the increasing profile of climate change had reinforced awareness of the importance of an effective global climate observing system;
- developed countries had improved many of their climate observation capabilities, but with little progress in ensuring long-term continuity for several important observing systems;
- developing countries had made only limited progress in filling gaps in their *in situ* observing networks, with some evidence of decline, and capacity building remained small in relation to needs;
- both operational and research networks and systems, established principally for other purposes, were increasingly responsive to climate needs including the need for timely data exchange;
- space agencies had improved mission continuity, observational capability, data reprocessing, product generation and access;
- GCOS had progressed significantly, but still fell short of meeting all the climate information needs of the UNFCCC and broader user communities.

The Third World Climate Conference (WCC-3) in 2009 underscored the importance of systematic observations (Manton et al. 2010; Karl et al. 2010). WCC3 recommended strengthening GCOS by:

- sustaining the established *in situ* and space-based components of GCOS;
- applying the GCOS Climate Monitoring Principles (GCMPs);
- improving the operation and planning of observing systems; identify deficiencies, achieving resilience, and assuring reliable and timely delivery of quality data, traceable to international standards;

- enhancing observing systems wherever feasible; filling gaps in spatial coverage and in the breadth of variables measured, improving measurement accuracy and frequency, increasing use of operational platforms for satellite sensors, monitoring urban and coastal conditions, and establishing reference networks;
- rescuing, exchanging, archiving and cataloging data, and recalibrating, reprocessing and reanalyzing long-term records, working towards full and unrestricted access to data and products;
- giving high priority to observational needs for adaptation planning, identifying country needs in National Adaptation Programs of Action;
- assisting developing countries to maintain and strengthen their observing networks through support for updating, refining and implementing the GCOS Regional Action Plans and other regional observational and service initiatives.

The 2010 update (GCOS 2010) also noted advances in observational science and technology, an increasing focus on adaptation, and the demand to optimize mitigation measures. It reaffirmed the importance of the GCMPs, emphasizing the need for and ways to achieve continuity and stability of measurements. Guidelines for operations including on-orbit calibration and validation, the need for global coverage, timeliness of data, and development of a maturity index for each ECV, were also included. It introduced a small number of new ECVs, and called for colocated measurement of ecosystem variables along with the ECVs that influence or are influenced by them. Table 1 provides details of the ECVs.

The 2010 GCOS update provided cost estimates for fully implementing and operating the climate observing system; around US$2.5 billion each year (in addition to the current annual global expenditure of some US$5–7 billion on global observing systems serving climate and related purposes). Around US$1.4 billion of this additional expenditure is needed for satellites or for *in situ* observation of the open ocean, in both cases for the benefit of all. In addition, around US$600 million per year are needed for *in situ* observations in developing countries (GCOS 2010). Consequently, the magnitude of the investment required is order $\frac{1}{3}$ to $\frac{1}{2}$ of the current expenditure (whose estimate depends on how costs are assigned when the observations serve multiple purposes).

A definition of a climate data record is, "…a time series of measurements of sufficient length, consistency, and continuity to determine climate variability and change" (NRC 2004). A challenge for climate observations is to have a consistent, well-understood framework for observations that is independent of a parameter's origin and observing approach, and, easily found and accessed.

2.2 *Building a System for Climate Observations*

The push to develop a systems approach to climate observations has been detailed in Trenberth et al. (2002, 2006). Trenberth (2008) outlined a framework for how observations, data and analyses feed into assimilation and modeling that support prediction and attribution. Assessments build on the products to inform stakeholders,

Table 1 Essential climate variables (ECVs) that are both currently feasible for global implementation and have a high impact on UNFCCC requirements (GCOS 2010)

Domain	Essential climate variables
Atmospheric (over land, sea and ice)	**Surface:** air temperature, wind speed and direction, water vapor, pressure, precipitation, surface radiation budget **Upper-air:** temperature, wind speed and direction, water vapor, cloud properties, earth radiation budget (including solar irradiance) **Composition:** carbon dioxide, methane, and other long-lived greenhouse gases; ozone and aerosol, supported by their precursors
Oceanic	**Surface:** Sea-surface temperature, sea-surface salinity, sea level, sea state, sea ice, surface current, ocean color, carbon dioxide partial pressure, ocean acidity, phytoplankton **Sub-surface:** temperature, salinity, current, nutrients, carbon dioxide partial pressure, ocean acidity, oxygen, tracers
Terrestrial	River discharge, water use, ground water, lakes, snow cover, glaciers and ice caps, ice sheets, permafrost, albedo, land cover (including vegetation type), fraction of absorbed photosynthetically active radiation (FAPAR), leaf area index (LAI), above-ground biomass, soil carbon, fire disturbance, soil moisture

users and decision makers. Because of the long time scales associated with climate variations and change, basic research and operational applied research are inherent parts of the entire system that ultimately feed into climate services. All elements are essential for a useful and robust climate information system.

Not all observing systems and datasets are suitable for climate studies. The evolution of data systems to support climate observations has been a multi-step process. Many *in situ* observations originated in a single investigator or team developing an approach, building a network and eventually moving to a systematized network, *e.g.*, meteorological variables followed such a path and transitioned to primarily nationally operated and internationally coordinated observing enterprises by the mid-twentieth century. While *in situ* ocean, land, and ice observing activities have moved along similar trajectories, they have been less mature for the most part. In contrast, space-based remotely sensed observations required significant investments from the outset, most of which were national in origin. Thus, these activities were subject to a systems engineering rigor from very early in their evolution due to their platform dependencies and expense. Nevertheless, the same rigor did not apply to calibration, and recalibration and reprocessing of the data has become essential. It is important to appreciate that there are differing strategies and maturities associated with each ECV.

A "maturity matrix" (Privette et al. 2008) translated NASA concepts on technology readiness into similar attributes for satellite observation maturity. It defines six levels of maturity as a function of sensor use, algorithm stability, metadata completeness, documentation, validation, availability of data, and science and applications. Such an approach provides a framework for defining the attributes and readiness of space-based observations for use in climate applications. While this approach was applied initially to space-based observations, more recently it has been suggested that it be applied to *in situ* observations as well. CEOS, GCOS, GOOS, GTOS and GEOSS are stewarding an integrated approach for Earth observations along with WCRP through its WCRP Observations and Assimilation Panel (WOAP), which is transitioning into the WCRP Data Advisory Council.

The history of space-based observations and currently funded initiatives gives a basis for looking at the state of each ECV (Fig. 2). Combining this information with similar information from *in situ* systems provides the basis for doing assessments of integrated observing system health, gaps, and so forth.

2.3 Developing Operational Components

No single agency, organization, or country has the resources to develop a robust operational end-to-end system for monitoring Earth's climate over the required spatial and temporal scales. By operational we mean regular and with a sustained institutional commitment to the observing system, as opposed to single principle investigator-led or one-of-a-kind research missions. The developing international Global Framework for Climate Services (GFCS) led by WMO (WMO 2011) is a key driver of the need for a more operational approach to climate observations.

There are examples, however, that could serve as models or starting points for an operational climate system. One such example is the operational system that has been built over the last 40 years for weather observations, research, modeling and forecasting. Lives and property are saved everyday as a result of this operational weather system.

The challenges for climate monitoring are more complex, and are compounded by the lack of international agreements and architecture for developing a sustained, integrated climate monitoring capability. GCOS certainly provides an overarching framework and key components, yet much more is needed. Building blocks for an operational system would, at a minimum, include the following components: requirements identification and analysis, observations, intercalibration, contingency planning, analysis and product generation, archiving, distribution and dissemination, and user engagement and training.

Figure 3 shows key components required for an operational capability, which includes satellites sensors and data, climate data records (CDRs), satellite products, and ultimately users of those products. This value chain, although originally employed for weather purposes by WMO, is being extended for climate purposes by using the requirements that GCOS has identified and articulated for climate monitoring, e.g., the ECVs. Many agencies and organizations contribute to components of this value chain.

The WMO Global Observing System (GOS) (Fig. 4), was originally comprised of geostationary and polar-orbiting meteorological satellites (early 1960s to early

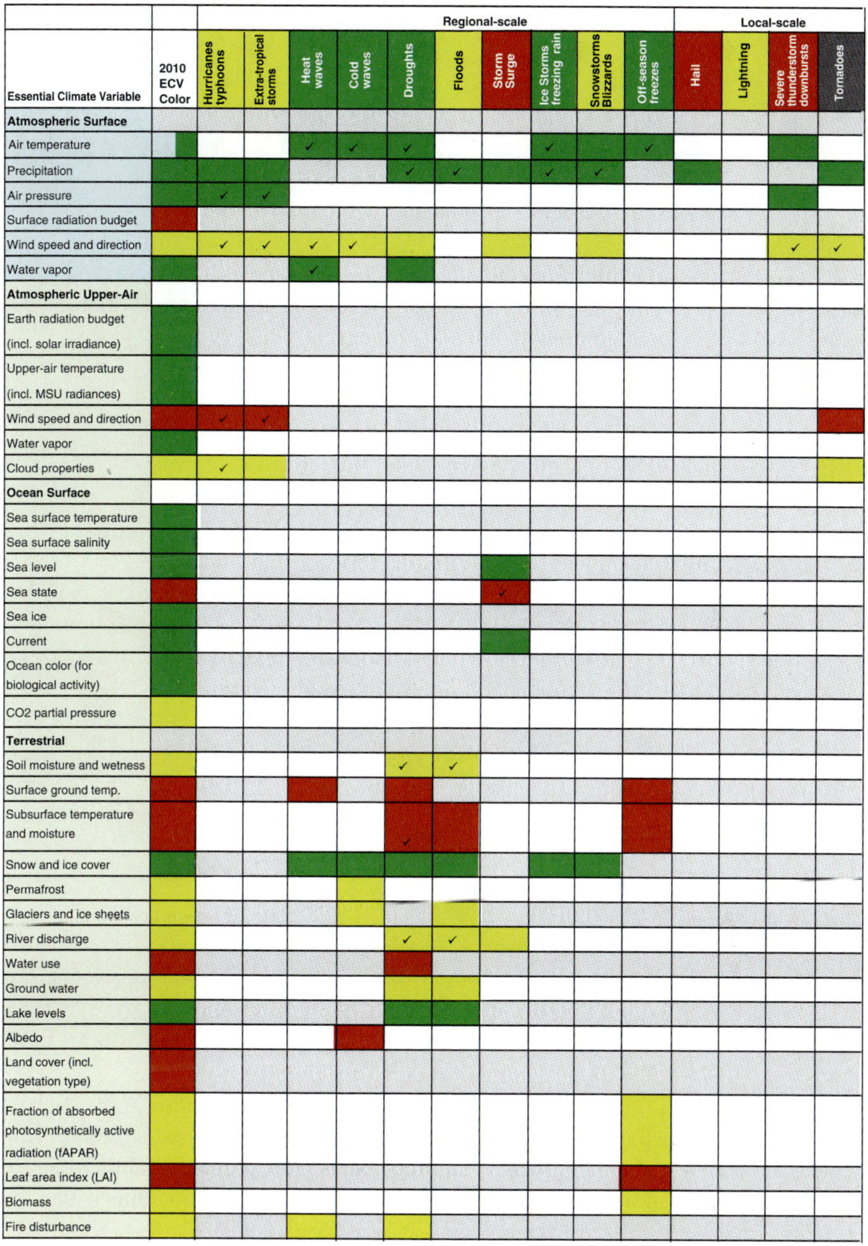

Fig. 2 Relationship of extreme phenomena to ECVs for monitoring. Both the phenomena and the ECVs are color coded to describe the adequacy of the current monitoring systems to capture trends on climate timescales set against alternating *grey* and *white* lines to enhance readability. *Green* indicates global coverage with a sufficient period of record, data quality, and metadata to make enable meaningful monitoring of temporal changes. *Yellow* indicates an insufficiency in one of those three factors. *Red* indicates insufficiency in more than one of the factors. The check mark in the colored ECV block indicates that the ECV is of primary importance to monitoring changes in the extreme event phenomenon

Fig. 3 Key components of an operational climate capability. Here *GSICS* is the global space-based intercalibration system, *IGDDS* WMO integrated global data dissemination service, *SCOPE-CM* sustained coordinated processing of environmental satellite data for climate monitoring, *VLab* virtual laboratory for training in satellite meteorology

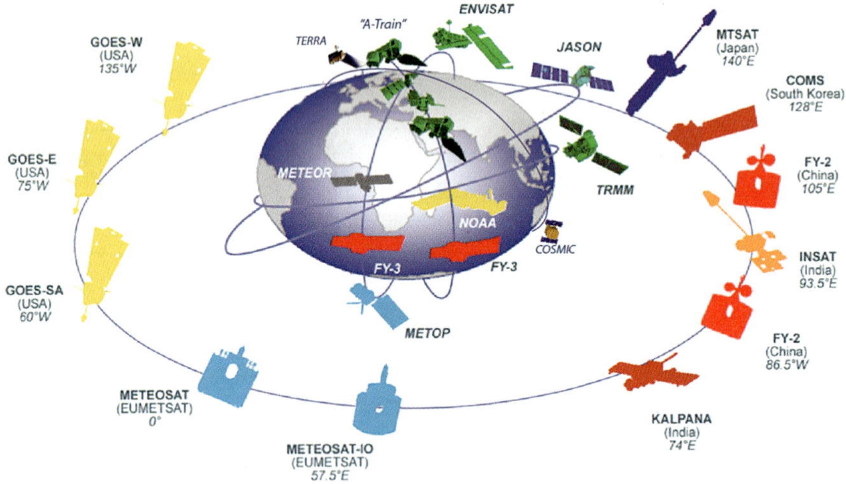

Fig. 4 Schematic of the space-based global observing system (*GOS*) as of about 2010

2000s) and has grown to include research and development satellites. This observing system, its underpinning architecture, and the results achieved illustrate the reliance on and importance of international collaboration. The GCOS reports suggest that the benefits countries receive from this global system far exceed the costs of their individual contributions. Additionally, the interplay between operational satellites and research and development satellites becomes more important to obtain the range of spatial and temporal scales and spectral resolutions needed for climate monitoring.

The Global Space-based Inter-Calibration System (GSICS) is an international program to improve the comparability of satellite measurements taken at different times and locations by different instruments operated by different satellite agencies

(Goldberg et al. 2011). GSICS inter-calibrates selected instruments of the GOS including operational low-Earth-orbit and geostationary Earth-orbit environmental satellites and, where possible, ties these measurements to common reference standards. The agencies participating in GSICS have developed a comprehensive calibration strategy involving inter-calibrating satellite instruments, tying measurements to absolute references and standards, and recalibrating archived data. GSICS corrections, initially for infrared channels and thereafter for visible and microwave sensors, are being performed and delivered operationally. GSICS results are used for CDR processing activities, as illustrated in Fig. 3, by the Sustained Co-Ordinated Processing of Environmental Satellite Data for Climate Monitoring (SCOPE-CM) effort. At present, GSICS reference observations (e.g., AIRS, IASI, MODIS) are SI traceable, but not at the absolute accuracy required for climate change. Planned observing systems (e.g., CLARREO) are designed to enable climate change accuracy requirements to be met if deployed.

A number of SCOPE-CM projects are underway, led by one of three space agencies (EUMETSAT and its Climate Monitoring Satellite Application Facility, JMA or NOAA). Structures are being established for the sustained generation of Fundamental CDRs and Thematic CDRs. Extension of the network is also being sought, as the existing projects are primarily target ECVs from the atmospheric domain; increased coverage of the oceanic and terrestrial domain ECVs is needed.

3 Lost in Space: Climate Observations?

The existence of GEOSS, its climate observing component (GCOS), its satellite observing component (CEOS) and their implementation plans (GEO 2005; GCOS 2010) are a strong initial step toward a true international climate observing system. Necessarily, there are both strong *in situ* and global orbiting satellite components. However, a comprehensive system remains more vision than reality, although very promising developments through GCOS, GSICS and SCOPE-CM are taking place. In addition WMO, with CGMS and CEOS, are drafting a climate monitoring from space architecture plan. This section highlights some of the key remaining challenges in observations, especially from space.

3.1 Current and Programmed Satellite Observations

Many new satellite remote sensing programs are under way. The Japan Aerospace Exploration Agency (JAXA) is developing and implementing a suite of climate monitoring satellites including ALOS (mainly for land), GOSAT (for carbon balance estimation among other applications), GCOM-W (for tasks including water circulation), and the EarthCARE platforms (cloud and aerosol observations).

From Europe, satellites flying today plus commissioned systems have the potential to generate 29 of the ECVs. The European Space Agency's Climate Change Initiative, EUMETSAT Satellite Application Facility on Climate Monitoring and the ECMWF ERA reanalysis already support production of some 40 % of the ECVs over the next 5–10 years (Wilson et al. 2010). The European Earth Observation program, GMES (Global Monitoring for Environment and Security), includes five new missions (the Sentinels, which include radar imaging of land and ocean, multi-spectral 10 m resolution land monitoring and a mission to measure sea-surface topography, sea- and land-surface temperature, ocean color, and terrestrial variables such as FAPAR). The first Sentinels are planned for launch in 2013 and each has a 7-year design lifetime.

NASA is developing and implementing a broad range of Earth space-borne remote sensing missions including the Decadal Survey (NRC 2007) and Climate Continuity series of satellites. NOAA operates operational weather satellites including the polar orbiters [Joint Polar Satellite System (JPSS) (previously called National Polar-Orbiting Environmental Satellite System NPOESS)] and two geostationary satellites [Geostationary Operational Environmental Satellite (GOES)]. The backbone of current global terrestrial monitoring for the U.S. are the NASA Earth Observing System platforms Terra, launched in 1999, Aqua, launched in 2002 and Aura, launched in 2004. At higher spatial resolution, the Landsat satellite series has operated since 1972, with the next satellite in the series planned for January, 2013. The Earth Observing System (EOS) platforms are currently likely to operate through about 2015 and possibly longer.

The first U.S. National Research Council (NRC) decadal survey for Earth sciences (hereafter the Decadal Survey; NRC 2007) reviewed the expected ongoing observations and recommended new observations over the next decade (roughly until 2020). It also provided an overview of translating satellite observations into knowledge and information for the benefit of society. NASA Earth Science has been responsive to and acted upon these recommendations, but significant issues have resulted in a much slower schedule than called for in the Decadal Survey (NRC 2012). CLARREO (Climate Absolute Radiance and Refractivity Observatory), DESDynl (Deformation, Ecosystem Structure, and Dynamics of Ice), SMAP (Soil Moisture Active/Passive) and ICESAT-II (Ice, Cloud, and land Elevation Satellite-II) all had follow up workshop reports (see http://science.nasa.gov/earth-science/decadal-surveys/) and the NASA Earth Science Data Systems has been pursuing a "system of systems" architecture in response to the report recommendations.

The Decadal Survey also recommended that NOAA carry out a fully operational follow-on mission to COSMIC (Constellation Observing System for Meteorology, Ionosphere and Climate). COSMIC (2006-), and other radio occultation missions such as GPS/MET (1993–1995), CHAMP (2001–2010) SAC-C (2000-) and METOP-A (2006-) have demonstrated the value of radio occultation in producing precise, accurate, climate quality observations in all weather (Anthes 2011). A follow-on mission (COSMIC-2) has been proposed (http://space.skyrocket.de/doc_sdat/formosat-7-cosmic-2.htm) and significant funding secured from Taiwan. Implementation is beginning with key U.S. support (DoD: US Air Force) but NOAA support has not yet been solidified.

Continuity of the key ECVs initiated in the EOS era is intended to transfer to the JPSS series over the next decade, beginning with the NPOESS Preparatory Project (NPP). However, three expected "foundation" missions have had a troubled history. OCO (Orbiting Carbon Observatory) and GLORY (carrying aerosol polarimetry and solar irradiance) both failed on launch and ended up in the Pacific Ocean, and JPSS replaced the cancelled NPOESS program, which has had rapidly rising costs. Hence several foundation missions have failed or been delayed. The NPP, originally intended to be a risk-reduction mission for a subset of the NPOESS sensors, slipped in time but was successfully launched late October 2011. NPP, now called Suomi NPP, now has an operational mandate for weather and climate applications, since the JPSS missions are delayed until late in the decade, and will serve as a gap filler. The Suomi NPP platform carries the ATMS, CERES, CriS, OMPS (nadir and limb) and VIIRS sensors. The latter is the successor to the widely used MODIS sensor on the Terra and Aqua platforms. The other relevant land sensor will be the SMAP (Soil Moisture Active/Passive) mission planned for an early 2015 launch, which will continue to monitor surface wetness and freeze/thaw conditions of the land surface, building on results from ESA's Soil Moisture and Ocean Salinity (SMOS) mission that was launched in November, 2009. There is a replacement for OCO (OCO-2) that has been supported and should launch later in the decade as well.

The overall impact of the above issues remains to be seen, but it is becoming clear that there is a significant probability of a lack of overlap between the EOS platforms, Suomi NPP, and the next generation operational system (JPSS). Cross-calibration from old to new sensors while both are still in orbit is essential for retaining ECV continuity for multiple decades. Lack of overlap provides challenges to continuity. Recent NOAA budget cuts have jeopardized the timely launch of the first full JPSS platform, originally planned for 2015, now possibly delayed to 2017–2018. An estimate of the likelihood of obtaining at least 1 year of intercalibration overlap as a function of instrument and spacecraft design lifetimes (Loeb et al. 2009) can be applied to 3 key climate sensors on EOS (CERES, MODIS and AIRS) with the follow-on sensors on NPP and JPSS-1 (CERES, VIIRS, CrIS). With a JPSS launch by late 2017 the probability of successful 1-year overlap for all three instruments is only about 37 %. Further delays in launching JPSS will lower this probability. However, some progress concerning cross-calibration of U.S. and European sensors, and the validation of products derived from them is being made (Zibordi et al. 2010).

Consistent measurement of the energy received from the sun is a case in point. There are considerable calibration issues with such measurements from space, but meaningful time series exist since 1979 only because of overlap between measurements. However, with the loss of Glory and because of cost constraints in JPSS that impact inclusion of a solar irradiance instrument, there is a distinct possibility of a gap that would break a more than three decades long record. Exploring alternative means of measuring solar output should be a high priority.

A number of emerging remote sensing programs are under development by other organizations and nations, including China, India and the Republic of Korea. Each of these contribute to the GOS and thus to GEOSS and, as the systems become operational, they are sharing increasingly more data and participating in GSICS in order to increase the quality of their observations.

3.2 Adequacy of In-Situ Observations

Many *in situ* measurements need to be combined with satellite measurements: for calibration/validation and for broader spatial coverage, and sometimes for temporal resolution. Examples of these synergies include greenhouse gases (many cannot yet be reliably measured from space), ozone (suborbital measurements can provide detailed vertical information), snow depth, cover and snow water equivalent. Other observations are of vital importance to understanding the physical climate system, including observations of the Earth surface radiation budget (such as the BSRN), temperature, greenhouse gases, leaf area index, land cover, surface albedo, precipitation, winds, and sea level. Other priority observing networks pertain to elements of the climate system and the important feedbacks therein: ocean color, biomass, fire disturbance, and water use.

Current *in situ* climate observations capabilities are diverse and contribute to both national needs and global partnerships. These capabilities make use of a broad range of airborne, terrestrial, and oceanic observations, some of which were designed primarily for climate, but many of which also serve other purposes. Overall, capabilities are most mature in the atmospheric domain, bolstered by observations made for weather forecasting, while needs and priorities are still emerging in the terrestrial, cryosphere, and oceanic domains. Gleick et al. (2013) provide examples of how some terrestrial *in situ* observations are evolving.

Unfortunately, many *in situ* networks have been in decline, as discussed more fully in Sect. 5.1 and, as noted in Sect. 2.1, hundreds of millions of dollar investments are needed to improve the adequacy of the in situ network.

An *in situ* climate observing component is highly desirable and is beginning to occur through the Global Reference Upper Air Network (GRUAN) (GCOS 2007). While other operational upper air observations exist they were not designed for climate purposes. A reference observation requires:

- traceability to SI or another commonly accepted standard
- comprehensively estimated uncertainty
- documentation of instrumentation, procedures and algorithms
- validation of the data products.

GRUAN will provide reference observations of upper-air ECVs, through a combination of *in situ* measurements made from balloon-borne instruments and ground-based remote sensing observations. The primary goals of GRUAN are to:

- Provide vertical profiles of reference measurements suitable for reliably detecting changes in global and regional climate on decadal time scales.
- Provide a calibrated reference standard for global satellite-based measurements of atmospheric essential climate variables.
- Fully characterize the properties of the atmospheric column.
- Ensure that gaps in satellite programs do not invalidate the long-term climate record.

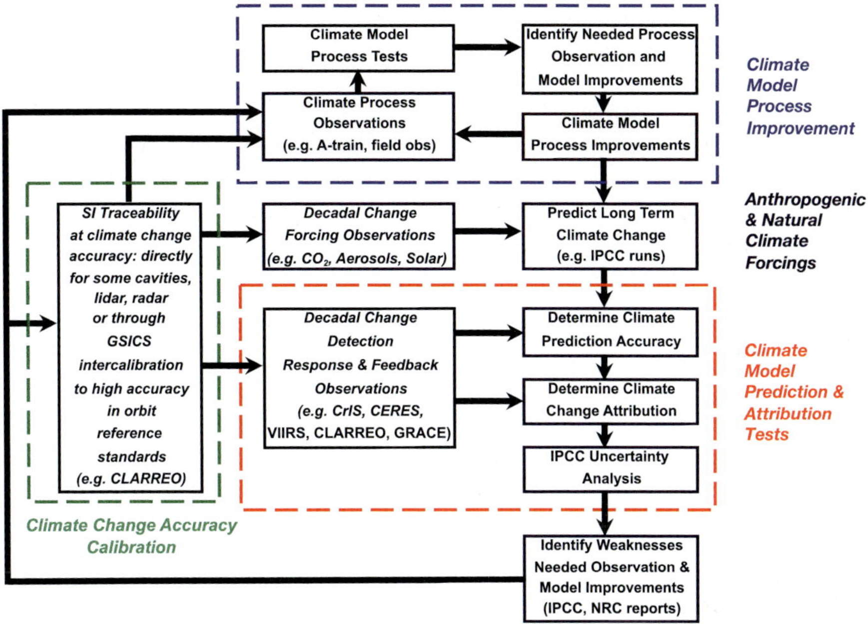

Fig. 5 The schematic shows the role of climate process and monitoring observations in climate change science: detection and attribution of climate change, climate model testing, and climate model improvements

The envisaged capabilities of a fully-implemented GRUAN (GCOS 2007) include plans to expand to include 30 or 40 sites worldwide. Strict site selection criteria and operating principles have been established, coordinated through the GRUAN Lead Centre, currently hosted by the Lindenberg Meteorological Observatory, Germany. Although GRUAN is a vital component for an adequate climate observing system, adequate support has been slow in developing.

3.3 *The Scope of the Challenge of Satellite Observations: Adequacy and Issues*

As noted in Sect. 1, an extreme range of scales, accuracy, and processes occurs across oceans, atmosphere, biosphere, cryosphere, and biogeochemistry. How scientists deal with this range is illustrated in Fig. 5. In general, climate process data are taken at small time/space scales more similar to weather data. These are critical to understanding underlying climate physics (blue box/text), but the accuracy of climate predictions of decadal change is primarily determined by decadal change in natural and anthropogenic radiative forcings (black) and decadal observations of the climate system response to those forcings (red box and text). The decadal change forcing and response observations drive the need for very high accuracy at large time/space scales. Resolving variability at finer spatial resolutions, however, is also required for many purposes such as extremes. To achieve high accuracy mandates a rigorously maintained link from satellite observations to metrological international physical standards, with a focus on traceability to SI standards at climate change accuracy in both ground calibration as well as in-orbit (green box); see Sect. 3.4.

3.3.1 The Missing Satellite Observing System Principles

The GCMPs include ten that are specifically directed at satellite observations (GCOS 2010). Two important additional principles have been proposed (USGCRP 2003):

- Provision for independent observations, especially to verify accuracy of other systems and to confirm and/or refute surprising climate change results.
- Provision for independent analysis of observations, especially satellite remote sensing data where analysis systems may involve ten thousand to a million lines of computer code.

The need for these two principles is well recognized in the metrological community. International standards are not accepted until they are independently verified, complete with an analysis of uncertainty in each step. A similar standard is required of fundamental tests of physical laws in research groups at particle accelerator laboratories around the world. Unfortunately, the need for independent scientific verification demands extensive resources especially for independent satellite observations. This may explain the absence of formal acceptance of these principles to date. But recent arguments over the accuracy of climate change observations reaffirm the need for the addition of these two key principles, as independent verification is the key to high confidence needed for societal decision making.

Independent analysis exists for some, but not most, current climate observations and processing. It also remains difficult to judge whether our current priorities will still be the same decades from now. However, a corollary advantage of the independence principles is to add reliability to the observing system when unexpected satellite failures occur such as the recent failures of Glory, OCO, and CryoSat missions, or premature loss in orbit of entire satellites, such as ADEOS and ADEOS 2.

3.3.2 Delays and Cost Increases

Technical development, schedule, and budget issues can also delay satellite observations as shown by the delays of JPSS, and the recent indefinite delay of the CLARREO and DESDynI missions, as well as a follow on copy of the Global Precipitation Mission radar. The delays of NPP and NPOESS/JPSS would already have had dire consequences had the Terra and Aqua missions not lasted a factor of 2 longer than design life. If those missions had only lasted the nominal 5 years planned, as did the recent ALOS satellite, the gap of a wide range of climate relevant observations would have begun in 2007 (Aqua 5 years old, Terra 7 years old), and continued until at least the end of 2011 with launch of the delayed NPP mission.

Delays and failures compromise the climate observing system's ability to deliver information concerning core UNFCCC needs and severely limit capacity to meet new demands. As emphasized in the introduction, we must have the ability to relate measurements of 20 or 50 years ago to those of today. This is equally the case for new demands for climate information, such as quantifying terrestrial source/sink dynamics of CO_2, and interchanges with the atmosphere (a need that is implicit in new policy instruments considered in the REDD++ framework). A mitigation example is testing different approaches with forests: by planting to enhance carbon sinks or reducing emissions from avoided deforestation and degradation. Therefore observations inherent in measures of disturbance are required as well as of land-cover and land-use change, from deforestation, wildfire, or other human activities which also influence albedo and water balance (Running 2008; Justice et al. 2011). Metrics to describe degradation require monitoring at spatial resolution of 30 m resolution and finer. These new demands are in danger of remaining unmet because of delays, and monitoring remains a challenge.

Accessible archives of historical observations are also fundamental to give that vital 20 or 50 year perspective on such changes – Landsat has been making observations since 1972 and significant progress has been made in cross calibrating the radiometry of the different sensors flown (Chander et al. 2009), but more than two thirds of the 7 million+ scenes acquired are held in largely inaccessible archives, which results in very uneven spatial and temporal coverage. Furthermore, the operational status of the Landsat system is still not fully secure. The unbroken record, secured since 1972, might not continue to grow. Landsat 5, which provided an unprecedented (and totally unexpected) 27 years of service suspended imaging mid-November 2011, Landsat 7 still flies, but with compromised sensor performance, and the launch of the next satellite in this series has been delayed. Gaps in the archive might yet be avoided if Landsat 7 survives until the follow-on mission's expected January 2013 launch date.

3.4 Decadal Change Accuracy: Unbroken Chain of Uncertainty to SI Standards

3.4.1 Accuracy and SI Standards

Observations of climate change require stability over decades, and unless overlapping observations are sustained, including verification of stability, absolute accuracy is required. Confidence in these observations depends on how accurately we can

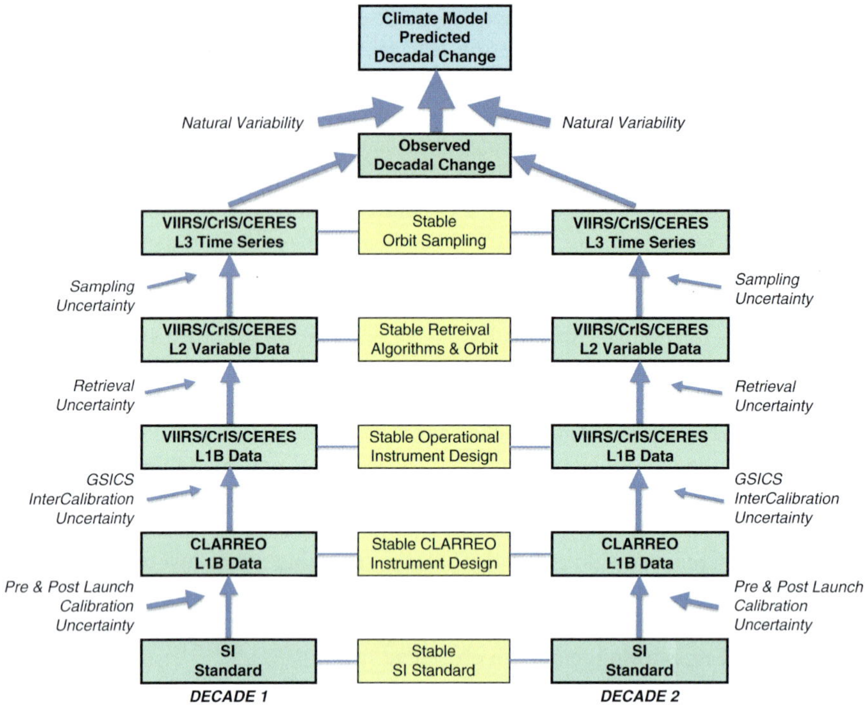

Fig. 6 Traceability of uncertainty in decadal change observations between two decades of data, followed by comparison of the observed decadal change with climate model predicted change. While the entire chain of uncertainty must be characterized, even perfect observations are limited by noise from natural variability of Earth's climate system itself (e.g., ENSO) when used to test climate models. The goal is to drive observation uncertainties to roughly a factor of 2 less than natural variability

relate satellite observations in one decade to those in another decade. However, few observations provide the rigorous on-board calibration and cross-calibration needed. Fortunately, progress is being made in cross calibration of U.S. and European sensors.

The schematic in Fig. 6 shows an example of the traceability required from SI standards as the anchor through instrument calibration, in-orbit intercalibration, retrieval of geophysical properties, orbit sampling, to final decadal change observations that could be used to test climate model predictions. The figure shows the goal of traceability to SI standards at the foundation that have absolute accuracy uncertainty much smaller than the signals expected from decadal change (NRC 2007; Ohring et al. 2007). In support of this, CEOS and GEO have led the development of a new internationally endorsed Quality Assurance Framework for Earth Observation (QA4EO) (CEOS 2008; GEO 2010). The framework concludes that "*All data and*

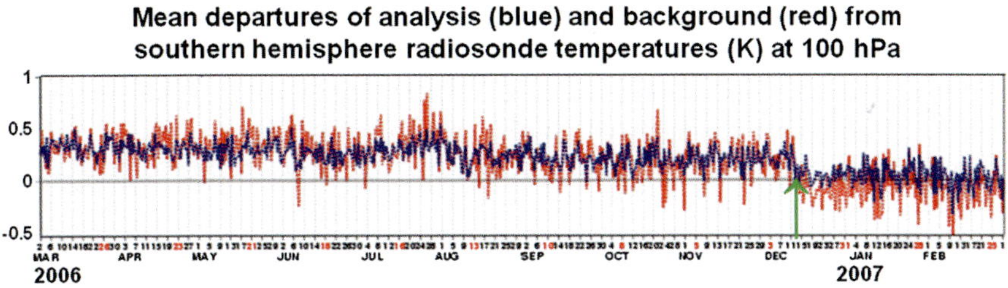

Fig. 7 Time series of the mean and standard deviations of the ECMWF background and analysis temperatures at 100 hPa showing a reduction in the bias errors on 12 December 2006 (*green arrow*) when COSMIC data began to be assimilated (After Luntama et al. 2008; courtesy Anthes 2011)

derived products must have associated with them a Quality Indicator (QI) based on documented quantitative assessment of its traceability to community agreed (ideally tied to SI) reference standards."

Some satellite observations can meet this goal: examples are GNSS radio occultation (e.g., Anthes 2011; Ho et al. 2010; Steiner et al. 2011), ocean altimeters and ice sheet or cloud elevation lidars which trace their accuracy in refractivity or height to SI standards in time measurement. Indeed, there have been marked improvements to atmospheric temperature and water vapor analyses through assimilation of COSMIC observations (see the bias reductions in Fig. 7 as an example and Poli et al. 2011). As another example, the diurnal heating of spacecraft platforms and instruments as they move into and out of the sun's shadow noticeably affects microwave and infrared soundings that can be corrected using radio occultation observations, as the latter are not so affected (Ho et al. 2009; Anthes 2011).

Most satellite instruments, including solar reflected and infrared emitted spectrometers and radiometers, as well as passive microwave instruments, do not currently achieve SI traceable in-orbit climate change accuracy. These instruments rely on less direct arguments of stability in orbit, and overlap of different instruments to remove calibration bias differences. This produces a fragile climate observing system with much weaker ties to SI standards than desired and severe vulnerability to any gaps in the overlap of instruments. While GSICS provides a very useful relative intercalibration of radiometers in orbit, we still lack a set of reference radiometers that could provide the absolute accuracy to serve as "metrology labs" in orbit and benchmarks for the GSICS activity (GSICS 2006; Goldberg et al. 2011).

Examples of designs of such platforms include NASA's CLARREO NRC Decadal Survey mission, and the TRUTHS mission proposed in 2010 to ESA. CLARREO is intended to provide the first observations of the full spectrum of reflected solar radiation and infrared emitted radiation, as well as radio occultation observations. TRUTHS would provide full reflected solar spectra as well as spectral solar irradiance observations. Because of the full spectrum and mission design, these missions serve as SI traceable transfer radiometers in orbit that can be used to increase the accuracy of orbiting operational sensors by matching them in time, space, angle, and wavelength. This includes future sensors covering a broad range of climate variables including temperature, water vapor, clouds, radiation, surface albedo, vegetation, and ocean color. In this sense CLARREO and TRUTHS could become anchors of the global climate observing system, but neither of these missions has an approved launch date.

3.4.2 Stability of Observations and Algorithms

A second key issue is the stability over decades of satellite geophysical retrieval algorithms which all have bias errors larger than decadal climate change signals. Moreover, the algorithms and ancillary data they depend on evolve with time. Current climate studies assume that these biases remain sufficiently stable to cancel out in observing decadal change anomalies, an assumption that should be verified. Otherwise, it would be essential to develop retrieval algorithms that are optimized for decadal change as opposed to optimization for instantaneous retrievals such as those from weather satellites. Another possibility to limit sensitivity to retrieval biases is the use of reflected solar and infrared spectral fingerprinting studies of climate change (Huang et al. 2010; Feldman et al. 2011; Jin et al. 2011). These climate Observing System Simulation Experiment (OSSE) studies have shown that infrared and solar reflected spectral fingerprints are very linear at the large time/space scales relevant to decadal climate change, unlike their highly nonlinear behavior for instantaneous retrievals.

Increasing attention to calibration and to algorithm performance is increasing the overall robustness of the global climate observing system. For example, structural and radiometric measures of plant canopies quantifying vegetation dynamics (terrestrial net primary productivity, FAPAR, Leaf Area Index) are being monitored with improving reliability using satellite observations from a range of polar orbiting platforms (Knyazikhin et al. 1998; Gobron et al. 2006, 2008; Zhao and Running 2010), but this has only been possible as greater attention has been paid to cross calibration and product validation. An example of rigorous intercomparison with reference data (Gobron et al. 2006, 2008) is given in Fig. 8 which shows how plant dynamics vary in both space and time as derived from daily observations from SeaWiFS (1998–2006) and MERIS (2002–2010).

3.4.3 Accuracy

Finally, the question arises as to what level of absolute accuracy is required to eliminate issues with gaps in climate data records, and to eliminate the uncertainties of changing instrument biases in orbit over time? Leroy et al. (2008) use mid-tropospheric temperature interannual variability to suggest an accuracy for infrared radiometers of 0.03 K for a 1 sigma confidence bound. Similar analyses could be performed for a wide range of climate variables and time/space scales.

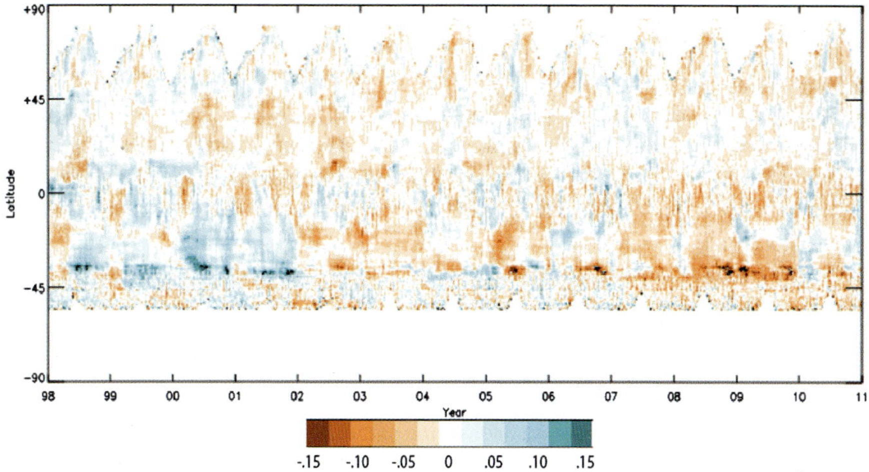

Fig. 8 Monthly zonal FAPAR anomalies relative to the period January 1998 to December 2010 estimated from decadal FAPAR products derived at a resolution of 0.5 × 0.5° from measurements acquired by the SeaWiFS (NASA) and MERIS (ESA) sensors. As rates of photosynthesis are affected by temperature and precipitation, FAPAR is an indicator of climate impacts on vegetation; favorable temperatures and soil moisture availability are accompanied by higher than average FAPAR values, drought and/or excessive temperature are accompanied by lower values (Gobron et al. 2010)

In summary, accuracy is not just about instrument calibration, but is the entire set of analysis steps required to move from SI standards at the foundation, to decadal change of a radiance or a geophysical variable at the other.

3.5 Improving Transitions Between Observing Systems

Arguably the biggest challenge to ensuring homogeneous time series is related to the timing of changes in observing systems and the critical need for continuity. Associated transitions in sampling (both in space and time), instrument accuracy (including biases), and processing methods are a major source of time-dependent biases in time series of Earth system observations. Nowhere is this more evident than in the satellite observing systems because of their relatively short lifetime of about 5 years, but *in situ* observing systems also have had a history of suboptimal transitions between old and new observing methods and systems. In some cases, information from other observations may help bridge gaps and constrain offsets.

Standard practice today either relies on launching a satellite on a planned date or launching in response to the loss of a satellite and/or specific instrument. In the former case, there may or may not be an adequate overlap, while the latter strategy does not comply with the GCMPs of planned overlaps. It inevitably leads to too short, or none-at-all observing overlaps between the old and new systems. Without absolute calibration and the use of exactly the same sampling strategy, undefined time-dependent observing system biases will likely be introduced into the time series. Poorly documented changes in processing systems can also introduce time-dependent biases. Similarly, for *in situ* observations, new observing methods and systems have been introduced with little consideration of the optimal overlap required with legacy systems.

Rule-of-thumb practices have resulted in seldom-adhered-to requirements of at least 1-year overlap between old and new observing systems to fully understand varying seasonal biases. It is unlikely that the overlap needed for a radiometer will be equivalent to that of a spectral irradiance sensor or an altimeter. Similarly, the overlap required for water vapor, precipitation measurements, and temperature are all likely to be different, especially when the sampling and accuracy changes.

Of course, to plan for an overlap, regardless of length, requires some prediction about the lifetime of the legacy observing system. For satellites, this includes the probability of failure of the satellite bus or the instruments. For some satellite research missions, Cramer (Remanifest of NASA's NPP and NOAA'S NPOESS instruments. Personal communication, 2008) and Loeb et al. (2009) have developed a few prototype probability density functions that help to understand the likelihood of failure of both instruments and the satellite bus.

For *in situ* observing systems, plans for a sufficient overlap must include an estimate of observing system degradation beyond which it cannot provide the sampling and accuracy needed to produce homogenous time series. Such analyses are needed for all climate-relevant observing systems. This would enable scientists to objectively communicate priorities for new observing systems. Optimization of observing system transitions could be based on climate risk assessments, which could then be evaluated in context with other requirements for multi-purpose observing systems.

3.6 How to Prioritize?

Observing system experiments (OSEs) have proven exceedingly useful in examining the impacts of a new set of observations (such as from a new satellite) by performing data assimilation with and without the new observations. This methodology also enables estimation of biases. The complexity of 50 ECVs, independent observations and analysis, and high accuracy traceability of all analysis steps to SI standards suggests that there is a need to also prioritize observation requirements within the climate observing system. This is fraught with difficulty because of the different and generally subjective underlying assumptions and the fact that observations are used for multiple purposes. The OSSE methodology (Sect. 3.4) can potentially be used to prioritize within the climate observing system but model errors currently limit their utility. However, as climate models become more accurate, OSSEs will become more effective and powerful, and needed to augment current dependence on scientific intuition "back of the envelope" estimates, and science committee voting approaches.

4 Analyses, Assessments and Reprocessing

Originally the task was getting a single time series of an ECV. Now there is a proliferation of multiple datasets purporting to be "the correct one". Many are created for specific purposes but all differ, often substantially, and the strengths and weaknesses or assumptions may not be well understood or well stated. Consequently, assessments are required to evaluate these aspects and to help improve the datasets. Moreover, continuous reprocessing is essential. Reprocessing can account for recalibration of satellite data from GSICS, take advantage of new knowledge and algorithms, and rectify problems and errors that have become evident. Repeat reprocessing and assessment should be hall-marks of a climate observing system.

Within the WCRP, the GEWEX Data and Assessments Panel is promoting the reprocessing of the GEWEX datasets so that they are globally consistent with regard to water and energy, complete with metadata and error estimates. The goal is to reduce errors, increase continuity, and improve homogeneity while comprehensively documenting uncertainties. The new processing will use calibrated and inter-calibrated satellite radiances for long time series of observations, and ensure that all products will "see" the same atmosphere especially in terms of temperature, water vapor, cloud and radiation. Surface radiative and turbulent fluxes are also included. ESA's Climate Change Initiative is also fostering reprocessing of individual variables to generate ECVs and take advantage of knowledge about problems and improved algorithms. At the same time, GEWEX is promoting the assessment of the variable products, not to rank the algorithms, because each often has a somewhat different application, but rather to adequately characterize each product as to its use in various ways. Some of these reprocessed data sets will provide the first long-term look at climate trends on a truly global basis for a number of climate variables. More generally, these reprocessing and assessment activities are promoted by WDAC and GCOS.

4.1 Reanalyses

Reanalysis is an activity to reprocess past observations in a fixed, state-of-the-art assimilation system. Most reanalysis activities have been for the atmosphere, but some exist for the ocean, sea ice and land variables. Reanalyses are based on data assimilation in numerical models, and are distinct from operational numerical weather prediction (NWP) as they can utilize data which were not received at the nominal analysis time as well as observations that have been more carefully processed than possible in real time. Freezing the analysis system removes the spurious variations that otherwise appear in the NWP analyses, and can potentially result in climate quality globally gridded products. However, the observing system changes as new sensors are developed and aging satellites expire (Fig. 1) thereby exposing different forecast model biases. As a result, some trends are not represented well in current reanalyses. Nevertheless, the model short-term predictions act as a powerful check on inconsistencies and errors in observations and model. The reanalysis process has become fairly mature and has developed variational techniques for bias correction of observations. The result can be an alternative source of an ECV record with an advantage that it is globally complete and associated variables are consistent with the ECV. A large user base is ensured by an open data policy and this enables scrutiny and evaluation of the results.

While reanalyses contain effects of both model and observation bias and error (see Fig. 9), there are some substantial strengths, such as their global scope. Simmons et al. (2010) show how the surface temperature record from reanalysis agrees with other analyses where overlapping data are available, but the reanalysis is able to extend the analysis into data sparse regions and provides a much better and more reliable record.

Uncertainty is important but difficult to quantify. A straightforward way to deal with it is to evaluate a multi-reanalysis collection of the variables of interest (e.g., Fig. 9). In addition, the imbalance of budgets (such as of mass of dry air or water, or energy) in reanalyses is representative of the forecast error (instantaneously) or the model and observation climate bias (long term). This needs to be better taken into account by users of reanalyses data. Lastly, reanalyses can provide the assimilated observations, as well as forecast error and analysis error for each observation.

4.2 Assessments

As well as assessments of datasets of individual variables, assessments of reanalyses are essential. The most comprehensive assessments with a focus on climate change are those of the IPCC that look at all aspects of the science. Nationally within the U.S. a series of Synthesis and Assessment Products (SAPs) has been carried out by the Climate Change Science Program (CCSP) and USGCRP, as well as Committee on the Environment and Natural Resources of the National Science and Technology Council.

The IPCC assessments are primarily based on peer-reviewed literature. But it is not just a review of the literature because conflicting claims and conclusions have to be reconciled to the extent possible. This means examining the methods, assumptions, and data used, and the logic behind the conclusions. The IPCC is convened by the United Nations jointly under UNEP and WMO. Its mandate is to provide policymakers with an objective assessment of the scientific and technical information available about climate change, its environmental and socio-economic impacts, and

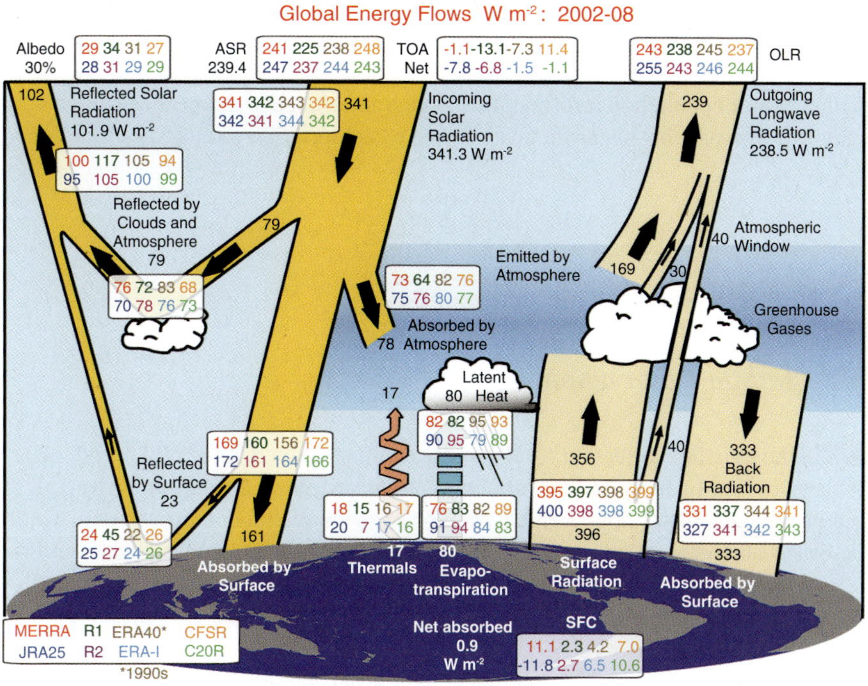

Fig. 9 The components of the global flow of energy through the climate system as given by Trenberth et al. (2009) as background values are compared with values from eight different reanalyses for 2002–2008 (except ERA-40 is for the 1990s), as given at lower left in the Figure, in W m^{-2}. From Trenberth et al. (2011). For example, the estimated imbalance at TOA and at the surface is 0.9 W m^{-2} for the 2002–2008 period, or 0.6 W m^{-2} for the 1990s, but values from reanalyses differ substantially at TOA and at the surface, and also differ between the two values implying a large source or sink in the atmosphere. Differences reveal assimilating model biases and the effects of analysis increments

possible response options. It has provided policymakers assessment reports since 1990, and the Fourth Assessment Report (AR4) was released in 2007. The IPCC assessments are produced through a very open and inclusive process. The volunteer authorship of the AR4 in Working Group I included 152 lead authors and over 400 contributing authors from over 130 countries. In addition, there were more than 30,000 comments from over 600 reviewers, as well as formal coordinated reviews by dozens of world governments. All review comments are addressed, and review editors are in place for each chapter of the report to ensure that this is done in a satisfactory and appropriate manner.

The IPCC assessments provide a snapshot of the state of the science every 6 or 7 years, but increasingly there is a need for yearly, monthly and even shorter–term assessments. The "*State of the Climate*" reports published annually in the *Bulletin of the American Meteorological Society* are a step towards meeting needs between IPCC reports. NOAA's National Climate Data Center (NCDC) also reports monthly on the observed state and provides some commentary on what is happening and why. However, near-real time information and attribution is increasingly in demand, especially when major events occur, such as the 2010 Russian heat wave. How to include model prediction information and guarantee quality and peer review of near real time assessments to ensure that they have "authority" are key issues for climate services.

5 Further Needed Improvements

5.1 In Situ Observations

While the existing collection of *in situ* observations covers most of the high priority and currently feasible measurements, their spatial and temporal coverage is incomplete and many improvements can be envisioned. Such improvements would be based on technical innovations in the measurement techniques, the recognition of new needs for observations, and improved integration of variables for societally-relevant topics, including providing a sound scientific basis for mitigation and adaptation efforts.

There is a general need for improved integration and synthesis of satellite and *in situ* observations beyond that provided by reanalysis. Observations from multiple sources complement each other and provide calibration and validation. It should not be assumed, therefore, that observations from multiple sources are redundant and unnecessary. Some observation systems are currently at risk because they require substantial investments that cannot be done incrementally; or because budget constraints and ageing equipment have gradually reduced capabilities or data quality to unacceptable levels.

Several networks in need of physical repair and maintenance to ensure data quality include stream gauge networks, surface sensors for Earth radiation budget, ground-based snow cover (including snow depth), especially in mountainous areas; gaps exist in observations for ice caps, ice sheets, glaciers, and permafrost, and temperature profiles of permafrost in bore holes that are being degraded or lost by warming. Some important measurements could provide a cost-effective way to enhance the information obtained. These include enhancement of greenhouse gas networks including sensor automation, expansion of the network of ground-based soil moisture measurements, increased measurement frequency/time resolution, and airborne sensor deployments. Accurate and precise ground-based GPS measurements of total column water vapor also contribute to climate-quality data sets, calibration of other instruments, and verification of reanalysis data sets (Wang and Zhang 2009; Vey et al. 2010).

Measurements of variables describing terrestrial fresh water in its liquid and solid phase are currently limited, as are the fluxes (see Jung et al. 2010). Satellite altimetry is used to monitor river and lake levels, but only for a few river basins and large lakes. Fresh water is considered in more detail by Gleick et al. (2013).

Snow-cover extent is mapped daily by satellites, but sensors change and continuing research and surface observations are needed to calibrate and verify satellite products for snow depth and snow water equivalent. Monitoring glaciers and ice caps is important for early detection of climate changes because their contraction indicates warming trends. Satellite observations of polar ice caps, continental mountain glaciers and ice shelves increasingly help provide a regular inventory. Satellite derived digital elevation maps of the ice surface for Greenland and Antarctica are available, though long term commitments to such monitoring are not in place.

One area where potential exists for cost savings, improved efficiency, and more comprehensive observations is through the consolidation and rationalization of the multitude of *in situ* networks that have grown up under different agencies and countries. For instance, the networks for radiosondes, ozonesondes, other atmospheric constituents (GAW), radiation (BSRN), flux towers (IGBP), and so on have been developed for specific purposes. By consolidating some of these measurements increased value accrues and the networks become more sustainable because they serve more purposes.

Numerous bilateral and multilateral international partnerships exist, providing highly productive avenues for coordination and cooperation. Partnership opportunities exist with communities other than the international framework: with defense agencies, the private sector, and non-governmental organizations, although sometimes with adverse consequences. Major strengths include the leveraging of individual national resources toward common goals, and the sharing of data and expertise. However, more effort is needed in overcoming differences in data and metadata standards, data sharing and data policy, and access to currently restricted data (this includes both *in situ* and satellite data).

In summary, WCRP should take a leadership role in an international coordination framework to perform a comprehensive assessment of the research priorities of an operational global *in situ* observation system. WCRP should also provide recommendations for transition from research to operational capability and identify where overlap is needed to prevent critical gaps in this extensive array of climate-relevant observations administered by many agencies from the international community. The challenge to WCRP is to recommend guidelines and identify specific ways that the international community can optimize this mix, across agencies and under consideration of international agreements and participation with other partners. Such a framework and set of guidelines could greatly serve the needs of the climate research community and yet exercise maximum fiscal responsibility for a global observation capability.

5.2 Data Documentation and Adequacy of Metadata

For several decades, metadata and data discovery have been inextricably intertwined because of the difficulty in keeping up with the explosion in observations and data products. Discovery alone, however, is not adequate for understanding observations and, more importantly, temporal variations in those observations. Excellent documentation of environmental observations and data, preferably in peer-reviewed literature, is more important today than ever before:

- Rapid evolution of the global climate adds requirements for understanding temporal variations in observed properties. Pertinent data must be documented so as to unambiguously recognize change and differentiate real change from observational, experimental or analytical error.
- The changing environment increases the importance of older observations that provide context but which may have been collected, processed or synthesized by scientists who are no longer available. Detailed documentation is essential to ensure that today's observations can contribute to answering tomorrow's questions.
- There are increased requirements for sharing data across broad communities with diverse expertise. Users include decision and policy makers, inter-disciplinary scientists, and the general public.
- The international environmental community is coming together in unprecedented collaborations.

A series of international metadata (International Organization for Standardization-ISO) standards have emerged recently, forming the foundation for effectively documenting observed and synthesized data. These standards include mechanisms for describing sensors, data quality assessments, provenance (sources and algorithms), and temporal variations in all these items. They also include mechanisms for creating metadata at many levels (sensor, platform, network, project…) and connecting to related documentation in standard or non-standard forms. The global scientific community needs to work together to:

- *Develop conventions for how standards will be used to describe important data types* to enable meaningful sharing of metadata. Like the Climate and Forecast Conventions for data, metadata conventions will include standard names and ontologies for shared concepts.
- *Extend high-quality documentation with increased emphasis on preservation and sharing of that documentation.* Adoption of the ISO standards supports both of these goals.
- *Participate in evolving the standards as documentation and sharing needs change.*

Considerable progress has been made towards supporting open data across a growing segment of the scientific community. Scientists around the world should share environmental observations along with their documentation, or risk undermining a basic scientific premise of independent verification of results that supports the credibility of the scientific process.

5.3 Tracking Climate Observing Performance

As we strive to be more effective in our climate observing and research activities, an important objective is the effective use of both operations and research for early

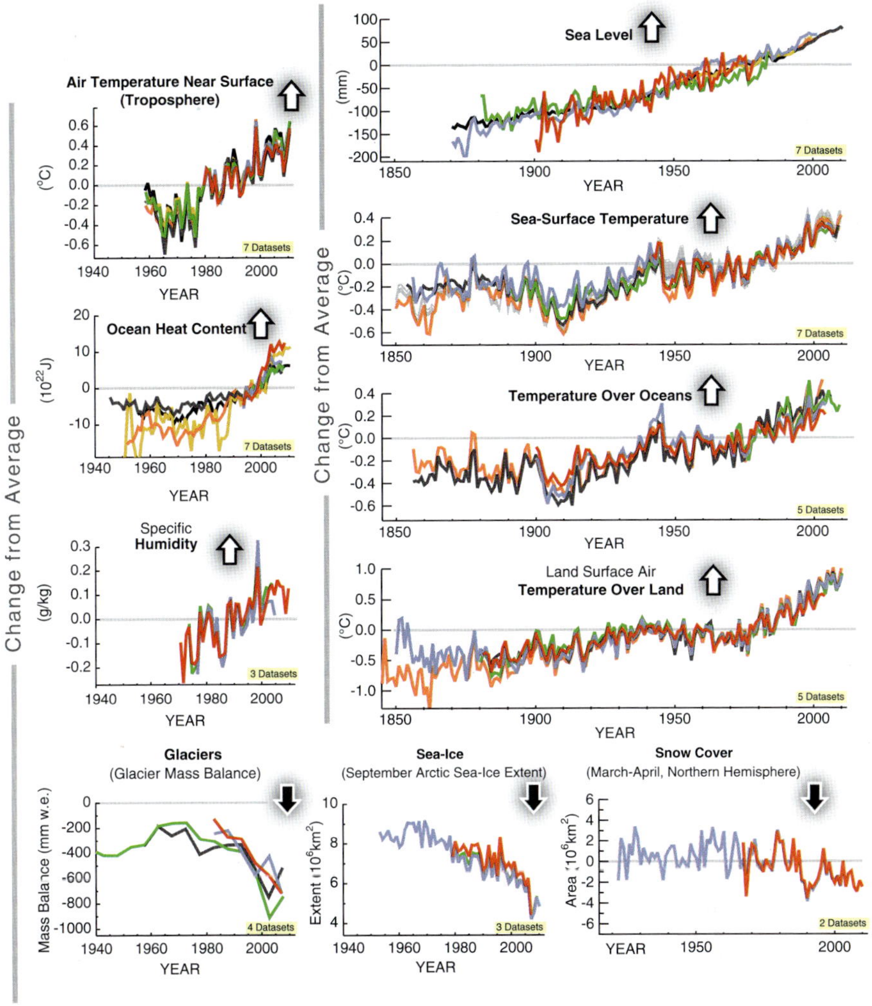

Fig. 10 Observations of the ten indicators over time (SOC 2009) (Adapted from figure courtesy NCDC, NOAA)

identification of time-dependent biases. The International State of the Climate Report and the subsequent special NOAA report (SOC 2009) focused on a set of nine indicators in a warming world. In SOC (2009), numerous indicators and indices representing ECVs were compared and contrasted to ensure that observing systems (satellite and *in situ*) were providing a physically consistent set of information about climate and global change (Fig. 10). These analyses demonstrate the value of collectively analyzing a broad set of essential climate variables across various observing systems using independent time series developed by various science teams.

Figure 10 shows time series from independent observing systems (satellite and *in situ*) and various independent analyses. This kind of display enables checks of consistency among datasets of the same variable and also the physical consistency among variables.

Consistency among other variables is being explored within the GEWEX Data and Assessments Panel for temperature, water vapor, cloud, precipitation, surface fluxes of sensible and latent heat, and surface radiation. This kind of display therefore also reveals changes in the climate that are extremely useful for many purposes.

Nonetheless, understanding differences among datasets, their strengths and weaknesses is also very important in order to properly utilize the most appropriate data for certain purposes. At NCAR a new Climate Data Guide http://climatedataguide.ucar.edu/ is being developed to provide this information about the multitudes of datasets.

5.4 Climate Observations at High Risk

The GCOS is designed to meet evolving national and international requirements for climate observations. Certainly our current observing system and the one in the foreseeable future (taking all planned U.S., European and Asian satellite missions into account), will lead to a lot of new information about our planet and the climate system. Many observations can be used for climate purposes although more so for some ECVs than others. But unless there is major progress on climate observations, we shall not see as much or as clearly as needed for effective climate research and applications. Moreover, progress is much needed to reduce the probability of being tripped up by something unexpected that we cannot grasp with our deficient vision. While the need for climate information has greatly increased, the effort to meet this need has not.

A recent mid-course assessment of the Decadal Survey (NRC 2012) supports our assessment. It notes that despite some successes (e.g., successful launches of the Ocean Surface Topography Mission (OSTM), Aquarius, and the Suomi NPP), a number of significant issues have had damaging effects on the U.S. satellite observing system. These include significant budget shortfalls in NASA and NOAA, launch failures, delays, changes in scope, and cost growth of missions. NOAA has made significant reductions in scope to the future operational Earth satellites, omitting observational capabilities assumed by the Decadal Survey to be part of NOAA's future capability and failing to implement the three new missions recommended for NOAA by the Survey (the Operational GPS Radio Occultation Mission, the Extended Ocean Vector Winds Mission, and the NOAA portion of CLARREO).

Furthermore, the U.S. Earth observing capability from space is in jeopardy as older missions fail faster than they are replaced; thus the number of NASA and NOAA Earth observing instruments in space is likely to decline to as little as 25 % of the current number by 2020 (Fig. 11, NRC 2012).

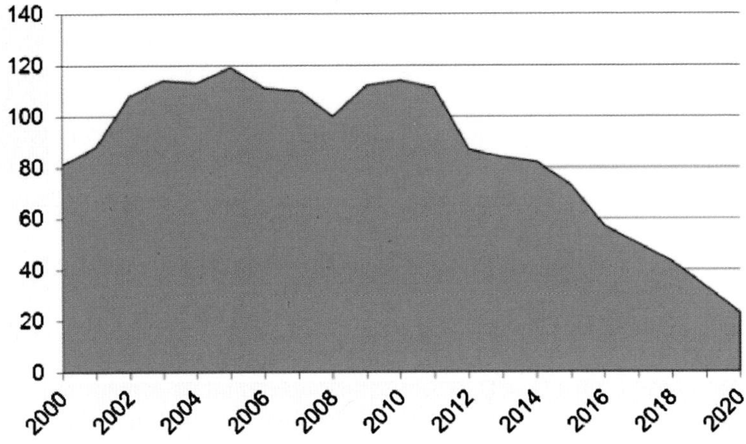

Fig. 11 Estimated number of NASA/NOAA Earth Observing instruments in space out to 2020 (NRC 2012)

While significant progress has been made in the last decade, we conclude that the climate observation architecture is still very much a work in progress, with a long way to go before we achieve a fully implemented climate observing system. Serious challenges remain in the areas of data accuracy, independence, continuity, and prioritization within the observing system. Comprehensive standard metadata is also missing for many observations. Much more complete spatial and temporal sampling is essential if we are to determine how extremes are changing; as an example the need for hourly data on precipitation has long been recognized because of its inherent intermittent nature. Changes in extremes are the main way climate change is perceived (Trenberth 2011) and of special interest are changes in hurricanes, storm surges, severe convection, tornadoes, hail, lightning, floods, droughts, heat waves and wild fires. All of these depend on detailed information about precipitation: its distribution, intensity, frequency, amount, type, and sequences in time. The evidence is increasing for changes in weather and climate extremes whereby, for example, 500-year events become 50-year events, but the information is not being made available and planning for those changes is wholly inadequate. The need to assess model capabilities from this standpoint is also clear.

Other needs are rearing up in the form of irreversible climate change and tipping points as thresholds are crossed, and whether it is possible to even recognize that we have passed such a point when we do, until decades or centuries later, when it is far too late to do anything about it (Solomon et al. 2009). A classic example is the increased melting of the Greenland and West Antarctic ice sheets. Are these reversible, or is it already too late?

Nations have continued to recognize the needs for a fully implemented climate observing system, for example through acceptance of the GCOS Implementation

Plans and other reports by the Parties to the UNFCCC: most recently GCOS (2010); and in the resolutions of the WMO Congresses relating to GCOS. But in many cases, funding commitments have not yet been made by GCOS member nations to provide or improve key components of the climate observing system. As we have seen with losses of ADEOS, Cryosat, OCO, Glory, inability to fully implement COSMIC-2, delays of NPP and JPSS, CLARREO, DESDynI, the GPM follow-on, limb soundings, as well as the TAO buoy network preventive maintenance, the stream gauge network and an integrated carbon-tower network; the risk of major satellite and *in situ* observing system gaps is already present, and will grow in the future.

Climate observations today contain many very good pieces, but are not yet well coordinated, understood, developed, maintained and preserved as a true global observing system. Satellite and *in situ* observations must be synthesized and analyzed and reanalyzed into usable and well documented integrated climate quality products. We must solve these challenges if we are not to walk blindly into our planet's future.

6 Appendix A: The GCOS Organizational Framework

The Global Climate Observing System activities are collectively sponsored by the (WMO), Intergovernmental Oceanographic Commission (IOC) of the United Nations Educational, Scientific and Cultural Organization (UNESCO), United Nations Environment Program (UNEP), and International Council of Science (ICSU) to meet national and international needs for climate-related observations of atmosphere, ocean and land. GCOS addresses the observations themselves, the transmission and management of data, the establishment of fundamental climate data records and the formation of products from these data records. In undertaking its review and advisory role, GCOS collaborates with other entities active in these fields, including the World Climate Research Program (WCRP).

GCOS functions through the contributions of nations to help implement:

- component comprehensive observing systems, principally the GOS and Global Atmosphere Watch (GAW), the IOC-led Global Ocean Observing System (GOOS) and the FAO-led Global Terrestrial Observing System (GTOS);
- baseline and reference networks designated or established for specific monitoring purposes;
- observing principles and guidelines for dataset production;
- operation of regional lead centers, network monitoring centers and lead centers for analysis/archiving and the reference upper-air measurement network;
- a cooperation mechanism and associated technical program for observing-system improvements in developing countries; and
- coordination of GCOS activities at national and regional levels across the atmospheric, oceanic and terrestrial domains.

GCOS is guided by a steering committee, and supported by co-sponsored panels, and by a secretariat working alongside those of WMO, GOOS and GTOS.

GCOS focuses on observations to support the United Nations Framework Convention on Climate Change (UNFCCC). Its activities include detailed assessments of the adequacy of the composite observing system, statements of required actions and reports on progress, and it interacts with the UNFCCC's Subsidiary Body for Scientific and Technological Advice (SBSTA) and open public reviews via responses and requests. Activities also cover many systematic observational needs for climate-change assessment, research and the provision of climate services, and serve many societal benefit areas of the GEOSS, including agriculture, biodiversity, climate, disasters, ecosystems, energy, health, water and weather.

The Second Adequacy Report (GCOS 2003) identified a set of ECVs judged to be the minimum required to support the work of the Convention and to be technically and economically feasible for systematic observation. It was followed by a 5–10 year implementation plan in 2004, which identified 131 specific actions. The response to the space-based actions was coordinated by the CEOS, with the CGMS – the international forum for the exchange of technical information on geostationary and polar orbiting meteorological satellite systems.

Acronyms

ALOS	Advanced Land Observing Satellite
ADEOS	Advanced Earth Observing Satellite
AIRS	Atmospheric Infrared Sounder
AR4	Fourth Assessment Report (IPCC)
ATMS	Advanced Technology Microwave Sounder
BSRN	Baseline Surface Radiation Network
CCSP	Climate Change System Program
CDR	Climate Data Record
CEOS	Committee on Earth Observation Satellites
CERES	Clouds and the Earth's Radiant Energy System
CF	Climate and Forecast
CGMS	Coordination Group for Meteorological Satellites
CLARREO	Climate Absolute Radiance and Refractivity Observatory
COSMIC	Constellation Observing System for Meteorology, Ionosphere and Climate
CrIS	Crosstrack Infrared Sounder
DESDynl	Deformation, Ecosystem Structure, and Dynamics of Ice
DoD	Department of Defense
EarthCARE	Earth, Cloud, Aerosol, Radiation and Energy
ECMWF	European Centre for Medium-range Weather Forecasts
ECV	Essential Climate Variable
ENSO	El Niño-Southern Oscillation

EOS	Earth Observing System
ERA	ECMWF Re-Analysis
ESA	European Space Agency
EUMETSAT	European Organisation for the Exploitation of Meteorological Satellites
FAPAR	Fraction of Absorbed Photosynthetically Active Radiation
GAW	Global Atmospheric Watch
GCOM	Global Change Observation Mission (JAXA)
GCOS	Global Climate Observing System
GCMPs	GCOS Climate Monitoring Principles
GEO	Group on Earth Observations
GEOSS	Global Earth Observation System of Systems
GEWEX	Global Energy and Water Exchanges (WCRP)
GFCS	Global Framework for Climate Services
GMES	Global Monitoring for Environment and Security
GNSS	Global Navigation Satellite System
GOES	Geosynchronous Operational Environmental Satellite
GOOS	Global Ocean Observing System
GOS	Global Observing System
GOSAT	Greenhouse Gases Observation Satellite (JAXA)
GPM	Global Precipitation Mission
GPS	Global Positioning System
GRUAN	GCOS Reference Upper-Air Network
GSICS	Global Space-based Intercalibration System
GTOS	Global Terrestrial Observing System
ICESAT	Ice, Cloud, and Land Elevation Satellite
ICSU	International Council for Science
IGBP	International Geosphere-Biosphere Programme
IGDDS	WMO Integrated Global Data Dissemination Service
IOC	Intergovernmental Oceanographic Commission
IPCC	Intergovernmental Panel on Climate Change
JAXA	Japan Aerospace Exploration Agency
JMA	Japanese Meteorological Agency
JPSS	Joint Polar Satellite System
LAI	Leaf Area Index
MERIS	Medium Resolution Imaging Spectrometer
MERRA	Modern Era Retrospective-Analysis for Research and Applications
MODIS	Moderate Resolution Imaging Spectro-radiometer (NASA)
NASA	National Aeronautics and Space Administration
NCAR	National Center for Atmospheric Research
NCDC	National Climatic Data Center (NOAA)
NOAA	National Oceanic and Atmospheric Administration
NPOESS	National Polar-Orbiting Operational Environmental Satellite System
NPP	NPOESS Preparatory Project
NRC	National Research Council (USA)

NWP	Numerical Weather Prediction
OCO	Orbiting Carbon Observatory
OMPS	Ozone Mapping and Profiler Suite
OSE	Observing System Experiment
OSSE	Observing System Simulation Experiment
REDD	Reducing Emissions from Deforestation and Forest Degradation
SAPS	Synthesis and Assessment Products
SBSTA	Subsidiary Body for Scientific and Technological Advice
SCOPE-CM	Sustained Co-Ordinated Processing of Environmental satellite data for Climate Monitoring
SI	International System of units (Système International)
SMAP	Soil Moisture Active/Passive
SOC	State of Climate
TOA	Top of Atmosphere
TRUTHS	Traceable Radiometry Underpinning Terrestrial- and Helio-Studies
UNEP	United Nations Environment Programme
UNFCCC	United Nations Framework Convention on Climate Change
USGCRP	United States Global Change Research Program
VIIRS	Visible/Infrared Imager/Radiometer Suite
WCC-3	World Climate Conference-3
WCRP	World Climate Research Programme
WDAC	WCRP Data Advisory Council
WG	Working Group
WMO	World Meteorological Organization
WOAP	WCRP Observation and Assimilation Panel

Acknowledgments We thank Adrian Simmons for substantial comments and suggestions. The National Center for Atmospheric Research is sponsored by the National Science Foundation.

References

Anthes RA (2011) Exploring Earth's atmosphere with radio occultation: contributions to weather, climate and space weather. Atmos Meas Tech 4:1077–1103. www.atmos-meas-tech.net/4/1077/2011/ doi:10.5194/amt-4-1077-2011

Brink AB, Eva HD (2009) Monitoring 25 years of land cover change dynamics in Africa: a sample based remote sensing approach. Appl Geogr 29:501–512

Brönnimann S, Ewen T, Luterbacher J, Diaz HF, Stolarski R, Neu U (2008) A focus on climate during the past 100 years. In: Brönnimann SJ, Luterbacher, Ewen T, Diaz HF, Stolarski R, Neu U (eds) Climate variability and extremes during the past 100 years. Adv Global Change Res 33:1–25

CEOS (2008) EO handbook 2008. Climate change special edition 2008. http://www.eohandbook.com/eohb2008/

Chander G, Markham BL, Helder DL (2009) Summary of current radiometric calibration coefficients for Landsat MSS, TM, ETM+, and EO-1 ALI sensors. Remote Sens Environ 113(5):893–903

Feldman DR, Algieri CA, Ong JR, Collins WD (2011) CLARREO shortwave observing system simulation experiments of the twenty-first century: simulator design and implementation. J Geophys Res 116:D10107. doi:10.1029/2010JD015350

GCOS (2003) The second report on the adequacy of the global observing systems for climate in support of the UNFCCC. GCOS-82, WMO-TD/No. 1143. WMO, Geneva, 74 pp

GCOS (2007) GCOS Reference Upper-Air Network (GRUAN): justification, requirements, siting and instrumentation options. GCOS-12, WMO-TD/No. 1379, WMO, Geneva, 25 pp

GCOS (2009) Progress report on the implementation of the global observing system for climate in support of the UNFCCC 2004–2008. GCOS-129, WMO-TD/No. 1489, GOOS-173, GTOS-70. http://www.wmo.int/pages/prog/gcos/Publications/gcos-129.pdf

GCOS (2010) Implementation plan for the global observing system for climate in support of the UNFCCC. GCOS-138, Geneva, 180 pp

GEO (Group on Earth Observations) (2005) The Global Earth Observation System of Systems (GEOSS) 10-Year Impl Plan. http://www.earthobservations.org/documents/10-Year%20Implementation%20Plan.pdf

GEO (2010) A quality assurance framework for Earth observation: Principles. V4, Jan 2010. http://qa4eo.org/docs/QA4EO_Principles_v4.0.pdf

Gleick P, Cooley H, Famiglietti J, Lettenmaier D, Oki T, Vörösmarty C, Wood E (2013) Improving understanding of the global hydrological cycle. In: Asrar GR, Hurrell JW (eds) Climate science for serving society: research, modelling and prediction priorities, Springer, in press

Gobron N, Pinty B, Aussedat O, Chen J, Cohen WB, Fensholt R, Gond V, Hummerich KF, Lavergne T, Mélin F, Privette JL, Sandholt I, Taberner M, Turner DP, Verstraete MM, Widlowski J-L (2006) Evaluation FAPAR products for different canopy radiation transfer regimes: methodology and results using JRC products derived from SeaWiFS against ground-based estimations. J Geophys Res 111:D13110. doi:10.1029/2005JD006511

Gobron N, Pinty B, Aussedat O, Taberner M, Faber O, Mélin F, Lavergne T, Robustelli M, Snoeij P (2008) Uncertainty estimates for the FAPAR operational products derived from MERIS – impact of top-of-atmosphere radiance uncertainties and validation with field data. Remote Sens Environ 112:1871–1883. doi:10.1016/j.rse.2007.09.011

Gobron N, Knorr W, Belward AS, Pinty B (2010) Fraction of Absorbed Photosynthetically Active Radiation (FAPAR). In: state of the climate in 2009. Bull Am Meteorol Soc 91:S50–S52

Goldberg M, Coauthors (2011) The global space-based inter-calibration system. Bull Am Meteorol Soc 92:467–475

GSICS (2006) Implementation plan for a Global Space-based Inter-calibration System (GSICS). WMO-CGMS, WMO, Geneva, 22 pp

Ho S-P, He W, Kuo Y-H (2009) Construction of consistent temperature records in the lower stratosphere using Global Positioning System radio occultation data and microwave sounding measurements. In: Steiner AK, Pirscher B, Foelsche U, Kirchengast G (eds) New horizons in occultation research. Springer, Berlin, pp 207–217

Ho S-P, Kuo Y-H, Schreiner W, Zhou X (2010) Using SI-traceable Global Positioning System Radio Occultation measurements for climate monitoring. In: state of the climate in 2009. Bull Am Meteorol Sci 91:S36–S37

Huang Y, Leroy S, Gero PJ, Dykema J, Anderson J (2010) Separation of longwave climate feedbacks from spectral observations. J Geophys Res 115:D07104. doi:10.1029/2009JD012766

IPCC (Intergovernmental Panel on Climate Change) (2007) In: Solomon S, Qin D, Manning M, Chen Z, Marquis M, Averyt KB, Tignor M, Miller HL (eds) Climate change 2007: the physical science basis, contribution of working group I to the fourth assessment report of the Intergovernmental Panel on Climate Change. Cambridge University Press, Cambridge/New York, 996 pp

Jin Z, Wielicki BA, Loukachine C, Charlock TP, Young D, Noël S (2011) Spectral kernel approach to study radiative response of climate variables and interannual variability of reflected solar spectrum. J Geophys Res 116:D10113. doi:10.1029/2010JD015228

Jung M, Reichstein M, Ciais P, Seneviratne SI, Sheffield J, Goulden ML, Bonan GB, Cescatti A, Chen J, Running S et al (2010) Recent decline in the global land evapotranspiration trend due to limited moisture supply. Nature 467:951–954

Justice CO, Giglio L, Roy D, Boschetti L, Csiszar I, Davies D, Korontzi S, Schroeder W, O'Neal K, Morisette J (2011) MODIS-derived global fire products. Remote Sens Digit Image Process 11(Pt. 5):661–679. doi:10.1007/978-1-4419-6749-7_29

Karl TR, Diamond HJ, Bojinski S, Butler JH, Dolman H, Haeberli W, Harrison DE, Nyong A, Rösner S, Seiz G, Trenberth KE, Westermeyer W, Zillman J (2010) Observation needs for climate information, prediction and application: capabilities of existing and future observing systems. In: World Climate Conference-3, Elsevier, Procedia Environmental Sciences 1, pp 192–205. doi:10.1016/j.proenv.2010.09.013

Knyazikhin Y, Martonchik JV, Myneni RB, Diner DJ, Running SW (1998) Synergistic algorithm for estimation of vegetation canopy leaf area index and fraction of absorbed photosynthetically active radiation from MODIS and MISR data. J Geophys Res 103(32):32,257–32,276

Leroy SS, Anderson JG, Ohring G (2008) Climate signal detection times and constraints on climate benchmark accuracy requirements. J Clim 21:841–846

Loeb NG, Wielicki BA, Wong T, Parker PA (2009) Impact of data gaps on satellite broadband radiation records. J Geophys Res 114:D11109. doi:10.1029/2008JD011183

Luntama J-P, Kirchengast G, Borsche M, Foelsche U, Steiner A, Healy S, von Engeln A, O'Clerigh E, Marquardt C (2008) Prospects of the EPS GRAS mission for operational atmospheric applications. Bull Am Meteorol Soc 89:1863–1875

Manton MJ, Belward A, Harrison DE, Kuhn A, Lefale P, Rösner S, Simmons A, Westermeyer W, Zillman J (2010) Observation needs for climate services and research. World Climate Conference-3, Elsevier, Procedia Environmental Sciences 1, pp 184–191. doi:10.1016/j.proenv.2010.09.012

National Research Council (NRC) (2004) Climate data records from Environmental Satellites. National Academies Press, Washington DC, 150 pp. ISBN:0-309-09168-3

NRC (2007) Earth science and applications from space: national imperatives for the next decade and beyond. National Academies Press, Washington DC, 428 pp. ISBN-10: 0-309-14090-0

NRC (2008) Earth observations from space-the first 50 years of scientific achievements. National Academies Press, Washington, DC, 129 pp

NRC (2012) Earth science and applications from space-a midterm assessment of NASA's implementation of the decadal survey. National Academies Press, Washington DC, 124 pp. ISBN-10: 0-309-25702-6

Ohring et al (2007) Achieving satellite instrument calibration for climate change. Eos Trans AGU 88(11):136. http://www.agu.org/eos_elec/eeshome.html

Poli P, Healy SB, Dee DP (2011) Assimilation of Global Positioning System radio occultation data in the ECMWF ERA–Interim reanalysis. Q J R Meteorol Soc 136:1970–1990

Privette J, Barkstrom B, Bates J, Bonadonna M, Boyd K, Cecil D-W, Cramer B, Davis G, Karl T, Kaye J, Koblinsky C, Tanner M, Young D (2008) Restoration of NPOESS climate capabilities–climate data records. 4th symposium future national operational environmental satellite systems, The 88th AMS Annual Meeting, New Orleans. http://ams.confex.com/ams/88Annual/techprogram/paper_131058.htm

Running SW (2008) Ecosystem disturbance, carbon, and climate. Science 321:652–653

Schneider AM, Friedl A, Potere D (2009) A new map of global urban extent from MODIS satellite data. Environ Res Lett 4:44003. doi:10.1088/1748-9326/4/4/044003

Simmons AJ, Willett KM, Jones PD, Thorne PW, Dee DP (2010) Low-frequency variations in surface atmospheric humidity, temperature, and precipitation: inferences from reanalyses and monthly gridded observational data sets. J Geophys Res 115:D01110. doi:10.1029/2009JD012442

SOC (State of the Climate) (2009) Report at a glance. NOAA. http://www1.ncdc.noaa.gov/pub/data/cmb/bams-sotc/2009/bams-sotc-2009-brochure-lo-rez.pdf

Solomon S, Plattner G-K, Knuttic R, Friedlingstein P (2009) Irreversible climate change because of carbon dioxide emissions. PNAS 106:1704–1709. doi:10.1073_pnas.0812721106

Steiner AK, Lackner BC, Ladstädter F, Scherllin-Pirscher B, Foelsche U, Kirchengast G (2011) GPS radio occultation for climate monitoring and change detection. Radio Sci 46:RS0D24. doi:10.1029/2010RS004614

Trenberth KE (2008) Observational needs for climate prediction and adaptation. WMO Bull 57:17–21
Trenberth KE (2011) Attribution of climate variations and trends to human influences and natural variability. Wiley Interdiscip Rev Clim Change 2(6):925–930. doi:10.1002/wcc.142
Trenberth KE, Karl TR, Spence TW (2002) The need for a systems approach to climate observations. Bull Am Meteorol Soc 83:1593–1602
Trenberth KE, Moore B, Karl TR, Nobre C (2006) Monitoring and prediction of the Earth's climate: a future perspective. J Clim 19:5001–5008
Trenberth KE, Fasullo JT, Kiehl J (2009) Earth's global energy budget. Bull Am Meteorol Soc 90:311–323
Trenberth KE, Fasullo JT, Mackaro J (2011) Atmospheric moisture transports from ocean to land and global energy flows in reanalyses. J Clim 24:4907–4924. doi:10.1175/2011JCLI4171.1
USGCRP (2003) Strategic plan for the U.S. climate change science program, July 2003, Chapter 12, observing and monitoring the climate system, Washington DC, 142 pp
Vey S, Dietrich R, Rülke A, Fritsche M, Steigenberger P, Rothatcher M (2010) Validation of precipitable water vapor within the NCEP/DOE reanalysis using global GPS observations from one decade. J Clim 23:1675–1695. doi:10.1175/2009JCLI2787.1
Wang J, Zhang L (2009) Climate applications of a global, 2-hourly atmospheric precipitable water dataset derived from IGS tropospheric products. J Geod 83:209–217. doi:10.1007/s00190-008-0238-5
Wilson J, Dowell M, Belward AS (2010) European capacity for monitoring and assimilating space-based climate change observations–Status and Prospects. Publ Off Eur Union, JRC54704, 40 pp. ISBN: 978-92-79-15154-5
World Meteorological Organization (2011) Climate knowledge for action: a global framework for climate services – empowering the most vulnerable, WMO/TD-No. 1065, Geneva, p 240
Zhao M, Running SW (2010) Drought induced reduction in global terrestrial net primary production from 2000 through 2009. Science 329:940–943
Zibordi G, Holben B, Melin F, D'Alimonte D, Berthon J-F, Slutsker I, Giles D (2010) AERONET-OC: an overview. Can J Remote Sens 36:488–497

On the Reprocessing and Reanalysis of Observations for Climate

Michael G. Bosilovich, John Kennedy, Dick Dee, Rob Allan, and Alan O'Neill

Abstract The long observational record is critical to our understanding of the Earth's climate, but most observing systems were not developed with a climate objective in mind. As a result, tremendous efforts have gone into assessing and reprocessing the data records to improve their usefulness in climate studies. The purpose of this paper is to both review recent progress in reprocessing and reanalyzing observations, and summarize the challenges that must be overcome in order to improve our understanding of climate and variability. Reprocessing improves data quality through more scrutiny and improved retrieval techniques for individual observing systems, while reanalysis merges many disparate observations with models through data assimilation, yet both aim to provide a climatology of Earth processes. Many challenges remain, such as tracking the improvement of processing algorithms and limited spatial coverage. Reanalyses have fostered significant research, yet reliable global trends in many physical fields are not yet attainable, despite significant advances in data assimilation and numerical modeling. Oceanic reanalyses have

M.G. Bosilovich (✉)
Global Modeling and Assimilation Office, Goddard Space Flight Center, NASA, 8800 Greenbelt Rd., Greenbelt, MD 20771, USA
e-mail: Michael.Bosilovich@nasa.gov

J. Kennedy • R. Allan
The Met Office, Hadley Centre, FitzRoy Road, Exeter, Devon EX1 3PB, UK
e-mail: John.Kennedy@metoffice.gov.uk; Rob.Allan@metoffice.gov.uk

D. Dee
European Centre for Medium Range Weather Forecasting (E.C.M.W.F), Shinfield Park, Reading RG2 9AX, UK
e-mail: Dick.Dee@ecmwf.int

A. O'Neill
Department of Meteorology, National Centre for Earth Observation, The University of Reading, Earley Gate Bldg 58, Reading RG6 6BB, UK
e-mail: Alan.ONeill@nceo.ac.uk

G.R. Asrar and J.W. Hurrell (eds.), *Climate Science for Serving Society: Research, Modeling and Prediction Priorities*, DOI 10.1007/978-94-007-6692-1_3,

made significant advances in recent years, but will only be discussed here in terms of progress toward integrated Earth system analyses. Climate data sets are generally adequate for process studies and large-scale climate variability. Communication of the strengths, limitations and uncertainties of reprocessed observations and reanalysis data, not only among the community of developers, but also with the extended research community, including the new generations of researchers and the decision makers is crucial for further advancement of the observational data records. It must be emphasized that careful investigation of the data and processing methods are required to use the observations appropriately.

Keywords Essential climate variables • Climate data records • Data rescue • Data provenance • Reanalysis • Uncertainty • Bias correction

1 Reprocessing Observations

A major difficulty in understanding past climate change is that, with very few exceptions, the systems used to make the observations that climate scientists now rely on were not designed with their needs in mind. Early measurements were often made out of simple scientific curiosity or needs other than for understanding climate or forecasting it; latterly, many systems have been driven by other needs such as operational weather forecasting, or by accelerating improvements in technology. This has two major consequences.

The first consequence is that although large numbers of observations are available in digital archives, many more still exist only as paper records, or on obsolete electronic media and are therefore not available for analysis. Measurements made by early satellites, whaling ships, missions of exploration, colonial administrators, and commercial concerns (to name only a few) are found in archives scattered around the world. Finding, photographing and digitizing observations from paper records and locating machines capable of reading old data tapes, punch cards, strip charts or magnetic tapes are each time-consuming and costly, but they are vital to improving our understanding of the climate. Furthermore, there is a growing need for longer, higher quality data bases of synoptic timescale phenomena in order to address questions and concerns about changing climate and weather extremes, risks and impacts under both natural climatic variability and anthropogenic climate change. Such demands are leading to a greater emphasis on the recovery, imaging, digitization, quality control and archiving of, plus ready access to, daily to sub-daily historical weather observations. These new data will ultimately improve the quality of the various reanalyses that rely on them. There is also a sense of urgency as many observations are recorded on perishable media such as paper and magnetic tapes which degrade over time. Without intervention, our ability to understand and reconstruct the past is disintegrating in a disturbingly literal sense.

The second major consequence is that current observation system requirements for climate monitoring and model validation such as those specified by GCOS (http://www.wmo.int/pages/prog/gcos/index.php?name=ClimateMonitoringPrinciples) – typically emphasizing continuity and stability over resolution and timeliness – are met by few historical observing systems. Changes in instrumentation, reporting times and station locations introduce non-climatic artifacts in the data necessitating consistent reprocessing to recover homogeneous climate records. Nevertheless, reliable assessments of changes in the global climate have been made such as the IPCC's statement that "warming of the climate system is unequivocal". This assessment relies on the many multi-decadal climate series which now exist.

Reprocessing of observations aims to improve the quality of the data through better algorithms and to understand and communicate the errors and consequent uncertainties in the raw and processed observations. Reanalyses differ from reprocessed observational data sets in that sophisticated data assimilation techniques are used in combination with global forecast models to produce global estimates of continuous data fields based on multiple observational sources (to be discussed in the following section).

1.1 Data Recovery and Archiving

A vital first step for the understanding of historical data and hence past climate is to digitize and make freely available the vast numbers of measurements, other observations and related metadata that currently exist only in hard copy archives or on inaccessible (or obsolete) electronic media. Some estimates suggest that the number of undigitized observations prior to the Second World War is larger than the number of observations currently represented in the largest digital archives.

Digitizing large numbers of observations that are printed or hand-written in a variety of languages is labor intensive: imaging fragile paper records is time consuming and optical character recognition (OCR) technology is not yet capable of dealing with handwritten log book or terrestrial registers entries, so they must be keyed by hand. Scientific projects such as CLIWOC (García-Herrera et al. 2005), RECLAIM (Wilkinson et al. 2011) and the international ACRE initiative (Atmospheric Circulation Reconstructions over the Earth, Allan et al. 2011) have worked to recover and make available these observations. More recently they have been supplemented by citizen science projects such as oldweather.org (http://www.oldweather.org) and Data.Rescue@Home (http://www.data-rescue-at-home.org/) which have reliably and rapidly digitized large numbers of meteorological observations online at the same time as increasing public engagement with science via lively e-communities. Such projects are not only of climatological interest but can also be of wider historical interest (Allan et al. 2012).

The international ACRE initiative (Allan et al. 2011) both undertakes data rescue and facilitates data recovery projects around the world and their integration with existing data archives. A number of these data archives exist. The International Comprehensive Ocean Atmosphere Data Set (ICOADS Woodruff et al. 2010) holds marine meteorological reports covering a wide range of surface variables.

The World Ocean Database (WOD, Showstack 2009) has large holdings of oceanographic measurements. The Integrated Surface Database (ISD, Lott et al. 2008) holds high-temporal resolution data for land stations. The International Surface Pressure Databank (ISPD, Yin et al. 2008) contains measurements of surface pressure from ICOADS and land stations, supplemented by information about tropical cyclones from the International Best Track Archive for Climate Stewardship (IBTrACS, Knapp et al. 2010). The Global Precipitation Climatology Centre (GPCC) has gathered precipitation observations from many different sources. The International Surface Temperature Initiative (ISTI, Thorne et al. 2011b) is bringing together temperature measurements from many different sources to provide a single, freely available databank of temperature measurements combined with metadata concerning the provenance of the data. Nevertheless, these various activities are very fragile, and often only exist as a result of 'grassroots' actions by the climate science community (Allan et al. 2011, 2012). These projects and initiatives urgently need to be imbedded in an overarching, sustainable, fully funded and staffed international infrastructure that oversees data rescue activities, and compliments the various implementation and strategy plans and documents on data through international coordinating bodies, such as GCOS, GEO, WMO and WCRP.

The consolidation of meteorological, hydrological and oceanographic reports and observations into large archives facilitates the creation of a range of 'summary' data sets which are widely used in climate science and can also act as a focus for an international community of researchers. However, further consolidation could bring greater benefits. A land equivalent of the ICOADS, for example, would bring together many of the elements needed to fully describe the meteorological situation and potentially reduce the efforts that are currently expended to maintain and grow a large number of different datasets. In fact, both the terrestrial and marine data efforts need to be integrated and better linked up under an international framework that supports their activities in a fully sustainable manner.

1.2 Data Set Creation and Evaluation

The difficulties of converting raw observations into data sets which are of use to climate researchers are well documented (e.g. Lyman et al. 2010; Thorne et al. 2011a; Kent et al. 2010; Lawrimore et al. 2011; Hossain and Huffman 2008). Systematic errors and inhomogeneities in data series caused by changes in instrumentation, time of observation and in the environment of the sensor are often as large, or larger than, the signals we hope to detect. Without reliable traceability back to international measurement standards, the problem of detecting and accounting for these errors is not easy. Before the satellite era, observations were often sparsely distributed. Various methods have been devised to impute the values of climatological variables at locations and times when no such observations were made. The problems are further compounded by the necessity of making approximations, using uncertain inputs (such as climatologies), the use of different data archives and having sometimes limited statistics with which to estimate important parameters. Three examples will help to illustrate some of these difficulties and the way that they have been tackled.

One long running example is seen in the different reprocessings of the data from the satellite-based Microwave Sounding Units (MSU) which can be used to derive vertical temperature profiles through the free atmosphere (Thorne et al. 2011a). The earliest processing by Spencer and Christy (1990) suggested a monthly precision of 0.01° C in the global average lower troposphere temperatures but the lack of a trend in the satellite data was not physically consistent with contemporary surface temperature estimates. However, when other teams (Prabhakara et al. 2000; Vinnikov and Grody 2003; Mears et al. 2003) processed the data they found quite different long term behavior. Successive iterations of the datasets have considered an increasingly broad range of confounding factors including orbital decay, hot target temperature and diurnal drift. Twenty years of analysis and reprocessing have undoubtedly improved the overall understanding of the MSU instruments (Christy et al. 2003; Mears and Wentz 2009a, b), the quality of the data sets and estimates of atmospheric temperature trends, but despite these improvements temperature trends from the different products still do not agree. This implies either the existence of unknown systematic effects, or significant sensitivity to data processing choices. Mears et al. (2011) used a monte-carlo approach to assess the uncertainty arising from data processing choices, but this did not fully bridge the gap between their analysis and others.

In the past decade, the view of ocean heat content has changed considerably. Early estimates of global ocean heat content (Levitus et al. 2000) showed marked decadal variability. Gouretski and Koltermann (2007) identified a time-varying bias in measurements made by eXpendable BathyThermographs (XBT). An XBT is a probe that is launched from the deck of a ship and falls down through the ocean trailing behind it a fine wire that relays water temperature measurements to the operator. The depths of the measurements are estimated from an equation that relates time-since-launch to depth. Gouretski and Koltermann (2007) found that there were time-varying differences between the actual and estimated depths. Since 2007, various groups (Wijffels et al. 2008; Ishii and Kimoto 2009; Levitus et al. 2009; Gouretski and Reseghetti 2010; Good 2011) have proposed adjustments for the XBT data based on a number of factors including, the make and model of the XBT, water temperature (which is related to viscosity) as well as a pure thermal bias of unknown origin. By running the different correction methods on a defined set of data, it has been possible to begin to assess the uncertainty arising from the different parts of the reprocessing e.g. bias adjustment, choice of climatology etc. (Lyman et al. 2010).

The third example provides contrasting depth to the problems at hand. A number of sea-surface temperature data sets extend back to the start of the twentieth century (and before). Because observations become fewer the further back in time one goes, statistical methods are used to estimate SSTs in data gaps. However, as before, the data sets differ. Trends in SSTs in the tropical Pacific show different behavior depending on the data set used. Some data sets show an El Niño-like pattern, others a La Niña-like pattern (Deser et al. 2010) indicating that uncertainty in long-term trends can arise from sources other than systematic instrumental error.

Because of the obvious difficulties with observationally-based data sets, it is dangerous to consider them as unproblematic data points which one can use to build and challenge theories and hypotheses regarding the climate. The reality is not so simple. The data sets are themselves based on assumptions and hypotheses concerning the means by which the observed quantity is physically related to the climatological variable of interest. In the first example given above, the MSUs are sensitive to microwave emissions from oxygen molecules in the atmosphere. To convert the measured radiances to atmospheric temperature requires knowledge of atmospheric structure, the physical state of the satellite, quantum mechanics and orbital geometry.

In the first two examples above, the earliest attempts to create homogeneous data series underestimated the uncertainties because they did not consider a wide enough range of systematic effects. The physical understanding of the system under study was incomplete. Such problems are not unique to the study of climate data; see for example, Kirshner (2004) on the difficulties of estimating the Hubble constant. The uncertainty highlighted by the differences between independently processed data sets is often referred to as *structural* uncertainty. It arises from the many different choices made in the processing chain from raw observations to finished product. Part of this difference will arise from the different systematic effects considered – implicitly and explicitly – by the groups, but part will also arise from the different ways independent groups tackle the same problems. In most cases there are a wide variety of ways in which a particular problem can be approached and no single method can be proved definitively to be correct. The uncertainty associated with small changes in method (for example, using a 99 % significance cutoff as opposed to 95 % for identifying station breaks) can be assessed using monte-carlo techniques (see e.g. Mears et al. 2011; Kennedy et al 2011; Williams et al. 2012) and is referred to as *parametric* uncertainty to differentiate it from the deeper – and often larger – uncertainties associated with more significant structural chances that can only be assessed by taking independent approaches.

This slow evolution underlies what drives improvements in the understanding of the data. It also highlights the fact that no reprocessing is likely to be final and definitive. These considerations show the ongoing importance of making multiple, independent data sets of the same variable and many analyses that rely on climate data sets use multiple data sets to show that their results are not sensitive to structural uncertainty.

Comparisons between different methods have been used to assess the relative strengths and weaknesses of different approaches. Side by side comparisons of existing data sets have been made (Yasunaka and Hanawa 2011) but the use of carefully designed tests datasets can be far more illuminating. Real observations can be used (e.g. Lyman et al. 2010), but in this case the ‘true’ value is unknown. By using synthetic data sets, where the truth is known, much more can be learned (e.g. Venema et al. 2012; Williams et al. 2012). The use of carefully designed test data sets has been used in metrology to understand uncertainties associated with software in the measurement chain. However, the National Physics Laboratory (NPL) best practice guide on validation of software in measurement systems (NPL report DEM-ES 014) excludes measurement systems where the physics is still being researched which is arguably the case for many climate data sets. The International Surface Temperature Initiative (ISTI Thorne et al. 2011b) is developing a sophisticated process for developing test data sets based on synthetic 'pseudo-observations' that have been constructed to contain errors and inhomogeneities thought to be representative of real world cases. By running the algorithms designed to homogenize station data on these analogues of the real world as well as on the real data, it will be possible to directly compare the performance of different methods. Tests like these have been used to study the effectiveness of paleo-reconstruction techniques (Mann and Rutherford 2002) and have long formed the basis of Observing System Simulation Experiments (OSSE's). Ideally, such processes need to be ongoing for two reasons. Firstly, benchmark tests become less useful over time because there is a danger that the methods will become tuned to their peculiarities. Secondly, because the benchmarks might not address novel uses of the data or reflect new understanding of the error structures present in real world data.

Such methods are less effective for assessing homogenization procedures where they are based on empirical studies (Brunet et al. 2011), or on physical reasoning (Folland and Parker 1995). However, they could be used to cross-check results if statistically-based alternatives can be developed. A more empirical approach to the problem of assessing data biases is to run observational experiments (Brunet et al. 2011) whereby different sensors, including historical sensors, are compared side by side over a period of years. Such comparisons can be used to estimate the biases and associated uncertainties that can be used to cross check other methods, and in periods with fewer observations they may be the only means of assessing the data uncertainties.

Greater emphasis is now being given to the importance of uncertainty in observationally-based data sets, but it is not always clear how a user of the data should implement or interpret published uncertainty estimates. The traditional approach of providing an error bar on a derived value is often unsatisfactory because it provides information only on the magnitude of the uncertainty, but not how uncertainties co-vary. For example in the schematic in Fig. 1, each of the red lines is consistent with the median and 95 % uncertainty range indicated by the black line and blue area. By providing only the black line and 'error bar', information concerning (in this case) the temporal covariance structure of the errors is lost. This has implications when the data are further processed, because the covariance is needed to correctly propagate the uncertainties.

Recent approaches have drawn representative samples (roughly equivalent to the red lines shown in Fig. 1) from the posterior distributions of statistically reconstructed fields (Karspeck et al. 2011; Chappell et al. 2012) or representative samples from a particular error model (Mears et al. 2011; Kennedy et al. 2011). Each sample, or realization, can then be run through an analysis to generate an ensemble of results that show the sensitivity of the analysis to observational uncertainty.

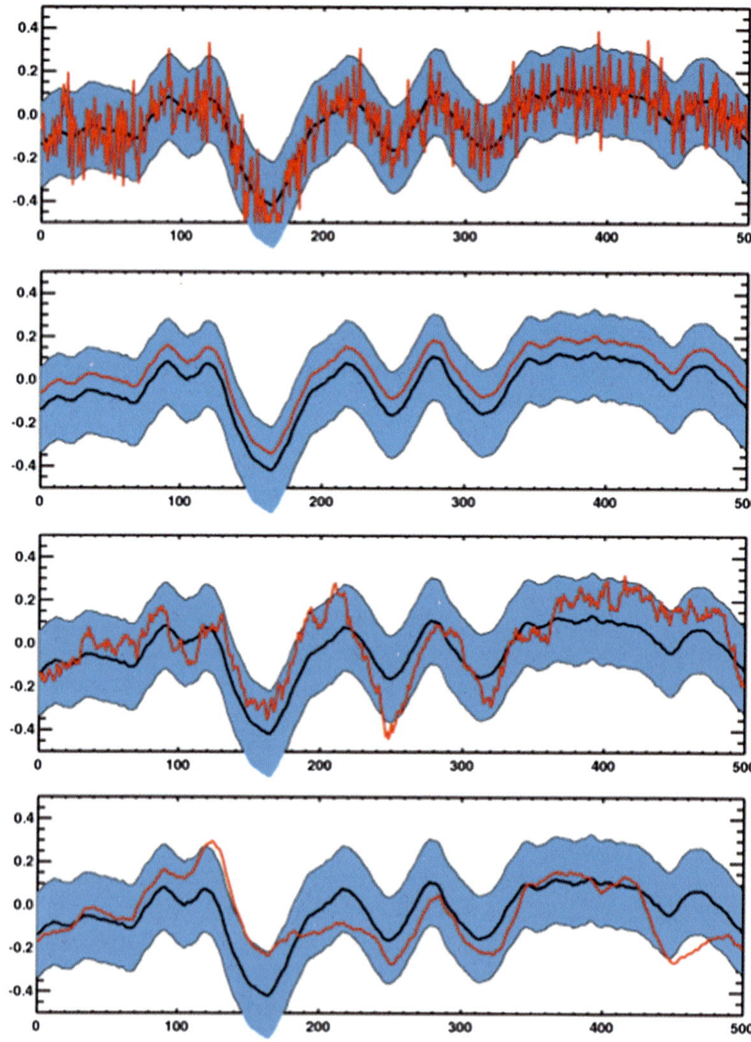

Fig. 1 Four examples showing that very different behaviors are consistent with the same 'error bars'. (*Top*) uncertainty range indicates that high-frequency variability is missing. (*second* from *top*) uncertainty range indicates a systematic offset. (*bottom* and *second* from *bottom*) uncertainty range indicates red-noise error variance

While these issues have been important for assessing large scale long term climate change, the challenges become even more formidable when data sets are used to assess climate change at higher resolution in time and in space. It is the extremes of weather that most often have the highest societal impacts and detecting and attributing changes in the statistics of these events is hampered by sparse data and poorly characterized uncertainties (see the OSC Community Paper on Extremes by Alexander et al.). The analysis of extremes demands more careful quality control – which in turn necessitates greater understanding of the underlying processes – because unusual events can sometimes resemble data errors and vice versa. In order to provide the data sets demanded by climate services the problems detailed above need to be resolved for a new generation of high resolution data set; from the discovery imaging and digitizing of paper records and metadata, through the management of appropriate archives, the generation of multiple independent data sets and their intercomparison to the wide dissemination and documentation of the final products.

Addressing the above concerns is vital for the creation of Climate Data Records (CDR http://www.ncdc.noaa.gov/cdr/guidelines.html), defined by the National Research Council (NRC) as "a time series of measurements of sufficient length, consistency, and continuity to determine climate variability and change". At the moment, the concept of a CDR has been associated with satellite processing, but a similar approach would be illuminating for in situ measurements of other geophysical variables. Of particular interest, from this point of view are the importance accorded to transparency of data and methods. Openness and transparency have many advantages over their opposites. They lay bare the assumptions made in the analysis: although methods sections in papers can adequately describe an algorithm, there is always the danger of ambiguity, or unstated assumptions. Where computer codes are provided, they unambiguously describe the methods used. In addition, the discovery and correction of errors in data and analysis are greatly facilitated, as is the reuse of methods in later analyses (Barnes 2010). The Climate Code Foundation (http://climatecode.org/) has been set up to help improve the visibility, availability and quality of code used in climate assessments and has recoded the NASA Goddard Institute of Space Studies global temperature data set, which has been developed over a number of years, in a single consistent package.

Assessing the quality of anything is a difficult task (Pirsig 1974) and CDRs are no exception. Indices attempting to measure the quality, or maturity of CDRs have been proposed (www1.ncdc.noaa.gov/pub/data/sds/cdrp-mtx-0008-v4.0-maturity-matrix.pdf). These include considerations of criteria such as scientific maturity, preservation maturity and metadata completeness as well as highlighting the importance of independent cross-checks and the provision of validated uncertainty estimates. A concept such as "maturity" is dangerous when applied to a single dataset: longevity and quality are not equivalent. As shown above, scientific maturity has typically developed by means of making *multiple independent* data sets. Even when considering the understanding of a variable across a range of data sets, difficulties arise because systematic errors in the data can go undetected for many years. "Immaturity" has only ever been obvious in hindsight.

Climate research encompasses a large range of studies, from process studies, overlapping more traditional research, that focus on large space-time scale interactions and coupling (i.e, feedbacks) to global, long-term monitoring (change detection) and attribution (change explanation). Planning for the needs of all of these uses is difficult. The need for greater transparency and traceability of raw data characteristics, analysis methods and data product uncertainties also have to help users judge whether a particular product is useful for a particular study. Given the large range of data products currently available—both raw and analyses—it is sometimes difficult for users to identify, locate and obtain what they need unless there is an organized set of information available. A number of approaches can help users find the data they need.

First, users need information about the various data sets. Journal papers and technical reports describing data set construction are often less useful as user guides, with technical details hidden behind journal paywalls or spread across a series of publications. Initiatives such as the Climate Data Guide project aims to provide expert and concise reviews of data and quality (http://climatedataguide.ucar.edu/). By comparing data sets side by side in a common setting, it should be easier for users to understand the relative strengths and weaknesses of different data sets.

Second, the users need to be able to find the data. This is easiest to do if there exists a common method for data discovery. At the basic level of individual meteorological reports, there exist a large number of archives (as mentioned before). At a higher level, there is no single repository for gridded and otherwise processed observational data sets that is analogous to the CMIP archive of model data (Meehl et al. 2000). Generating such an archive would have the dual effect of giving users easy access to the data in a standard format while allowing data producers to get their work more widely recognized. Presenting different data sets side by side will also serve to highlight the uncertainties in the observations themselves. A problem common to all data sets is that of accurate citation. Where data sets are regularly updated, a citation to a journal paper might not be sufficient to allow full reproducibility. Data archives could allow systematic version control of data set through a common mechanism allowing future users to extract a particular data set downloaded at any time. There is a growing concern about archiving and ready access to all of these data under a viable system that can easily handle the storage and access to an ever expanding volume of data. By combining such an archive with detailed provenance information, as anticipated by ISTI, would allow users to use data of a kind that is appropriate for their particular analysis. In gathering together observational data, thought must also be given to archiving and systematizing metadata and documentation. Such things as, quality flags, stations histories, calibration records, reanalysis innovations and feedback records, observer instructions, and so on, provide valuable information for analysts. Ideally, archives of metadata should coexist with the archives of data to which they refer.

Third, the information and data sets need to be integrated. There is not as yet a systematic way to gather value that has been added by a community that works with the data. The Climate Data Guide points to the data, but the data exist in a variety of formats. Collections of data sets exist, but they are sometimes divorced from the expert guidance necessary to understand them. A number of initiatives are addressing these problems. The ICOADS does incorporate some information concerning quality control, or bias identification and adjustment, but the IVAD (ICOADS Value-Added Data http://icoads.noaa.gov/ivad/) data base plans to add a layer which will give users access to a range of value-added data. The ISTI (International Surface Temperature Initiative) plans to create an archive of air temperature data and go further by planning to include other variables, as well as full provenance information for each observation in the archive allowing users to drill down from fully analyzed products to the original handwritten note made by the observer. Other projects, such as Group for High Resolution Sea Surface Temperature (GHRSST, www.ghrsst.org; Donlon et al. 2007), have produced alternative models for their own user communities that give access to greater detail allowing them to make their own evaluations of uncertainty.

1.3 Recommendations

1. Projects and initiatives concerning data digitization and archiving of basic observations urgently need to be imbedded in an overarching, sustainable, fully funded and staffed international infrastructure that oversees data rescue activities, and compliments the various implementation and strategy plans and documents on data coming out of GCOS, GEO, WMO, WCRP and the like.
2. Terrestrial and marine data efforts need to be integrated and better linked up under an international framework that supports their activities in a fully sustainable manner.
3. An archive of observational data sets analogous to the CMIP archive of model data, should be set up and integrated with user-oriented information such as the Climate Data Guide.

2 Reanalysis of Observations

Reanalyses differ from reprocessed observational data sets in that sophisticated data assimilation techniques are used in combination with global forecast models to produce global estimates of continuous data fields based on multiple observational sources (Bengtsson and Shukla 1988; Trenberth and Olson 1988). One advantage of this approach is that reanalysis data products are available at all points in space and time, and that many ancillary variables, not easily or routinely observed, are generated by the forecast model subject to the constraints provided by the observations. An important disadvantage of the reanalysis technique, however, is that the effect of model biases on the reanalyzed fields depends on the strength of the observational constraint, which varies both in space and time. This needs to be taken into account when reanalysis data are used for weather and climate research (e.g. Kalnay et al. 1996). Nevertheless, recent developments in data assimilation techniques, combined with improvements in models and observations (e.g. due to reprocessing of satellite data) have led to increasing use of modern reanalyses for monitoring of the global climate (Dee and Uppala 2009; Dee et al. 2011b; Blunden et al. 2011).

With multiple reanalyses now available for weather and climate research, investigators must consider the strengths and weaknesses of each reanalysis. Estimates of the basic dynamic fields in modern reanalyses are increasingly similar, especially in the vicinity of abundant observations (Rienecker et al. 2011). The physics fields (e.g. precipitation and longwave radiation) are more uncertain due to shortcomings in the assimilating model and its parameterizations. Understanding the effect of model errors is important both for users and developers of reanalyses, and ultimately needed to further improve the representation of climate signals in reanalysis. Observations provide the essential information content of reanalysis products; their quality and availability ultimately determines the accuracy that can be achieved. The types of observations assimilated span the breadth of remotely sensed and instrumental in-situ observations. Dealing with the complexities and uncertainties in the observing system, including data selection, quality control and bias correction, can have a crucial effect on the quality of the resulting reanalysis data.

Given the importance of reanalysis for weather and climate research and applications, successive generations of advanced reanalysis products can be anticipated. In the near future, coupling ocean, land and atmosphere will allow an integrated aspect of the reanalysis of historical observations, but may also increase the presence of model uncertainty. However, with the complexity of all the components of the Earth system, realizing the true potential of such advancements will require coordination, not only among developers of future reanalyses but also with the research community.

2.1 Current Status

The most used and cited reanalysis is the NCEP/NCAR reanalysis, which includes data going back to 1948 (Kalnay et al. 1996). The 45 year ECMWF reanalysis (ERA-40, Uppala et al. 2005), which stops in August 2002, has also been extensively used in weather and climate studies. Both of these reanalyses span the transition from a predominantly conventional observing system (broadly referring to in situ observations and retrieved observations that are assimilated) to the modern period with abundant satellite observations, marked by the introduction of TOVS radiance measurements in 1979. Many spurious variations in the climate signal have been identified in these early-generation reanalyses (Bengtsson et al. 2004; Andersson et al. 2005; Chen et al. 2008a, b), mainly resulting from inadequate bias corrections of the satellite data and modulated effects of model biases related with changes in the observing system. There now exist several atmospheric reanalyses covering the post-1979 period that are being continued forward in near-real time. The Japanese 25-year Reanalysis (JRA-25), released for use in March 2006 (Onogi et al. 2007) is the first effort by the JMA, and their second, JRA-55 is underway (Ebita et al. 2011). The National Centers for Environmental Prediction (NCEP) second reanalysis (NCEP-DOE, Kanamitsu et al. 2002) improved upon the NCEP/NCAR reanalysis data. More recently, ECMWF has produced the ERA-Interim reanalysis based on a 2006 version of their data assimilation system (Dee et al. 2011a), in preparation for a new climate reanalysis to be produced starting in 2014. NASA's Modern Era Retrospective-analysis for Research and Applications (MERRA) was developed as a tool to better understand NASA's remote sensing data in a climate context (Rienecker et al. 2011). The NCEP Climate Forecast System Reanalysis (Saha et al. 2010) became available in early 2010, produced with a data assimilation system that includes precipitation assimilation over land, and a semi-coupled ocean/land/atmosphere model and intended for seasonal prediction initialization. This is a brief description of the latest atmospheric reanalyses. The basic information about the data can be found at http://reanalyses.org/atmosphere/comparison-table, along with similar information for the latest oceanic reanalyses.

While the fundamental strength in resolving dynamical processes remains, recent reanalyses have improved on many aspects of the earlier-generation systems. Direct assimilation of the remotely-sensed satellite radiances, rather than assimilation of retrieved state estimates, has become the norm. Variational bias correction of the satellite radiances effectively anchors these data to high-quality observations from radiosondes and other sources (Dee and Uppala 2009; used in ERA-Interim, MERRA, and CFSR as well as the forthcoming JRA-55). The recently completed CFSR is the first reanalysis to use a weakly-coupled ocean/atmosphere model, and also assimilates precipitation data over land. In addition to the technical and scientific improvements of the reanalysis systems, increased computational resources allow the use of higher-resolution models that better resolve the observations. These advances combined have lead to improved representations of many physical parameters and processes in reanalyses, for example improved skill of the large-scale global and tropical precipitation (Bosilovich et al. 2008, 2011). In addition, the need for reanalyses to contribute to climate change studies has prompted significant innovations. For example, the twentieth century Reanalysis (20CR) project carried out by NOAA in collaboration with CIRES uses the available global surface pressure observations and sea surface temperature record reconstructed through the 1870s in an ensemble-based global analysis method. The resulting analysis is able to produce weather patterns with the quality of a modern 3-day numerical forecast (Compo et al. 2011).

Even with substantial improvements, assessment of the uncertainties in reanalysis output, especially in the physical processes needed to study climate variations and change, remains a significant concern. For a more complete picture of the climate system, as represented by reanalyses, the impact of the observations on the resulting data should be captured in the analysis of the physical processes (as in Schubert and Chang 1996; Roads et al. 2002). Even the most recent reanalyses demonstrate, to varying degrees, shifts in the time series that can be related to changes in the observing systems being assimilated (Dee et al. 2011a; Saha et al. 2010; Bosilovich et al. 2011). These shifts, which may be due to changing biases in the observations, systematic errors in the assimilating model, or both, interfere with the ability to detect reliable climate trends from the reanalyses. While there are some post-processing techniques that may address these spurious features (Robertson et al. 2011), dealing with biases in models and observations remains the most difficult challenge for the reanalysis and data assimilation community in developing future generations of climate reanalyses.

The number of global reanalyses has increased greatly in recent years, as computing improves, and various entities have need for specific missions to support.

Furthermore, spanning the various Earth system disciplines shows that uncoupled ocean and land reanalyses are being performed as regularly as those for the atmosphere (Guo et al. 2007; Xue et al. 2012; an evolving list of reanalyses is maintained at *reanalysis.org*). Regional reanalyses attempt to improve upon the local representation of climate and processes that must be handled more generally in global systems (Mesinger et al. 2006; Verver and Klein Tank 2012). While this increase in new reanalyses can cause additional work for the research community in understanding the various strengths and weaknesses, it does provide opportunity to more quantitatively investigate the uncertainties of the reanalysis data. For example, in studying the global water and energy budgets Trenberth et al. (2011) characterized the range of values for each term. In addition, collections of analyses have been used to derive a super ensemble mean and variance for the ocean (Xue et al. 2012), land (Guo et al. 2007) and atmosphere (Bosilovich et al. 2009). While the ensembles can expose biases in the character of various reanalyses, there is some evidence that the ensemble itself can also provide reasonable data from weather to monthly timescales. Despite the difficulties in dealing with a large amount of data, a researcher will find more advantage to have multiple data sets available for study. Just as several coupled model integrations are required for present day and future climate projections, multiple reanalyses will better contribute to the characterization of present day climate. Reanalyses may well benefit from common data standards that facilitate evaluation and analysis of the IPCC climate change experiments.

2.2 *Integrating Earth System Analyses*

Observations are the critical resource for a reanalysis, which needs as many as possible to characterize the state of the Earth system. As decadal predictions begin to play a role in understanding near-term climate variations, the Earth system ocean/land/atmosphere needs to be initialized in a balanced state. Newer measurements, such as aerosols, sea ice and ocean salinity contribute to the need for reanalyses that encompass the broad Earth system. Therefore, Integrated Earth Systems Analysis (IESA) encompasses the connections of these disparate observations, and have become an important challenge for data assimilation development.

NCEP CFSR provides a reanalysis produced with a semi-coupled ocean/land/atmosphere model, along with an analysis of land precipitation gauge measurements (Saha et al. 2010). Development of the next reanalysis from NASA includes aerosols, ocean (temperature and salinity), land (soil water) and ocean color (biology) analysis. While there are significant difficulties in both the modeling and assimilation of the integrated Earth system, extending these more complex reanalyses to historic periods, when little or none of the diversity in observations is available will require even more effort on addressing the impact of changes in the observing systems. Likewise, maintaining and expanding many of the Earth observations forward in time is also a critical issue (Trenberth et al. OSC position paper on observing system), and reference networks can provide stable benchmarks for reanalyses and their data assimilation. Consistency and overlap of newer systems will help maintain the consistency in the integrated reanalyses.

2.3 Reanalysis Input Observations

Essentially, reanalyses without input observations revert to model products, hence the importance of the observing system emphasized here. As discussed previously, there are numerous value added advantages from reanalysis, but they cannot replace observed data. It is very important, especially for new reanalysis users, to understand that reanalyses are *not* observations, but rather, an observation-based data product. Since reanalyses combine many types of observations, their relative comparison should be valuable in assessing the quality of the observation as well. However, it is not always easy to determine which observations are included in the reanalysis at specific spatio-temporal coordinates. Any given observation will be weighted with other nearby observations and the model forecast in the assimilation process. It may be accepted or rejected, and if accepted will contribute to the overall analysis including other accepted observations. The degree to which an observation influences an analysis can be determined from the output background model forecast error and the analysis error (as discussed in Rienecker et al. 2011).

Such output data have been available from reanalysis and data assimilation products for some time, but generally only used by developers or those closely familiar with the data assimilation methodology. However, these assimilated observations represent a key component in the output of the reanalyses, and can show which observations are used and how. For example, Haimberger (2007) used feedback information from ERA40 to better characterize inhomogeneities in the radiosonde time series, and this information was, in turn, used to improve the input observations to both ERA-Interim and MERRA. To facilitate broader access, assimilated observations need to be provided in a format easily accessible to the reanalysis users, so that users can more appropriately identify the agreement between observed features (including all sources of a given state variable) and reanalysis features at any specific point in space and time. Even just the capability of easily determining the presence (or lack thereof) of assimilated observations during a given event would be useful in many research studies. Typically, the data is produced in "observation-space", in that, it is an ascii record including space and time coordinates. To facilitate comparisons with the gridded reanalysis output, the GMAO has processed MERRA's assimilated observations to its native grid (Rienecker et al. 2011) called the MERRA Gridded Innovations and Observations (GIO). It includes each observation, its forecast error and analysis error (as well as the count of observations and variance within the grid box). Similarly, recent efforts at ECMWF aim to make assimilated observations and the "feedback" files available through a WWW interface. With these data, researchers can quickly identify the observation assimilated at each of the reanalysis grid points.

Of course, reanalyses rely on the broad and open availability of increasing numbers of observing systems and variables. Regarding in situ (or sometimes referred to as conventional) observing networks, reanalysis projects have been able to coordinate and update data holdings to reflect the latest quality assessments and reprocessing of the data. For the remote sensing data, however, there remains much less organization of the data and how it is used in reanalyses. As part of preparations for a new comprehensive climate reanalysis, an inventory of satellite radiances potentially available for reanalysis is currently being compiled at ECMWF. Some remotely sensed data is still assimilated as retrieved state fields, instead of radiances, and is therefore a function of the algorithm or radiative transfer model and its version, as well as the version of the input radiance.

There is significant work progressing on the radiances themselves that should affect their use in reanalyses. For example, intercalibrated MSU (channels 2–4) (Zou et al. 2006) were newly available and assimilated from the start of MERRA production, but this was not an option for reanalyses beginning prior to it. The satellite data input is generally handled by the reanalysis center, which must maintain contacts with the data community to be informed on all the latest information and updates. Presently, each center documents its own data usage, but there is no central information about this for research users to access and intercompare among reanalyses. As discussed earlier, observations are the key resource for reanalysis, reanalysis are sensitive to the assimilated observations and so, it is vitally important for reanalysis projects to have the latest information and reprocessing of the input data type, and also convey that information to the research community. The series of international reanalyses conferences have provided a focal point for discussions on the accomplishments, challenges and future directions of reanalyses (e.g. jra.kishou.go.jp/3rac_en.html and icr4.org). Additionally, a grass roots effort to open communication among reanalysis developers and the research community leveraging internet communication technology has begun and is gaining momentum (reanalysis.org).

2.4 Recommendations

1. The research community and reanalysis developers benefit from the availability of multiple international reanalysis products. Researchers should be encouraged to use as many as possible to better define the uncertainty of reanalyses. Data management practices and utilities should be developed to facilitate intercomparison among reanalyses.
2. Given the criticality of observations and their quality in reanalyses, efficient and open communications among the reanalyses developers and observation developers/stewards needs to be enhanced. Likewise, information on how the observations are used in the reanalysis can be used by the observation developers and research community. Reanalysis developers should be encouraged to provide the assimilated observations and innovations alongside the characteristic reanalysis data.
3. Interdisciplinary coupled modeling and assimilation across the atmosphere (including aerosols and the stratosphere), ocean, land and cryosphere needs significant advancement and communications to accomplish the long-term goals of integrated reanalyses.

3 Future Directions

Global data products and their further refinement will continue to be a critical resource for understanding the Earth's climate, variability and change. Not only is reduction of uncertainty for any individual product important, through improved algorithms and processing, but also, global data must be physically integrated and consistent in their use of ancillary information and consistency in assumptions. These considerations are leading to more formal assessments of global data products, such as those put forward by the GEWEX Data and Assessment Panel (e.g. Gruber and Levizzani 2008).

A substantial amount of observations are not regularly analyzed in present day research projects because it has yet to be digitized. Projects and initiatives concerning data digitization and archiving of basic observations urgently need to be imbedded in an overarching, sustainable, fully funded and staffed international infrastructure that oversees data rescue activities, and compliments the various implementation and strategy plans and documents on data coming out of international coordinating agencies. Terrestrial and marine data efforts need to be integrated and better linked up under an international framework that supports their activities. An archive of observational data sets analogous to the CMIP archive of model data, should be established and integrated with user-oriented information such as the Climate Data Guide.

The reanalysis developer and user community has increased substantially over the last decade, mostly due to the broad utility of the data. This paper has addressed some of the most pressing challenges facing the international reprocessing and reanalysis communities. WCRP has been an integral partner in the development of reprocessing and reanalyses, fostering communications within the community through workshops, conferences and its scientific panels. Recently, reanalyses data have been discussed and considered in the derivation of Essential Climate Variables (ECVs), as well as using the data for climate monitoring and information services (Dee et al. 2011b). Assessment of global data products is also a major issue for ECVs.

As can be easily seen in the overview summary of reanalyses, the reanalysis systems are evolving and growing. There will be newer, more advanced and comprehensive reanalysis data products available in coming years. Regarding the most recent reanalysis data products, there are many questions on their relative performance for the many uses and regions covered. It is not feasible for any one institution to be able to fully address the exact quality among all the reanalyses, simply because there are too many applications of reanalyses. While this does put the burden of intercomparison on the individual researcher, in quite a few instances, communication and sharing of knowledge between users and developers will have become critically important. In a grass roots effort to address the communications issues, an effort to utilize the internet and live documents has begun, to provide a forum that facilitates communication within the reanalysis community. It is considered a pilot project, and is called *reanalyses.org*. At this site, developers can contribute to a central knowledge-base regarding all issues of reanalyses.

In addition, reanalyses.org provides a function to allow users to compare reanalyses. In the long run, users are encouraged to summarize their results with pointers to detailed information and ultimately publications on the ongoing efforts. While this should not be the sole effort to facilitate communications, it does provide an outlet and focal point for anyone in the community. The Climate Data Guide (climatedataguide.ucar.edu) provides concentrated information and expert analysis of many reprocessed data set, data sources for reanalysis and the reanalyses themselves. Another platform, the Earth System Grid (ESG) is under development and will allow users to easily compare the existing reanalyses with observations and also CMIP present day simulations. While significant challenges remain, the active communities of users and developers have numerous avenues of information and interaction to pursue the solutions.

Acknowledgments William Rossow provided tremendous insight and refinement of the final form of this manuscript. Likewise, Gil Compo and Peter Thorne's thoughtful reviews and insights strengthened the recommendations and the reasoning behind them. Two anonymous reviewers exposed several key points that also solidified the final manuscript.

References

Allan R, Brohan P, Compo GP, Stone R, Luterbacher J, Brönnimann S (2011) The international Atmospheric Circulation Reconstructions over the Earth (ACRE) initiative. Bull Am Meteorol Soc 92:1421–1425. http://dx.doi.org/10.1175/2011BAMS3218.1

Allan R, Brohan P, Leduc S, Willett KM (2012) Climate data rescue and stewardship: filling gaps in 21st century needs. In "State of the Climate in 2011". Bull Am Meteorol Soc 93(7):S12–S13. doi:10,1175/2012BAMSStateoftheClimate.1 and http://www.ncdc.noaa.gov/bams-state-of-the-climate/

Andersson E, Bauer P, Beljaars A, Chevallier F, Hólm E, Janisková M, Kållberg P, Kelly G, Lopez P, Mcnally A, Moreau E, Simmons AJ, Thépaut JN, Tompkins AM (2005) Assimilation and modeling of the atmospheric hydrological cycle in the ECMWF forecasting system. Bull Am Meteorol Soc 86:387–402

Barnes N (2010) Publish your computer code: it is good enough. Nature 467:753. doi:10.1038/467753a

Bengtsson L, Shukla J (1988) Integration of space and in situ observations to study global climate change. Bull Am Meteorol Soc 69(10):1130–1143

Bengtsson L, Hagemann S, Hodges KI (2004) Can climate trends be calculated from reanalysis data? J Geophys Res 109:D11111. doi:10.1029/2004JD004536

Blunden J, Arndt DS, Baringer MO (2011) State of the Climate in 2010. Bull Amer Meteorol Soc 92:S1–S236. doi: http://dx.doi.org/10.1175/1520-0477-92.6.S1

Bosilovich MG, Chen J, Robertson FR, Adler RF (2008) Evaluation of global precipitation in reanalyses. J Appl Meteorol Climatol 47:2279–2299

Bosilovich MG, Mocko D, Roads JO, Ruane A (2009) A multimodel analysis for the Coordinated Enhanced Observing Period (CEOP). J Hydrometeorol 10:912–934

Bosilovich MG, Chen JY, Robertson FR, Adler RF (2011) Evaluation of global precipitation in reanalyses. J Appl Meteorol Climatol 47:2279–2299. doi:10.1175/2011JCLI4175.1

Brunet M, Asin J, Sigró J, Bañón M, García F, Aguilar E, Palenzuela JE, Peterson TC, Jones P (2011) The minimization of the screen bias from ancient Western Mediterranean air temperature records: an exploratory statistical analysis. Int J Climatol 31(12):1879–1895

Chappell A, Renzullo L, Haylock M (2012) Spatial uncertainty to determine reliable daily precipitation maps. J Geophys Res 117:D17115. doi:10.1029/2012JD017718

Chen J, Del Genio AD, Carlson BE, Bosilovich MG (2008a) The spatiotemporal structure of long term climate variations in the 20th century based on observed and reanalysis data. Part I: The global warming trend. J Clim 21:2611–2633

Chen J, Del Genio AD, Carlson BE, Bosilovich MG (2008b) The spatiotemporal structure of long term climate variations in the 20th century based on observed and reanalysis data. Part II: Pacific pan-decadal variability. J Clim 21:2634–2650

Christy JR, Spencer RW, Norris WB, Braswell WD, Parker DE (2003) Error estimates of version 5.0 of MSU-AMSU bulk atmospheric temperatures. J Atmos Ocean Technol 20:613–629

Compo GP and Co-authors (2011) The twentieth century reanalysis project. Q J R Meteorol Soc 137:1–28. doi:10.1002/qj.776

Dee D and Coauthors (2011a) The ERA-Interim reanalysis: configuration and performance of the data assimilation system. Q J R Meteorol Soc 137:553–597

Dee D and Coauthors (2011b) The use of reanalysis data for monitoring the state of the climate. In: "The State of the Climate in 2010". Bull Am Meteorol Soc 92(6):S34–S35

Dee DP, Uppala S (2009) Variational bias correction of satellite radiance data in the ERA-Interim reanalysis. Q J R Meteorol Soc 135:1830–1841. doi:10.1002/qj.493

Deser C, Phillips AS, Alexander MA (2010) Twentieth century tropical sea surface temperature trends revisited. Geophys Res Lett 37:L10701. doi:10.1029/2010GL043321

Donlon C and Co-authors (2007) The Global Ocean Data Assimilation Experiment high-resolution sea surface temperature pilot project. Bull Am Meteorol Soc 88(8):1197–1213. doi:10.1175/BAMS-88-8-1197

Ebita A and Co-authors (2011) "The Japanese 55-year reanalysis "JRA-55": an Interim Report". SOLA **7**:149–152

Folland CK, Parker DE (1995) Correction of instrumental biases in historical sea surface temperature data. Q J R Meteorol Soc 121(522):319–367

García-Herrera R, Jones P, Wheeler D, Können G, Prieto MR (Guest Eds) (2005) CLIWOC: Climatology of the World's Oceans, pp 1750–1850 (Special Issue). Climatic Change 73:1–194

Good SA (2011) Depth biases in XBT data diagnosed using bathymetry data. J Atmos Ocean Technol 28:287–300. http://dx.doi.org/10.1175/2010JTECHO773.1

Gouretski VV, Koltermann KP (2007) How much is the ocean really warming? Geophys Res Lett 34:L01610. doi:10.1029/2006GL027834

Gouretski V, Reseghetti F (2010) On depth and temperature biases in bathythermograph data: development of a new correction scheme based on analysis of a global ocean database. Deep Sea Res I 57:812–833. doi:10.1016/j.dsr.2010.03.011

Gruber A, Levizzani V (2008) Assessment of global precipitation products: a project of the World Climate Research Programme Global Energy and Water Cycle Experiment (GEWEX) radiation panel, Bologna, WCRP-128 WMO/TD-No. 1430

Guo Z, Dirmeyer PA, Gao X, Zhao M (2007) Improving the quality of simulated soil moisture with a multi-model ensemble approach. Q J R Meteorol Soc 133:731–747. doi:10.1002/qj.48

Haimberger L (2007) Homogenization of radiosonde temperature time series using innovation statistics. J Clim 20:1377–1403

Hossain F, Huffman GJ (2008) Investigating error metrics for satellite rainfall at hydrologically relevant scales. J Hydrometeorol 9(3):563–575

Ishii M, Kimoto M (2009) Reevaluating of historical ocean heat content variations with time-varying XBT and MBT depth bias corrections. J Oceanogr 65:287–299

Kalnay E and Coauthors (1996) The NCEP/NCAR 40-year reanalysis project. Bull Amer Meteorol Soc 77(3):437–472. doi:10.1175/1520-0477(1996)0772.0.CO;2

Kanamitsu M, Ebisuzaki W, Woollen J, Yang S-K, Hnilo JJ, Fiorino M, Potter GL (2002) NCEP-DOE AMIP-II reanalysis (R-2). Bull Am Meteorol Soc 83:1631–1643

Karspeck AR, Kaplan A, Sain SR (2011) Bayesian modelling and ensemble reconstruction of mid-scale spatial variability in North Atlantic sea-surface temperatures for 1850–2008. Q J R Meteorol Soc 138(662):234–248. doi:10.1002/qj.900

Kennedy JJ, Rayner NA, Smith RO, Parker DE, Saunby M (2011) Reassessing biases and other uncertainties in sea surface temperature observations measured in situ since 1850: 2. Biases and homogenization. J Geophys Res 116:D14104. doi:10.1029/2010JD015220

Kent EC, Kennedy JJ, Berry DI, Smith RO (2010) Effects of instrumentation changes on sea surface temperature measured in situ. Wiley Interdiscip Rev Clim Change 1(5):718–728. doi:10.1002/wcc.55

Kirshner RP (2004) Hubble's diagram and cosmic expansion. Proc Natl Acad Sci 101(1):8–13. doi:10.1073/pnas.2536799100

Knapp KR, Kruk MC, Levinson DH, Diamond HJ, Neumann CJ (2010) The International Best Track Archive for Climate Stewardship (IBTrACS): unifying tropical cyclone best track data. Bull Am Meteorol Soc 91:363–376. doi:10.1175/2009BAMS2755.1

Lawrimore JH, Menne MJ, Gleason BE, Williams CN, Wuertz DB, Vose RS, Rennie J (2011) An overview of the Global Historical Climatology Network monthly mean temperature data set, version 3. J Geophys Res 116:D19121. doi:10.1029/2011JD016187

Levitus S, Antonov JI, Boyer TP, Stephens C (2000) Warming of the World Ocean. Science 287(5461):2225–2229. doi:10.1126/science.287.5461.2225

Levitus S, Antonov JI, Boyer TP, Locarnini RA, Garcia HE, Mishonov AV (2009) Global ocean heat content 1955–2008 in light of recently revealed instrumentation problems. Geophys Res Lett 36:L07608. doi:10.1029/2008GL037155

Lott N, Del Greco SA, Ross T, Vose R (2008) The integrated surface database: partnerships and progress, 88th AMS Annual Meeting, 20–25 January 2008, New Orleans, combined preprints [CD-ROM], American Meteorological Society, Boston, MA (January 2008)

Lyman JM, Good SA, Gouretski VV, Ishii M, Johnson GC, Palmer MD, Smith DM, Willis JK (2010) Robust warming of the global upper ocean. Nature 465:334–337. doi:10.1038/nature09043

Mann ME, Rutherford S (2002) Climate reconstruction using 'Pseudoproxies'. Geophys Res Lett 29:1501. doi:10.1029/2001GL014554

Mears CA, Wentz FJ (2009a) Construction of the Remote Sensing Systems V3.2 atmospheric temperature records from the MSU and AMSU microwave sounders. J Atmos Ocean Technol 26:1040–1056

Mears CA, Wentz FJ (2009b) Construction of the Remote Sensing Systems V3.2 lower-tropospheric temperature dataset from the MSU and AMSU microwave sounders. J Atmos Ocean Technol 26:1493–1509

Mears CA, Schabel MC, Wentz FJ (2003) A reanalysis of the MSU channel 2 tropospheric temperature record. J Clim 16:3650–3664

Mears CA, Wentz FJ, Thorne P and Others (2011) Assessing uncertainty in estimates of atmospheric temperature changes from MSU and AMSU using a Monte-Carlo estimation technique. J Geophys Res 116:D08112. doi:10.1029/2010JD014954

Meehl GA, Boer GJ, Covey C, Latif M, Stouffer RJ (2000) The Coupled Model Intercomparison Project (CMIP). Bull Am Meteorol Soc 81:313–318. doi:10.1175/1520-0477(2000)081<0313:TCMIPC>2.3.CO;2

Mesinger F and Co-authors (2006) North American regional reanalysis. Bull Am Meteorol Soc 87:343–360

Onogi K and Co-authors (2007) The JRA-25 reanalysis. J Meteorol Soc Japan 85:369–432

Pirsig RM (1974) Zen and the Art of Motorcycle Maintenance. William Morrow and Company, New York, ISBN 0-688-00230-7

Prabhakara C, Iacovazzi R, Yoo J-M, Dalu G (2000) Global warming: evidence from satellite observations. Geophys Res Lett 27:3517–3520

Rienecker MR and Co-authors (2011) NASA's Modern-Era Retrospective Analysis for Research and Applications. MERRA. J Clim 24(14):3624–3648

Roads JO, Kanamitsu M, Stewart R (2002) CSE water and energy budgets in the NCEP-DOE reanalysis II. J Hydrometeorol 3:227–248

Robertson FR, Bosilovich MG, Chen J, Miller TL (2011) The effect of satellite observing system changes on MERRA water and energy fluxes. J Climate 24:5197–5217. doi:10.1175/2011JCLI4227.1

Saha S and Coauthors (2010) The NCEP climate forecast system reanalysis. Bull Am Meteorol Soc 91:1015–1057

Schubert SD, Chang Y (1996) An objective method for inferring sources of model error. Mon Wea Rev 124:325–340

Showstack R (2009) World Ocean Database. Eos Trans AGU 90(49):472. doi:10.1029/2009EO490003

Spencer RW, Christy JR (1990) Precise monitoring of global temperature trends from satellites. Science 247(4950):1558–1562. doi:10.1126/science.247.4950.1558

Thorne PW, Lanzante JR, Peterson TC, Seidel DJ, Shine KP (2011a) Tropospheric temperature trends: history of an ongoing controversy. WIREs Clim Change 2(1):66–88. doi:10.1002/wcc.80

Thorne PW and Coauthors (2011b) Guiding the creation of a comprehensive surface temperature resource for twenty-first-century climate science. Bull Am Meteor Soc 92:ES40–ES47. doi:10.1175/2011BAMS3124.1

Trenberth KE, Olson JG (1988) An evaluation and intercomparison of global analyses from NMC and ECMWF. Bull Am Meteor Soc 69:1047–1057

Trenberth KE, Fasullo JT, Mackaro J (2011) Atmospheric moisture transports from ocean to land and global energy flows in reanalyses. J Climate 24:4907–4924. doi:10.1175/2011JCLI4171.1

Uppala SM and Co-authors (2005) The ERA-40 re-analysis. QJRMS 131(Part B):2961–3012

Venema VKC et al (2012) Benchmarking homogenization algorithms for monthly data. Clim Past 8:89–115. doi:10.5194/cp-8-89-2012

Verver G, Klein Tank A (2012) European reanalysis and observations for monitoring (EURO4M). Presented at the 4th WCRP international conference on reanalyses, Silver Spring, 7–11 May 2012. http://icr4.org/posters/Verver_UA-43.pdf

Vinnikov KY, Grody NC (2003) Global warming trend of mean tropospheric temperature observed by satellites. Science 302:269–272

Wijffels SE, Willis J, Domingues CM, Barker P, White NJ, Gronell A, Ridgway K, Church JA (2008) Changing expendable bathythermograph fall rates and their impact on estimates of thermosteric sea level rise. J Clim 21:5657–5672

Wilkinson C, Woodruff SD, Brohan P, Claesson S, Freeman E, Koek F, Lubker SJ, Marzin C, Wheeler D (2011) Recovery of logbooks and international marine data: the RECLAIM project. Int J Climatol Special Issue Achiev Mar Climatol 31(7):968–979. doi:10.1002/joc.2102

Williams CN, Menne MJ, Thorne PW (2012) Benchmarking the performance of pairwise homogenization of surface temperatures in the United States. J Geophys Res 117:D05116. doi:10.1029/2011JD016761

Woodruff S, Worley S, Lubker S, Ji Z, Freeman J, Berry D, Brohan P, Kent E, Reynolds R, Smith S, Wilkinson C (2010) ICOADS release 2.5: extensions and enhancements to the surface marine meteorological archive. Int J Climatol. doi:10.1002/joc.2103

Xue Y, and Coauthors (2012) A comparative analysis of upper-ocean heat content variability from an ensemble of operational ocean reanalyses. J Climate 25:6905–6929. doi:10.1175/JCLI-D-11-00542.1

Yasunaka S, Hanawa K (2011) Intercomparison of historical sea surface temperature datasets. Int J Climatol 31(7):1056–1073

Yin X, Gleason BE, Compo GP, Matsui N, Vose RS (2008) The International Surface Pressure Databank (ISPD) land component version 2.2. National Climatic Data Center, Asheville, pp 1–12. Available from ftp://ftp.ncdc.noaa.gov/pub/data/ispd/doc/ISPD2_2.pdf

Zou C-Z, Goldberg MD, Cheng Z, Grody NC, Sullivan JT, Cao C, Tarpley D (2006) Recalibration of microwave sounding unit for climate studies using simultaneous nadir overpasses. J Geophys Res 111:D19114. doi:10.1029/2005JD006798

Climate Processes: Clouds, Aerosols and Dynamics

Steven C. Sherwood, M. Joan Alexander, Andy R. Brown, Norm A. McFarlane, Edwin P. Gerber, Graham Feingold, Adam A. Scaife, and Wojciech W. Grabowski

Abstract Physical processes not well resolved by climate models continue to limit confidence in detailed predictions of climate change. The representation of cloud and convection-related processes dominates the model spread in global climate sensitivity, and affects the simulation of important aspects of the present-day climate

S.C. Sherwood (✉)
Climate Change Research Centre (CCRC) and Australian Research Council (ARC), Centre of Excellence for Earth Systems Science, University of New South Wales, Sydney, NSW 2052, Australia
e-mail: ssherwood@alum.mit.edu

M.J. Alexander
NorthWest Research Associates, 3380 Mitchell Lane, Boulder, CO 80301, USA
e-mail: alexand@cora.nwra.com

A.R. Brown • A.A. Scaife
Met Office, Hadley Centre, FitzRoy Road, Exeter, Devon EX1 3PB, UK
e-mail: andy.brown@metoffice.gov.uk; adam.scaife@metoffice.gov.uk

N.A. McFarlane
Canadian Centre for Climate Modelling and Analysis, University of Victoria, PO Box 3065 STN CSC, Victoria, BC V8W 3V6, Canada
e-mail: Norm.McFarlane@ec.gc.ca

E.P. Gerber
Center for Atmosphere Ocean Science, Courant Institute of Mathematical Sciences, New York University, 251 Mercer Street, New York, NY 10012, USA
e-mail: gerber@cims.nyu.edu

G. Feingold
Chemical Sciences Division, NOAA Earth System Research Laboratory, 325 Broadway R/CSD2, Boulder, CO 80305, USA
e-mail: Graham.Feingold@noaa.gov

W.W. Grabowski
Mesoscale and Microscale Meteorology Division, National Center for Atmospheric Research, PO Box 3000, Boulder, CO 80307-300, USA
e-mail: grabow@ucar.edu

G.R. Asrar and J.W. Hurrell (eds.), *Climate Science for Serving Society: Research, Modeling and Prediction Priorities*, DOI 10.1007/978-94-007-6692-1_4,

especially in the tropics. Uncertainty in aerosol radiative effects complicates the interpretation of climate changes in the observational and paleoclimate records, in particular limiting our ability to infer climate sensitivity. Dynamical uncertainties, notably those involving teleconnections and troposphere-stratosphere interaction, also affect simulation of regional climate change especially at high latitudes. In response, targeted field programs, new satellite capabilities, and new computational approaches are promoting progress on these problems. Advances include recognition of the likely importance of non-greenhouse gas forcings in driving recent trends in the general circulation, compensating interactions and emergent phenomena in aerosol-cloud-dynamical systems, and the climatic importance of cumulus entrainment. Continued progress will require, among other things, more integrative analysis of key processes across scales, recognizing the complexity at the local level but also the constraints and possible buffering operating at larger (system) scales.

Keywords Clouds • Atmospheric convection • Aerosols • Cloud-aerosol interaction • Atmospheric dynamics • Climate feedbacks • Climate modeling

1 Introduction

Cloud, aerosol, and dynamical processes remain at the core of uncertainties about atmospheric aspects of climate and continue to be the subject of detailed research. This research encompasses observations, process modeling, and the analysis of global climate models (GCMs) to examine the possible broader consequences of the processes. While aerosols play an important role in air quality and visibility, this paper will consider only their climatic consequences; similarly, our discussion of cloud and dynamical issues will be oriented toward WCRP science objectives rather than purely weather-related or highly localized phenomena.

Anthropogenic aerosols are now cooling the climate by an amount that remains difficult to quantify accurately, but could be comparable to the warming effect of anthropogenic carbon dioxide. Moreover, because aerosols are highly nonuniform and therefore warm the atmosphere and cool the surface non-uniformly over the Earth, they can drive changes to the atmospheric circulation that may affect patterns of rainfall (Rotstayn and Lohmann 2002) or cloud (e.g., Allen and Sherwood 2010) independently of any impact on global-mean temperature.

Clouds remain the greatest source of spread in model predictions of future climate. Much of this spread comes from low clouds, but other cloud types also contribute and/or may be more important than suggested by their contribution to this among present models. Cirrus clouds, for example, are not well represented in models and exert a net warming effect that is comparable to the net cooling effect of low clouds; models are beginning to hint at the potential importance of this for climate change. Convective clouds interact with the circulation and tend to amplify or organize many tropospheric circulations, playing a central role, for example, in tropical intra-seasonal variability and helping to drive the general circulation at low latitudes (Slingo and Slingo 1991). Polar clouds interact not only with atmospheric dynamics, but also with sea ice. See Heintzenberg and Charlson (2009) for a thorough review of our understanding of how clouds respond to both aerosols and climate changes, and Rosenfeld et al. (this volume) for a more focused perspective on current ideas about aerosol impacts on clouds.

Dynamical processes at all scales modulate how global heat inputs are expressed regionally, and affect global-mean climate indirectly through their role in transporting energy to where it can be radiated to space. The dynamical processes considered here are not comprehensive but include motions from the cloud-system scale upward, that appear to be important for climate or inadequately understood. While it is often assumed that global-scale circulations are fully captured by existing climate models, this is not necessarily the case as shown by recent examinations of varying circulations in different model designs as described in Sect. 2.3. Also, even if global models do capture a phenomenon correctly there are typically intellectual and practical advantages to achieving a more fundamental or heuristic understanding (see, e.g., Held 2005). Rosenlof et al. (this volume) discuss global-scale dynamical changes more extensively, including their ocean and surface components.

2 Recent Scientific Advances

2.1 Clouds and Convection

The representation of clouds in climate models continues to exhibit mean biases that have been brought into sharper focus by the data from active remote sensors on board the CloudSat and CALIPSO satellites. These sensors reveal more clearly the vertical distribution of cloudiness, confirming that many climate models generate too much cloud in upper levels and too little at middle and low levels (e.g., Chepfer et al. 2008).

2.1.1 Boundary Layer Clouds and Dynamics

Field programs have shed new light on the strong and varied dynamical and microphysical interactions in maritime shallow convection and marine stratus clouds (Wood 2012). In many cases these systems are remarkably robust, but occasionally exhibit rapid transitions from open-celled to closed-celled morphologies, with substantially different albedos and rainfall characteristics. The role of aerosol-cloud interactions in these transitions is discussed further in Sect. 2.2.3.

Recent progress in the representation of boundary layer clouds in climate models has been brought about through both parameterization improvements and in many cases the use of higher vertical resolution. Other recent parameterization developments include: (i) Non-local boundary layer schemes with explicit entrainment, which typically lead to improved stratocumulus (e.g. Lock et al. 2000); (ii) Eddy diffusion mass flux schemes, which seek to unify turbulence and cumulus parameterizations (e.g. Siebesma et al. 2007).

Improved community coordination through groups that bring together observationalists, process modelers and parameterization developers, such as GCSS (Global Cloud System Studies group, now being subsumed into a new program called GASS that also includes land processes), has been a positive development in recent years. GCSS and CFMIP (Cloud Feedback Model Intercomparison Project) efforts have additionally engaged members of the climate feedback community. Observation sites that monitor detailed surface and remotely sensed information on turbulent fluxes, boundary layer depth, and cloud properties have been linked to create improved networks through programs like CLOUDNET and ARM.

2.1.2 Deep Convection and Its Dynamical Coupling to Larger Scales

There is now evidence that phenomena such as the Madden Julian Oscillation (MJO) and other tropical wavelike phenomena are sensitive to aspects of convective behavior (Hannah and Maloney 2011; Raymond and Fuchs 2009). Raising barriers to deep convection, either through more stringent triggering conditions or greater entrainment, generally improves the representation of the MJO. However these changes usually affect other aspects of simulations adversely, and are not a modeling panacea. It now appears that the eastward propagation of the MJO, previously attributed either to dynamical/wavelike propagation or to a wind-surface flux feedback, may actually arise from simple advection of mid-level moisture (Maloney et al. 2010). This accounts for the importance of convective sensitivity to this variable in reproducing the phenomenon in models.

After a long period of relative apathy since the early 1990s, the last few years have seen renewed interest in developing new parameterizations for deep convection and in cloud dynamics generally. This has been motivated partly by negative drivers such as the significant failure of many existing schemes to properly respond to atmospheric humidity variations (Derbyshire et al. 2004) or simulate realistic diurnal and intraseasonal variations, but also by positive drivers such as the advent of new computational approaches and the spread of cloud-resolving models. Some recent studies have questioned the centrality of thermodynamic, parcel-based reasoning in theories of convection, emphasizing the additional role of mesoscale dynamical constraints in influencing convective growth (Robinson et al. 2008, 2011). At the same time climate models with "superparameterizations," or explicit convection models in place of the usual convective and cloud parameterizations (Randall et al. 2003), have also come into wider use and global models have appeared at resolutions better than 10 km (Satoh et al. 2008). These models are too expensive to run as conventional climate models themselves, but are beginning to provide insights that may help improve standard parameterizations; for example, convective mass fluxes from these simulations can be used in parameterizations of aerosol physics (Gustafson et al. 2008; Wang et al. 2003).

As model grid sizes decrease, traditional assumptions of grid independence and statistically equilibrated cloud fields used in convective parameterizations appear increasingly unjustifiable. Two alternative strategies gaining attention are the inclusion of evolving mesoscale structure, and some elements of stochasticity. While only one convective scheme (Donner 1993) accounts for mesoscale motions explicitly, several new strategies capture in other ways the qualitative evolution of convective events, and seem to improve both diurnal and intraseasonal variability. One such strategy is to add prognostic parameters representing the evolving degree of convective organization (Mapes and Neale 2011) or boundary-layer forcings (Rio et al. 2009), while another is to represent transitions between convective stages or regimes in a population of clouds (e.g. Frenkel et al. 2011a, b; Khouider and Majda 2008). Stochastic parameterizations are also being tested for many model physical schemes, the basic idea being to predict a range of possible outcomes (or one chosen at random) from the inputs to the scheme. One advantage of this is to create a more physical way of generating ensemble forecasts; another is to "smooth" the behavior of the physical scheme with respect to resolved state variables. It is as yet unclear whether stochastic physics will improve climate simulations, or whether any of these strategies will systematically improve the simulated mean climate or cloud feedbacks.

2.1.3 Microphysics

More climate models are beginning to include multiple-moment cloud microphysical schemes to represent both liquid and ice particles. This allows prediction of cloud droplet sizes as well as bulk condensate amounts, and makes possible the computation of more aerosol indirect effects.

However, the fundamental problem with applying more sophisticated cloud microphysics schemes in models that rely on cloud parameterizations is that microphysics is tightly coupled to the cloud dynamics, with the latter unresolved when clouds are parameterized. Arguably, some bulk aspects of convective clouds (such as their total water content profiles) may be well constrained by the mass flux quantities that convective schemes predict. However, predicting sizes of cloud and precipitation particles requires additional assumptions. For instance, in shallow convective clouds in the tropics and subtropics, activation of cloud condensation nuclei strongly depends not only on aerosol characteristics, but also on the vertical velocity field. Some recent cloud parameterizations include information about the vertical velocity in order to provide an estimate of the droplet concentration (Chen et al. 2010; Golaz et al. 2011; Ghan et al. 2011).

2.1.4 Trends, Variations and Feedbacks

While absolute trends in cloud cover have always been difficult to verify due to calibration difficulties, Bender et al. (2012) found evidence in multiple observing systems of a poleward shift of storm-track clouds, that is relative increases at high latitudes and decreases in the subtropics. This shift is qualitatively consistent with poleward shifts of the general circulation reported on the basis of other indices (Sects. 2.3.1 and 2.3.4), and on its own would imply a significant increase in net radiative heating of the planet in recent decades. This phenomenon contributes strongly to a net positive cloud-amount feedback in GCMs (Zelinka and Hartmann 2010).

Climate models, process models, and observations show that upper-level clouds at a given latitude rise or fall roughly in accord with upper-tropospheric isotherms, as predicted by Hartmann and Larson (2002) (Zelinka and Hartmann 2011). This produces a positive feedback on global temperature that accounts for most of the overall mean positive cloud feedback in the CMIP3 collection of climate models (Zelinka and Hartmann 2010).

In general, cloud fields in models change in roughly the same way that the relative humidity field changes (Sherwood et al. 2010). However the exception is boundary-layer clouds, which are crucial to the spread in model predictions. Boundary-layer relative humidity changes are small generally in models. Instead these clouds appear to be sensitive to subtle perturbations in radiation, subsidence and surface fluxes (Zhang and Bretherton 2008; Colman and McAvaney 2011).

2.2 *Aerosols and Aerosol-Cloud Interactions*

2.2.1 Sources, Ageing and Sinks of Aerosols in the Atmosphere

Volkamer et al. (2006) identified evidence that the natural production of secondary organic aerosol (SOA) is much larger than expected, perhaps by an order of magnitude. This aerosol forms from organic precursor gases such as VOCs (volatile organic compounds) emitted from vegetation and other sources. Recent studies have explored this discrepancy and are suggesting that it is not quite as large as previously thought, but still evident in model-observation comparisons (Spracklen et al. 2011; Hodzic et al. 2009). It is not yet clear whether the main problem is insufficient sources, or incorrect sinks in models.

Aerosol sinks are not as well understood as sources, but some progress is being made. The crucial importance of wet scavenging of CCN aerosols in the dynamics of shallow cloud systems is now recognized (see Sect. 2.2.3). Sinks of organic aerosols are not fully understood, and may include unexpected processes such as fragmentation (Kroll et al. 2009). Aerosol ageing is a complex process especially for organics, but recent work suggests possible simplifications in how this can be described (Heald et al. 2010).

A significant problem affecting aerosol-cloud interactions is that currently IN concentrations are poorly quantified, and we still don't have a very good idea which substances are the most important IN, or what fraction of IN are anthropogenic. An important factor determining IN concentrations in the atmosphere appears to be the overall number concentration of aerosol particles at sizes greater than 0.5 μ diameter (Demott et al. 2010), but there are still large variations in the ratio of IN to other aerosol. While primary organic aerosol such as pollen do not appear to be dominant sources of IN in clouds, organic residues on dust and in soils do appear to contribute significantly to the ice-nucleating ability of these substances (Conen et al. 2011) but in ways that vary mysteriously from one region to another. Most IN are undoubtedly natural; the most likely anthropogenic IN would either be black carbon (whose ability to nucleate ice is still in question) or additional dust emissions arising from human land use changes or other activity (which are hard to isolate from the much greater quantities of natural dust).

2.2.2 Direct and Indirect Radiative Effects of Aerosols on Climate

Aerosols exert a direct cooling effect on climate by reflecting sunlight to space, although dark carbonaceous aerosols can exert either warming or cooling effects because they absorb as well as scatter sunlight. Quantifying these effects from observations alone is difficult, as some type of model is needed to establish the radiative balance that would have occurred in the absence of whatever aerosol is present. Some kind of model is also needed to establish how much of the observed aerosol is anthropogenic, given that global observations are unable to distinguish aerosol types sufficiently for this purpose, except via crude assumptions. Interest in aerosol effects on climate has been enhanced by proposals to disperse aerosols in boundary layer clouds and in the stratosphere as a geoengineering strategy for cooling the planet.

The most straightforward and long-established aerosol impact on cloud albedo comes through the so-called Twomey (sometimes known as cloud-albedo) effect, whereby more droplets are nucleated by greater aerosol counts, increasing the surface area and thus albedo of a given total cloud water content. Model estimates of the magnitude of this forcing over time have changed little. Additional indirect effects due to changes in cloud lifetime or cover, or arising from changes to atmospheric circulations arising from aerosol thermal and microphysical effects, are increasingly being considered but are much more difficult to quantify. There is some suggestion in recent studies that as new effects are added, compensation occurs with existing effects such that the total impact on cloud albedo and/or precipitation doesn't change as much as might have been expected (see Sect. 2.2.3). However, rapid transitions can be triggered in stratocumulus such that changes in cloud amount and thickness strongly amplify the Twomey effect (see Rosenfeld et al. this volume).

A number of GCMs equipped with aerosol physics now predict the radiative effects of anthropogenic aerosol. Model predictions of both the direct (Myhre 2009; Bellouin et al. 2008) and aerosol-cloud related (Storelvmo et al. 2009) cooling effects have decreased somewhat in more recent studies, with estimates of total forcing (not including ice processes) now near $-1.5\ W\ m^{-2}$; a few models with ice effects tend to show greater cooling. Considering only the albedo effect, estimates of forcing constrained by satellite observations show significantly less cooling than those predicted by models alone: from $-0.5\ W\ m^{-2}$ to near zero. This may mean models are still overestimating the albedo effect, though it is also possible that observations of aerosol in the vicinity of clouds, and methodologies for averaging data from the satellite pixel scale to model grid-box scale, bias the strength of the cloud-aerosol relationships used to constrain climate models (McComiskey and Feingold 2012). Inter-model estimates of aerosol-cloud forcing that allow for dynamical feedbacks tend to be more variable than estimates of the albedo effect alone because of the greater range of processes considered. However there are some indications, from both observations and small-scale models, that compensating factors may be at play in real cloud systems, and that the higher negative forcing estimates are a result of the inability of climate models to resolve small spatiotemporal scale cloud, and aerosol-cloud interaction processes (see Sect. 2.2.3). This is an active area of research.

There are several reasons why model estimates of aerosol forcing have dropped. Perhaps the most important is increased estimates of the absorbing effect of black carbon (Myhre 2009; Chung et al. 2005), which offsets the cooling effect of aerosol scattering and can warm climate further by settling on ice surfaces where it is a particularly efficient absorber. Also, new observations are showing somewhat greater natural contributions to the observed aerosol burden (see Sect. 2.2.1).

There is growing evidence that decadal changes in aerosols may be responsible for the observed phenomenon of global dimming (the reduction of sunlight observed at the surface) prior to about 1990 and global brightening since, although changes in cloudiness (whether due to aerosols or not) play a large role especially on a regional basis (Wild 2009). Background stratospheric aerosol and water vapor may also vary on decadal or longer time scales, making some contribution to radiative forcing (Solomon et al. 2010, 2011). Aerosols may also drive interdecadal climate variations in the Atlantic basin (Booth et al. 2012).

New research highlights the possibility of IN effects on cirrus or mixed-phase cloud properties, which has even been suggested as another geoengineering strategy (Mitchell and Finnegan 2009). The main anticipated mechanism for IN to affect clouds is by causing the earlier nucleation of smaller numbers of ice particles at temperatures between −10 and −40° C in deep convective clouds. These early-initiators would grow rapidly and become efficient collectors, leading (in principle) to optically thinner deep-cloud outflows. However the complexity of mixed-phase cloud systems means that currently such mechanisms are hypothetical; indeed some simulations show IN leading to increased cirrus (Zeng et al. 2009). See Rosenfeld et al. (this volume) for more details.

2.2.3 Microphysical Effects of Aerosols on Precipitation and Vice Versa

A long history of efforts to ascertain the influence of CCN aerosol on warm clouds (Gunn and Phillips 1957; Warner 1968) have indicated a likely suppression of rainfall, although there exists no definitive, statistically-sound, observational proof of this. The proposed mechanism is that by nucleating more droplets, droplets do not grow as fast, fall speeds are reduced, and the formation of rain by collision and coalescence is delayed or prevented. However this suppression of precipitation will lead to more evaporation in the free troposphere, destabilization and deepening of subsequent clouds, and the potential for more rain. Dynamical feedbacks of this kind make it particularly difficult to untangle aerosol effects on precipitation (e.g., Stevens and Feingold 2009). The net effect of aerosol on cloud albedo is a complex function of small-scale processes and feedbacks that occur at a range of scales. As a result it is likely cloud-regime-dependent. When averaged over multiple regimes, it may be significantly less than would be expected from consideration of the simple microphysical response in isolation (Stevens and Feingold 2009).

Recent work shows that the knock-on effects from the initial modification of clouds are sometimes "absorbed" by the cloud system, but other times are more profound. Observations of shallow convective cloud layers confirm strong connections between aerosol loading, precipitation and cloud morphology, with precipitating portions of marine cloud decks appearing nearly devoid of aerosols (Sharon et al. 2006; Wood 2012). This suggests a strong positive feedback where precipitation removes aerosol, leading to more efficient formation of precipitation, a feedback thought to shift closed-cellular to open-cellular convection, in sub-regions that are non-raining and raining respectively (Stevens et al. 2005; Sharon et al. 2006). Both A-Train observations (Christensen and Stephens 2011) and large eddy simulation (e.g., Wang et al. 2003; Ackerman et al. 2004; Xue et al. 2008; Wang and Feingold 2009) show that the aerosol increases cloud amount and cloud water in clean, open-cell regions and decreases cloud amount in non-precipitating, closed-cell regions.

It is now argued that as coupled cloud systems evolve, they tend to prefer certain modes (e.g., non-precipitating closed cells and precipitating open cells) that are resilient to change due to internal compensating processes (Stevens and Feingold 2009; Koren and Feingold 2011). However under certain conditions, e.g., very low aerosol concentrations, instability sets in and the closed-cell, stable system may transfer to the precipitating open-cell system. The open cells appear to constantly rearrange themselves as precipitation-driven outflows collide and drive new convection, which forms new precipitation, and so on (Feingold et al. 2010).

A weakness of the detailed process-level large eddy simulation is that it is rather idealized. Cloud resolving and regional models allow for a much broader range of scale interactions and timescales and are increasingly being used to explore aerosol-cloud interactions (e.g., Grabowski 2006). Modeling of deep convective cloud systems suggests that the average impact of added aerosol is very short-lived, with a slight delay in the initial development of rainfall but no effect on the integrated rainfall amounts over times approaching a day or longer (Morrison and Grabowski 2011; Seifert et al. 2012). Similarly, under conditions of radiative-convective equilibrium van den Heever et al. (2011) have shown that aerosol perturbations have little influence on domain-averaged precipitation and cloud fraction. However this is a result of compensation between the responses of shallow and deep convective clouds, in keeping with the idea that while average aerosol influences may be small, local influences may be significant.

In addition to their potential to study aerosol-cloud interactions, cloud resolving and regional models show that gradients in the aerosol may generate changes in circulation patterns via changes in heating rates (Lau et al. 2006), radiative properties of cloud anvils (van den Heever et al. 2011), or in the spatial distribution of precipitation (Lee 2012).

2.2.4 Advances in Parameterizing Aerosols

Aerosol treatments in global climate models remain fairly crude, although this could be said of all model parameterizations. Studies using chemical transport models driven by observational estimates of wind fields have proven useful in constraining and refining the schemes for predicting poorly-constrained natural sources of aerosols such as sea-salt and organic aerosol precursors (Lapina et al. 2011).

Aerosol effects on clouds are being treated in more models, and are beginning to include effects on convective clouds including secondary effects although this involves massive uncertainties. Mass fluxes obtained from explicit simulations are being used to implement aerosol effects on convective clouds (see Wang et al. 2003).

2.3 *Dynamics from Small to Global Scales*

2.3.1 Gravity Waves

Small scale atmospheric gravity waves (or internal waves), produced by flow over topography, convection, and imbalances in the geostrophic flow, influence climate through their effects on the large-scale circulation, which in turn affect synoptic and planetary wave propagation and dissipation (e.g. Alexander et al. 2010). With important horizontal and vertical scales as small as 5 km and 1 km, respectively, much of the gravity wave spectrum remains unresolved at current climate model resolution. Mountain wave drag reduces westerly biases in zonal winds near the tropopause, and parameterized mountain wave drag settings in climate models can affect high-latitude climate change response patterns in surface pressure (Sigmond and Scinocca 2010). The changes in wind shear that occur with tropospheric warming and stratospheric cooling alter the altitude and strength of mountain wave drag; this affects planetary wave propagation and associated surface pressure patterns, strengthening aspects of the Brewer-Dobson circulation such as poleward stratospheric transport and upwelling and downwelling near the tropical and polar tropopause respectively.

Trends in upwelling near the tropical tropopause have been related to changes in stratospheric water vapor, an important greenhouse gas (Solomon et al. 2010). An increasing trend in twenty-first century upwelling is predicted in models that resolve the stratospheric Brewer-Dobson circulation (Butchart et al. 2006). This wave-driven transport circulation responds to changes in forcing by planetary-scale and gravity waves, and many models ascribe a large fraction of the trend to changes in parameterized orographic gravity wave drag (Li et al. 2008; McLandress and Shepherd 2009; Butchart et al. 2010). Cooling in the stratosphere and warming in the troposphere associated with greenhouse gas (GHG) trends lead to stronger subtropical jets, and these changes in the winds explain the changes in the parameterized drag.

An early focus on different dissipation mechanisms within non-orographic gravity wave parameterizations has given way in recent years to a focus on defining wave sources and the properties of the waves emitted. This has followed from research demonstrating effective equivalence of different parameterization methods in climate model applications (McLandress and Scinocca 2005). For climate prediction, the sources of non-orographic gravity waves should respond to climate changes, but in most current models wave sources are simply prescribed. A few models do include multiple wave sources like convection and fronts in addition to orography (e.g. Richter et al. 2010; Song et al. 2007). However, the underlying processes remain rather poorly understood and the parameterizations are largely based on two-dimensional theoretical models.

Recent global simulations at very-high resolution capable of resolving many (though not all) scales of gravity waves have advanced our understanding of the processes important for improving parameterizations (e.g. Sato et al. 2009; Watanabe et al. 2008), and comparisons of these with observations are assessing their ability to realistically represent the resolvable portions of the wave spectrum (Shutts and Vosper 2011).

2.3.2 Blocking Events

Atmospheric blocking is characterized by abnormally persistent (i.e. time scales of 1–2 weeks) high pressure systems which steer, or "block," the usual propagation of midlatitude cyclones, and thus play a critical role in intraseasonal variability and extreme events in the extratropics. Limitations in the ability of climate-models to capture these important synoptic scale features were described in the IPCC's AR4, and appear to persist in more recent models. Since the 1980s many authors reported an upscale feedback of eddy vorticity that helps to maintain blocking highs (e.g. Shutts 1986; Lau 1988). Recently this has been verified in models and analyses, and the self-maintaining nature of blocking eddies has been confirmed (e.g. Kug and Jin 2009).

Despite this, it is not yet clear what resolution is required to successfully model enough of the vorticity flux to give reasonable blocking statistics. Traditionally, models have under-represented the frequency of blocking (D'Andrea et al. 1998) in a way consistent with their limited resolution. Some studies have shown an increase in blocking when either horizontal resolution (Matsueda et al 2009) or vertical resolution (Scaife and Knight 2008) is increased. This is consistent with the idea of an upscale feedback from poorly resolved eddies. Evidence has also emerged that climate models are systematically westerly biased (Kaas and Branstator 1993), which can greatly bias blocking frequencies diagnosed via standard measures (Doblas-Reyes et al. 1998), even if the simulated variability appears adequate (Scaife et al. 2010). In coupled models, the westerly bias and blocking deficit over the Atlantic may be associated with errors in the simulated Gulf Stream (Scaife et al. 2011).

2.3.3 Widening of the Tropics

On planetary scales, evidence for a widening of the Hadley circulation, or tropical belt, in the last decades of the twentieth century has been deduced from various data sources, and model simulations show that GHG increases cause widening (e.g., Schneider et al. 2010). This has potential connections to important changes in global precipitation patterns and other climate variables (e.g. Seidel et al. 2008). How the width of the Hadley cell is controlled is however unclear. Both thermodynamic changes at low latitudes and eddy flux changes in the subtropics and extratropics likely play a role. Indeed, Son et al. (2008) show that changes in polar stratospheric ozone influence the width of the Hadley Cell, most likely by displacing the midlatitude jets and so modifying eddy momentum fluxes in the subtropics. Based on model simulations, the expansion of the Hadley cell has been ascribed to radiative forcing associated with changes in GHG and stratospheric ozone depletion (Lu et al. 2007) or absorbing aerosols or ozone in the troposphere (Allen et al. 2012), and is consistent with poleward shifts of the subtropical jet streams (Yin 2005). However changes in tropical tropopause heights that have been associated with the Hadley cell widening (Seidel and Randel 2007) are also strongly affected by changes in the Brewer-Dobson circulation (Birner 2010) and therefore coupled to changes in the extra-tropical circulation in the stratosphere.

2.3.4 Impact of the Stratosphere on the Large-Scale Circulation

Observational evidence for a significant impact of stratospheric ozone loss on the tropospheric circulation emerged prior to the IPCC's AR4 (e.g., Thompson and Solomon 2002). To date, the largest change in the midlatitude jet streams and storm tracks is observed in the Southern Hemisphere in summer, following the annual formation of the ozone hole, and climate model studies have verified the critical role of ozone in these changes (e.g. Arblaster and Meehl 2006; Polvani et al. 2011). However some of the CMIP3 models used in the last assessment ignored ozone changes, and most represented the stratosphere poorly in general. Understanding of the connection between twenty-first century ozone recovery and SH climate projections has advanced very recently. Son et al. (2008) showed that models with realistic ozone recovery predict a weak equatorward shift in the summertime extra-tropical jet in the twenty-first century, while models with constant ozone predict a poleward shift in the jet due to GHG increases. These trends in jet position project strongly onto the Southern Annular Mode (SAM). While GHG trends lead to a year-round positive trend in the SAM, some models including ozone recovery with a well-resolved stratosphere predict a large negative trend in the SAM in summer (e.g. Perlwitz et al. 2008). Seasonally dependent trends in SAM could influence carbon uptake in the Southern Ocean (Lenton et al. 2009) and may further couple with Antarctic sea ice trends (Turner et al. 2009).

New work shows the stratosphere plays another important role in climate change independent of ozone changes. In models with good representation of the stratosphere, regional climate changes, particularly those associated with ENSO teleconnection to European winter climate, can propagate through a stratospheric pathway (Ineson and Scaife 2009; Cagnazzo and Manzini 2009), and even long-term predictions of precipitation and wind patterns in models lacking a well-resolved stratosphere can suffer from first order errors compared to those of models that better resolve the stratosphere (Scaife et al. 2012). These changes often project onto the North Atlantic Oscillation (NAO) and the Northern Annular Mode (NAM), a primary mode of northern hemisphere climate variability. Gerber et al. (2012) review the current understanding of stratospheric effects on surface weather and climate. Roughly ten models in the CMIP5 will include a better represented stratosphere, compared to almost no models in CMIP3, so these issues should become clearer in the IPCC's AR5 report.

2.3.5 Impact of Warming on Rainfall Extremes, Cyclones, and Severe Storms

Infrequent, intense weather events are part of a stable climate system, and involve many scales, from isolated convective cells on the order of kilometers to planetary scale features such as the Madden Julian Oscillation. Evidence of increases in certain extremes is beginning to emerge in the observational record (Zwiers et al. this volume), though attribution to specific aspects of climate change is difficult, especially for individual events (Stott et al. this volume). While model predictions of extremes remain dubious, certain expectations follow from our understanding of basic physical processes and are being investigated by process models.

Dynamical responses in the atmosphere to the warming climate lie behind changes in likelihood of some "extreme" weather events and therefore understanding and quantifying these is a basic step in determining changes in extremes. Poleward shifts of the extra-tropical jet stream with associated migrations of storm tracks and changes in the intensity of the storms may be accompanied by changes in weather patterns and associated extremes (Gastineau and Soden 2009, 2011). Expansion of sub-tropical dry zones at the edges of the widening Hadley circulation may be accompanied by pronounced changes in precipitation patterns and associated desertification (Johanson and Fu 2009).

Assessing the response of tropical circulations and associated weather extremes to changes in GHG forcing using climate models has proved to be difficult because of the lack of agreement among models (Kharin et al. 2007) and their general inability to consistently represent some key physical features such as the observed mean precipitation regimes of the Asian summer monsoon (Stowasser et al. 2009). Such deficiencies are in large part associated with resolution constraints and associated inadequate parameterization of unresolved small scale processes. Large-scale increases in tropical sea surface temperatures (SSTs) associated with a warming climate do not necessarily translate directly into local increases in precipitation intensity associated with enhanced deep moist convection. In fact model results suggest that precipitation may decrease in regions such as the equatorial Indian

Ocean in association with uniform increases in SSTs. However modeling results do indicate that intensified deep convection with higher precipitation is more likely to occur where SSTs are locally larger than their surroundings (Stowasser et al. 2009; Neelin and Held 1987). Only a few of the coupled models used in AR4 simulate a qualitatively realistic climatology of the Asian monsoon (Annamalai et al. 2007; Stowasser et al. 2009); under global warming, these models predict an increase in monsoon rainfall over southern India, despite weakened cross-equatorial flow (Stowasser et al. 2009).

3 Current Scientific Gaps and Open Questions

3.1 Clouds and Convection

Observational capabilities for clouds have improved significantly with the launch of MODIS, CloudSat/CALIPSO and other satellite sensors. However we lack good data on the detailed motions at the convective scale that would be beneficial for testing the assumptions of cloud models and in particular for constraining processes such as entrainment. Also, observations of precipitation still have large errors even from the best spaceborne sensors, particularly for light rain.

Many GCMs still have difficulty in successfully simulating transitions between different cloud regimes (e.g., stratocumulus to cumulus). Most deep convective schemes used in global models appear to make the transition from shallow to deep convection much too quickly, which among other problems leads to inaccurate diurnal cycles. A possibly related problem is that convection in models is insufficiently sensitive to humidity above the cloud base (Derbyshire et al. 2004). This problem is well-recognized by model developers but a fundamental basis for redeveloping the convective schemes is currently lacking, such that most approaches to address the problem have so far been convenient fixes that don't come to grips with underlying problems.

While recent research (e.g. through GEWEX) has focused particularly on low clouds due to their role as a "known unknown," (e.g., Soden and Vecchi 2011), the representation of upper-level and cirrus clouds in GCMs is a source of concern as it is highly simplified, and models currently underpredict mid-level cloud which begs the question of whether feedbacks by these clouds might be missing or underrepresented. Cirrus clouds have also been hypothesized as playing a role in polar amplification of warmer past climate states (Sloan and Pollard 1998) but this has not been reproduced by climate models so far.

Models still have difficulty representing tropical variability (Lin et al. 2006). Convective parameterizations tend to well represent either the mean climate or the variability, but not both. Convectively coupled equatorial waves (CCEWs) control a substantial fraction of tropical rainfall variability. CCEWs have broad impacts within the tropics, and their simulation in general circulation models is still problematic, although progress has been made using simpler models. A complete understanding of CCEWs remains a challenge in tropical meteorology (Kiladis et al. 2009).

Cloud microphysics remains a great challenge, with most work so far limited to liquid clouds, which have still proven difficult to model. For ice clouds the situation is even more difficult because of complications of ice initiation (i.e., homogeneous versus heterogeneous activation) and subsequent growth. Only about 1 in 10^5 aerosol particles are active as heterogeneous ice nuclei, they are hard to measure, and the detailed nature of the freezing mechanisms is uncertain. Cloud physics has struggled with representation of ice processes in detailed models for decades, so it should not be surprising that representation of such processes in large-scale models remains highly uncertain. In summary, parameterizing cloud microphysics in models with parameterized clouds is extremely difficult. Arguably explicitly cloud-resolving approaches are a significant improvement, but often not at an affordable cost for many applications.

The modeling of clouds is badly hampered by the poor state of understanding of basic cloud physics and dynamics, and the inability to represent all scales of cloud motion and entrainment. Fundamental uncertainties about entrainment and mixing may significantly affect our ability to quantify aerosol impacts on cloud radiative forcing (e.g., Jeffery 2007).

Some researchers are calling for greater emphasis on basic cloud physics in the context of aerosol effects (e.g. Stevens and Feingold 2009), on the grounds that we cannot fully understand or quantify how clouds are modified by aerosols before we are able to predict what clouds do in the absence of aerosol perturbations. While that article focuses mainly on warm boundary layer clouds, an equally or stronger case can be made for mixed-phase stratus clouds (Morrison et al. 2011) or cirrus clouds, where even the relative importance of homogeneous vs. heterogeneous nucleation is still unknown let alone the cloud dynamics or evolution of ice particles after they have formed. An alternative view however, is advanced by Rosenfeld (this volume) on the basis that aerosol impacts on clouds can be observed even if we don't have complete theories of cloud behavior.

3.2 Aerosols and Aerosol-Cloud Interactions

The discrepancy between model and observational estimates of aerosol cloud-mediated forcings (Sect. 2.2.2) is a significant issue. It is not yet clear whether biases lie predominantly with the observations or with the models. If satellite-derived estimates are correct, most GCMs are probably overestimating the cooling effect of aerosols during the twentieth century.

The quantitative study of aerosols is greatly hampered by the complexity of aerosol structures in the atmosphere and the limited compositional information provided by most observing systems, especially satellite sensors. It is evident that most aerosols are inhomogeneous mixtures, with optical and hygroscopic properties that depend on how they are mixed. One upshot is that particles not normally thought to be effective CCN may become effective after a modification through the deposition of other materials while the particle is airborne (Ervens et al. 2010). The reverse may be true for IN because their effectiveness is reduced by the addition of soluble material. There are also many forms of organic aerosol with different source and deposition properties. Economically describing or categorizing such a rich spectrum of possible aerosol types, mixtures, and sizes is a significant observational and modeling challenge.

Relatively little research has gone into quantifying aerosol sinks, in comparison to sources (e.g., Lee and Feingold 2010). The measurement of dry deposition of aerosols is difficult in many cases, and measurements are currently too scarce to constrain models. The processing of secondary organic aerosols through aqueous chemistry is also not well understood. It is possible that poor representation of sinks may be affecting model simulations of aerosol distribution as much as inaccurate sources.

Aerosol modeling is also affected by transport issues. Models typically make naive assumptions about vertical redistribution of aerosols by boundary layer motions and deep convective mixing. Aerosol effects on clouds are quite sensitive to mixing assumptions and the science is currently hampered by basic questions of how to model turbulent entrainment and mixing within clouds noted above. Vertical distributions of aerosol vary significantly with region and aerosol type, and are of concern in interpreting both satellite observations and in-situ near-surface observations.

Observational studies of aerosol impacts on clouds have long been plagued by a problem of correlation vs. causality, since clouds strongly affect aerosols as well as the reverse, and both are affected by meteorology. Satellite-based aerosol observations are mainly provided by polar orbiters, but these only give snapshots, providing little traction against the causality dilemma. Geostationary satellites can provide crucial temporal information but produce relatively poor aerosol and cloud products compared to polar orbiting satellites.

It continues to be difficult to unambiguously distinguish aerosol and cloud in remote sensing observations, because of a combination of factors, including aerosols becoming hydrated and growing in size with decreasing distance to clouds, cloud fragments, and enhanced scattering of photons between clouds (Wen et al. 2007). Since even in principle there is no clear distinction between a hydrated CCN aerosol and an incipient cloud droplet, it may for some purposes be better not to attempt to distinguish aerosol and clouds at all (Koren et al. 2007; Charlson et al. 2007).

Ice nuclei remain a particularly puzzling aspect of the global aerosol burden. Progress in predicting IN concentrations appears to be hampered by the incomplete understanding of why some substances nucleate ice well and others poorly. It is hard to see how aerosol-cloud radiative effects modulated by deep convection, and subsequently affecting anvils and cirrus, will be properly understood or quantified while issues surrounding ice nucleation and growth remain so unresolved.

Aerosol-cloud related forcings remain poorly quantified. Even in the relatively well-studied case of shallow clouds, it remains unclear whether secondary effects globally tend to cancel (e.g., Stevens and Feingold 2009) or reinforce (e.g., Rosenfeld et al. this volume) the primary ("Twomey") effect, since both outcomes are possible depending on circumstances. The prevalence and areal coverage of the sign and magnitude of these responses would seem to be an important line of enquiry. Aerosol effects on ice-containing clouds are likely in opposition to those on shallow clouds, and climate model simulations suggest that radiative forcings involving these are potentially larger than those of liquid-phase clouds, and involve large infrared forcing effects. While this result is highly uncertain, it highlights the need for progress on mixed-phase cloud microphysics, and points to large uncertainties in model-based "forward" estimates of indirect forcing; it also leaves open the possibility that a modest net aerosol-cloud forcing represents a near-balance between opposing large ones from deep and shallow clouds (Rosenfeld et al. this volume).

Studies attempting to back out aerosol forcing from the observed temperature record ("inverse estimates") must consider not only uncertainties in climate sensitivity and ocean heat uptake, but also the role of other forcings such as tropospheric ozone, stratospheric water vapor, and land use changes. Recent studies also show that aerosol impacts on surface temperature can be highly non-local, nonlinear, and can include impacts on the general circulation. This complicates attribution efforts, as for example changes in tropical aerosol may have affected the extratropical temperatures in either hemisphere and may not be strictly additive with other forcings.

3.3 *Dynamics from Small to Global Scales*

The push toward higher horizontal resolution leads to resolution of more gravity waves in climate and NWP models. Observational verification of these waves and their effects on general circulation is needed. Evidence in the tropics suggests that higher vertical resolution is more urgently needed to properly simulate large-scale equatorially trapped modes (e.g. Evan et al. 2012) important to driving the QBO (e.g. Scaife et al. 2000; Giorgetta et al. 2002). Even at NWP resolutions, short horizontal wavelength gravity waves with substantial momentum fluxes and inferred large effects on circulation remain unresolved (e.g. Alexander et al. 2009). Improvements in the parameterization of gravity wave sources is needed to properly simulate gravity wave effects in future climate scenarios.

Higher resolution also impacts the representation of synoptic scale variability in climate models. It is still unclear what resolution is required to accurately represent atmospheric blocking. Further work is needed to understand the role of mean state errors in blocking statistics and how blocking might be improved in models. The organization of synoptic scale heat and momentum fluxes in the planetary scales generates the midlatitude jet streams. There are substantial biases in the location of austral jets in almost all CMIP3 models, which are associated with errors in their intraseasonal variability and sensitivity to climate forcing (e.g. Kidston and Gerber 2010). While these processes are nominally resolved by all CMIP3 models, simply increasing the resolution appears to help correct (but not eliminate) biases (Arakelian and Codron 2012). Further work is need to understand how errors in marginally resolved mesoscale processes are scattering back and biasing the resolved variability.

The issue of resolved vs. unresolved scales is a more pressing problem in tropical meteorology, where key processes must be parameterized. The interactions of unresolved cloud and convective processes with resolved waves and vortices is a critical area of current research (e.g. Khouider et al. 2013). This coupling across scales (or lack thereof) is likely behind the most persistent problems in climate model's representation of tropical variability, including convective coupled waves and the Madden–Julian oscillation (e.g. Lin et al. 2006). Poor tropical variability in turn affects both the mean climate (i.e. the double inter-tropical convergence zone problem; Lin 2007) and the frequency of high- and low-intensity rainfall events (e.g., Stephens et al. 2011).

Although the simulated pattern of sea-surface temperature response to global warming includes an El Nino-like component, the extratropical atmospheric responses occur in a somewhat opposite fashion to the El Nino teleconnection pattern (Lu et al. 2008). Understanding the difference between the response to El Nino (jets shift equatorward) and global warming (jets shift poleward) may provide important clues to understanding mechanisms for the poleward shift of the jet and widening of the Hadley cell in climate change scenarios.

A common theme in many of these gaps in our understanding is the relationship between natural, or internal variability, and the mean climate. One can view the climate as a stochastically forced system, and formulate the questions: what does climate "noise" tell us about the system and its response to external forcing, and how does noise at unresolved scales scatter back to resolved scales? To account for unresolved variability, new stochastic parameterizations are being developed to explicitly introduce uncertainty in subgrid scale processes (e.g. in the sources of non-orographic gravity waves; Berner et al. 2009; Eckermann 2011). To account for resolved variability, modeling groups are turning to large ensemble forecasts, as is routinely done in numerical weather prediction. Properly accounting for natural variability is also extremely important for predicting changes in the extremes and making regional climate forecasts, where the signal to noise ratio is smaller (e.g. Deser et al. 2012).

Another general issue which affects all research areas covered in this article is the limited size of the community involved in model development (e. g., Jakob 2010). A relatively large community of researchers use global and regional climate models, or study the processes that are not well represented. Some of this work gets as far as proposing parameterization improvements. However, there is a large and separate task of improving the GCMs, which is crucial, but in which there are only a relatively small number of people participating. The problem is exacerbated by current funding models which tend to separate basic research (largely at universities) from model development (largely at big modeling centers) with too little support or incentive to link these activities. Further, scientific achievement is measured by counting papers, which may be harder for hands on-model developers to do in quantity. Finally, model development is a challenging undertaking for a postgraduate student or short-term postdoc, really requiring longer-term support and a team environment; this will become more true as models become more complex and parameterizations more interconnected.

4 Strategic Opportunities and Recommendations

After decades of effort it remains evident that no current model can reliably simulate both individual clouds and the climate at the same time. Yet the cloud and climate scales cannot be decoupled. One question that then arises is how to best harness high-resolution computations, and whether they can ultimately bridge the gap and render parameterization unnecessary? Second, how can observations be used to help make progress? The complexity of the system makes it very difficult either to durably improve models by haphazard experimentation, or to diagnose their problems directly from discrepancies with observations, although these activities must continue. Nor is there evidence that numerical cloud models, even at extreme resolutions, converge to solutions that are insensitive to parameterizations. These difficulties highlight the need for better fundamental understanding. We believe this applies equally to aerosol and dynamical research.

4.1 Research Foci, Strategies and Resources

While there is a wide array of diverging views on the best paths forward, we see several promising opportunities, as well as important assets that must be protected and nourished.

4.1.1 Confront Two-Way Integration Across Scales

A recurring theme in cloud, aerosol and dynamics research is the tight connections between behavior across scales. It is becoming evident for example that the immediate response of a cloud to an aerosol perturbation, in the absence of any interactions or feedbacks from the larger environment, may differ dramatically from what happens in a more realistic setting where the cloud interacts with others dynamically. Thus role of clouds in climate may be as difficult to discern from traditional small-scale (e.g. cloud-scale) studies—where dynamical adjustments and feedbacks from remote processes cannot occur—as from global studies that cannot resolve the clouds. Numerical (e.g. LES) simulations may capture some, but not all of these adjustments. A similar limitation affects observational analyses based on local relationships between variables that do not account for the fact that the putative causal agent (e.g., aerosol) can effect the target quantity (e.g., clouds) nonlocally.

A key research priority should be the development and implementation of strategies to couple large-scale responses into process modeling efforts, and the application of this to interpretation of observations. One approach is simply to perform extremely large and expensive computations; another has been "superparameterization/" The latter approach could for example be extended to resolve gravity wave propagation into the stratosphere. However, other, more affordable and widely adoptable strategies are needed.

A useful prototype strategy is to run process models in a "weak temperature gradient" setup (Sobel and Bretherton 2000) that allows some idealized feedback from larger scales in a Tropical setting. Development and standardized use of a small set of analogous strategies or testbeds, perhaps involving the coupling of multiple process models, would fill a crucial gap. Another strategy for combining models and observations is to exploit emergent behavior or other non-traditional measures of the behavior of a tightly coupled aerosol-cloud-dynamical system, rather than trying to isolate deterministic impacts of one part of the system on the others (e.g., Harte 2002; Koren and Feingold 2011; Bretherton et al. 2010; Morrison et al. 2011). A prototype for this strategy is the longstanding effort to explain convectively-coupled wave activity in the tropics, with models of varying complexity and design, to see what is needed to get it right.

4.1.2 Emphasize Fundamental Science and Model Development

Our perception is that the amount of effort being expended toward the proper development of atmospheric model "physics" (cumulus and other parameterizations) is too small relative to the expanding use of the models for predictions and demands from users for greater regional accuracy, which in most cases the models cannot yet deliver (Jakob 2010). While there are significant model development efforts at some centers, more often the development is driven toward short-term model improvement rather than identifying and resolving fundamental problems. A larger, vibrant community working on the development of more solid theory through basic research into poorly understood processes and, crucially, the transfer of this to practical applications in more comprehensive models, is essential to sustained improvement in global and regional simulations. This probably requires more durable institutional support for broadly engaged model development teams, as well as promotion of stronger links between basic research and model development.

4.1.3 Explore Hierarchical Modeling Approaches

While adding new processes to models has value, there is equal value (but currently less effort) in simplifying models—even in highly idealized ways—in order to reveal deeper aspects of system behavior, narrow down possible explanations for phenomena or for model differences, or identify misconceptions (see Bony et al. this volume). One specific example could be the use of aquaplanets or other even more idealized configurations to explore the cloud-mediated effects of aerosols or other forcings; another could be switching off selected processes in GCMs systematically as part of future intercomparisons. Single-column versions of GCMs are a potentially valuable resource that is currently underutilized outside model development centers.

4.1.4 Integrate the Whole Atmosphere, Ocean and Surface

The recent reorientation of SPARC toward troposphere-stratosphere coupling is already a good development in light of new awareness that such interactions may be more important than previously thought. This accompanies a growing development of "high-top" atmosphere models. However, as the stratosphere, cryosphere and ocean each have more "memory" than the troposphere, they may be capable of interactions (through the troposphere) that would only be resolved by fully coupled high-top models. Such models barely exist at present; more should be pursued. One area of attention would be the impact of solar variability on climate.

4.1.5 Plan for the High-Resolution Future

Advancing computer power will inevitably lead to higher resolution global and process models, a potential boon for atmospheric physics research but one not without problems. First, performance does not always increase, and can even drop, when resolution rises beyond those for which parameterizations were optimized. It is thus becoming clear that physical parameterizations in models should be "scale aware"—their behavior should depend on the grid size, and in particular, they should gradually stop acting if and when the grid size shrinks to where it can explicitly resolve the parameterized phenomenon. Second, data transfer and storage technologies are not keeping pace with CPU power, and data analysis software is typically not parallelized, with the result that the analyses needed to take full advantage of large simulations will continue to become more difficult. Traditional practices of dumping output and then analyzing it may become increasingly impractical. Modeling, IT and theory communities should together devise strategies to maximize the practical scientific utility of state-of-the-art computations.

Similar issues exist for more modest but more numerous CRM and LES computations, which have entered a rapid-growth phase, and could benefit from the adoption of canonical test cases (analogous to CO_2-doubling, 1 %/year and twentieth century hindcasts for GCMs) and standardized output quantities and formats. Moves in this direction are already occurring in GEWEX and e.g. CGILS. These studies are often based on observed cases, but simpler, idealized cases also have a role to play in testing hypotheses and understanding key processes and how best to represent them in larger-scale models.

4.1.6 Bring Weather to Climate

The experience of the weather forecasting community, which routinely runs at high resolution, could be better utilized by climate modelers. Efforts to examine the behavior of climate models on short time scales in a variety of different environments, and the climatic behavior of forecast models, should be encouraged as possible pathways to better understanding. For example, idealized studies with simplified GCMs suggest a connection between the internal variability and the response to external forcing (Ring and Plumb 2008; Gerber et al. 2008). Other evidence is that strong connections are found between biases in the time-averaged position of the extratropical jets in different GCMs, the time scales of their natural weather variability, and biases in blocking (e.g. Kidston and Gerber 2010; Barnes and Hartmann 2010). The similarity of short-term and long-term errors in model forecasts from a specified initial state also suggests the utility of this approach for climate (Brown et al. 2012). Related to this is a need for more statistical rigor, and perhaps opportunities from new statistical approaches, in many aspects of climate and climate-process research.

4.1.7 Sustain and Improve Observations

Last but not least, new observational capabilities are needed to address key weaknesses, and existing capabilities should be protected and kept as homogeneous and continuous as possible. Experience has shown the importance of sustained observations in order to capture crucial variability on decadal and multi-decadal time scales, and how sensitive this can be to gaps or too-short overlaps in satellite records. Continuation of existing cloud- and aerosol-observing capabilities is not assured, as few new missions are in the pipeline; plans to incorporate process- and climate-oriented observations into operational satellites in the US in particular have largely fallen by the wayside.

New observables that would be particularly useful include better fine-scale observations of clouds on a range of scales, better information on vertical velocities in clouds (promised by the EarthCare satellite scheduled to launch in 2015), measurements of aerosols and water vapor underneath clouds, better characterization of cloud microphysics and water content, more accurate global measurement of light and/or shallow precipitation, and better monitoring of spectral solar variability (Harder et al. 2009). Some of these could potentially be provided from space by multiangular, multispectral sensors, by GPS technologies or by new active sensors.

New observational opportunities need not be limited to big satellite missions or traditional aircraft observations, but could also include unattended aerial observations that can dwell over a single scene (Stevens and Feingold 2009). Expansion of inexpensive radar networks or cameras, perhaps combined with advanced data-mining/reduction techniques to cope with the large amount of information potentially available, is another possibility. The network of DOE ARM (Atmospheric Radiation Measurement) and similar European sites will prove the more valuable as record lengths grow, and their value could be further augmented by expanding the network to new sites and/or better integrating modeling and observations at such sites, as described by Neggers et al. (2012).

4.2 Research Coordination

Existing projects under the WCRP are well structured to improve the problem associated with lack of resources for model development. Examples include WGNE/WGCM model development and testing; GCSS/GABLS (now GASS) looking at details of boundary layer/clouds/convection; SPARC DynVar for defining necessary improvements in representation of the stratosphere (Gerber et al. 2012); CFMIP for representation of cloud feedbacks. In addition, recent efforts to improve the links between the groups (and the proposed new modeling council) should provide further support. Important links to THORPEX (subseasonal prediction) and WGSIP and WGCM (seasonal to centennial prediction) and through WGNE to the numerical weather prediction (NWP) community will also assist in the effort to achieve 'seamless science'.

Similar programs or efforts would be very useful, however, for aerosol and aerosol-cloud interactions. While all GCMs include similar cloud types and processes, different models include different types of aerosol-cloud effects (lifetime, semi-direct, cumulus, IN etc.) and this makes it difficult to compare these effects between models, or distinguish the impacts of different aerosol predictions from those of different aerosol sensitivities (e.g., Quaas et al. 2009). It is also difficult to distinguish the impacts of aerosol physics and cloud microphysical assumptions in assessing behavioral differences among models. Finally, although the AEROCOM program evaluates global models (Textor et al. 2006), no systematic program is in place to use available field data from observational case studies to evaluate detailed aerosol process models in the manner analogous to GCSS intercomparisons of cloud process models. Such a program could be helpful in identifying the root causes of model-observation discrepancies and could draw on the testbed established by Fast et al. (2011) for this purpose.

5 Summary

In this paper we have attempted to summarize a broad sweep of issues relating to atmospheric physical processes and their impact on our understanding and simulation of climate. Significantly, recent work has highlighted that some important aspects of climate change, including global cloud feedbacks and regional climate changes, may be modulated by shifts of the atmospheric general circulation that are not thought to depend in particular on small-scale processes. These shifts are evident in observations and qualitatively in models, but not all are fundamentally understood or well simulated. Some involve interactions with the stratosphere, which may be more important to tropospheric climate than previously assumed, and was given short shrift in most climate models until very recently. These findings represent a real advance in terms of confidence in model predictions, but do not resolve longstanding problems in how to model the smaller-scale processes, which remain broadly important.

Progress on smaller-scale processes, as well as the larger-scale issues, is being driven by results of new observing campaigns, growing awareness of key unexplained phenomena, targeted research initiatives e.g. through the WCRP, and advancing computational resources. We have identified key problems and presented a number of suggestions for emphasis in coming years. Chief among these is the need for research approaches that confront the interactions on a wide array of scales from the process scale out to (potentially) near-global scales. Such approaches must treat the complexity at the local process level but also account for feedbacks from remote dynamical adjustments, which may occur at any scale, and which could either buffer, enhance, or qualitatively modify local changes. This requires novel modeling, theoretical or observational analysis approaches because traditional numerical models will not be able to span the full range of scales required in the foreseeable future, for many key applications.

The evolution of scientific efforts will continue to be shaped by rapidly advancing information technology. Applications of this should not be limited to bigger computations alone, although these will be carried out. Equally important is facilitating intercomparison and hypothesis-testing efforts via greater accessibility of the complete spectrum of modeling approaches and results to the greater scientific community, members of which are always generating the new ideas that may eventually become the basis for new and deeper understanding of atmospheric physical phenomena.

Acknowledgments AAS was supported by the Joint DECC/Defra Met Office Hadley Centre Climate Programme (GA01101). EPG is supported by the US National Science Foundation under grant AGS- 0938325. GF acknowledges NOAA's Climate Goal.

References

Ackerman AS, Kirkpatrick MP, Stevens DE, Toon OB (2004) The impact of humidity above stratiform clouds on indirect aerosol climate forcing. Nature 432(7020):1014–1017

Alexander MJ, Eckermann SD, Broutman D, Ma J (2009) Momentum flux estimates for South Georgia Island mountain waves in the stratosphere observed via satellite. Geophys Res Lett 36:L12816. doi:10.1029/2009GL038587

Alexander MJ, Coauthors (2010) Recent developments in gravity-wave effects in climate models and the global distribution of gravity-wave momentum flux from observations and models. Q J R Meteorol Soc 136:1103–1124

Allen RJ, Sherwood SC (2010) The aerosol-cloud semi-direct effect and land-sea temperature contrast in a GCM. Geophys Res Lett 37:L07702

Allen RJ, Sherwood SC, Norris JR, Zender CS (2012) Recent Northern Hemisphere tropical expansion primarily driven by black carbon and tropospheric ozone. Nature 485:350–354

Annamalai H, Hamilton K, Sperber KR (2007) South Asian Monsoon and its relationship with ENSO in the IPCC-AR4 simulations. J Clim 20:1071–1092

Arakelian A, Codron F (2012) Southern hemisphere jet variability in the IPSL GCM at varying resolutions. J Atmos Sci 69(12):3788–3799

Arblaster JM, Meehl GA (2006) Contributions of external forcings to Southern annular mode trends. J Clim 19:2896–2905

Barnes EA, Hartmann DL (2010) Dynamical feedbacks and the persistence of the NAO. J Atmos Sci 67:851–865

Bellouin N et al (2008) Updated estimate of aerosol direct radiative forcing from satellite observations and comparison against the Hadley Centre climate model. J Geophys Res Atmos 113:D10205
Bender FA et al (2012) Changes in extratropical storm track cloudiness 1983–2008: observational support for a poleward shift. Clim Dyn 38:2037–2053
Berner J, Shutts GJ, Leutbecher M, Palmer TN (2009) A spectral stochastic kinetic energy backscatter scheme and its impact on flow-dependent predictability in the ECMWF ensemble prediction system. J Atmos Sci 66:603–626
Birner T (2010) Recent widening of the tropical belt from global tropopause statistics: sensitivities. J Geophys Res 115:D23109. doi:10.1029/2010JD014664
Booth BBB, Dunstone NJ, Halloran PR, Andrews T, Bellouin N (2012) Aerosols implicated as a prime driver of twentieth-century North Atlantic climate variability. Nature 484:228–232. doi:10.1038/nature10946
Bretherton CS, Uchida J, Blossey PN (2010) Slow manifolds and multiple equilibria in stratocumulus-capped boundary layers. J Adv Model Earth Syst 2, Art.#14. doi:10.3894/JAMES.2010.2.14
Brown A, Milton S, Cullen M, Golding B, Mitchell J, Shelly A (2012) Unified modeling and prediction of weather and climate: a 25 year journey. Bull Am Meteorol Soc 93:1865–1877. doi:10.1175/BAMS-D-12-00018.1
Butchart N, Scaife AA, Bourqui M, de Grandpre J, Hare SHE, Kettleborough J, Langematz U, Manzini E, Sassi F, Shibata K, Shindell D, Sigmond M (2006) Simulations of anthropogenic change in the strength of the Brewer-Dobson circulation. Clim Dyn 27:727–741
Butchart N, Cionni I, Eyring V, Waugh DW, Akiyoshi H, Austin J, Brühl C, Chipperfield MP, Cordero E, Dameris M, Deckert R, Frith SM, Garcia RR, Gettelman A, Giorgetta MA, Kinnison DE, Li F, Mancini E, McLandress C, Pawson S, Pitari G, Plummer DA, Rozanov E, Sassi F, Scinocca JF, Shepherd TG, Shibata K, Tian W (2010) Chemistry–climate model simulations of 21st century stratospheric climate and circulation changes. J Clim 23:5349–5374
Cagnazzo C, Manzini E (2009) Impact of the stratosphere on the winter tropospheric teleconnections between ENSO and the North Atlantic and European Region. J Clim 22:1223–1238
Charlson RJ, Ackerman AS, Bender FA-M, Anderson TL, Liu Z (2007) On the climate forcing consequences of the albedo continuum between cloudy and clear air. Tellus B 59:715–727. doi:10.1111/j.1600-0889.2007.00297.x
Chen WT, Nenes A, Liao H, Adams P, Seinfeld JH (2010) Global climate response to anthropogenic aerosol indirect effects: present day and year 2100. J Geophys Res 115:D12207. doi:10.1029/2008JD011619
Chepfer H, Bony S, Winker D, Chiriaco M, Dufresne, JL, Seze G (2008) Use of CALIPSO lidar observations to evaluate the cloudiness simulated by a climate model. Geophys Res Lett 35, Art. L15 704. doi:10.1029/2008GL034207
Christensen MW, Stephens GL (2011) Microphysical and macrophysical responses of marine stratocumulus polluted by underlying ships: evidence of cloud deepening. J Geophys Res Atmos 116:D03201. doi:10.1029/2010JD014638
Chung CE, Ramanathan V et al (2005) Global anthropogenic aerosol direct forcing derived from satellite and ground-based observations. J Geophys Res Atmos 110(D24):D24207
Colman R, McAvaney BJ (2011) On tropospheric adjustment to forcing and climate feedbacks. Clim Dyn 36:1649–1658
Conen F et al (2011) Biological residues define the ice nucleation properties of soil dust. Atmos Chem Phys Discuss 11:16585
D'Andrea F et al (1998) Northern hemisphere atmospheric blocking as simulated by 15 atmospheric general circulation models in the period 1979–1998. Clim Dyn 14:385–407. doi:10.1007/s003820050230
Demott P et al (2010) Predicting global atmospheric ice nuclei distributions and their impacts on climate. Proc Natl Acad Sci 107(25):11217–11222. doi:10.1073/pnas.0910818107
Derbyshire SH, Beau I, Bechtold P, Grandpeix J-Y, Piriou J-M, Redelsperger J-L, Soares PMM (2004) Sensitivity of moist convection to environmental humidity. Q J R Meteorol Soc 130:3055–3079
Deser D, Knutti R, Solomon S, Phillips AS (2012) Communication of the role of natural variability in future North American climate. Nat Clim Change 2:775–779

Doblas-Reyes FJ, Deque M, Valero F, Stephenson DB (1998) North Atlantic wintertime intraseasonal variability and its sensitivity to GCM horizontal resolution. Tellus Ser A 50:573–595

Donner LJ (1993) A cumulus parameterization including mass fluxes, vertical momentum dynamics, and mesoscale effects. J Atmos Sci 50:889–906

Eckermann SD (2011) Explicitly stochastic parameterization of nonorographic gravity wave drag. J Atmos Sci 68:1749–1765

Ervens B, Cubison MJ, Andrews E, Feingold G, Ogren JA, Jimenez JL, Quinn PK, Bates TS, Wang J, Zhang Q, Coe H, Flynn M, Allan JD (2010) CCN predictions using simplified assumptions of organic aerosol composition and mixing state: a synthesis from six different locations. Atmos Chem Phys 10:4795–4807. doi:10.5194/acp-10-4795-2010

Evan S, Alexander MJ, Dudhia J (2012) Model study of intermediate-scale tropical inertia-gravity waves and comparison to TWP-ICE campaign observations. J Atmos Sci 69:591–610. doi:10.1175/JAS-D-11-051.1

Fast JD, Gustafson WI Jr, Chapman EG, Easter RC Jr, Rishel JP, Zaveri RA, Grell G, Barth M (2011) The Aerosol Modeling Testbed: a community tool to objectively evaluate aerosol process modules. Bull Am Meteorol Soc 92(3):343–360. doi:10.1175/2010BAMS2868.1

Feingold G, Koren I, Wang H, Xue H, Brewer WA (2010) Precipitation-generated oscillations in open cellular cloud fields. Nature 466:849. doi:10.1038/nature09314

Frenkel Y, Khouider B, Majda A (2011a) Simple multicloud models for diurnal cycle of precipitation. Part I.: Formulation and the tropical ocean. J Atmos Sci 68(10 (October 2011)):2169–2190

Frenkel Y, Khouider B, Majda A (2011b) Simple multicloud models for diurnal cycle of precipitation. Part II: The continental regime. J Atmos Sci 61(17 (September 2004)):2188–2205

Gastineau G, Soden BJ (2009) Model projected changes of extreme wind events in response to global warming. Geophys Res Lett 36:L10810

Gastineau G, Soden BJ (2011) Evidence for a weakening of tropical surface wind extremes in response to atmospheric warming. Geophys Res Lett 38:L09706

Gerber EP, Coauthors (2012) Assessing and understanding the impact of stratospheric dynamics and variability on the Earth system. Bull Am Meteorol Soc 93(6):845–859

Gerber EP, Voronin S, Polvani LM (2008) Testing the annular mode autocorrelation timescale in simple atmospheric general circulation models. Mon Wea Rev 136:1523–1536

Ghan SJ, Abdul-Razzak H, Nenes A, Ming Y, Liu X, Ovchinnikov M (2011) Droplet nucleation: physically-based parameterizations and comparative evaluation. J Adv Model Earth Syst 3:M10001

Giorgetta MA, Manzini E, Roeckner E (2002) Forcing of the quasi-biennial oscillation from a broad spectrum of atmospheric waves. Geophys Res Lett 29(8):1245. doi:10.1029/2002GL014756

Golaz J-C, Salzmann M, Donner LJ, Horowitz LW, Ming Y, Zhao M (2011) Sensitivity of the aerosol indirect effect to subgrid variability in the cloud parameterization of the GFDL atmosphere general circulation model AM3. J Clim 24:3145–3160

Grabowski WW (2006) Indirect impact of atmospheric aerosols in idealized simulations of convective–radiative quasi equilibrium. J Climate 19:4664–4682

Gunn R, Phillips BB (1957) An experimental investigation of the effect of air pollution on the initiation of rain. J Meteorol 14:272–280

Gustafson WI Jr, Berg LK, Easter RC, Ghan SJ (2008) The Explicit-Cloud Parameterized-Pollutant hybrid approach for aerosol-cloud interactions in multiscale modelling framework models. Environ Res Lett 3:Art. 025005. doi:10.1088/1748-9326/3/2/025005

Hannah WM, Maloney ED (2011) The role of moisture-convection feedbacks in simulating the Madden-Julian oscillation. J Clim 245:2754–2770

Harder JW, Fontenla JM, Pilewskie P, Richard EC, Woods TN (2009) Trends in solar spectral irradiance variability in the visible and infrared. Geophys Res Lett 36:L07801

Harte J (2002) Towards a synthesis of the Newtonian and Darwinian worldviews. Phys Today 55:29–34

Hartmann DL, Larson K (2002) An important constraint on tropical cloud – climate feedback. Geophys Res Lett 29(20):1951–1954

Heald CL, Kroll JH, Jimenez JL, Docherty KS, DeCarlo PF, Aiken AC, Chen Q, Martin ST, Farmer DK, Artaxo P (2010) A simplified description of the evolution of organic aerosol composition in the atmosphere. Geophys Res Lett 37:L08803. doi:10.1029/2010GL042737

Heintzenberg J, Charlson RJ (eds) (2009) Clouds in the perturbed climate system: their relationship to energy balance, atmospheric dynamics, and precipitation. Struengmann Forum Report. MIT Press, Cambridge, pp 197–215

Held IM (2005) The Gap between simulation and understanding in climate modeling. Bull Am Meteorol Soc 86:1609–1615. doi:10.1175/BAMS-86-11-1609

Hodzic A et al (2009) Modeling organic aerosols during MILAGRO: importance of biogenic secondary organic aerosols. Atmos Chem Phys 9:6949–6982

Ineson S, Scaife AA (2009) The role of the stratosphere in the European climate response to El Nino. Nat Geosci 2:32–36

Jakob C (2010) Accelerating progress in global atmospheric model development through improved parameterizations: challenges, opportunities and strategies. Bull Am Meteorol Soc 91: 869–875

Jeffery CA (2007) Inhomogeneous cloud evaporation, invariance, and Damkohler number. J Geophys Res 112:D24S21. doi:10.1029/2007JD008789

Johanson CM, Fu Q (2009) Hadley cell widening: model simulations versus observations. J Clim 22(10):2713–2725

Kaas E, Branstator G (1993) The relationship between a zonal index and blocking activity. J Atmos Sci 50:3061–3077. doi:10.1175/1520-0469(1993)050<3061:TRBAZI>2.0.CO;2

Kharin VV, Zwiers FW, Zhang X, Hegerl GC (2007) Changes in temperature and precipitation extremes in the IPCC ensemble of global coupled model simulations. J Clim 20:1419–1444

Khouider B, Majda AJ (2008) Equatorial convectively coupled waves in a simple multicloud model. J Atmos Sci 65:3376–3397

Khouider B, Majda AJ, Stechmann SN (2013) Climate science in the tropics: waves, vortices, and PDEs, Nonlinearity 26:R1–R68

Kidston J, Gerber EP (2010) Intermodel variability of the poleward shift of the austral jet stream in the CMIP3 integrations linked to biases in 20th century climatology. Geophys Res Lett 37:L09708. doi:10.1029/2010GL042873

Kiladis GN, Wheeler MC et al (2009) Convectively coupled equatorial waves. Rev Geophys 47:RG2003

Koren I, Feingold G (2011) Aerosol-cloud-precipitation system as a predator-prey problem. Proc Natl Acad Sci U S A 108(30):12227–12232

Koren I, Remer LA, Kaufman YJ, Rudich Y, Martins JV (2007) On the twilight zone between clouds and aerosols. Geophys Res Lett 34:L08805. doi:10.1029/2007GL029253

Kroll JH, Smith JD et al (2009) Measurement of fragmentation and functionalization pathways in the heterogeneous oxidation of oxidized organic aerosol. Phys Chem Chem Phys 11(36): 8005–8014

Kug J-S, Jin F-F (2009) Left-hand rule for synoptic eddy feedback on low-frequency flow. Geophys Res Lett 36:L05709. doi:10.1029/2008GL036435

Lapina K, Heald CL, Spracklen DV, Arnold SR, Allan JD, Coe H, McFiggans G, Zorn SR, Drewnick F, Bates TS, Hawkins LN, Russell LM, Smirnov A, O'Dowd CD, Hind AJ (2011) Investigating organic aerosol loading in the marine environment. Atmos Chem Phys 11:8847–8860. doi:10.5194/acp-11-8847-2011

Lau N-C (1988) Variability of the observed midlatitude storm tracks in relation to low-frequency changes in the circulation pattern. J Atmos Sci 45:2718–2743

Lau KM, Kim MK, Kim KM (2006) Asian summer monsoon anomalies induced by aerosol direct forcing: the role of the Tibetan Plateau. Clim Dyn 26(7–8):855–864

Lee S-S (2012) Effect of aerosol on circulations and precipitation in deep convective clouds. J Atmos Sci 69:1957–1974. http://dx.doi.org/10.1175/JAS-D-11-0111.1

Lee S, Feingold G (2010) Precipitating cloud-system response to aerosol perturbations. Geophys Res Lett 37:L23806. doi:10.1029/2010GL045596

Lenton A, Codron F, Bopp L, Metzl N, Cadule P, Tagliabue A, Le Sommer J (2009) Stratospheric ozone depletion reduces ocean carbon uptake and enhances ocean acidification. Geophys Res Lett 36:L12606. doi:10.1029/2009GL038227

Li F, Austin J, Wilson J (2008) The strength of the Brewer–Dobson circulation in a changing climate: coupled chemistry – climate model simulations. J Clim 21:40–57

Lin J-L (2007) The double-ITCZ problem in IPCC AR4 coupled GCMs: ocean–atmosphere feedback analysis. J Clim 20:4497–4525. http://dx.doi.org/10.1175/JCLI4272.1

Lin J-L, Kiladis GN, Mapes BE, Weickmann KM, Sperber KR, Lin W, Wheeler MC, Schubert SD, Del Genio A, Donner LJ, Emori S, Gueremy J-F, Jourdin F, Rasch PJ, Roeckner E, Scinocca JF (2006) Tropical intraseasonal variability in 14 IPCC AR4 climate models. Part I: Convective signals. J Clim 19:2665–2690

Lock AP, Brown AR, Bush MR, Martin GM, Smith RNB (2000) A new boundary layer scheme for the unified model. Part I: Scheme description and single-column model tests. Mon Wea Rev 128:3187–3199

Lu J, Vecchi GA, Reichler T (2007) Expansion of the Hadley cell under global warming. Geophys Res Lett 34:L06805. doi:10.1029/2006GL028443

Lu J, Chen G, Frierson DMW (2008) Response of the zonal mean atmospheric Circulation to El Nino versus global warming. J Clim 21:5835–5851

Maloney ED, Sobel AH, Hannah WM (2010) Intraseasonal variability in an aquaplanet general circulation model. J Adv Model Earth Syst 2, Art. #5. doi:10.3894/JAMES.2010.2.5

Mapes BE, Neale RB (2011) Parameterizing convective organization to escape the entrainment dilemma. J Adv Model Earth Syst 3:M06004. doi:10.1029/2011MS000042

Matsueda M, Mizuta R, Kusunoki S (2009) Future change in wintertime atmospheric blocking simulated using a 20-km-mesh atmospheric global circulation model. J Geophys Res 114:D12114. doi:10.1029/2009JD011919

McComiskey A, Feingold G (2012) The scale problem in quantifying aerosol indirect effects. Atmos Chem Phys 12:1031–1049. doi:10.5194/acp-12-1031-2012

McLandress C, Scinocca JF (2005) The GCM response to current parameterizations of nonorographic gravity wave drag. J Atmos Sci 62:2394–2413

McLandress C, Shepherd TG (2009) Simulated anthropogenic changes in the Brewer–Dobson circulation, including its extension to high latitudes. J Clim 22:1516–1540

Mitchell DL, Finnegan W (2009) Modification of cirrus clouds to reduce global warming. Environ Res Lett 4(4):045102

Morrison H, Grabowski WW (2011) Cloud-system resolving model simulations of aerosol indirect effects on tropical deep convection and its thermodynamic environment. Atmos Chem Phys 11:10503–10523

Morrison H, DeBoer G, Feingold G, Harrington JY, Shupe M, Sulia K (2011) Resilience of persistent Arctic mixed-phase clouds. Nat Geosci 5:11–17. doi:10.1038/ngeo1332

Myhre G (2009) Consistency between satellite-derived and modeled estimates of the direct aerosol effect. Science 325(5937):187–190

Neelin JD, Held IM (1987) Modeling tropical convergence based on the moist static energy budget. Mon Wea Rev 115(1):3–12

Neggers RAJ, Siebesma AP, Heus T (2012) Continuous single-column model evaluation at a permanent meteorological supersite. Bull Am Meteorol Soc 93(9):1389–1400

Perlwitz J, Pawson S, Fogt R, Nielsen JE, Neff W (2008) The impact of stratospheric ozone hole recovery on Antarctic climate. Geophys Res Lett 35:L08714. doi:10.1029/2008GL033317

Polvani LM, Waugh DW, Correa GJP, Son S-W (2011) Stratospheric ozone depletion: the main driver of twentieth-century atmospheric circulation changes in the Southern Hemisphere. J Clim 24:795–812

Quaas J, Ming Y, Menon S, Takemura T, Wang M, Penner JE, Gettelman A, Lohmann U, Bellouin N, Boucher O, Sayer AM, Thomas GE, McComiskey A, Feingold G, Hoose C, Kristjansson JE, Liu X, Balkanski Y, Donner LJ, Ginoux PA, Stier P, Grandey B, Feichter J, Sednev I, Bauer SE, Koch D, Grainger RG, Kirkevag A, Iversen T, Seland O, Easter R, Ghan S, Rasch P, Morrison H, Lamarque J, Iacono M, Kinne S, Schulz M (2009) Aerosol indirect effects – general

circulation model intercomparison and evaluation with satellite data. Atmos Chem Phys 9:8697–8717
Randall D, Khairoutdinov M, Arakawa A, Grabowski W (2003) Breaking the cloud-parameterization deadlock. Bull Am Meteorol Soc 84:1547–1564
Raymond DJ, Fuchs Z (2009) Moisture modes and the Madden-Julian oscillation. J Clim 22:3031–3046
Richter JH, Sassi F, Garcia RR (2010) Toward a physically based gravity wave source parameterization in a general circulation model. J Atmos Sci 67:136–156
Ring MJ, Plumb RA (2008) The response of a simplified GCM to axisymmetric forcings: applicability of the fluctuation–dissipation theorem. J Atmos Sci 65:3880–3898
Rio C, Hourdin F, Grandpeix J-Y, Lafore J-P (2009) Shifting the diurnal cycle of parameterized deep convection over land. Geophys Res Lett 36:L07809. doi:10.1029/2008GL036779
Robinson FJ, Sherwood SC, Li Y (2008) Resonant response of deep convection to surface hot spots. J Atmos Sci 65:276–286
Robinson FJ, Sherwood SC, Gerstle D, Liu C, Kirshbaum DJ (2011) Exploring the land-ocean contrast in convective vigor using islands. J Atmos Sci 68:602–618
Rotstayn LD, Lohmann U (2002) Tropical rainfall trends and the indirect aerosol effect. J Clim 15(15):2103–2116
Sato K, Watanabe S, Kawatani Y, Tomikawa Y, Miyazaki K, Takahashi M (2009) On the origins of gravity waves in the mesosphere. Geophys Res Lett 36:L19801. doi:10.1029/2009GL039908
Satoh M, Matsuno T, Tomita H, Miura H, Nasuno T, Iga S (2008) Nonhydrostatic icosahedral atmospheric model (NICAM) for global cloud resolving simulations. J Comput Phys 227:3486–3514
Scaife AA, Knight JR (2008) Ensemble simulations of the cold European winter of 2005/2006. Q J R Meteorol Soc 134:1647–1659
Scaife AA, Butchart N, Warner CD, Stainforth D, Norton WA, Austin J (2000) Realistic Quasi-Biennial Oscillations in a simulation of the global climate. Geophys Res Lett 27:3481–3484
Scaife AA, Woollings T, Knight J, Martin G, Hinton T (2010) Atmospheric blocking and mean biases in climate models. Bull Am Meteorol Soc 23:6143–6152. doi:10.1175/2010JCLI3728.1
Scaife AA, Copsey D, Gordon C, Harris C, Hinton T, Keeley S, O'Neill A, Roberts M, Williams K (2011) Improved Atlantic winter blocking in a climate model. Geophys Res Lett 38:L23703. doi:10.1029/2011GL049573
Scaife AA, Spangehl T, Fereday D, Cubasch U, Langematz U, Akiyoshi H, Bekki S, Braesicke P, Butchart N, Chipperfield M, Gettelman A, Hardiman S, Michou M, Rozanov E, Shepherd TG (2012) Climate change and stratosphere-troposphere interaction. Clim Dyn 38:2089–2097. doi:10.1007/s00382-011-1080-7
Schneider T, O'Gorman PA, Levine XJ (2010) Water vapor and the dynamics of climate changes. Rev Geophys 48:RG3001. doi:10.1029/2009RG000302
Seidel DJ, Randel WJ (2007) Recent widening of the tropical belt: evidence from tropopause observations. J Geophys Res 112:D20113. doi:10.1029/2007JD008861
Seidel DJ, Fu Q, Randel WJ, Reichler TJ (2008) Widening of the tropical belt in a changing climate. Nat Geosci 1:21–24
Seifert A, Köhler C, Beheng KD (2012) Aerosol-cloud-precipitation effects over Germany as simulated by a convective-scale numerical weather prediction model. Atmos Chem Phys 12:709–725
Sharon TM, Albrecht BA, Jonsson HH, Minnis P, Khaiyer MM, Van Reken TM, Seinfeld J, Flagan R (2006) Aerosol and cloud microphysical characteristics of rifts and gradients in maritime stratocumulus clouds. J Atmos Sci 63:983–997
Sherwood SC, Ingram W, Tsushima Y, Satoh M, Roberts M, Vidale PL, O'Gorman PA (2010) Relative humidity changes in a warmer climate. J Geophys Res 115:D09104
Shutts GJ (1986) A case study of eddy forcing during an Atlantic blocking episode. Adv Geophys 29:135–162. doi:10.1016/S0065-2687(08)60037-0
Shutts GJ, Vosper SB (2011) Stratospheric gravity waves revealed in NWP model forecasts. Q J R Meteorol Soc 137:303–317

Siebesma AP, Soares PM, Teixeira J (2007) A combined eddy-diffusivity mass-flux approach for the convective boundary layer. JAS 64:1230–1248
Sigmond M, Scinocca JF (2010) The influence of the basic state on the Northern Hemisphere circulation response to climate change. J Clim 23:1434–1446
Slingo JM, Slingo A (1991) The response of a general circulation model to cloud longwave radiative forcing. II: Further studies. Q J R Meteorol Soc 117:333–364
Sloan LC, Pollard D (1998) Polar stratospheric clouds: a high latitude warming mechanism in an ancient greenhouse world. Geophys Res Lett 25(18):3517–3520
Sobel AH, Bretherton CS (2000) Modeling tropical precipitation in a single column. J Clim 13:4378–4392
Soden BJ, Vecchi GA (2011) The vertical distribution of cloud feedback in coupled ocean-atmosphere models. Geophys Res Lett 38:L12704
Solomon SK et al (2010) Contributions of stratospheric water vapor to decadal changes in the rate of global warming. Science 327(5970):1219–1223
Solomon S et al (2011) The persistently variable "background" stratospheric aerosol layer and global climate change. Science 333(6044):866–870
Son S-W, Coauthors (2008) The impact of stratospheric ozone recovery on the Southern Hemisphere westerly jet. Science 320:1486–1489
Song I-S, Chun H-Y, Garcia RR, Boville BA (2007) Momentum flux spectrum of convectively forced internal gravity waves and its application to gravity wave drag parameterization: Part II: Impacts in a GCM (WACCCM). J Atmos Sci 64:2286–2308
Spracklen V, Jimenez JL, Carslaw KS, Worsnop DR, Evans MJ, Mann GW, Zhang Q, Canagaratna MR, Allan J, Coe H, McFiggans G, Rap A, Forster P (2011) Aerosol mass spectrometer constraint on the global secondary organic aerosol budget. Atmos Chem Phys Discuss 11:5699–5755
Stephens GL, Ecuyer TL, Forbes FR, Gettelman A, Golaz JC, Bodas-Salcedo A, Suzuki K, Gabriel P, Haynes J (2011) Dreary state of precipitation in global models. J Geophys Res 115:D24211
Stevens B, Feingold G (2009) Untangling aerosol effects on clouds and precipitation in a buffered system. Nature 461(7264):607–613
Stevens B, Vali G, Comstock K, Wood R, van Zanten MC, Austin PH, Bretherton CS, Lenschow DH (2005) Pockets of open cells and drizzle in marine stratocumulus. Bull Am Meteorol Soc 86(1):51–57
Storelvmo T, Lohmann U, Bennartz R (2009) What governs the spread in shortwave forcings in the transient IPCC AR4 models? Geophys Res Lett 36, L01806
Stowasser M, Annamalai H, Hafner J (2009) Response of Asian summer monsoon to global warming: mean and synoptic systems. J Clim 22:1014–1036
Textor C, Schulz M, Guibert S, Kinne S, Balkanski Y, Bauer S, Berntsen T, Berglen T, Boucher O, Chin M, Dentener F, Diehl T, Easter RC Jr, Feichter H, Fillmore D, Ghan SJ, Ginoux P, Gong S, Grini A, Hendricks J, Horrowitz L, Huang P, Isaksen I, Iversen T, Kloster S, Koch D, Kirkevag A, Kristjansson JE, Krol M, Lauer A, Lamarque JF, Liu X, Montanaro V, Myhre G, Penner JE, Pitari G, Reddy S, Seland O, Stier P, Takemura T, Tie X (2006) Analysis and quantification of the diversities of aerosol life cycles within AeroCom. Atmos Chem Phys 6(7):1777–1813
Thompson DWJ, Solomon S (2002) Interpretation of recent southern hemisphere climate change. Science 296:895–899
Turner J, Coauthors (2009) Nonannular atmospheric circulation change induced by stratospheric ozone depletion and its role in the recent increase of Antarctic sea ice extent. Geophys Res Lett 36:L08 502. doi:10.1029/2009GL037524
van den Heever SC, Stephens GL, Wood NB (2011) Aerosol indirect effects on tropical convection characteristics under conditions of radiative-convective equilibrium. J Atmos Sci 68(4):699–718
Volkamer R, Jimenez JL, San Martini F, Dzepina K, Zhang Q, Salcedo D, Molina LT, Worsnop DR, Molina MJ (2006) Secondary organic aerosol formation from anthropogenic air pollution: rapid and higher than expected. Geophys Res Lett 33:L17811. doi:10.1029/2006GL026899

Wang H, Feingold G (2009) Modeling Mesoscale Cellular Structures and Drizzle in Marine Stratocumulus. Part I: Impact of drizzle on the formation and evolution of open cells. J Atmos Sci 66(11):3237–3256
Wang M, Ghan S, Ovchinnikov M, Liu X, Easter R, Kassianov E, Qian Y, Marchand R, Morrison H (2003) Aerosol indirect effects in a multi-scale aerosol-climate model PNNL-MMF. Atmos Chem Phys 11:5431–5455. doi:10.5194/acp-11-5431-2011
Wang S, Wang Q, Feingold G (2003) Turbulence, condensation and liquid water transport in numerically simulated nonprecipitating stratocumulus clouds. J Atmos Sci 60:262–278
Warner J (1968) A reduction in rainfall associated with smoke from sugar-cane fires: an inadvertent weather modification? J Appl Meteor 7:247–251
Watanabe S, Kawatani Y, Tomikawa Y, Miyazaki K, Takahashi M, Sato K (2008) General aspects of a T213L256 middle atmosphere general circulation model. J Geophys Res 113:D12110. doi: 10.1029/2008JD010026
Wen G, Marshak A, Cahalan RF, Remer LA, Kleidman RG (2007) 3-D aerosol-cloud radiative interaction observed in collocated MODIS and ASTER images of cumulus cloud fields. J Geophys Res 112:D13204. doi:10.1029/2006JD008267
Wild M (2009) Global dimming and brightening: a review. J Geophys Res 114:D00D16. doi:10.1029/2008jd011470
Wood R (2012) Stratocumulus clouds, Mon Wea Rev, 140:2373–2423
Xue H, Feingold G, Stevens B (2008) Aerosol effects on clouds, precipitation, and the organization of shallow cumulus convection. J Atmos Sci 65:392–406
Yin JH (2005) A consistent poleward shift of the storm tracks in simulations of 21st century climate. Geophys Res Lett 32:L18701. doi:10.1029/2005GL023684
Zelinka MD, Hartmann DL (2010) Why is longwave cloud feedback positive? J Geophys Res 115:D16117. doi:10.1029/2010jd013817
Zelinka MD, Hartmann DL (2011) The observed sensitivity of high clouds to mean surface temperature anomalies in the tropics. J Geophys Res 116, D23103
Zeng XP et al (2009) An indirect effect of ice nuclei on atmospheric radiation. J Atmos Sci 66:41–61
Zhang M, Bretherton CS (2008) Mechanisms of low cloud climate feedback in idealized single-column simulations with the Community Atmospheric Model (CAM3). J Clim 21:4859–4878

Aerosol Cloud-Mediated Radiative Forcing: Highly Uncertain and Opposite Effects from Shallow and Deep Clouds

Daniel Rosenfeld, Robert Wood, Leo J. Donner, and Steven C. Sherwood

Abstract Aerosol cloud-mediated radiative forcing, commonly known as the aerosol indirect effect (AIE), dominates the uncertainty in our ability to quantify anthropogenic climate forcing and respectively the climate sensitivity. This uncertainty can be appreciated based on the state of our understanding as presented in this chapter.

Adding aerosols to low clouds generally causes negative radiative forcing by three main mechanisms: redistributing the same cloud water in larger number of smaller drops, adding more cloud water, and increasing the cloud cover. Aerosols affect these components sometimes in harmony but more often in opposite ways. These processes can be highly non-linear, especially in precipitating clouds in which added aerosol can inhibit rain. There is probably little overall sensitivity in most clouds but hyper sensitivity in some, where the processes become highly nonlinear with positive feedbacks, causing changes of cloud regimes in marine stratocumulus under anticyclones. This leads to a complicated and uneven AIE. Process

D. Rosenfeld (✉)
Institute of Earth Sciences, The Hebrew University of Jerusalem,
Edmond J. Safra Campus, Givat Ram, Jerusalem 91904, Israel
e-mail: daniel.rosenfeld@huji.ac.il

R. Wood
Atmospheric Sciences, University of Washington, Box 351640,
Seattle, WA 98195-1640, USA
e-mail: robwood@atmos.washington.edu

L.J. Donner
Geophysical Fluid Dynamics Laboratory/NOAA, Princeton University Forrestal Campus,
201 Forrestal Road, Princeton, NJ 08540, USA
e-mail: Leo.J.Donner@noaa.gov

S.C. Sherwood
Climate Change Research Centre (CCRC) and Australian Research Council (ARC),
Centre of Excellence for Climate System Science, University of New South Wales,
Sydney, NSW 2052, Australia
e-mail: s.sherwood@unsw.edu.au

G.R. Asrar and J.W. Hurrell (eds.), *Climate Science for Serving Society: Research, Modeling and Prediction Priorities*, DOI 10.1007/978-94-007-6692-1_5,

models at high resolution (LES) have reached the stage that they can capture much of this complicated behavior of shallow clouds. The implementation of the processes of cloud aerosol interactions into GCMs is rudimentary due to severe computational limitations and the current state of cloud and aerosol parameterizations, but intense research efforts aimed at improving the realism of cloud-aerosol interaction in GCMs are underway.

Aerosols added to deep clouds generally produce an additional component of positive radiative forcing due to cloud top cooling, expanding, and detraining vapor to the upper troposphere and lower stratosphere. The level of scientific understanding of the AIE on deep clouds is even lower than for the shallow clouds, as mixed phase and ice processes play an important role. Respectively, the parameterization of these processes for GCMs is further away than for the low clouds.

Crucially, the AIE of both shallow and deep clouds must be considered for quantifying anthropogenic climate forcing and inferring climate sensitivity from observations.

While our objective is reducing the uncertainty, it appears that the recently acquired additional knowledge actually increased the uncertainty range of the AIE, as we learn of additional effects that should be quantified.

Keywords Cloud-aerosol interactions • Aerosol indirect radiative forcing

1 Introduction

Aerosols are thought to have exerted a net cooling effect on earth's climate that have grown over the last century or two due to aerosol added by anthropogenic activities, influencing climate. This negative radiative forcing must have offset some of the warming that would otherwise have occurred due to greenhouse gases. The magnitude of this however remains highly uncertain; indeed aerosols represent the most uncertain climate forcing over the last 150 years (IPCC 2007), due to the complex ways aerosols can directly and indirectly affect radiation.

First, aerosols scatter sunlight to space that would otherwise have been absorbed, causing a so-called direct radiative forcing especially for aerosols over dark surfaces (oceans and forests). This negative forcing is offset somewhat by the absorption of outgoing infrared radiation (e.g., Myhre 2009) and by the absorption of sunlight by dark (primarily carbonaceous) aerosols, both of which cause net warming, though nearly all studies find the cooling effects of the non-absorbing aerosols to be larger. This chapter will not discuss direct radiative forcing in detail, but chapters elsewhere in this volume touch on some of the issues (Sherwood et al. 2013, Chap. 4).

Second, aerosols serve as the nuclei (CCN or ice nuclei IN) for cloud droplets and can alter the albedo of clouds. As this component contributes the greatest uncertainty to our knowledge of Earth energy budget, it is the focus of our article. Adding CCN typically produces more droplets in a cloud, although this depends on details of the aerosols. Indeed the opposite can occur if the added particles are large enough compared to those already present, for example if sea salt is introduced into polluted continental air (Rosenfeld et al. 2002), although anthropogenic particles are generally too small for this to happen. All other things being equal (in particular, the cloud's size and condensed water content), more numerous droplets result in a so-called "Twomey" or droplet radius effect whereby the increased droplet surface area increases the cloud albedo, producing a negative indirect radiative forcing by the added CCN (Twomey 1977).

All other things are however not generally equal: aerosols can also alter the subsequent fate of condensed water, and can drive circulations that alter the formation of clouds. These impacts lead to "adjusted" aerosol forcings analogous to those following the stratospheric adjustment to added greenhouse gases (e.g., Hansen et al. 2005). Both direct (radiative) and indirect (CCN-based) pathways produce such adjustments. For example, heating of the air by absorbing aerosols can alter local stability and/or drive circulations that alter local or remote cloud amounts, producing a "semi-direct forcing" on regional or global radiative balances (e.g., Allen and Sherwood 2011). Smaller droplets may cause a cloud to dissipate either more quickly (by reducing fall speeds and increasing cloud break-up by increasing evaporative and radiatively driven entrainment) or more slowly (by decreasing droplet lifetimes in subsaturated air and the rate at which cloud is depleted by precipitation) – so called "lifetime" or "cloud amount effects" (Albrecht 1989). They also typically delay the formation of precipitation, which alters the latent heat release and therefore the dynamics of the cloud. Impacts can include invigoration and deepening of already deep clouds that would have rained anyway (e.g., Rosenfeld et al. 2008b), or the suppression of rain in weaker, shallower and more susceptible cloud systems (e.g., Rosenfeld 2000). Either implies changes to cloud water content, hence albedo; to cloud top height, hence greenhouse effect; to cloud amount, which affects both of these; and to net rainfall, hence the larger-scale circulation. It is in these "adjustments" where most of the uncertainty lies in quantifying the net climate forcing due to anthropogenic aerosols. Understanding of these has been sufficiently poor that the IPCC has not attempted to assess them up until now, but will do so to a limited degree in the upcoming AR5 report.

Model calculations of the aerosol indirect effect (AIE) have yielded radiative forcings of about −0.5 to −2.0 Wm^{-2} (e.g., Forster et al. 2007); these values overlap recent estimates based on satellite observations, which range from −0.2 to −1.2 Wm^{-2} (Quaas et al. 2009). Quaas et al. (2009) argued that models overestimate the AIE compared to satellite observations in present-day climate, while Penner et al. (2011) argue that flawed assumptions used in interpreting satellite data can cause several-fold underestimation of AIE between pre-industrial and present-day climate. Another possible reason for the discrepancy could be that additional effects not yet included in models offset the Twomey effect. Such an effect might be positive radiative forcing due to aerosol impacts on deep convective clouds.

Since other anthropogenic radiative forcings are known better than the AIE, and since temperature changes over the last century or so are relatively well-measured, the total net forcing due to aerosols (including also any semi-direct effects of greenhouse gases) can be constrained based on the energetics of recent global climate, yielding a so-called "inverse" or "top-down estimate." Anderson et al. (2003) compiled similar inverse calculations and concluded that total (direct and indirect) aerosol forcing near −1.0 Wm^{-2} but without taking the ocean heat uptake into account. Murphy et al. (2009) obtained a 68 % range of −1.5 to −0.7 Wm^{-2} based purely on observations since 1950, but with no direct estimate of contributions from cloud and other feedbacks. Forest et al. (2006) obtained a 90 % range of −0.74 to −0.14 Wm^{-2} by fitting a simple climate model (including feedbacks and ocean heat storage) to the spatiotemporal distribution of observed twentieth-century temperature changes.

Stronger (more negative) aerosol forcings correspond to higher climate sensitivity (Kiehl 2007). Values stronger than −1.5 Wm^{-2} would negate the impact of CO_2 since 1850, as a lag of the oceans of even −1.0 Wm^{-2} would imply implausibly high climate sensitivities (Forest et al. 2006). Since these estimates include the direct effect of aerosols, which is already about −0.6 to −0.1 Wm^{-2}, the Forest et al. (2006) numbers imply an AIE near zero while the Murphy et al. (2009) numbers would leave room for an AIE of weaker than −1.0 Wm^{-2}. These numbers are hard to reconcile with the estimates from GCMs. General circulation models (GCMs) began to estimate AIE in the middle 1990s. Early estimates ranged from about −0.5 Wm^{-2} to nearly −4.0 Wm^{-2}, but more recently constructed GCMs do not cool more than about −2.6 Wm^{-2} (Isaksen et al. 2009; Quaas et al. 2009). Quaas et al. (2009) used satellite observations, which generally indicate weaker interactions between clouds and aerosols than GCMs, to scale GCM estimates, finding that an average AIE estimate from ten GCMs of −1.1 Wm^{-2} was reduced to −0.7 Wm^{-2} when scaled by satellite observations. These lower numbers are presented in the radiative forcing chart of Isaksen et al. (2009), shown here as Fig. 1. When considering the high uncertainty range, especially for the cloud lifetime effect, a net forcing of zero or even negative values are included in the range of possibilities. Net zero or negative forcings are unlikely, of course, because it is hard to understand how the climate has warmed with zero or negative overall forcing, and this situation exemplifies the difficulty in estimating forcing due to cloud-aerosol interactions.

The metrics of the effect of the aerosols on cloud properties are often defined in logarithmic formulations (e.g., McComiskey and Feingold 2008; Koren et al. 2008). This means that the clouds respond to the fractional change in CCN concentrations. This means, in turn, that large impacts can be expected when small amounts of aerosols are added to pristine air. Therefore, the background to which the aerosols are emitted is at least as important as the amount of emissions.

This chapter addresses the main sources of uncertainty in AIE in the various kinds of clouds and aerosols, the way that they might be working together or at opposite directions, and suggests possible ways to address these questions. Section 1 (this section) introduces the uncertainty of the AIE and the motivation for its reduction. Section 2 addresses the processes that determine the AIE from low clouds, whereas Sects. 3 and 4 do the same for deep clouds and for supercooled layer clouds, respectively. Section 5 contrasts the mostly negative radiative forcing caused by the AIE of low clouds to the mostly positive forcing due to the deep and supercooled layer clouds. It also discusses the implications with respect to GCMs. In Sect. 6 we provide some recommendations for ways to address the formidable challenges that were discussed in this chapter. An overall summary is provided in Sect. 7.

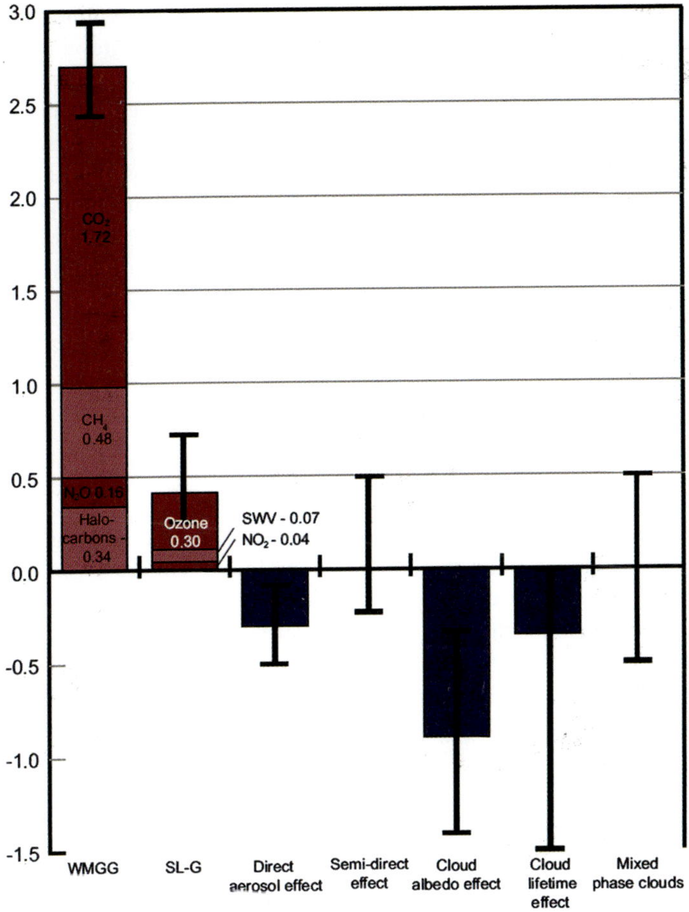

Fig. 1 Radiative forcing estimates of atmospheric compounds from the pre-industrial period 1750–2007 (From Isaksen et al. 2009) in W/m^2

2 Aerosol-Induced Radiative Forcing by Boundary-Layer Warm Clouds

2.1 The Fundamental Physical Processes

The CCN supersaturation activation spectrum, CCN(S), along with the updraft at cloud base, determines the maximum super saturation at cloud base, S, and hence the number of activated cloud drops, N_d. In a rising adiabatic non-precipitating cloud parcel the liquid water content, *LWC*, is determined exclusively by thermodynamic considerations and is highly linear with the vertical distance z above cloud base. In general, however, mixing processes (lateral and cloud top entrainment) cause the liquid water profile to be subadiabatic. Under most circumstances, mixing is predominantly inhomogeneous and causes the observed growth of the mean volume radius r_v with z in boundary layer clouds to follow closely the theoretical value of an adiabatic cloud parcel (Brenguier et al. 2000; Freud et al. 2011). It follows that, at any given height, r_v is inversely proportional to $N_d^{1/3}$, as long as the development of the cloud drop size distribution is dominated by diffusional growth, i.e., before drop coalescence advances and initiates warm rain, unless rain is already falling from above into the cloud. The same applies to the cloud drop effective radius, r_e, as it is very highly linearly correlated with r_v, where $r_e = 1.08\ r_v$ (Freud and Rosenfeld 2012). The r_e is a useful measure because it can be directly retrieved from satellite observations (Arking and Childs 1985). We can write the aerosol indirect effect as the sensitivity of the albedo α to changes in N_d as

$$\frac{d\alpha}{dN_d} = \left(\frac{\partial \alpha}{\partial N_d}\right)_C + \sum_i \left(\frac{\partial \alpha}{\partial C_i}\right)\left(\frac{\partial C_i}{\partial N_d}\right) \tag{1}$$

where C_i are radiatively important cloud macrophysical properties (e.g. liquid water path, cloud thickness, cloud cover, etc.). The first term on the RHS of (1) represents the change in albedo caused only by changes in microphysics, in the absence of changes in cloud macrophysical properties. This is generally referred to as the Twomey effect, or the first aerosol indirect effect. The second term on the RHS represents the changes in albedo associated with aerosol-induced changes in cloud macrophysical properties. Equation 1 is very general since C_i can represent *any* changes to the system induced by aerosols. Examples for such properties are cloud liquid water path, precipitation content, geometrical depth, cloud top height, cloud cover and organization. The Twomey term is called the albedo susceptibility (Platnick and Twomey 1994), and is well-approximated (e.g. Twomey 1991) by

$$\left(\frac{\partial \alpha}{\partial N_d}\right)_C \approx \frac{\alpha(1-\alpha)}{N_d} \tag{2}$$

Equation 2 indicates that aerosol-induced cloud albedo increases are greatest for clouds with low initial N_d. Further compounding the impact of aerosols on the albedo of clean clouds with low N_d is the fact that in this aerosol-limited cloud regime, almost all accumulation mode aerosols are activated to form cloud drops, i.e. $N_d \approx N_a$. As aerosol concentrations increase, the limiting factor on N_d increasingly becomes the updraft speed (updraft limited regime), and $N_d < N_a$, leading to much weaker sensitivity of albedo to aerosol increases (Pöschl et al. 2010).

In addition to the Twomey effect, observations and modeling results indicate that, in this aerosol-limited regime, cloud macrophysical properties (i.e. the second term on the RHS in (1)) are also particularly sensitive to aerosols. Cloud macrophysical responses to aerosols are more challenging to understand than the purely microphysical effect and are mediated via changes in the precipitation, sedimentation and evaporation of hydrometeors. These changes induce macrophysical responses in turbulent dynamics, entrainment rate, and, in some cases, mesoscale reorganization. Many of these processes remain poorly understood (Wood 2012). This issue will be discussed later where it will be shown that when CCN is decreased below a certain concentration a full cloud cover can no longer be sustained.

Aerosol effects on the microphysical properties of boundary layer clouds (i.e., cloud drop size distribution and precipitation forming processes) may affect the macrophysical properties of the same clouds (i.e., cloud LWP, geometrical depth, cover and organization). The microphysical impacts of aerosol changes on boundary layer cloud macrophysical properties can be partitioned into precipitation/sedimentation mediated impacts and those that do not involve precipitation changes. Precipitation impacts are non linear due to internal mechanisms of feedbacks (some positive and some negative), which under some circumstances may lead to changes in cloud regime (e.g. closed to open cells, or stratocumulus-to-cumulus transition) that are associated with drastic jumps in the cloud cover and the respective radiative effect (Ackerman et al. 1995; Rosenfeld et al. 2006a; Wang and Feingold 2009; Wang et al. 2010, 2011a). Because precipitation can play an important role in these transitions, it is critical to understand the processes controlling transitions between lightly or non-precipitating marine stratocumulus (MSC) and heavily drizzling MSC.

Marine stratocumulus that form in stable atmosphere and maintained by radiative cooling of their tops persist under anticyclones and subtropical highs over the ocean, and occupy nearly 25 % of the ocean surface. The radiative properties of these clouds represent large sensitivity to CCN concentrations, and might have a substantial global impact. While having a globally important cloud radiative effect, the overall actual radiative forcing from these clouds is a subject of intense debate due to complicated feedback mechanisms that are positive in some cases, mostly in precipitating MSC, and negative in others, mostly in non precipitating MSC.

Rain intensity in stratocumulus depends on N_d and cloud thickness h (Fig. 2). Aircraft measurements (Van Zanten et al. 2005), supported by physical considerations (Kostinski 2008), showed that cloud base rain rate $R \sim h^3/ N_d$. Since effective radius $r_e^3 \sim LWC/N_d \sim h/N_d$, then $R \sim h^2\ r_e^3$. This was also reproduced by the simulations of Wang and Feingold (2009), but only for clouds with $N_d < 100\ \mathrm{mg}^{-1}$, and h of about 600 m. For clouds with similar h but $N_d \approx 150\ \mathrm{mg}^{-1}$ the surface rain rate was zero. This implies cloud top r_e of about 15 μm. Wang and Feingold (2009), Wang et al. (2011a) found similar results of complete suppression of surface precipitation at high N_d and respectively small r_e. The relation of $R \sim h^3/N_d$ depends on the existence of rain embryos, but their scarcity in clouds with very small drops, as expressed by cloud top significantly smaller than 15 μm, causes R to become practically zero for any h and N_d. The dependence of R on liquid water path (*LWP*) and h was replicated by bulk microphysics models (Kubar et al. 2009; Wood et al. 2009), but they could not capture the complete suppression of R at high N_d and low r_e that was simulated with the explicit bin microphysics models. Aircraft measurements in MSC (Gerber 1996) showed that when r_e exceeds 16 μm most cloud water already resides in the drizzle mode, and that this can occur due to diffusional growth in the convective elements when N_d is sufficiently small. Interestingly, this height for onset of heavy drizzle increases linearly with N_d, A similar linear relationship

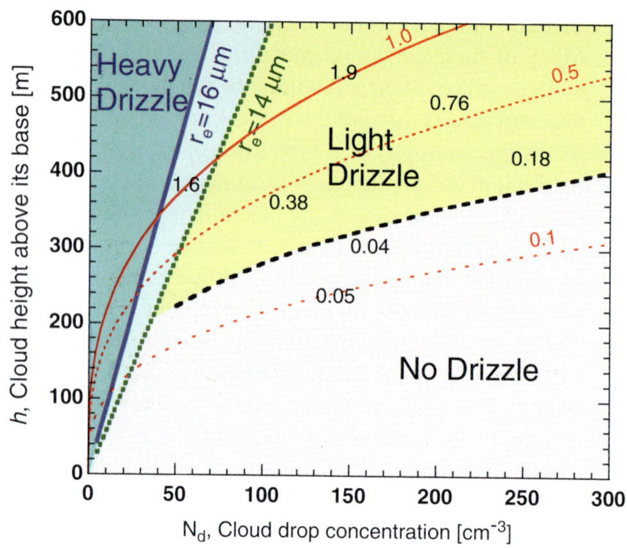

Fig. 2 The dependence of drizzling regimes in marine stratocumulus clouds on drop number concentration and cloud depth. Heavy drizzle is defined where most water resides in the drizzle drops. Light drizzle is defined where most water resides in the cloud drops. The cloud drop effective radius of $r_e = 16$ µm was shown to be the minimal size for the heavy drizzle regime (Gerber 1996). Transition to light drizzle occurs between r_e of 14–16 µm. The *dashed line* separates between negligible drizzle and light drizzle of $R > 0.2$ mm day^{-1} is based on DYCOMS-II observations. The *red lines* show the approximation of $R \sim h^3/N_d$, for R of 0.1, 0.5 and 1 mm/day. The individual points and their R values are posted (From Table 3 of van Zanten et al. 2005. After Rosenfeld et al. (ACP 2006a))

between N_d and cloud depth for initiation of rain was observed by Freud and Rosenfeld (2012) in convective clouds over land. The validity of this threshold cloud top r_e as separating between the logarithmic response of the Twomey effect (Eq. (2)) and the highly non-linear response to aerosols by regime change is supported by satellite observations, which show consistently that a cloud top r_e of 16 µm separates the closed and open cell regimes (Rosenfeld et al. 2006a). Aircraft-measurements show that this change in r_e is also manifested in changes of N_d. An average N_d of 21 cm^{-3} was measured near cloud base of the open cells and 70 cm^{-3} in the closed cells (Wood et al. 2011).

2.2 *Aerosol Effects on Non-precipitating and Modestly Precipitating Clouds*

The aerosol indirect effect on cloud albedo was introduced by Twomey (1977) and Eq. (2) expresses its dependence upon cloud albedo and droplet concentration N_d. However, changes in aerosols rarely affect only N_d without changing cloud macrophysical properties such as cloud thickness and *LWP*. One might expect *LWP* to increase with CCN because less water is lost to precipitation (Albrecht 1989). This is true for some meteorological conditions (Ackerman et al. 2004; Wood 2007). Certainly, there is good modeling and observational evidence that added aerosols can suppress precipitation (Ackerman et al. 2004; Lu and Seinfeld 2006; Sandu et al. 2008; Feingold and Seibert 2009; Sorooshian et al. 2009, 2010; Wang et al. 2010, 2011a; Chen et al. 2011; Terai et al. 2012). However, besides influencing the moisture budget of the clouds, precipitation also impacts the turbulent mixing, which can alter the moisture and energy budget of the boundary layer by changing entrainment (Ackerman et al. 2004; Wood 2007). Aerosol-suppressed precipitation results in increased cloud top entrainment that can warm and dry the boundary layer and thin the cloud, an effect that works in the opposite direction to the effects of precipitation on the surface moisture budget (Wood 2007). The overall effect on *LWP* therefore depends upon the ratio of the surface moistening (suppression of precipitation) compared with the entrainment drying/warming. When significant precipitation reaches the surface (usually heavily drizzling cases), or when the free-troposphere is relatively moist, precipitation suppression tends to increase *LWP*. In weakly precipitating cases, where there is little surface precipitation, the entrainment drying may dominate, leading to aerosol-induced reductions in *LWP* (Chen et al. 2011). Indeed, many ship track cases appear to show such a response (Coakley and Walsh 2002; Christensen and Stephens 2011).

Increasing N_d can also enhance mixing due to faster evaporation of the smaller drops at the border of the clouds and resultant enhanced mixing with the dry ambient air (Wang et al. 2003; Lu and Seinfeld 2006; Hill et al. 2008, 2009; Chen et al. 2011; Small et al. 2009). Increased N_d also reduces the sedimentation of cloud droplets which can increase entrainment rate (Bretherton et al. 2007). Large eddy modeling shows that increases in CCN shorten the life time and reduce the size of small trade wind cumuli (Jiang et al. 2009a).

Overall, the macrophysical responses to aerosols in weakly precipitating and non precipitating clouds appear to reduce their solar reflecting capabilities, which counteracts the brightening associated with the Twomey effect itself.

2.3 Aerosol Effects on the Transition to Precipitating Clouds

The dependence of precipitation rate in marine stratocumulus clouds on N_d and h is shown in Fig. 2. The strong dependence on aerosols is evident by the dependence of N_d on CCN. The relationship between CCN and N_d is approximately linear at the low concentrations characterizing the aerosol-limited regime (Martin et al. 1994; Hegg et al. 2011), where the transition from heavy to lightly or not drizzling clouds occurs (Fig. 2). Deeper clouds transition at greater N_d.

Upon the transition to heavy drizzle the fast loss of cloud water can no longer be compensated by evaporation, and a net loss of cloud water from the domain occurs.

The precipitation also scavenges efficiently the aerosols (Feingold et al. 1996; Wood 2006), hence reducing CCN and N_d even more, increasing r_e and causing even faster coalescence and precipitation in a positive feedback loop. In the extreme this process progresses all the way until there are insufficient CCN for sustaining the growth of new clouds. Because of the essential role of the clouds in determining the lapse rate of the marine boundary layer, the suppression of their formation due to dearth of CCN suppresses also the vertical mixing of air from sea surface to very shallow heights, thus in fact causing the collapse of the marine boundary layer to a thin layer of sea fog composed of drizzle drops. The precipitation scavenging feedback leads in some cases to the collapse of the cloudy boundary layer (Ackerman et al. 1993) and in other cases to a deep boundary layer with open cellular convection.

This runaway feedback effect is a basis for a situation of bi-static stability (Baker and Charlson 1990; Gerber 1996), where once the atmosphere has reached a very clean situation the highly efficient rainout mechanisms keeps it clean until it will be overwhelmed by a strong aerosol source such as anthropogenic emissions.

The full cloud cover of closed cells is maintained by the strong radiative cooling from the cloud tops that causes top-down convection and entrainment of air from the free troposphere just above the clouds (Agee et al. 1973). This replenishes the CCN that may have lost by the cloud processes (Randall 1980; Clarke et al. 1997; Jiang et al. 2002; Stevens et al. 2005).

A mechanism for the transition between the closed- and open-cell regimes was proposed by Rosenfeld et al. (2006a, b). This mechanism is illustrated in Fig. 3. Based on this mechanism, Rosenfeld et al. (2006a) hypothesized that dynamically the closed cells are inverse Benard convection, where the cooling at the top causes polygons of sinking cool air with compensating rising air at the center of the polygons. The rising centers are manifested as patches of polygonal clouds, with narrow regions of dry downward moving air at the cell fringes (see Figs. 3 and 4). The onset of heavy precipitation that occurs when the cloud top r_e exceeds 16 μm, due to decrease in N_d and/or increase in h, breaks the full cloud cover by depleting the cloud water and by decoupling it from the surface due to the low level evaporation of the precipitation. With reduced cloud cover at the top of the boundary layer the radiative cooling there decreases respectively, and allows thermal radiation to be emitted upward from the vapor within the boundary layer and the lower cloud fragments. This reverses the driving of the convection, from inverse convection due to the radiative cooling at the top, to normal convection of Benard cells that is triggered by weak surface heating, where the air rises along the walls of the polygons and sinks in the centers. The rising polygons are manifested as the polygons of the clouds (see Figs. 3 and 4). This picture is complicated by the evaporative cooling of the rain shafts, which form mini gust fronts at the surface that regenerate the convergence lines away from the rain cells, especially where several such fronts collide (Feingold et al. 2010). When the original rain cell decays new clouds and rain showers form at the convergence along the old gust fronts. This, in turn, produces new gust fronts and so on, leading to regular oscillations of the locations of the low level convergence lines and the respective polygonal cloud and rain patterns.

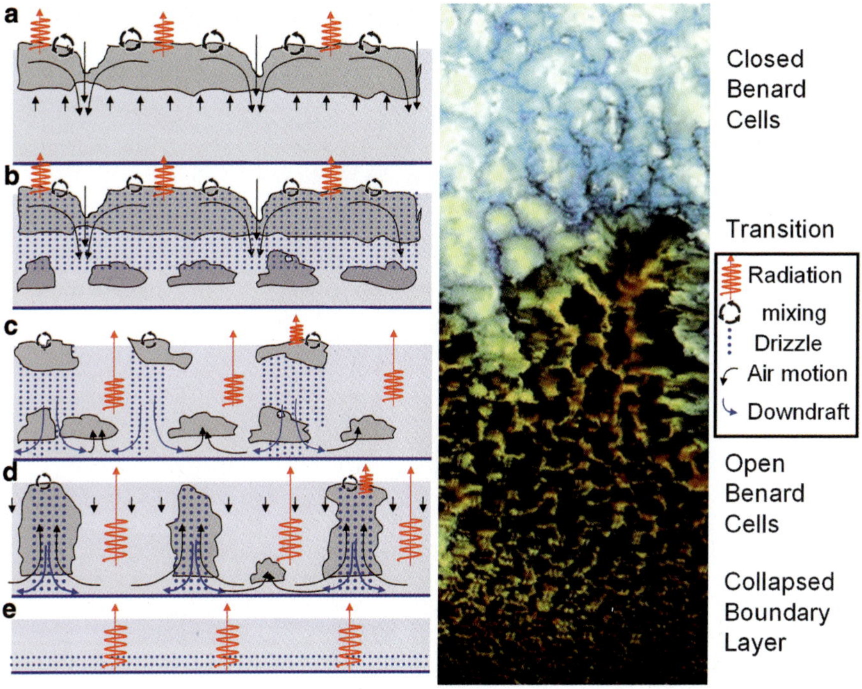

Fig. 3 A schematic illustration of the mechanism for transition from non precipitating closed Benard cells to precipitating open cells and onward to nearly complete rainout and elimination of the clouds (After Rosenfeld et al. 2006a). In the closed cells (**a**) the convection is propelled by thermal radiative cooling from the tops of the extensive deck of clouds with small drops. The clouds mix aerosols and vapor with the free troposphere from above. The onset of drizzle depletes the water from the cloud deck and cools the sub-cloud layer (**b**). This leads to decoupling of the cloud cover and to its subsequent breaking. The downdrafts due to the evaporational cooling starts triggering new convection (**c**). The propulsion of the convection undergoes transition from radiative cooling at the top of the fully cloudy MBL to surface heating at the bottom of the partly cloudy MBL. This causes a reversal of the convection from closed to open Benard cells, that develop, rainout and produce downdrafts that trigger new generations of such rain cells (**d**). The mixing with of aerosols with the free troposphere at cloud tops is much reduced. Therefore, the process can continue to a runaway effect of cleansing by the CCN and direct condensation into drizzle that directly precipitates and prevents the cloud formation altogether (**e**). The satellite strip is a 300 km long excerpt from Fig. 4

The self organization of clouds into the three distinct regimes was described by Koren and Feingold (2011) by simple principles of prey (cloud water) and predator (rain process):

1. The non or weakly precipitating clouds, where the rain-forming process is too slow for large depletion of cloud water. This corresponds to the closed cells regime, with suppressed rain due to high aerosol concentration or a very shallow cloud with little *LWP*.

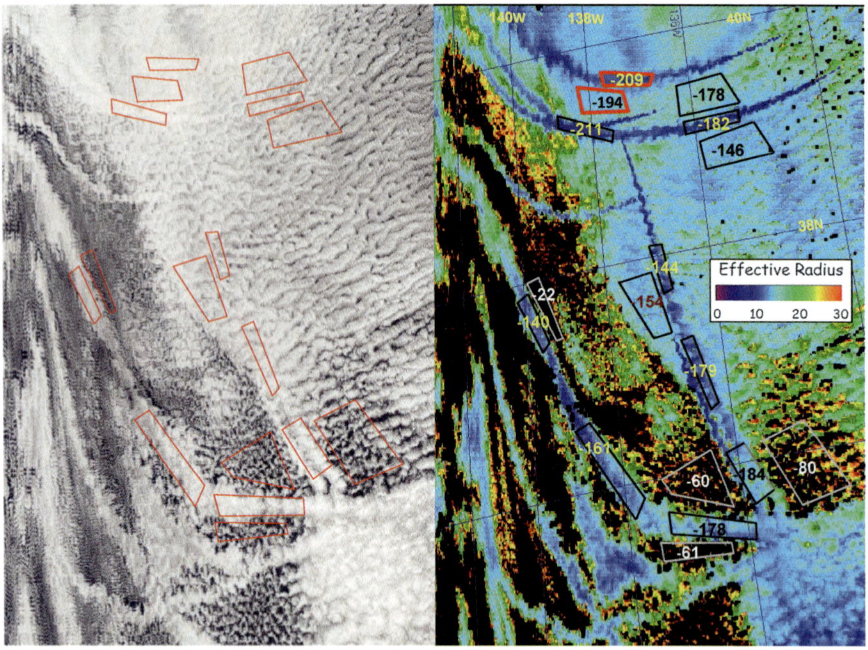

Fig. 4 MODIS satellite image of open and closed cells in marine stratocumulus with ship tracks in an area lying between about 35–40 North and 134–142 West, off the coast of California on 26 July 2003 19:40 UTC. The *left panel* is a true color image, and the *right panel* is the MODIS-retrieved cloud top r_e. The ship tracks appear as a marked decrease in cloud drop effective radius (r_e in μm) on the right panel. The ship tracks are barely discernible in the true color image on the left panel, except for areas where $r_e > 16$μm, above which significant drizzle occurs (Gerber 1996) and open the closed cells. The cloud radiative effect (CRE, Wm^{-2}) is given for the *marked rectangles*. The difference in CRE between the open and ship track induced closed cells well exceeds 100 Wm^{-2}, whereas the RF of the ship tracks within the closed cells is an order of magnitude smaller. The image is the same as in from Rosenfeld et al. (2006a) with added calculated CRE

2. The heavily drizzling regime, where rain can deplete the cloud water, but the supply of new aerosols is able to replenish the cloud water after a while, so that cycles of clouds building and raining out occur. This corresponds to the regime of oscillating and raining open cells.
3. The heavily precipitating clouds, where all incipient cloud water effectively precipitates along with the aerosols on which it condensed, probably due to insufficient rate of replenishment of aerosol. This corresponds to the situation of the ultra-clean collapsed boundary layer.

The value in this highly simplified description is in elucidating these different cloud patterns as fundamentally different regimes. It is of particular importance on the background that internal processes can buffer the aerosol effects within the regimes (Stevens and Feingold 2009), but not between the regimes. The buffering should not be regarded as a full compensation, but rather as a negative feedback that attenuates the results of the initial microphysical effect of the aerosols on the cloud microstructure. An example for the buffering in the closed cell regime is the opposite effects of aerosols increasing the cloud albedo for a given *LWP*, but decreasing the *LWP* at the same time. This is evident in the red rectangles in Fig. 4, where half of the albedo changes due to N_d (Twomey effect) of -31 Wm^{-2} was offset by a decrease in LWP that incurred $+16$ Wm^{-2}, leaving a net effect of -15 Wm^{-2}. An example for the buffering in the open cell precipitating regime is that an increase in aerosols would delay, but not completely shut off, the onset of rain in a convective cell, causing it to grow more, and when it eventually precipitates it would rain more.

Based on the above consideration, we have to consider the hypothesis that most of the cloud-mediated aerosol forcing is manifested by changes between cloud regimes. Such transitions are associated with change in cloud radiative effect (CRE) of the order of 100 Wm^{-2}, whereas the aerosol net effect within the cloud regimes are 1–2 orders of magnitude smaller.

It is difficult to ascribe the changes of CRE between regimes to aerosol cloud-mediated RF, because the aerosol amounts are interactive with the clouds, especially in the open and collapsed BL regimes, so that they are not independent of the cloud forms. Another major difficulty in ascribing satellite-measured aerosols to their effects on the clouds is the fact that the greatest effect occurs in the regime where $N_d < 100$ cm^{-3} (see Fig. 2), where on average AOD is <0.05, which is at the low boundary of the measurement capability, and its conversion to CCN is highly uncertain (Andreae 2009). Therefore, using the retrieved N_d instead of AOD as proxy for the CCN provides a more sensitive metric of the aerosol cloud-mediated effects on MSC. Therefore, it is argued here that assessment of the differential CRE between MSC regimes with respect to N_d captures an important element of the aerosol cloud-mediated radiative forcing. The remaining challenge will be quantifying the extents of the attribution of the regime changes to anthropogenic causes.

A case where the regime changes could be ascribed to anthropogenic aerosols from ship tracks is reproduced here from Fig. 3 of Rosenfeld et al. (2006a), with the added CRE, and presented here in Fig. 4. It is shown for this case that the negative CRE over the closed cells is on average higher by well over 100 Wm^{-2} than the adjacent open cells or collapsed boundary layer. This forcing is calculated for the averaged 24 h diurnal cycle. The r_e in the closed cells of this example is smaller than 16 μm, very close to the heavy drizzle threshold of 15 μm (see Fig. 2), whereas the r_e is considerably larger than 16 μm in the open cells. The appearance of heavy drizzle after cloud top r_e exceeds this threshold appears to be the main cause for opening the closed cells (e.g., Rosenfeld et al. 2006a; Koren and Feingold 2011). The cloud top r_e is determined mostly by N_d and h (Freud et al. 2011). Therefore, the combination of N_d and h is required for explaining the transitions from closed to open cells.

The ship tracks within the closed cells obviously did not incur a regime change and hence the associated change in CRE was about 10–15 Wm^{-2}, which is lower by an order of magnitude than the change associated with regime change.

A consistent picture emerges from the study of George and Wood (2010) who quantified the dependence of the variance in albedo over the southeastern Pacific Ocean on the variances in the controlling variables (i.e., cloud fraction, *LWP* and N_d).

The variability in cloud fraction, *LWP* and N_d explained on average roughly 1/2, 1/3 and 1/10 of the spatial variance of the area-mean albedo that was accounted for by these variables, respectively. It is interesting that despite a strong gradient in N_d within the analyzed region, N_d does not explain more than 10 % in the variance of the area-mean albedo. Is it because h and hence *LWP* increases along with the decrease in N_d with distance from land? These results should be treated with caution, because part of this variability could be explained by meteorological factors that are correlated with the cloud fraction, *LWP* and N_d.

Does it mean that much, if not most of the variability in the cloud RF in the southeastern Pacific is not contributed by MSC regime changes? It appears that this partition of the CRE components is not limited to areas where MSC regime changes occur frequently, because these results are in agreement with the previous global studies that separated the contributions of RF. Sekiguchi et al. (2003) showed based on AVHRR data that the N_d effect could not have contributed more than 25 % of the total cloud RF over the global oceans. Kaufman et al. (2005) analyzed MODIS data over the Atlantic Ocean and showed that only 10–20 % of the enhanced cloud RF that was associated with increased τ_a was contributed from N_d. The dominance of cloud cover effect over ocean was also supported qualitatively by several other satellite studies (Matheson et al. 2006; Myhre et al. 2007b; Menon et al. 2008). Lebsock et al. (2008) used CLOUDSAT for showing that the LWP effect dominated the Twomey effect, being positive with added τ_a in precipitating clouds and negative in non-precipitating clouds.

How much of the aerosol indirect effect on climate can be explained globally by regime changes, and how much by net radiative changes within regimes? It is possible that a large fraction occurs through the latter. Buffering (Stevens and Feingold 2009) and cancellation (Wood 2007) mechanisms have been shown to work within regimes, but between the regimes it is not so clear that this is the case (Koren and Feingold 2011). A possible mechanism to communicate information that may cause some buffering between regimes pertains to the determination of the inversion height.

The two regimes have two different equilibrium states. Weakly precipitating closed cells have large inversion heights at the top of a well-mixed boundary layer and strong entrainment at the top of the inversion. Very pristine drizzling clouds or a thin layer of very low clouds in equilibrium state are topped by a very low inversion height, also defined as a "collapsed" boundary layer (Bretherton et al. 2010b). However, this does not result in a step change in PBL height at the boundary between the regions, but instead the inversion tends to "homogenize" due to the strong buoyancy forcing at a scale in the order of at least 100 km, thus inducing a shallow secondary circulation above the PBL top (Berner et al. 2011) so that, in effect, open cell regions keep the adjacent closed cell region's PBL from deepening as fast as it would in the absence of the open cell region. From the other side, the closed cells regions keep the open cell PBL from collapsing in their vicinity. We don't yet know what the consequences of this interaction are for cloudiness, but they are likely to be important for determining AIEs associated with regime change in MSC.

These questions will have to be answered quantitatively by future research. In particular, an emphasis should be placed on the role that aerosols play in mediating regime changes in marine low clouds. This might require some experiments with controlled dispersion of aerosols into MSC.

2.4 The Frequency of Occurrence of Aerosol-Starved Cloud Regimes

The regime of open cells cannot inherently sustain full cloud cover, and water that does condense is depleted quickly by precipitation. Therefore, it is appropriate to describe this as a situation where scarcity of aerosol limits the cloud cover and *LWP*, i.e., an aerosol starved cloud regime (Van Zanten and Stevens 2005; Petters et al. 2006; Sharon et al. 2006; Wood et al. 2008, 2011). The regime of the collapsed boundary layer was not yet analyzed for its differential CRE with respect to the other regimes, but given the mechanism of its creation, it can be considered even more strongly as an aerosol starved cloud regime.

How frequent are these conditions where clouds are starved for aerosols, such that the depletion of aerosols can incur a regime change from closed to open cells with decreased radiative forcing in the order of $-100\ Wm^{-2}$? The addition of aerosols has been observed to close the open cells, at least in the regime of collapsed boundary layer (Christensen and Stephens 2011). Simulations of added aerosols to open cells stopped their precipitation, but failed to convert them back to closed cells (Wang et al. 2011a). The ability of aerosols to close relatively deep open cells requires additional research. Figure 5 presents global maps of the occurrence of mesoscale cellular convection, partitioned into closed cells, open cells that are organized in Benard convection, and disorganized open cellular convection. The organization of the first two regimes can be ascribed clearly to the aerosols and N_d as discussed above, but this is not obvious for the latter regime. These three regimes cover a large part of the eastern subtropical and tropical oceans. The frequency of the open cells increases with the distance westward away from land. This occurs due to a combination of decreasing N_d (Fig. 6) and increasing cloud thickness (see e.g. George and Wood 2010), the combination of which increases precipitation dramatically (Fig. 6, see also Bretherton et al. 2010a). Open cells observed during the VOCALS Regional Experiment tended to be associated with aerosol-starved conditions (e.g. Wood et al. 2011), but it is not yet clear the extent to which this is the case for all open cell regions in the subtropics.

Open cells are also frequent in midlatitudes, but here they can occur due to cold advection of air (e.g. cold air outbreaks), which provide strong surface forcing in subsiding conditions which dominates the dynamics of open cells regardless of possible aerosol effects. The extent to which these open cell systems modulate their own microphysical state and become aerosol-starved is currently poorly known.

Some light can be shed on this question from the shape of the functional dependence of cloud cover on aerosol amounts, as represented by AOD. Globally, almost

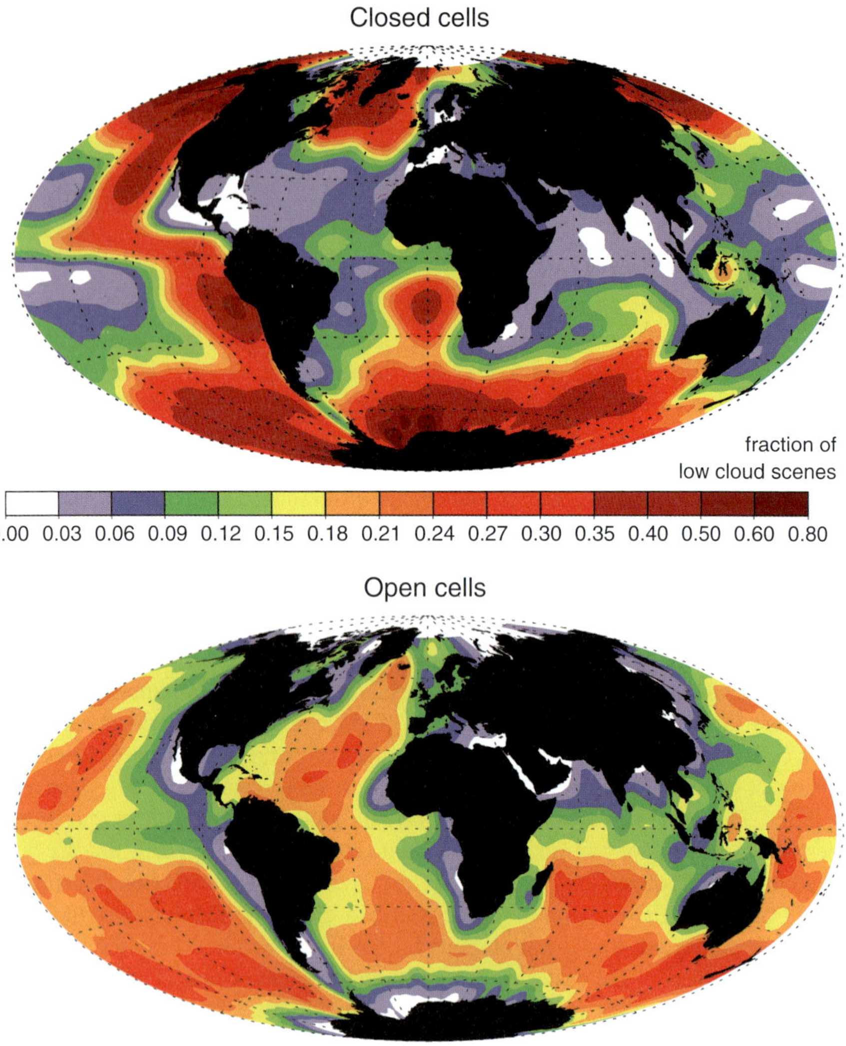

Fig. 5 Frequency of occurrence of closed (*top*) and open (*bottom*) mesoscale cellular convection (*MCC*), based on all available MODIS data from 2008, using method of Wood and Hartmann (2006)

all of the increase in cloud cover f_c with AOD occurs at AOD < 0.2 (see Fig. 7). For AOD ≤ 0.75 the $\ln(f_c)/\ln(\mathrm{AOD}) = 0.57$. This shows that the sensitivity of f_c to AOD is much greater than logarithmic at the lowest AOD, and that the behavior is consistent with the aerosol changes with the MSC regimes responsible for a large part of the dependence of f_c on AOD.

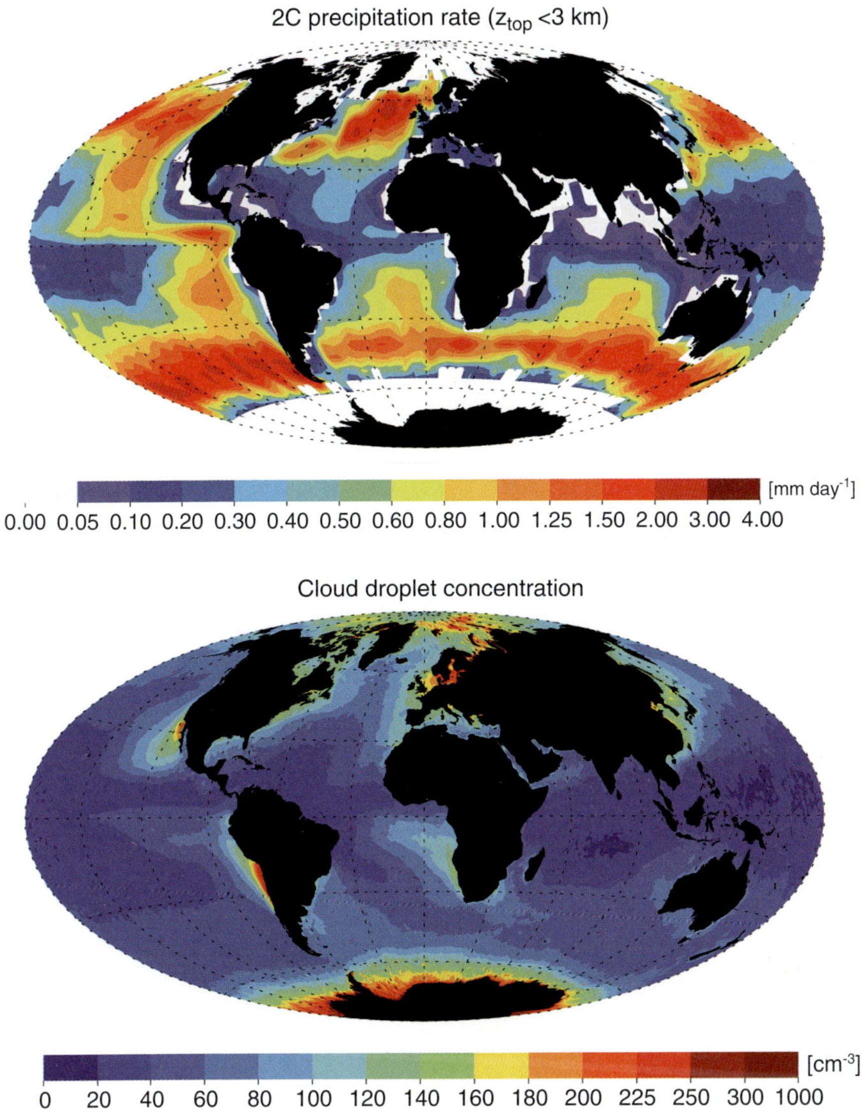

Fig. 6 *Top* Annual mean map of column maximum precipitation rate from clouds with tops below 3 km altitude, from the CloudSat Precipitation Radar (Lebsock et al. 2011). *Bottom* Annual mean cloud droplet number concentration for horizontally extensive (instantaneous cloud cover exceeding 0.8 for 1 × 1° boxes) liquid clouds N_d, using data from MODIS, following the method of Bennartz (2007)

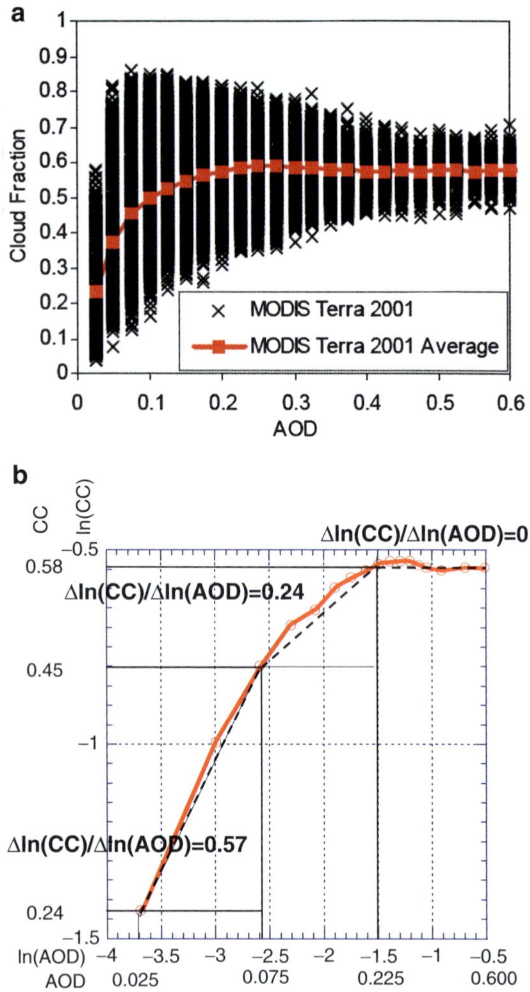

Fig. 7 Annual global MODIS retrieved cloud cover as a function of AOD. (**a**) as presented by Myhre et al. (2007b); (**b**), presented on a logarithmic scale, with ln(CC)/ln(AOD) calculated for three AOD intervals

2.5 *The Attribution of the Regime Changes to Anthropogenic Aerosols*

Open cellular convection is more frequent over the Southern Hemisphere subtropical and midlatitude oceans than over the corresponding regions of the Northern Hemisphere (Fig. 5). It is interesting to ask the extent to which this might be attributable to anthropogenic aerosol influence. Mean N_d values for low clouds in polluted regions are higher than for clean regions (e.g. Quaas et al. 2009), and the ability of increased cloud droplet concentrations to keep large areas of MSC at the closed regime is evident in the observations of Goren and Rosenfeld (2012), where the large areas of closed cells are shown to have been shaped by old ship emissions. Other mechanisms that can transport aerosols from land to the remote ocean areas are pollution plumes in the free troposphere that subside in the anticyclones to the underlying MSC (Wilcox et al. 2006).

It is hypothesized that the greater amount of aerosols from the northern hemisphere continents are responsible for the hemispheric differences in open cell frequency, but more understanding of factors controlling this frequency, including the large- scale meteorology, is required to test this hypothesis. If the reduction in open cells is a manifestation of the added anthropogenic aerosols it implies a huge negative radiative forcing, because the differential RF between the closed and open cells can exceed 100 Wm^{-2} (see Fig. 4).

2.6 The Possible Underestimation of the Radiative Forcing Via Low Clouds

As we have discussed in Sect. 2.3, it is possible that the Twomey effect is 1/4 the overall AIE from low clouds, or less (Sekiguchi et al. 2003; Kaufman et al. 2005; Lebsock et al. 2008). Yet, the IPCC AR4 found a cloud drop radius effect of –0.7 Wm^{-2} with the large uncertainty range of –0.3 to –1.8 Wm^{-2}. If indeed the cloud-cover effect is much larger than the clouddrop-radius effect, the AR4 range has to be increased by a large factor to account for other effects. Even if not all cloud types respond in the same way as our example of MSC, we face the possibility of a very large and highly uncertain net forcing from low clouds, especially once adjustments involving dynamics occur.

This should be contrasted with the inverse calculations showing that the overall net cloud-mediated RF should likely be even lower than the IPCC-estimated albedo effect alone (see Sect. 1). To resolve this apparent contradiction, there are two likely possibilities:

1. The aerosols that are involved in regime changes and the respective RF are predominantly natural, or,
2. Most of the strong negative RF is balanced by another similarly strong positive RF, particularly by anthropogenic aerosols interacting with deep and high clouds.

While at least part of the aerosols involved in the regime changes are natural, based on some of the evidence presented here, we cannot discard the second possibility, especially in view of its far-reaching consequences, The second possibility, that a strong negative RF is partially countered by a positive RF from less-studied cloud types, is explored next.

3 Aerosol Induced Radiative Forcing by Deep Convective Clouds

If indeed the forcing of low- level cloud is large to the extent that the climate should have been cooling, the constraints described in Sect. 1 would be difficult if not impossible to satisfy without a similarly large positive radiative forcing to balance most of this cooling effect. We hypothesize here that such a positive forcing may indeed exist, in the form of aerosol effects on deep and/or high clouds, through several possible mechanisms that are presented in this section.

3.1 Aerosol Invigoration of Deep Clouds in Warm and Moist Atmosphere

Most of the condensed cloud water in deep tropical convective clouds in pristine air masses is precipitated as warm rain before reaching the freezing level. Adding CCN to the clouds causes N_d to increase, and respectively the height for onset of warm rain to increase as well. This effect was quantified in several aircraft field campaigns in the Amazon tropical clouds (Andreae et al. 2004), Argentina hail storms (Rosenfeld et al. 2006b), California winter storms (Rosenfeld et al. 2008a), Israel winter clouds and India summer monsoon clouds (Freud and Rosenfeld 2012). As seen for the case of MSC (Fig. 2), and for the same fundamental physical reasons, the number of activated cloud drops near cloud base scales linearly with the cloud depth required to grow droplets to the threshold r_e of ~14 μm for rain initiation in deep convective clouds (Freud et al. 2011; Freud and Rosenfeld 2012). Increasing the number of activated aerosols by 100 cm^{-3} increases h for the onset of rain by ~280 m. Therefore, in deep tropical clouds with freezing level of 3–4 km above cloud base, an adiabatic concentration of nearly 1,000 drops cm^{-3} would delay the onset of precipitation to above the freezing level, thus preventing warm rain formation. Observations from the Amazon (Andreae et al. 2004) and India (Freud and Rosenfeld 2012; Konwar et al. 2012) support this conclusion.

It has been hypothesized (e.g., Khain et al. 2004, Rosenfeld et al. 2008b) that delaying the precipitation to above the freezing level would cause the cloud water to freeze first onto ice hydrometeors and so release the latent heat of freezing, which would not have been realized had rain at lower levels not been prevented by the aerosols (see illustration at Fig. 8). The released added latent heat adds buoyancy to the cloud, increases the updraft speed, and causes the cloud top to grow higher and the anvil to expand over a larger area. The melting of the ice hydrometeors while falling cools lower levels, with a net result of more low-level cooling and high-level warming for the same surface rainfall amount. This means consumption of more static gravitational energy and its conversion into respectively more kinetic energy, which is the essence of the invigoration of the storm. The invigoration, along with enhanced ice precipitation processes, enhance also the cloud electrification (Molinie and Pontikis 1995; Williams et al. 2002; Andreae et al. 2004; Rosenfeld et al. 2008b). Set against this possibility, however, is the added gravitational loading of the retained condensate. The net result of these competing factors is not obvious a priori.

Cloud simulation studies have generally confirmed the net invigoration hypothesis for deep warm- base clouds with weak wind shear in moist environments. For other conditions no invigoration was obtained, and for cool-base clouds, dry environment and/or strong wind shear the precipitation amount was even decreased

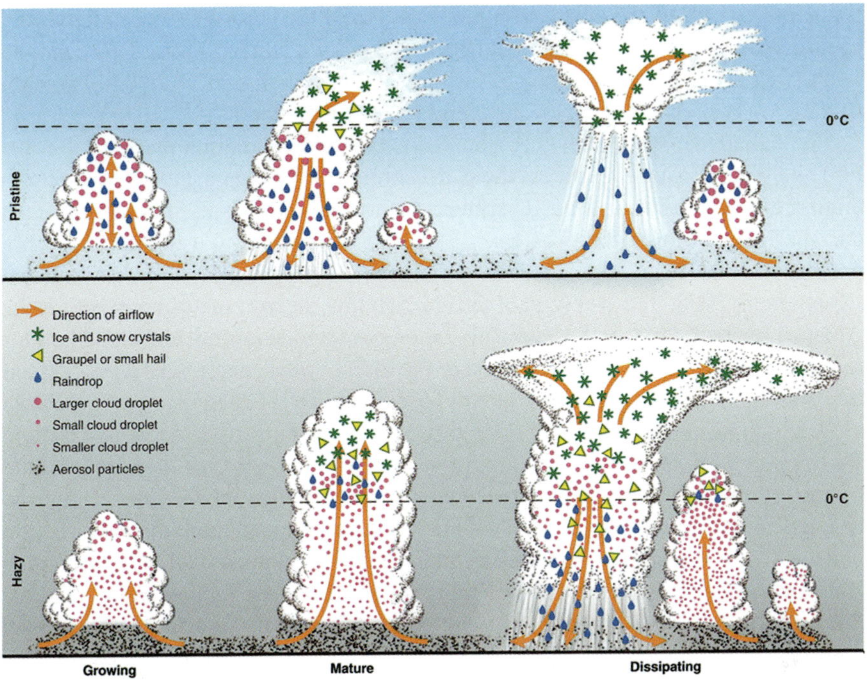

Fig. 8 Illustration of the aerosol cloud invigoration hypothesis. *Top* Clouds in pristine air rain-out their water before reaching the freezing level. *Bottom* The aerosols delay the rain until the cloud reaches the freezing level, where the water freezes into ice hydrometeors and releases more intensely the latent heat of freezing, which invigorates the cloud. The cloud tops grow to greater heights and expand to larger anvils (From Rosenfeld et al. 2008b)

(Khain and Pokrovsky 2004; Khain et al. 2004, 2005, 2008a; Wang 2005; Seifert and Beheng 2006; van den Heever et al. 2006; Fan et al. 2007, 2009, 2012). In some of the simulations, the greater low-level evaporative cooling of the enhanced rainfall produced stronger gust fronts that triggered more new clouds and invigorated them (Tao et al. 2007; Lee et al 2010). Morrison and Grabowski (2011) argue that invigoration is counteracted in radiative-convective equilibrium by large-scale feedbacks, but clouds still become deeper.

Satellite observations using MODIS showed deeper and more expansive convective clouds associated with greater aerosol optical depth over the tropical Atlantic Ocean (Koren et al. 2005, 2010a, b). The reality of such associations has been questioned due to possible errors in the retrieved aerosols due to cloud contamination and other artifacts that are caused by the proximity to the clouds, but Koren et al. (2010a) showed that this was not the cause in a study of the North Atlantic region, because the cloud invigoration was detected with a similar magnitude when comparing the retrieved cloud properties to the results of an aerosol transport model. They also partitioned their analysis to different meteorological conditions that control the depth of the convection, and still found the aerosol invigoration effect having a similar magnitude for the different meteorological partitions. However, the average measured cloud top height in the study of Koren et al. (2010a) was only about 3 km, which is well below the height of an anvil cloud.

The radiative effects of the aerosols reduce the solar radiation reaching the surface and therefore act to suppress the convection, working against the aerosol invigoration effect, at least on land. Therefore the aerosol effect is not monotonic, such that the invigoration effect was calculated by Rosenfeld et al. (2008b) to reach a maximum at AOD of ~0.3. This was confirmed observationally over the Amazon (Koren et al. 2008; Ten Hoeve et al. 2012). Satellite measurements reported for the Amazon region by Lin et al. (2006), also showed that total rainfall and cloud heights increased on average with AOD and that the effects weakened above an AOD near 0.3, although these observations did not specifically target deep convective clouds.

Anvil clouds associated with deep convection exert a substantial longwave and shortwave cloud forcing, and the longwave component dominates in clouds that are optically thin, including the cirrus clouds produced by the anvil. Aerosol-induced changes in anvil clouds associated with deep convection and more distant cirriform clouds whose ice is partly supplied by convective detrainment can therefore act as warming mechanisms. Lee et al. (2009) found in a deep-convection simulation that 28 % of the increased shortwave cloud forcing (cooling) associated with higher aerosol concentrations was offset by increased longwave cloud forcing (warming). The corresponding offset for stratocumulus clouds was only 2–5 %.

Critical supporting observational evidence to the validity of the invigoration hypothesis was obtained very recently, where volcanic aerosols, whose variability was completely independent on meteorology, were observed to invigorate deep convective clouds over the northwest Pacific Ocean and more than double the lightning activity (Yuan et al. 2011; Langenberg 2011). This lends credibility to the suggestion of Zhang et al. (2007b) that the trend of increasing emissions of air pollution from East Asia caused their observed trend of increasing deep convection and intensification of the storm tack at the North Pacific Ocean.

The aerosol-induced invigoration on the peripheral clouds of tropical cyclones was hypothesized to occur at the expense of the converging air to the eye wall, and hence decrease maximum wind speeds (Rosenfeld et al. 2007b). This aerosol effect was simulated extensively (Rosenfeld et al. 2007b; Cotton et al. 2007; Khain et al. 2008b, 2010; Khain and Lynn 2011; Zhang et al. 2007a, 2009). The variability in aerosols was also observed to explain about 8 % of the variability in the intensity of Atlantic hurricanes (Rosenfeld et al. 2011a). The aerosol effects on the microphysics and intensity of tropical cyclones are reviewed in Rosenfeld et al. (2012a).

A weekly cycle in the anthropogenic aerosols, peaking during mid-week, was shown to be associated with a similar cycle in the rain intensity and cloud top heights (Bell et al. 2008), on the lightning frequency (Bell et al. 2009), and even on the probability of severe convective storms that produce large hail and tornadoes

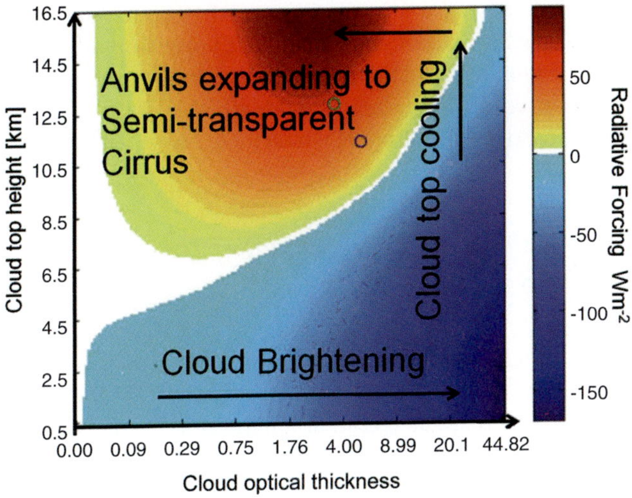

Fig. 9 The net TOA radiative forcing of a cloud in a tropical atmosphere, as a function of its cloud top height and optical thickness (After Koren et al. 2010b)

(Rosenfeld and Bell 2011) in the eastern USA during summer. These findings are supported by a recent study analyzing 10 years of surface measurements of clouds and aerosols over the ARM site at the Southern Great Plains in Oklahoma, showing clearly the cloud invigoration effect, associated with decreasing probabilities of light rain matched by similar increasing probability for heavy rain (Li et al. 2011). It is important to note that the invigoration effect is anticipated mainly in the intensity of rain events or vigor of convection, and not necessarily in the average rainfall (e.g., Storer and van den Heever 2013), which may explain why studies examining total rainfall (e.g., Morrison and Grabowski 2011) sometimes do not find it.

All these findings, and especially the long-term measurements of Li et al. (2011), suggest that the aerosol invigoration is a robust effect in the atmosphere. The aerosol invigoration of deep convective clouds could exert a cloud-modulated radiative forcing in several ways, as illustrated in Fig. 9:

- Brightening of the clouds at a fixed cloud top, increasing their albedo and cooling effect. However, for already thick convective cloud, where the albedo effect is nearly saturated, the negative RF is expected to be rather small.
- Higher cloud tops, which emit less thermal radiation to space and hence induce a warming effect.
- More extensive anvils and/or more semi transparent ice clouds. Such cirrus clouds have small albedo in the visible, but still have large emissivity in the

thermal IR, thus causing a strong positive RF. They could also reduce the radiative cooling and air subsidence rates in the upper troposphere, which would increase relative humidity and therefore the atmospheric greenhouse effect.

- Deep convection could more frequently reach the lower stratosphere adding more water vapor to the stratosphere which increases the greenhouse effect.

3.2 Aerosols Enhancing Detrainment of Ice and Vapor in the UTLS

Aerosols may well enhance the amount of ice contained in and detrained from anvils into the upper troposphere and lower stratosphere (UTLS), even without having any dynamic effects (e.g., invigoration) on the clouds. Analyzing deep convective clouds in satellite data, Sherwood (2002a) and Jiang et al. (2009b, 2011) found that biomass burning aerosols were associated with smaller ice particle r_e at the anvils of tropical deep convective clouds. These storms also were more intense, as indicated by their colder cloud tops, though these studies found that proxies for intensity were too small to explain the smaller r_e. The effect could be due to CCN nucleating small cloud drops that freeze homogeneously into respectively small ice particles (e.g., Phillips et al. 2002), or (in part) due to invigoration of the storms activating more aerosols aloft, or to meteorological factors not accounted for in those studies. The clouds with smaller ice particle r_e produce significantly more lightning, supporting the hypothesis that aerosols played a role in reducing the r_e of the ice particle (Sherwood et al. 2006). Satellite measurements of pyro-cumulonimbus showed that the extreme CCN concentrations in the dense smoke keep the cloud drops extremely small up to the homogeneous ice nucleation level, where they become similarly small ($r_e \sim 10$ μm) ice particles, whereas ice in the ambient clouds formed mostly by mixed-phase processes producing particles in the anvils with $r_e > 30$ μm (Rosenfeld et al. 2007a). Tracking the life cycle of such anvils showed that they lived twice as long as anvils from ambient clouds and expanded to much larger areas (Lindsey and Fromm 2008). This is likely due to the smaller fall speeds and/or slower aggregation of the particles.

Aircraft measurements and model simulations show that aerosols from Africa indeed nucleate small cloud drops aloft that freeze homogeneously into small ice crystals in the anvils of clouds over southern Florida (Fridlind et al. 2004). In simulating this process, Jensen and Ackerman (2006) showed that the detrainment of small ice crystals was responsible for creating long-lived cirrus clouds. The simulations of deep tropical clouds by Fan et al. (2010) show that added CCN can lead to such enhancement of small ice particles in the anvils, and nearly double the extent of the resulting clouds; similar results were obtained by Morrison and Grabowski (2011).

A cloud-system resolving model simulation of the aerosol effect at a regional scale with bin microphysics for tropical and midlatitude summertime convective cloud situations (Fan et al. 2012) found invigoration in the tropical case with weak wind shear, but not with strong wind shear. However, the positive RF from the anvil

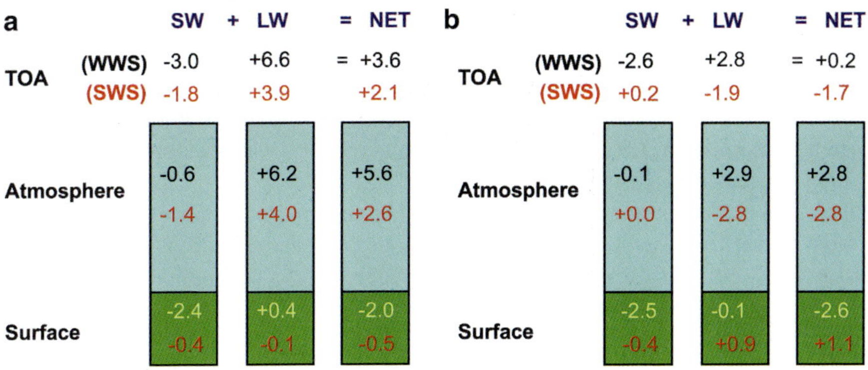

Fig. 10 Short wave (*SW*), long wave (*LW*), and net radiative forcing of aerosol cloud mediated effect at the top of atmosphere (*TOA*), atmosphere, and surface (*SFC*) for the China tropical (**a**) and SGP temperate (**b**) cases of deep convective cloud system, with weak wind shear (*WWS*) and strong wind shear (*SWS*). Values in red are for the stronger wind shear condition. Values are averaged over the last day of simulations (From Fan et al. 2012)

expansion with added CCN dominated the negative RF due to cloud brightening in both cases, as shown in Fig. 10a. In the temperate case the net RF were weaker and of opposite signs for the different wind shears (Fig. 10b).

Another possible pathway for aerosol indirect effects is through altering stratospheric water vapor, a strong greenhouse gas. Solomon et al. (2010) found that decadal variability in lower stratospheric water vapor was contributing to decadal climate variability, following previous calculations showing that increases in stratospheric water vapor over the latter part of the twentieth century contributed a radiative climate forcing of order 0.2 Wm^{-2} (Forster and Shine 1999, 2002; Myhre et al. 2007a). While the decadal humidity variations can largely be explained by those of tropopause temperature through a simple freeze-drying model (e.g., Notholt et al. 2010), radiosonde data do not show a longer-term warming trend, and the source of the moistening trend is still unknown. The radiative forcing is significantly larger than accounted for by the IPCC in 2007, which only included the part attributable to methane oxidation.

Two plausible mechanisms have been suggested linking this trend to anthropogenic aerosols. First, smaller ice particles lofted in polluted storms could cause overshooting clouds to re-evaporate more quickly when mixing with dry stratospheric air, delivering more water vapor to levels where it can reach the lower stratosphere as shown by satellite and in-situ observations and simulated by models (Sherwood 2002b; Chen and Yin 2011; Wang et al. 2011b; Nielsen et al. 2011). Back-of-the-envelope calculations suggest this mechanism, which is observed to affect stratospheric humidity independently of tropopause temperature, could account for the observed trend since 1950 even discounting any invigoration effect (Sherwood 2002b), but this has not been comprehensively modeled; isotopic data do not suggest any trend in ice re-evaporation since 1991 (Notholt et al. 2010) but most of the humidity trend occurred before 1991. A similar microphysical effect from ice nuclei could also occur for cirrus clouds formed near the tropopause (Notholt et al 2005). The second possibility is suggested by observations (Su et al. 2011; Wu et al. 2011) and models (Liu et al. 2009) indicating that pollution particles lofted in deep convection elevate cirrus cloud height and water vapor mixing ratios, which would increase water transport into the stratosphere (Liu et al. 2009). Observations do not show a corresponding temperature trend since 1958, but this could be due to biased trends in the radiosondes which are difficult to correct (JS Wang et al. 2012). In summary, aerosols probably exert a second indirect warming effect through lower stratospheric water vapor, and this could be of nontrivial magnitude.

4 Aerosol Induced Radiative Forcing by Supercooled Layer Clouds

New satellite remote sensing data, especially from active sensors, are revealing that supercooled layer clouds are more common than previously suspected (Hogan et al. 2004; Zhang et al. 2010; Hu et al. 2010; Morrison et al. 2011). Almost all clouds with tops warmer than –20° C contain supercooled liquid water (Hu et al. 2010), and supercooled liquid can exist in many stratiform clouds with temperatures down to –30° C (Hu et al. 2010, Shupe et al. 2006). In the midlatitude storm tracks and high latitudes supercooled liquid layers can occur 10–15 % of the time (Zhang et al. 2010), making this a climatologically important category of clouds.

Often supercooled layer clouds, both at low and mid troposphere, are maintained by radiative cooling at their tops that induces inverse convection in a layer within a stable atmosphere, similar to the mechanism that sustains decks of marine stratocumulus. What is particularly remarkable about such clouds is their apparent sensitivity to small changes in ice nuclei (IN) ingested (e.g., Prenni et al. 2007; Morrison et al. 2011), with the result of increased IN being a rapid glaciation, the loss of liquid condensate and a suppression of longwave cooling that weakens the turbulent mixing sustaining the supercooled layer (H. Morrison et al. 2011). Thus, the hypothesis has been raised in recent studies that Arctic clouds exhibit bistability: they consist either of turbulent supercooled layers with minimal ice, or rarefied clouds containing only ice particles (Morrison et al. 2011, 2012). Because the ice-only clouds tend to be optically thin (perhaps even just diamond snow under some circumstances), such bistability could permit a particularly strong aerosol cloud-mediated radiative forcing. In addition to IN-mediated impacts, observational data indicate that a CCN-starved regime is often present in the Arctic over sea ice (Mauritsen et al. 2011). This change between the two stable regimes of supercooled water and ice clouds is illustrated in Fig. 11.

Adding CCN to supercooled layer clouds may delay their transition to glaciated clouds for a given IN concentration. With more CCN cloud drops are smaller and less likely to accreted to ice crystals or coalesce and precipitate, leading to smaller

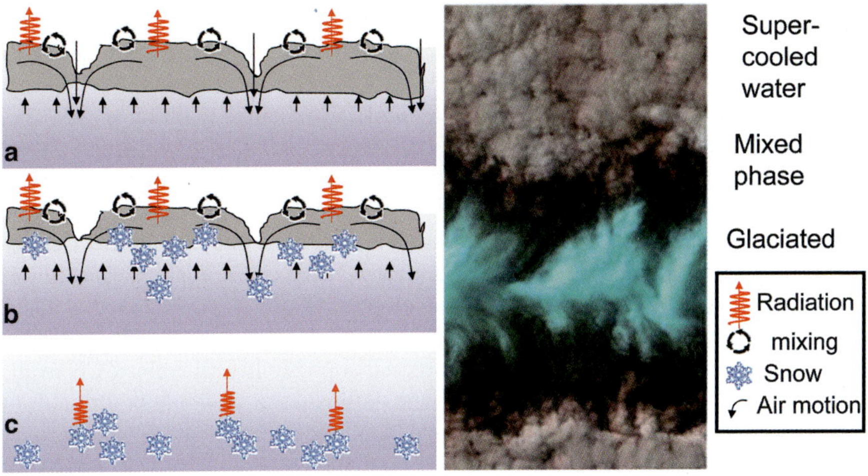

Fig. 11 A schematic illustration of the mechanism for transition from persistent supercooled layer clouds to a stable situation of glaciated or no clouds. The mechanism that maintains the supercooled layer clouds is the radiative cooling and mixing at their tops, the same as for closed marine stratocumulus. The inverted convection replenishes the cloud water that is lost to ice precipitation (**a**). An increased loss of cloud water to ice precipitation, due to increased concentrations of IN and/or decrease in CCN, makes the cloud thinner and broken with less water (**b**). When the cloud water is fully consumed by the ice crystals that precipitate, there is nothing that will keep the radiative cooling that regenerated and maintained the cloud at the first place, and the cloud dissipates and leaves some falling ice crystals or no cloud at all (**c**). The image was taken by LANDSAT over the eastern USA on 11 December 2009. The glaciation in this case occurred probably due to addition of IN by aircraft exhaust to clouds at temperatures of –30 to –35° C (*Source*: NASA Visible Earth)

loss of cloud water. In such case, added CCN would have a negative forcing, whereas added IN have a positive forcing.

The remarkable sensitivity of supercooled layer clouds seems at odds with their frequent occurrence. This has led to the search for mechanisms limiting the formation of ice in these clouds. It appears that most of the IN in supercooled layer clouds, particularly in the Arctic, arrive via long-range transport in the freetroposphere. They enter the cloud by subsidence and are then entrained into the cloud. This process does not supply IN at a particularly high rate, with typical replenishment timescales of several days after entrainment removal (Fridlind et al. 2012). This results in a slower rate of IN activation than would occur if all the free-tropospheric IN were activated at once, and limits the desiccation of supercooled liquid clouds.

Similarly, the effect of aerosols on cirrus clouds could result in either a positive or negative forcing on climate. GCM simulations of the effect of added IN show a reduction in cirrus water content and optical thickness, and sometimes a reduction in relative humidity, producing a negative forcing of uncertain magnitude (Penner et al. 2009; Hendricks et al. 2011). Opposing this, the introduction of CCN aerosol can enhance cirrus coverage; the radiative forcing of this is fairly small for aircraft emissions (Burkhardt and Kärcher 2011) but could be larger when including all anthropogenic aerosol. As already stated, such calculations should be regarded as provisional at this time.

There are still major unresolved questions. Chief among these is that ice-nucleating mechanisms of aerosols are poorly understood. In some supercooled liquid clouds the concentration of ice crystals appears to be significantly greater than the IN concentration (Fridlind et al. 2007). We do not yet fully understand how pollution aerosols affect the IN concentration in the Arctic and midlatitude freetroposphere where supercooled clouds are common (McFarquhar et al. 2011). Aircraft observations suggest that high ice crystal concentrations in supercooled shallow cumulus in maritime polar airmasses tended to occur in the presence of large drizzle drops (Rangno and Hobbs 1991). This suggests a possibility whereby CCN availability might be important for ice formation. At this stage, it would be difficult to attribute even a sign to the potential aerosol cloud-mediated effects on supercooled water clouds, but there nevertheless exist important possibilities that warrant further exploration.

5 Discussion and Implications for GCMs

Based on the previous section, AIE on deep convective clouds appear to induce positive radiative forcing of yet unknown global magnitude by invigorating clouds, expanding their anvils, and enriching the lower stratosphere with water vapor. Air pollution aerosols were also observed to glaciate mid- and upper-tropospheric supercooled clouds (Rosenfeld et al. 2011b), and thus adding positive radiative forcing.

This compensates to an unknown extent the negative forcing due to the AIE on low clouds.

Even if the net effect is very small on a global average, the cooling occurs mainly over the subtropical highs and migratory anticyclones over ocean, whereas the warming occurs mainly at the areas of deep tropical convection. The spatial separation can propel atmospheric circulation systems that would modify the weather patterns. GCMs do not yet treat AIEs in both deep convection and shallow clouds comprehensively enough to ascertain the nature of these changes, but studies focusing on direct effects of aerosols and/or indirect effects on shallow clouds suggest aerosol-induced circulation changes are possible in the tropical Atlantic climate (Chang et al. 2011), Sahel rainfall (Ackerley et al. 2011), south Asian monsoon circulations (Bollasina et al 2011), the Hadley circulation (Ming and Ramaswamy 2011; Allen et al. 2012), North Atlantic (Booth et al. 2012), and the boreal winter extra-tropical circulation (Ming et al. 2011).

As noted elsewhere in this chapter, observational and process studies suggest that aerosols and clouds interact through a range of radiative, microphysical, thermodynamic, and dynamic mechanisms. With increasing aerosol concentrations, these mechanisms all recognize an initial response taking the form of smaller cloud particles, delayed precipitation formation, and larger water contents. The instantaneous radiative forcing is comprised of increased shortwave reflection (cooling) and increased longwave emission (possible warming from high clouds) and can be described as a radiative indirect effect. Several subsequent competing mechanisms resulting from smaller cloud particles, delayed precipitation formation, and larger water contents are possible. In the absence of mechanisms responding to larger water contents, cloud lifetimes and areas increase, enhancing the instantaneous radiative forcing (included in "adjusted" radiative forcing). Numerous counteracting mechanisms have been identified. Increased water contents near cloud top enable evaporation resulting from entrained dry air to break up clouds, reducing water content, cloud lifetime, and cloud areas. The "adjusted" radiative forcing by this mechanism opposes that described above. Increased water content near cloud top can enhance radiative cooling and generate instabilities, leading to a similar set of consequences. Increased water content can also lead to changes in the heights and thicknesses of clouds. Changes in the sizes of drizzle particles below cloud base can change evaporation and stability below cloud base. In some cases, aerosol-induced changes can alter the cloud regime, changing significantly cloud areas and lifetimes. Microphysical changes in deep convection can change distributions of latent heating and induce evaporatively driven downdrafts, increasing the intensity of convection. Effects related to ice nucleation are likely, and absorbing aerosols can heat the atmosphere around clouds, altering clouds in what is referred to as a semi-direct effect.

While observational and process studies suggest this wide range of cloud-aerosol interactions capable of both warming and cooling the earth-atmosphere system, scaling these interactions to global scale and inferring their impacts on climate and climate change requires synthesis provided by climate models. On the other hand, state-of-the-science atmospheric general circulation models (GCMs) treat processes relevant for cloud-aerosol interactions in a highly simplified manner, limiting the confidence with which conclusions can be drawn.

Quaas et al. (2009) compared ten GCMs which treat cloud-aerosol interactions with satellite observations. All of the GCMs in that study, as well as those summarized in Isaksen et al. (2009), are cooled by their cloud-aerosol interactions. To the extent underlying relationships between clouds and aerosols in GCMs can be evaluated using satellite observations, present-day positive relationships between aerosol optical depths and cloud liquid in GCMs seem to be too strong, while positive relationships between aerosols and drop number are comparatively well simulated (Quaas et al. 2009). Penner et al. (2011) note that GCMs suggest present-day relationships between cloud and aerosol properties may differ from their pre-industrial counterparts, with the latter stronger than the former. Quaas et al. (2009) had noted that present-day aerosol optical depths and their variations with cloud properties are related in GCMs to AIEs between pre-industrial and present-day climates in those GCMs. By replacing the modeled aerosol optical depths and their variations with cloud properties with the corresponding satellite observations, they infer GCM AIEs are larger than would be consistent with satellite observations. Quaas et al. (2009) also found most GCMs had difficulty simulating reductions in cloud-top temperature with increasing aerosol optical depth, especially over oceans, consistent with the absence of interactions between deep convection and aerosols in most GCMs.

The complexity with which GCMs treat aerosol processes varies widely, from empirical methods relating aerosol concentrations to drop number (e.g., Lin and Leaitch 1997) to physically based methods using aerosol activation theory (e.g., Abdul-Razzak and Ghan 2000; Ming et al. 2007). Aerosol size distributions are specified (e.g., in terms of aerosol concentration, Donner et al. (2011)) in some models but calculated from prognostic aerosol modal equations (e.g., Liu at al. 2012) in others.

The chief limitation in GCM representations of aerosol-cloud interactions arises from simplifications in their cloud macrophysics (the processes governing the environments for activating cloud liquid and ice particles and their subsequent microphysical evolution) and the absence of aerosol interactions with deep convection in most GCMs. GCM cloud macrophysics also dominates the interactions between radiation, microphysics, thermodynamics, and dynamics; these interactions are quite restricted in current GCM macrophysics relative to the interactions identified by process studies. As an example, in GFDL CM3, a normal distribution whose variance is related to large-scale eddy diffusivity is used to characterize the small-scale variations in vertical velocity, which is a major control on aerosol activation (Golaz et al. 2011). CM3 treats cloud-aerosol interactions only in stratiform and shallow cumulus clouds. CM3 macrophysics can straightforwardly capture microphysics interactions which increase cloud water paths as aerosol concentrations increase but is much less able to represent processes discussed in the preceding paragraph in which increasing aerosol concentrations could reduce water paths. Indeed, GFDL CM3 exhibits an annual global-mean temperature increase of 0.32 °C between the period from 1980 to 2000 and the period from 1880 to 1920 (Donner et al. 2011). The corresponding increase for GFDL CM2.1, which does not include cloud-aerosol interactions, is 0.66 °C (Knutson et al. 2006). Observed estimates of this difference from the Climate Research Unit (Brohan et al. 2006) and the Goddard Institute for Space Studies (http://data.giss.nasa.gov/gistemp/tabledata/GLB.Ts+dSST.txt) are 0.56 and 0.52 °C, respectively. Changes other than incorporation of cloud-aerosol interactions between CM2.1 and CM3 preclude attributing the change in temperature increase solely to these interactions. Six of the ten models analyzed in Quaas et al. (2009) impose lower limits on cloud drop number concentration, which arbitrarily restricts cooling by cloud-aerosol interactions. An important research priority is for GCMs to improve their parameterization of aspects of cloud-aerosol interactions which are poorly represented currently, many of which limit cooling by aerosols.

The simulation of temperature increases in climate models between pre-industrial and present times depends on their adjusted forcings, climate sensitivities, and transient climate responses. Since climate sensitivity is not known, the extent to which a climate model (e.g., CM3) simulates this temperature increase would not strongly constrain the adjusted forcing due to anthropogenic cloud-aerosol interactions, even if greenhouse gas forcing and aerosol direct forcing were known. Related to the latter, it is important that climate models simulate aerosol distributions and properties realistically. Global observation networks for aerosols and surface downward shortwave radiative fluxes are available for evaluating climate models, e.g., as in Donner et al. (2011).

Advanced cloud macrophysics parameterizations offer a prospect for improving representation of cloud-aerosol interactions in climate models. For example, Guo et al. (2010) show that a parameterization using multi-variate probability distribution functions for vertical velocity, liquid water potential temperature, and total water mixing ratio can capture a range of responses of liquid water path to increasing aerosol concentrations. Guo et al. (2011) find that a key mechanism in these responses is cloud entrainment, as discussed above and modeled by large-eddy simulation. These methods to date have been used successfully in simulating single columns in field experiments. Incorporating them in climate models is an ongoing activity, e.g., at GFDL and NCAR. Droplet activation and ice nucleation in deep convection depends on vertical velocities therein. Since most GCMs parameterize deep cumulus convection in terms of mass flux only, they are not able to represent the interactions between deep convection and aerosols described elsewhere in this chapter. Examples of promising prospective developments include the use of deep cumulus parameterizations based on ensembles of cumulus clouds with vertical velocities in GFDL CM3 (Donner 1993; Donner et al. 2001), the use of double-moment microphysics in deep convection in experimental versions of GFDL AM3 (Salzmann et al. 2010), the NCAR Community Atmosphere Model (Song and Zhang 2011), and the ECHAM5-HAM model (Lohmann 2008).

In summary, assessing the role of cloud-aerosol interactions in the climate system requires studying these interactions in climate models to integrate them to global scales. Current macrophysical aspects of cloud-aerosol interactions in climate models remain rudimentary, however, with process studies suggesting a more nuanced picture of these processes than encompassed by current GCM parameterizations. In particular, a number of processes which may limit cooling by cloud-aerosol interactions are not well parameterized at present. High priority should be given to addressing the challenge of more realistically representing cloud-aerosol interactions in climate models.

6 What Should We Do Next?

A key obstacle to better understanding aerosol indirect effects is our poor ability to model cloud macrophysics. As noted in Sect. 5, high priority should be given to improving the realism with which cloud macrophysical processes governing cloud-aerosol interactions are represented in GCMs. Only recently have physically based approaches to aerosol activation been used in GCMs, and their usefulness is limited by incomplete representations of the full set of processes which govern cloud-aerosol interactions in GCMs and by the lack of resolution at the cloud scale. New approaches to parameterizing cloud macrophysics for both shallow and deep cloud systems are emerging. Evaluating and further developing these parameterizations will require extensive collaboration between GCM developers and scientists studying cloud macrophysics using process models, large-eddy and cloud-system simulation, and field observations. Satellite observations will also be critical in assessing cloud-aerosol interactions on a global scale.

More realistic physics has to be parameterized into both cloud resolving and global circulation models, and their results need to be validated against actual observations. A limiting factor in the present Earth observations is the ability to separate the aerosol from thermodynamic and meteorological effects. Doing so requires measuring of the CCN and cloud microphysical, thermodynamic and dynamic properties simultaneously from space at the necessary spatial and vertical resolution, which is in the order of 50–100 m. This requires a new generation of satellites with multi-spectral and multi-angle sensors. A way to do that is described by Rosenfeld et al. (2012b) in a proposed satellite mission. High resolution multi-angle imager (as in MISR) will be able to map the topography of the cloud surfaces and their vertical motions. A multi-spectral imager can map the microstructure and temperature of the cloud surfaces at various heights above cloud base, which will allow retrieving N_a from the vertical evolution of N_d in convective elements (Freud et al. 2011). The vertical development rate of the cloud surface just above its base will provide a measure of cloud-base updraft, which when combined with N_a yields the supersaturation and the CCN concentrations. Multi-angular near-infrared observations can also provide information on ice particle habit and microphysical history not obtainable at visible wavelengths (Sherwood 2005). Such a mission does not represent a major technological challenge, but requires the recognition to be of high priority in addressing the large uncertainties in RF that are the subject of this chapter.

Field campaigns are necessary for performing case studies of simultaneous measurements of the CCN and cloud microphysical, thermodynamic and dynamic properties in a way that will allow reaching closure of the aerosol, water and energy budgets, at a scale of a box of several hundred km on the side. This needs to done both in the shallow and deep clouds, as much as possible in similar meteorological conditions but with very different aerosols. The outlines for such campaigns are given by Andreae et al. (2009).

7 Summary

The aerosol indirect effect on radiative forcing (AIE) is the main source of uncertainty in the overall anthropogenic climate forcing and climate sensitivity. The uncertainties are summarized in Table 1. The AIE can be generally divided into negative forcing from low clouds, which is at least partially countered by positive forcing from deep and high clouds and by the IN effects on glaciating supercooled water clouds. The quantification of both opposite and possibly large effects is highly uncertain, to the extent that even the sign of the overall net effect cannot be determined with any degree of certainty.

Table 1 Aerosol cloud-mediated radiative forcing: status of current understanding

Process	Current understanding
Activation of liquid droplets	For aerosols with known solubility properties and size distributions, understanding is well-established
Primary nucleation of ice crystals	Although some ice nuclei have been identified, significant uncertainty remains as to the nucleating abilities of black carbon, biogenics, and mixtures
Aerosol size distributions for cloud condensation and ice nuclei	Size distributions can be modeled reasonably accurately in detailed process models, but considerable simplifications, the consequences of which are not fully understood, are required for computational efficiency in GCMs
Aerosol-induced changes in cloud regimes and organization	Conceptual and numerical models have identified basic issues. Field and satellite observations have been limited and will remain so in the absence of simultaneous characterization of dynamics, microphysics, and aerosols, enabling closure of aerosol budgets. GCM parameterizations have not explicitly been developed, and capabilities of current GCM parameterizations to capture these changes are likely to be severely limited
Aerosol-induced changes in cloud entrainment, dynamics, and microphysics	Large-eddy simulations with advanced microphysics have identified key issues. Observations have been limited. Current GCM parameterizations are very limited regarding these processes, but multi-variate probability distribution functions with dynamics have been able to capture entrainment-aerosol interactions
Aerosol-induced changes in dynamics, radiation, and microphysics of deep convection	Cloud-system models have identified basic processes Observations have been limited. Most GCM cumulus parameterizations lack the physical basis to simulate these processes, but GCM cumulus parameterizations with vertical velocities and advanced microphysics have recently been developed

Aerosols added to low clouds generally incur negative radiative forcing, because they can cause cloud brightening by three main mechanisms: redistributing the water in larger number of smaller drops; adding more cloud water, and increasing the cloud cover. Aerosols affect these components some times in harmony and quite often in opposite ways that cancel each other at least partially. These processes can be highly non-linear, especially in precipitating clouds that added aerosol can inhibit from raining. This amounts to behavior of little overall sensitivity in most of the clouds, and hyper-sensitivity in some of the clouds where the processes become highly non linear with positive feedbacks, leading to very complicated and uneven AIE. Present observations assume a logarithmic relation between aerosol amount and cloud response. This hides the physics of much more complicated behavior, whose state-of–the-art understanding is described in this chapter. Key processes that are involved in the AIE are the precipitation-forming processes and the response of the cloud properties to the precipitation, which have profound impacts on the clouds and their environment. Some of these impacts are the formation of downdrafts and cold pools that alter the dynamics of the clouds, change the vertical diabatic heating profiles and the atmospheric instability, and scavenging the aerosols that affect the clouds at the first place. Process models at high resolution (LES) have reached very recently to the development stage that they can capture much of this complicated behavior, but the implementation into a GCM has been rudimentary due to severe computational limitations and the present state of cloud and aerosol parameterizations in GCMs. The latter deficiencies are an active research area at present.

Aerosols added to deep clouds generally incur positive radiative forcing, where to the effects that are operative in low clouds (cloud drop size, cloud water path and cloud cover) are added the effects of cloud top cooling, expanding, and detraining vapor to the upper troposphere and lower stratosphere. The latter three factors generally incur positive radiative forcing. The level of scientific understanding of the AIE on deep clouds is even lower than for the shallow clouds, as the deep clouds are much more complicated, where mixed phase and ice processes play an important role. Process models still have a major void in the knowledge in mixed-phase and ice processes, for both layer and deep convective clouds, both low and high level, in the arctic and lower latitudes. Respectively, the parameterization of these processes for GCMs is further away than for the low clouds.

Future efforts must address the AIE of both shallow and deep clouds for obtaining the net effect, which is required so much for quantifying the anthropogenic climate forcing, climate sensitivity and climate predictions. Furthermore, the cooling occurs mainly over the subtropical highs and migratory anticyclones over ocean, whereas the warming occurs mainly at the areas of deep tropical convection. The spatial separation can propel atmospheric circulation systems that would modify the weather patterns at all scales and the hydrological cycle. Therefore, the AIE must be quantified correctly not only for understanding climate, but also for improving weather and precipitation forecasts.

As a limiting factor in our understanding and quantification of the weather-forming processes and its integration into the climate system, we recommend coordinated field campaigns and satellite missions for addressing this problem, with the objective to describe and parameterize correctly these complex processes, and to measure these processes from space and quantify their effects at a global coverage and climate time scales. Present day satellite missions (CLOUDSAT, CALIPSO, GPM) focus at measuring the precipitation and large cloud particles and aerosols, but lack the critical measurements of CCN and detailed cloud microstructure. An example of a proposed satellite mission that is being designed to address the issues described here is given by Rennó et al. (2013). An example of the concept of field campaigns that are designed to address is issues is given by the Aerosols-Clouds-Precipitation-Climate initiative (Andreae et al. 2009), which provides the template for the design of a closure box experiment for quantifying all the energy and mass fluxes within a region of several 100 km on the side. The recommendations are summarized in Table 2.

This position chapter can be summarized in the following points:

1. While many of the clouds have little sensitivity, some of the clouds are hypersensitive, especially when the mechanism of regime change is involved.
2. The sign of the effects are of opposite signs for different kinds of clouds and aerosols.

Table 2 Aerosol cloud-mediated radiative forcing: key uncertainties and recommendations for increased understanding

Process	Issues and uncertainties	Recommendations
Activation of liquid drops	Solubility properties of organics	Characterization and lab studies
Primary nucleation of ice Crystals	Identification of ice nuclei, especially roles of biogenics	Field and lab studies of nuclei candidates
Aerosol size distributions for cloud condensation and ice nuclei	Accurate, computationally efficient parameterizations, especially for GCMs	Interactions between process-level studies and GCM development
Aerosol-induced changes in cloud regimes and organization	Process-level understanding, parameterization for GCMs	Model development; analysis of satellite and field observations to evaluate conceptual models, numerical models, and parameterizations
Aerosol-induced changes in cloud entrainment, dynamics, and microphysics	Improved process-level understanding, parameterization for GCMs	Interactions between process- level studies and GCM development; analysis of field and satellite observations
Aerosol-induced changes in dynamics, radiation, and microphysics of deep convection	Convective invigoration, changes in UTL clouds and vapor, parameterization for GCMs	Interactions between process- level studies and GCM development; closure box experiment at a regional scale; analysis of field and satellite observations

3. We have little quantitative knowledge on the AIE of any of these cloud and aerosol types.
4. We have even much less knowledge on the combined effect, even as far as its sign
5. We propose certain ways to address it.

Finally, we have shown here that the recently acquired additional knowledge actually increased the uncertainty bar in the chart of the radiative forcing, while everyone strives to reduce it. How large is this uncertainty? Do we know now all what we should know that we don't know yet? When we will be there the uncertainty range will peak, and start to be reduced from there on.

References

Abdul-Razzak H, Ghan S (2000) A parameterization of aerosol activation. 2. Multiple aerosol types. J Geophys Res 105:6837–6844

Ackerley D, Booth BBB, Knight SHE, Highwood EJ, Frame DJ, Allen MR, Rowell DP (2011) Sensitivity of twentieth-century rainfall to sulfate aerosol and CO2 forcing. J Climate 24:4999–5014

Ackerman AS, Toon OB, Hobbs PV (1993) Dissipation of marine stratiform clouds and collapse of the marine boundary layer due to the depletion of cloud condensation nuclei by clouds. Science 262:226–229

Ackerman AS, Toon OB, Hobbs PV (1995) Numerical modeling of ship tracks produced by injections of cloud condensation nuclei into marine stratiform clouds. J Geophys Res 100:7121–7133

Ackerman AS, Kirkpatrick MP, Stevens DE, Toon OB (2004) The impact of humidity above stratiform clouds on indirect aerosol climate forcing. Nature 432:1014–1017

Agee EM, Chen TS, Dowell KE (1973) A review of mesoscale cellular convection. Bull Amer Meteorol Soc 54:1004–1012

Albrecht BA (1989) Aerosols, cloud microphysics and fractional cloudiness. Science 245:1227–1230

Allen RJ, Sherwood SC (2011) The impact of natural versus anthropogenic aerosols on atmospheric circulation in the Community Atmosphere Model. Clim Dyn 36:1959–1978

Allen RJ, Sherwood SC, Norris JR, Zender CS (2012) Recent Northern Hemisphere tropical expansion primarily driven by black carbon and tropospheric ozone. Nature 485:350–354. doi:10.1038/nature11097

Anderson TL, Charlson RJ, Schwartz SE, Knutti R, Boucher O, Rodhe H, Heintzenberg J (2003) Climate forcing by aerosols–a hazy picture. Science 300:1103–1104

Andreae MO (2009) Correlation between cloud condensation nuclei concentration and aerosol optical thickness in remote and polluted regions. Atmos Chem Phys 9:543–556

Andreae MO, Rosenfeld D, Artaxo P, Costa AA, Frank GP, Longo KM, Silva-Dias MAF (2004) Smoking rain clouds over the Amazon. Science 303:1337–1342

Andreae MO, Stevens B, Feingold G, Fuzzi S, Kulmala M, Lau WK, Lohmann U, Rosenfeld D, Siebesma P (2009) Aerosols, clouds, precipitation and climate (ACPC) – science plan & implementation strategy. http://www.ileaps.org/index.php?option=com_phocadownload&view=category&id=8:p

Arking A, Childs JD (1985) Retrieval of cloud cover parameters from multispectral satellite images. J Clim Appl Meteor 24:322–333

Baker MB, Charlson RJ (1990) Bistability of CCN concentrations and thermodynamics in the cloud-topped boundary layer. Nature 345:142–145

Bell TL, Rosenfeld D, Kim KM, Yoo JM, Lee MI, Hahnenberger M (2008) Midweek increase in U.S. summer rain and storm heights suggests air pollution invigorates rainstorms. J Geophys Res 113:D02209. doi:10.1029/2007JD008623

Bell TL, Rosenfeld D, Kim KM (2009) Weekly cycle of lightning: evidence of storm invigoration by pollution. Geophys Res Lett 36:L23805. doi:10.1029/2009GL040915

Bennartz R (2007) Global assessment of marine boundary layer cloud droplet number concentration from satellite. J Geophys Res 112:D02201. doi:10.1029/2006JD007547

Berner AH, Bretherton CS, Wood R (2011) Large-eddy simulation of mesoscale dynamics and entrainment around a pocket of open cells observed in VOCALS-REx RF06. Atmos Chem Phys 11:10525–10540. doi:10.5194/acp-11-10525-2011

Bollasina MA, Ming Y, Ramaswamy V (2011) Anthropogenic aerosols and the weakening of the South Asian summer monsoon. Science 334:502–505

Booth BBB, Dunstone NJ, Halloran PR, Andrews Bellouin TN (2012) Aerosols implicated as a prime driver of twentieth-century North Atlantic climate variability. Nature 484:228–232. doi:10.1038/nature10946

Brenguier JL, Pawlowska H, Schuller L, Preusker R, Fischer J, Fouquart Y (2000) Radiative properties of boundary layer clouds: droplet effective radius versus number concentration. J Atmos Sci 57:803–821

Bretherton CS, Blossey P, Uchida J (2007) Cloud droplet sedimentation, entrainment efficiency, and subtropical stratocumulus albedo. Geophys Res Lett 34(3):L03813

Bretherton CS, George R, Wood R, Allen G, Leon D, Albrecht B (2010a) Southeast Pacific stratocumulus clouds, precipitation and boundary layer structure sampled along 20S during VOCALS-REx. Atmos Chem Phys 10:15 921–15 962

Bretherton CS, Uchida J, Blossey PN (2010b) Slow manifolds and multiple equilibria in Stratocumulus-Capped boundary layers. J Adv Model Earth Syst 2(12)

Brohan P, Kennedy J, Harris I, Tett S, Jones P (2006) Uncertainty estimates in regional and global observed temperature changes: a new dataset from 1850. J Geophys Res 111:D12106. doi:10.1029/2005JD006 548

Burkhardt U, Kärcher B (2011) Global radiative forcing from contrail cirrus. Nat Clim Change 1:54–58

Chang CY, Chiang JCH, Wehner MF, Friedman AR, Ruedy R (2011) Sulfate aerosol control of tropical Atlantic climate over the twentieth century. J Climate 24:2540–2555

Chen B, Yin Y (2011) Modeling the impact of aerosols on tropical overshooting thunderstorms and stratospheric water vapor. J Geophys Res 116(D19):D19203

Chen TC, Xue L, Lebo ZJ, Wang H, Rasmussen RM, Seinfeld JH (2011) A comprehensive numerical study of aerosol-cloud-precipitation interactions in marine stratocumulus. Atmos Chem Phys 11:9749–9769. doi:10.5194/acp-11-9749-2011

Christensen MW, Stephens GL (2011) Microphysical and macrophysical responses of marine stratocumulus polluted by underlying ships: evidence of cloud deepening. J Geophys Res 116:D03201

Clarke AD, Uehara T, Porter JN (1997) Atmospheric nuclei and related aerosol fields over the Atlantic: clean subsiding air and continental pollution during ASTEX. J Geophys Res 102(D21):25,281–25,292

Coakley JA, Walsh CD (2002) Limits to the aerosol indirect radiative effect derived from observations of ship tracks. J Atmos Sci 59:668–680

Cotton WR, Zhang H, McFarquhar GM, Saleeby SM (2007) Should we consider polluting hurricanes to reduce their intensity? J Weath Modif 39:70–73

Donner L (1993) A cumulus parameterization including mass fluxes, vertical momentum dynamics, and mesoscale effects. J Atmos Sci 50:889–906

Donner L, Seman C, Hemler R (2001) A cumulus parameterization including mass fluxes, convective vertical velocities, and mesoscale effects: thermodynamic and hydrological aspects in a general circulation model. J Climate 14:3444–3463

Donner LJ, Wyman BL, Hemler RS, Horowitz LW, Ming Y, Zhao M, Golaz JC, Ginoux P, Lin SJ, Schwarzkopf MD, Austin J, Alaka G, Cooke WF, Delworth TL, Freidenreich SM, Gordon CT, Griffies SM, Held IM, Hurlin WJ, Klein SA, Knutson TR, Langenhorst AR, Lee HC, Lin Y, Magi BI, Malyshev SL, Milly PCD, Naik V, Nath MJ, Pincus R, Ploshay JJ, Ramaswamy V, Seman CJ, Shevliakova E, Sirutis JJ, Stern WF, Stouffer RJ, Wilson RJ, Winton M, Wittenberg AT, Zeng F (2011) The dynamical core, physical parameterizations, and basic simulation characteristics of the atmospheric component of the GFDL global coupled model CM3. J Climate 24:3484–3519. doi:10.1175/2011JCLI3955.1

Fan J, Zhang R, Li G, Tao W-K (2007) Effects of aerosols and relative humidity on cumulus clouds. J Geophys Res 112:D14204. doi:10.1029/2006JD008136

Fan J, Yuan T, Comstock JM, Ghan S, Khain A, Leung LR, Li Z, Martins VJ, Ovchinnikov M (2009) Dominant role by vertical wind shear in regulating aerosol effects on deep convective clouds. J Geophys Res 114:D22206. doi:10.1029/2009JD012352

Fan J, Comstock JM, Ovchinnikov M, McFarlane SA, McFarquhar G, Allen G (2010) Tropical anvil characteristics and water vapor of the tropical tropopause layer: impact of heterogeneous and homogeneous freezing parameterizations. J Geophys Res 115:D12201. doi:10.1029/2009JD012696

Fan J, Rosenfeld D, Ding Y, Leung LR, Li Z (2012) Potential aerosol indirect effects on atmospheric circulation and radiative forcing through deep convection. Geophys Res Lett 39(9):L09806

Feingold G, Siebert H (2009) Cloud-aerosol interactions from the micro to the cloud scale. In: Heintzenberg J, Charlson R (eds) Perturbed clouds in the climate system: their relationship to energy balance, atmospheric dynamics, and precipitation. MIT Press, Cambridge, MA

Feingold G, Stevens B, Cotton WR, Frisch AS (1996) On the relationship between drop in-cloud residence time and drizzle production in stratocumulus clouds. J Atmos Sci 53:1108–1122

Feingold G, Koren I, Wang H, Xue H, Brewer WA (2010) Precipitation-generated oscillations in open cellular cloud fields. Nature 466:849–852. doi:10.1038/nature09314

Forest CE, Stone PH, Sokolov AP (2006) Estimated PDFs of climate system properties including natural and anthropogenic forcings. Geophys Res Lett 33:L01705. doi:10.1029/2005GL023977
Forster PMdF, Shine KP (1999) Stratospheric water vapor changes as a possible contributor to observed stratospheric cooling. Geophys Res Lett 26:3309–3312
Forster PMdF, Shine KP (2002) Assessing the climate impact of trends in stratospheric water vapor. Geophys Res Lett 29(6):1086. doi:10.1029/2001GL013909
Forster P, Ramaswamy V, Artaxo P, Berntsen T, Betts R, Fahey DW, Haywood J, Lean J, Lowe DC, Myhre G, Nganga J, Prinn R, Raga G, Schulz M, Van Dorland R (2007) Radiative forcing of climate change. In: Solomon S, Qin D, Manning M, Chen Z, Marquis M, Averyt KB, Tignor M, Miller HL (eds) Climate change 2007: the physical science basis. Contribution of Working Group I to the Fourth Assessment Report of the Intergovernmental Panel on Climate Change. Cambridge University Press, Cambridge/New York, pp 129–234
Freud E, Rosenfeld D (2012) Linear relation between convective cloud drop number concentration and depth for rain initiation. J Geophys Res 117:D02207. doi:10.1029/2011JD016457
Freud E, Rosenfeld D, Kulkarni JR (2011) Resolving both entrainment-mixing and number of activated CCN in deep convective clouds. Atmos Chem Phys 11:12887–12900. doi:10.5194/acp-11-12887-2011
Fridlind AM, Ackerman AS, Jensen EJ, Heymsfield AJ, Poellot MR, Stevens DE, Wang D, Miloshevich LM, Baumgardner D, Lawson RP, Wilson JC, Flagan RC, Seinfeld JH, Jonsson HH, VanReken TM, Varutbangkul V, Rissman TA (2004) Evidence for the predominance of mid-tropospheric aerosols as subtropical anvil cloud nuclei. Science 304:718–722
Fridlind AM, Ackerman AS, McFarquhar G, Zhang G, Poellot MR, DeMott PJ, Prenni AJ, Heymsfield AJ (2007) Ice properties of single-layer stratocumulus during the Mixed-Phase Arctic Cloud Experiment: 2. Model results. J Geophys Res 112:D24202. doi:10.1029/2007JD008646
Fridlind AM, van Diedenhoven B, Ackerman AS, Avramov A, Mrowiec A, Morrison H, Zuidema P, Shupe MD (2012) A FIRE-ACE/SHEBA case study of mixed-phase arctic boundary layer clouds: entrainment rate limitations on rapid primary ice nucleation processes. J Atmos Sci 69:365–389
George RC, Wood R (2010) Subseasonal variability of low cloud radiative properties over the southeast Pacific Ocean. Atmos Chem Phys 10:4047–4063. doi:10.5194/acp-10-4047-2010
Gerber H (1996) Microphysics of marine stratocumulus clouds with two drizzle modes. J Atmos Sci 53(12):1649–1662
Golaz JC, Salzmann M, Donner L, Horowitz L, Ming Y, Zhao M (2011) Sensitivity of the aerosol indirect effect to subgrid variability in the cloud parameterization of the GFDL atmosphere general circulation model AM3. J Climate 24:3145–3160. doi:10.1175/2010JCLI3945.1
Goren T, Rosenfeld D (2012) Satellite observations of ship emission induced transitions from broken to closed cell marine stratocumulus over large areas. J Geophys Res. doi:10.1029/2012JD017981
Guo H, Golaz JC, Donner LJ, Larson V, Schanen D, Griffin B (2010) Multivariate probability density functions for cloud droplet activation in large-scale models: single column tests. Geosci Model Dev 3:475–486. doi:10.5194/gmd-3-475-2010
Guo H, Golaz JC, Donner LJ (2011) Aerosol effects on stratocumulus water paths in a PDF-based parameterization. Geophys Res Lett 38(17):L17807. doi:10.1029/2011GL048 611
Hansen J, Sato M, Ruedy R, Nazarenko L, Lacis A, Schmidt GA, Russell G, Aleinov I, Bauer M, Bauer S, Bell N, Cairns B, Canuto V, Chandler M, Cheng Y, Del Genio A, Faluvegi G, Fleming E, Friend A, Hall T, Jackman C, Kelley M, Kiang N, Koch D, Lean J, Lerner J, Lo K, Menon S, Miller R, Minnis P, Novakov T, Oinas V, Perlwitz J, Perlwitz J, Rind D, Romanou A, Shindell D, Stone P, Sun S, Tausnev N, Thresher D, Wielicki B, Wong T, Yao M, Zhang S (2005) Efficacy of climate forcings. J Geophys Res 110(D18):D18104. doi:10.1029/2005JD005776
Hegg DA, Covert DS, Jonsson HH, Woods RK (2011) A simple relationship between cloud drop number concentration and precursor aerosol concentration for the regions of earth's large marine stratocumulus decks. Atmos Chem Phys Discuss 11:28663–28687. doi:10.5194/acpd-11-28663-2011

Hendricks J, Karcher B, Lohmann U (2011) Effects of ice nuclei on cirrus clouds in a global climate model. J Geophys Res 116:D18206

Hill AA, Dobbie S, Yin Y (2008) The impact of aerosols on non-precipitating marine stratocumulus. Q J R Meteorol Soc 134:1143–1154. doi:10.1002/qj.278

Hill AA, Feingold G, Jiang H (2009) The influence of entrainment and mixing assumption on aerosol-cloud interactions in marine stratocumulus. J Atmos Sci 66:1450–1464

Hogan RJ, Behera MD, O'Connor EJ, Illingworth AJ (2004) Estimate of the global distribution of stratiform supercooled liquid water clouds using the LITE lidar. Geophys Res Lett 31:L05106. doi:10.1029/2003GL018977

Hu Y, Rodier S, Xu K, Sun W, Huang J, Lin B, Zhai P, Josset D (2010) Occurrence, liquid water content, and fraction of supercooled water clouds from combined CALIOP/IIR/MODIS measurements. J Geophys Res 115:D00H34. doi:10.1029/2009JD012384

IPCC (2007) Summary for policymakers. In: Solomon S, Qin D, Manning M, Chen Z, Marquis M, Averyt KB, Tignor M, Miller HL (eds) Climate change 2007. The physical science basis. Contribution of Working Group I to the Fourth Assessment Report of the Intergovernmental Panel on Climate Change. Cambridge University Press, Cambridge

Isaksen ISA, Granier C, Myhre G, Berntsen TK, Dalsøren SB, Gauss M, Klimont Z, Benestad R, Bousquet P, Collins W, Coxk T, Eyring V, Fowlerm D, Fuzzi S, Jöckel P, Laj P, Lohmann U, Maione M, Monks P, Prevot ASH, Raes F, Richter A, Rognerud B, Schulz M, Shindell D, Stevenson DS, Storelvmo T, Wang WC, van Weele M, Wild M, Wuebbles D (2009) Atmospheric composition change: climate-chemistry interactions. Atmos Environ 43:5138–5192

Jensen EJ, Ackerman AS (2006) Homogeneous aerosol freezing in the tops of high-altitude tropical cumulonimbus clouds. Geophys Res Lett 33:L08802. doi:10.1029/2005GL024 928

Jiang H, Feingold G, Cotton WR (2002) A modeling study of entrainment of cloud condensation nuclei into the marine boundary layer during ASTEX. J Geophys Res 107(D24):4813. doi:10.1029/2001JD001502

Jiang H, Feingold G, Koren I (2009a) Effect of aerosol on trade cumulus cloud morphology. J Geophys Res 114:D11209. doi:10.1029/2009JD011750

Jiang JH, Su H, Massie ST, Colarco PR, Schoeberl MR, Platnick S (2009b) Aerosol-CO relationship and aerosol effect on ice cloud particle size: analyses from Aura Microwave Limb Sounder and Aqua Moderate Resolution Imaging Spectroradiometer observations. J Geophys ResAtmos 114:D20207

Jiang JH, Su H, Zhai C, Massie ST, Schoberl M, Colarco PR, Platnick S, Gu Y, Liou KN (2011) Influence of convection and aerosol pollution on ice cloud particle effective radius. Atmos Chem Phys 11:457–463

Kaufman YJ, Koren I, Remer L, Rosenfeld D, Rudich I (2005) The effect of smoke, dust, and pollution aerosol on shallow cloud development over the Atlantic Ocean. Proc Natl Acad Sci 102:11207–11212

Khain A, Lynn B (2011) Simulation of tropical cyclones using a mesoscale model with spectral bin microphysics. In: Lupo AR (ed) Recent hurricane research – climate, dynamics, and societal impacts. Intech Open Access Publisher, Rijeka, pp 197–227

Khain AP, Pokrovsky A (2004) Simulation of effects of atmospheric aerosols on deep turbulent convective clouds using a spectral microphysics mixed-phase cumulus cloud model, Part II: Sensitivity study. J Atmos Sci 61:2983–3001

Khain AP, Pokrovsky A, Pinsky M, Seifert A, Phillips V (2004) Effects of atmospheric aerosols on deep convective clouds as seen from simulations using a spectral microphysics mixed-phase cumulus cloud model Part 1: Model description. J Atmos Sci 61:2963–2982

Khain A, Rosenfeld D, Pokrovsky A (2005) Aerosol impact on the dynamics and microphysics of deep convective clouds. Q J R Meteorol Soc 131:1

Khain AP, BenMoshe N, Pokrovsky A (2008a) Factors determining the impact of aerosols on surface precipitation from clouds: an attempt at classification. J Atmos Sci 65:1721

Khain A, Cohen N, Lynn B, Pokrovsky A (2008b) Possible aerosol effects on lightning activity and structure of hurricanes. J Atmos Sci 65:3652–3667

Khain A, Lynn B, Dudhia J (2010) Aerosol effects on intensity of landfalling hurricanes as seen from simulations with the WRF model with spectral bin microphysics. J Atmos Sci 67:365–384

Kiehl JT (2007) Twentieth century climate model response and climate sensitivity. Geophys Res Lett 34:L22710. doi:10.1029/2007GL031383

Knutson TR, Delworth TL, Dixon KW, Held IM, Lu J, Ramaswamy V, Schwarzkopf MD, Stenchikov G, Stouffer RJ (2006) Assessment of twentieth-century regional surface trends using the GFDL CM2 coupled models. J Climate 19:1624–1651

Konwar M, Maheskumar RS, Kulkarni JR, Freud E, Goswami BN, Rosenfeld D (2012) Aerosol control on depth of warm rain in convective clouds. J Geophys Res 117(D13):D13204

Koren I, Feingold G (2011) Aerosol–cloud–precipitation system as a predator-prey problem. PNAS 108:12,227–12,232. doi:10.1073/pnas.1101777108

Koren I, Kaufman YJ, Rosenfeld D, Remer LA, Rudich Y (2005) Aerosol invigoration and restructuring of Atlantic convective clouds. Geophys Res Lett 32:L14828. doi:10.1029/2005GL023187

Koren I, Martins JV, Remer LA, Afargan H (2008) Smoke invigoration versus inhibition of clouds over the Amazon. Science 321:946. doi:10.1126/science.1159185

Koren I, Feingold G, Remer LA (2010a) The invigoration of deep convective clouds over the Atlantic: aerosol effect, meteorology or retrieval artifact? Atmos Chem Phys 10:8855–8872. doi:10.5194/acp-10-8855-2010

Koren I, Remer LA, Altaratz O, Martins JV, Davidi A (2010b) Aerosol-induced changes of convective cloud anvils produce strong climate warming. Atmos Chem Phys 10:5001–5010

Kostinski AB (2008) Drizzle rates versus cloud depths for marine stratocumuli. Environ Res Lett 3(4):045019

Kubar TL, Hartmann DL, Wood R (2009) Understanding the importance of microphysics and macrophysics for warm rain in marine low clouds. Part I: Satellite observations. J Atmos Sci 66:2953–2972

Langenberg H (2011) Triggered lightning. In "news and views". Nat Geosci 4:140

Lebsock MD, Stephens GL, Kummerow C (2008) Multisensor satellite observations of aerosol effects on warm clouds. J Geophys Res 113(D15):D15205. doi:10.1029/2008jd009876

Lebsock MD, L'Ecuyer TS, Stephens GL (2011) Detecting the ratio of rain and cloud water in low-latitude shallow marine clouds. J Appl Meteorol Clim 50:419–432

Lee SS, Donner LJ, Phillips VTJ (2009) Sensitivity of aerosol and cloud effects on radiation to cloud types: comparison between deep convective clouds and wam stratiform clouds over one-day period. Atmos Chem Phys 9:2555–2575

Lee SS, Donner LJ, Penner JE (2010) Thunderstorm and stratocumulus: how does their contrasting morphology affect their interactions with aerosols? Atmos Chem Phys 10:6819–6837. doi:10.5194/acp-10-6819-2010

Li Z, Niu F, Fan J, Liu Y, Rosenfeld D, Ding Y (2011) Long-term impacts of aerosols on the vertical development of clouds and precipitation. Nat Geosci 4(12):888–894. doi:10.1038/ngeo1313

Lin H, Leaitch W (1997) Development of an in-cloud aerosol activation for climate modeling. In: Proceedings of the WMO Workshop on Measurement of cloud properties for forecasts of weather, air quality, and climate, Geneva, pp 328–335

Lin JC, Matsui T, Pielke RA Sr, Kummerow C (2006) Effects of biomass-burning-derived aerosols on precipitation and clouds in the Amazon Basin: a satellite-based empirical study. J Geophys Res 111:D19204

Lindsey DT, Fromm M (2008) Evidence of the cloud lifetime effect from wildfire-induced thunderstorms. Geophys Res Lett 35:L22809. doi:10.1029/2008GL035680

Liu X, Penner JE, Wang M (2009) Influence of anthropogenic sulfate and black carbon on upper tropospheric clouds in the NCAR CAM3 model coupled to the IMPACT global aerosol model. J Geophys ResAtmos 114:D03204

Liu X, Easter RC, Ghan SJ, Zaveri R, Rasch P, Shi X, Lamarque J-F, Gettelman A, Morrison H, Vitt F, Conley A, Park S, Neale R, Hannay C, Ekman AML, Hess P, Mahowald N, Collins W, Iacono MJ, Bretherton CS, Flanner MG, Mitchell D (2012) Toward a minimal representation of aerosols in climate models: description and evaluation in the Community Atmosphere Model CAM5. Geosci Model Dev 5:709–739. doi:10.5194/gmd-5-709-2012

Lohmann U (2008) Global anthropogenic aerosol effects on convective clouds in ECHAM5-HAM. Atmos Chem Phys 8(7):2115–2131

Lu M, Seinfeld JH (2006) Effect of aerosol number concentration on cloud droplet dispersion: a large-eddy simulation study and implications for aerosol indirect forcing. J Geophys Res 111:D02207. doi:10.1029/2005JD006419

Martin GM, Johnson DW, Spice A (1994) The measurement and parameterization of effective radius of droplets in warm stratocumulus clouds. J Atmos Sci 51:1823–1842

Matheson MA, Coakley JA, Tahnk WR (2006) Multiyear Advanced Very High Resolution Radiometer observations of summertime stratocumulus collocated with aerosols in the northeastern Atlantic. J Geophys Res 111:D15206. doi:10.1029/2005JD006890

Mauritsen T, Sedlar J, Tjernström M, Leck C, Martin M, Shupe M, Sjogren S, Sierau B, Persson POG, Brooks IM, Swietlicki E (2011) An Arctic CCN-limited cloud-aerosol regime. Atmos Chem Phys 11:165–173. doi:10.5194/acp-11-165-2011

McComiskey A, Graham Feingold G (2008) Quantifying error in the radiative forcing of the first aerosol indirect effect. Geophys Res Lett 35:L02810. doi:10.1029/2007GL032667

McFarquhar GM, Ghan S, Verlinde J, Korole A, Strapp JW, Schmid B, Tomlinson JM, Wolde M, Brooks SD, Cziczo D, Dubey MK, Fan J, Flynn C, Gultepe I, Hubbe J, Gilles MK, Laskin A, Lawson P, Leaitch WR, Liu P, Liu X, Lubin D, Mazzoleni C, Macdonald AM, Moffet RC, Morrison H, Ovchinnikov M, Shupe MD, Turne DD, Xie S, Zelenyuk A, Bae K, Freer M, Glen A (2011) Indirect and semi-direct aerosol campaign. Bull Amer Meteorol Soc 92:183–201

Menon S, Del Genio AD, Kaufman YJ, Bennartz R, Koch D, Loeb N, Orlikowski D (2008) Analyzing signatures of aerosol-cloud interactions from satellite retrievals and the GISS GCM to constrain the aerosol indirect effect. J Geophys Res 113:D14S22. doi:10.1029/2007JD009442

Ming Y, Ramaswamy V (2011) A model investigation of aerosol-induced changes in tropical circulation. J Climate 24:5125–5133. doi:10.1175/2011JCLI4108.1

Ming Y, Ramaswamy V, Donner LJ, Phillips VTJ, Klein SA, Ginoux PA, Horowitz LW (2007) Modeling the interactions between aerosols and liquid water clouds with a self-consistent cloud scheme in a general circulation model. J Atmos Sci 64:1189–1209

Ming Y, Ramaswamy V, Chen G (2011) A model investigation of aerosol-induced changes in boreal winter extratropical circulation. J Climate 24:6077–6091. doi:10.1175/2011JCLI4111.1

Molinie J, Pontikis C (1995) A climatological study of tropical thunderstorm clouds and lightning frequencies on the French Guyana coast. Geophys Res Lett 22:1085–1088

Morrison H, Grabowski WW (2011) Cloud-system resolving model simulations of aerosol indirect effects on tropical deep convection and its thermodynamic environment. Atmos Chem Phys 11(20):10503–10523

Morrison H, Zuidema P, Ackerman AS, Avramov A, de Boer G, Fan J, Fridlind AM, Hashino T, Harrington JY, Luo Y, Ovchinnikov M, Shipway B (2011) Intercomparison of cloud model simulations of Arctic mixed-phase boundary layer clouds observed during SHEBA/FIRE-ACE. J Adv Model Earth Syst 3:M06003. doi:10.1029/2011MS000066, 23 pp

Morrison H, de Boer G, Feingold G, Harrington J, Shupe MD, Sulia K (2012) Resilience of persistent Arctic mixed-phase clouds. Nat Geosci 5:11–17. doi:10.1038/NGE01332

Morrison AE, Siems ST, Manton MJ (2011) A three-year climatology of cloud-top phase over the Southern Ocean and North Pacific. J Climate 24(9):2405–2418

Murphy DM, Solomon S, Portmann RW, Rosenlof KH, Forster PM, Wong T (2009) An observationally based energy balance for the Earth since 1950. J Geophys Res 114:D17107

Myhre G (2009) Consistency between satellite-derived and modeled estimates of the direct aerosol effect. Science 325:187–190

Myhre G, Nilsen JS, Gulstad L, Shine KP, Rognerud B, Isaksen ISA (2007a) Radiative forcing due to stratospheric water vapour from CH(4) oxidation. Geophys Res Lett 34(1):L01807

Myhre G, Stordal F, Johnsrud M, Kaufman YJ, Rosenfeld D, Storelvmo T, Kristjansson JE, Berntsen TK, Myhre A, Isaksen ISA (2007b) Aerosol-cloud interaction inferred from MODIS satellite data and global aerosol models. Atmos Chem Phys 7:3081–3101

Nielsen JK, Foster M, Heidinger A (2011) Tropical stratospheric cloud climatology from the PATMOS-x dataset: an assessment of convec- tive contributions to stratospheric water. Geophys Res Lett 38:L18801

Notholt J, Luo BP, Fueglistaler S, Weisenstein D, Rex M, Lawrence MG, Bingemer H, Wohltmann I, Corti T, Warneke T, von Kuhlmann R, Peter T (2005) Influence of tropospheric SO2 emissions on particle formation and the stratospheric humidity. Geophys Res Lett 32:L07810. doi: 10.1029/2004GL022159

Notholt J, Toon GC, Fueglistaler S, Wennberg PO, Irion FW, Mc-Carthy M, Scharringhausen M, Rhee TS, Kleinboehl A, Velazco V (2010) Trend in ice moistening the stratosphere – constraints from isotope data of water and methane. Atmos Chem Phys 10(1):201–207

Penner JE, Chen Y, Wang M, Liu X (2009) Possible influence of anthropogenic aerosols on cirrus clouds and anthropogenic forcing. Atmos Chem Phys 9:879–896

Penner JE, Xu L, Wang M (2011) Satellite methods underestimate indirect climate forcing by aerosols. Proc Nat Acad Sci 108:13404–13408. doi:10.1073/pnas.1018526108

Petters MD, Snider JR, Stevens B, Vali G, Faloona I, Russell L (2006) Accumulation mode aerosol, pockets of open cells, and particle nucleation in the remote subtropical pacific marine boundary layer. J Geophys Res 111:D02206

Phillips VTJ, Choularton TW, Blyth AM, Latham J (2002) The influence of aerosol concentrations on the glaciation and precipitation of a cumulus cloud. Q J R Meteorol Soc 128:951–971

Platnick S, Twomey S (1994) Determining the susceptibility of cloud albedo to changes in droplet concentration with the Advanced Very High Resolution Radiometer. J Appl Meteorol 33:334–347

Pöschl U, Martin ST, Sinha B, Chen Q, Gunthe SS, Huffman JA, Borrmann S, Farmer DK, Garland RM, Helas G, Jimenez JL, King SM, Manzi A, Mikhailov E, Pauliquevis T, Petters MD, Prenni AJ, Roldin P, Rose D, Schneider J, Su H, Zorn SR, Artaxo P, Andreae MO (2010) Rainforest aerosols as biogenic nuclei of clouds and precipitation in the Amazon. Science 329:1513–1516

Prenni AJ, Harrington JY, Tjernstrom M, DeMott PJ, Avramov A, Long CN, Kreidenweis SM, Olsson PQ, Verlinde J (2007) Can ice-nucleating aerosols affect arctic seasonal climate? Bull Amer Meteorol Soc 88(4):541–550. doi:10.1175/BAMS-88-4-541

Quaas J, Ming Y, Menon S, Takemura T, Wang M, Penner JE, Gettelman A, Lohmann U, Bellouin N, Boucher O, Sayer AM, Thomas GE, McComiskey A, Feingold G, Hoose C, Kristjánsson JE, Liu X, Balkanski Y, Donner LJ, Ginoux PA, Stier P, Grandey B, Feichter J, Sednev I, Bauer SE, Koch D, Grainger RG, Kirkeåg A, Iversen T, Seland Ø, Easter R, Ghan SJ, Rasch PJ, Morrison H, Lamarque J-F, Iacono MJ, Kinne S, Schulz M (2009) Aerosol indirect effects general circulation model intercomparison and evaluation with satellite data. Atmos Chem Phys 9:8697–8717

Randall DA (1980) Conditional Entrainment into a stratocumulus layer with distributed radiative cooling. J Atmos Sci 37:148–159

Rangno AL, Hobbs PV (1991) Ice particle concentrations and precipitation development in small polar maritime cumuliform clouds. Q J R Meteorol Soc 117:207–241

Rennó NO, Williams E, Rosenfeld D, Fischer DG, Fischer J, Kremic T, Agrawal A, Andreae MO, Bierbaum R, Blakeslee R, Boerner A, Bowles N, Christian H, Cox A, Dunion J, Horvath A, Huang X, Khain A, Kinne S, Lemos MC, Penner JE, Pöschl U, Quaas J, Seran E, Stevens B, Walati T, Wagner T (2013) CHASER: an innovative satellite mission concept to measure the effects of aerosols on clouds and climate. Bull Amer Meteorol Soc 94:e-View. doi:10.1175/BAMS-D-11-00239

Rosenfeld D (2000) Suppression of rain and snow by Urban and industrial air pollution. Science 287(5459):1793–1796

Rosenfeld D, Bell TL (2011) Why do tornados and hailstorms rest on weekends? J Geophys Res 116:D20211. doi:10.1029/2011JD016214

Rosenfeld D, Lahav R, Khain AP, Pinsky M (2002) The role of sea-spray in cleansing air pollution over ocean via cloud processes. Science 297:1667–1670

Rosenfeld D, Kaufman Y, Koren I (2006a) Switching cloud cover and dynamical regimes from open to closed Benard cells in response to aerosols suppressing precipitation. Atmos Chem Phys 6:2503–2511
Rosenfeld D, Woodley WL, Krauss TW, Makitov V (2006b) The structure of severe convective storms in Mendoza, Argentina. J Appl Meteor 45:1261–1281, September 2006
Rosenfeld D, Fromm M, Trentmann J, Luderer G, Andreae MO, Servranckx R (2007a) The Chisholm firestorm: observed microstructure, precipitation and lightning activity of a pyro-cumulonimbus. Atmos Chem Phys 7:645–659
Rosenfeld D, Khain A, Lynn B, Woodley WL (2007b) Simulation of hurricane response to suppression of warm rain by sub-micron aerosols. Atmos Chem Phys 7:3411–3424
Rosenfeld D, Woodley WL, Axisa D, Freud E, Hudson JG, Givati A (2008a) Aircraft measurements of the impacts of pollution aerosols on clouds and precipitation over the Sierra Nevada. J Geophys Res 113:D15203. doi:10.1029/2007JD009544
Rosenfeld D, Lohmann U, Raga GB, O'Dowd CD, Kulmala M, Fuzzi S, Reissell A, Andreae MO (2008b) Flood or drought: how do aerosols affect precipitation? Science 321:1309–1313
Rosenfeld D, Clavner M, Nirel R (2011a) Pollution and dust aerosols modulating tropical cyclones intensities. Atmos Res 102:66–76
Rosenfeld D, Yu X, Liu G, Xu X, Zhu Y, Yue Z, Dai J, Dong Z, Dong Y, Peng Y (2011b) Glaciation temperatures of convective clouds ingesting desert dust, air pollution and smoke from forest fires. Geophys Res Lett 38:L21804. doi:10.1029/2011GL049423
Rosenfeld D, Woodley WL, Khain A, Cotton WR, Carrió G, Ginis I, Golden JH (2012a) Aerosol effects on microstructure and intensity of tropical cyclones. Bull Amer Meteorol Soc 93(7):987–1001
Rosenfeld D, Williams E, Andreae MO, Freud E, Pöschl U, Rennó NO (2012b) The scientific basis for a satellite mission to retrieve CCN concentrations and their impacts on convective clouds. Atmos Meas Tech 5:2039–2055. doi:10.5194/amt-5-2039-2012
Salzmann M, Ming Y, Golaz JC, Ginoux PA, Morrison H, Gettelman A, Krämer M, Donner LJ (2010) Two-moment bulk stratiform cloud microphysics in the GFDL AM3 GCM: description, evaluation, and sensitivity tests. Atmos Chem Phys 10:8037–8064
Sandu I, Brenguier J, Geoffroy O, Thouron O, Masson V (2008) Aerosol impacts on the diurnal cycle of marine stratocumulus. J Atmos Sci 65:2705–2718
Seifert A, Beheng KD (2006) A two-moment cloud microphysics parameterization for mixed-phase clouds. Part 2: Maritime vs. continental deep convective storms. Meteorol Atmos Phys 92:67–82
Sekiguchi M, Nakajima T, Suzuki K, Kawamoto K, Higurashi A, Rosenfeld D, Sano I, Mukai S (2003) A study of the direct and indirect effects of aerosols using global satellite datasets of aerosol and cloud parameters. J Geophys Res 108(D22):4699. doi:10.1029/2002JD003359
Sharon TM, Albrecht BA, Jonsson H, Minnis P, Khaiyer MM, VanReken TM, Seinfeld J, Flagan R (2006) Aerosol and cloud microphysical characteristics of rifts and gradients in maritime stratocumulus clouds. J Atmos Sci 63:983–997
Sherwood SC (2002a) Aerosols and ice particle size in tropical cumulonimbus. J Climate 15:1051–1063
Sherwood SC (2002b) A microphysical connection among biomass burning, cumulus clouds, and stratospheric moisture. Science 295:1272–1275
Sherwood SC (2005) Detection of faceted crystals in deep convective clouds via the antisolar peak. J Geophys Res 110:D14205
Sherwood SC, Phillips VTP, Wettlaufer JS (2006) Small ice crystals and the climatology of lightning. Geophys Res Lett 33:L05804. doi:10.1029/2005GL025242
Sherwood S, Alexander MJ, Brown AR, McFarlane NA, Gerber EP, Feingold G, Scaife A, Grabowski W (2013) Climate processes: clouds, aerosols and dynamics (B6). In: Asrar GR, Hurrell JW (eds) Climate science for serving society. Springer, Dordrecht, pp 73–103
Shupe MD, Matrosov SY, Uttal T (2006) Arctic mixed-phase cloud properties derived from surface-based sensors at SHEBA. J Atmos Sci 63:697–711
Small JD, Chuang PY, Feingold G, Jiang H (2009) Can aerosol decrease cloud lifetime? Geophys Res Lett 36:L16806. doi:10.1029/2009GL038888

Solomon S, Rosenlof KH, Portmann RW, Daniel JS, Davis SM, Sanford TJ, Plattner GK (2010) Contributions of stratospheric water vapor to decadal changes in the rate of global warming. Science 327(5970):1219–1223. doi:10.1126/science.1182488
Song X, Zhang G (2011) Microphysics parameterization for convective clouds in a global climate model: description and single-column model tests. J Geophys Res 116:D02201. doi:10.1029/2010JD014 833
Sorooshian A, Feingold G, Lebsock M, Jiang H, Stephens G (2009) On the precipitation susceptibility of clouds to aerosol perturbations. Geophys Res Lett 36(L13):803
Sorooshian A, Feingold G, Lebsock MD, Jiang H, Stephens G (2010) Deconstructing the precipitation susceptibility construct: improving methodology for aerosol-cloud-precipitation studies. J Geophys Res 115:D17201. doi:10.1029/2009JD013426
Stevens B, Feingold G (2009) Untangling aerosol effects on clouds and precipitation in a buffered system. Nature 461:607. doi:10.1038/nature08281
Stevens B, Vali G, Comstock K, van Zanten MC, Austin PH, Bretherton CS, Lenschow DH (2005) Pockets of open cells (POCs) and drizzle in marine stratocumulus. Bull Amer Meteorol Soc 86:51–57
Storer RL, van den Heever SC (2013) Microphyiscal processes evident in aerosol forcing of tropical deep convective clouds. J Atmos Sci 70:430–446
Su H, Jiang JH, Lu X, Penner JE, Read WG, Massie S, Schoeberl MR, Colarco P, Livesey NJ, Santee ML (2011) Observed increase of TTL temperature and water vapor in polluted clouds over Asia. J Climate 24(11):2728–2736
Tao W-K, Li X, Khain A, Matsui T, Lang S, Simpson J (2007) Role of atmospheric aerosol concentration on deep convective precipitation: cloud-resolving model simulations. J Geophys Res 112:D24S18
Ten Hoeve JE, Jacobson MZ, Remer LA (2012) Comparing results from a physical model with satellite and in situ observations to determine whether biomass burning aerosols over the Amazon brighten or burn off clouds. J Geophys Res 117:D08203
Terai CR, Wood R, Leon DC, Zuidema P (2012) Does precipitation susceptibility vary with increasing cloud thickness in marine stratocumulus? Atmos Chem Phys 12:4567–4583. doi:10.5194/acp-12-4567-2012
Twomey S (1977) The influence of pollution on the short wave albedo of clouds. J Atmos Sci 34:1149–1152
Twomey S (1991) Aerosols, clouds and radiation. Atmos Environ 254:2435–2442
van den Heever SC, Carrio GG, Cotton WR, Demott PJ, Prenni AJ (2006) Impact of nucleating aerosol on Florida storms, part 1: mesoscale simulations. J Atmos Sci 63:1752
Van Zanten MC, Stevens B (2005) Observations of the structure of heavily precipitating marine stratocumulus. J Atmos Sci 62:4327–4342
Van Zanten MC, Stevens B, Vali G, Lenschow DH (2005) Observations in nocturnal marine stratocumulus. J Atmos Sci 62:88–106
Wang C (2005) A modeling study of the response of tropical deep convection to the increase of cloud condensation nuclei concentration: 1. dynamics and microphysics. J Geophys Res 110:D21211
Wang H, Feingold G (2009) Modeling mesoscale cellular structures and drizzle in marine stratocumulus, part I: impact of drizzle on the formation and evolution of open cells. J Atmos Sci 66:3237–3256
Wang SP, Wang Q, Feingold G (2003) Turbulence, condensation, and liquid water transport in numerically simulated nonprecipitating stratocumulus clouds. J Atmos Sci 60:262–278
Wang H, Feingold G, Wood R, Kazil J (2010) Modeling microphysical and meteorological controls on precipitation and cloud cellular structures in Southeast Pacific stratocumulus. Atmos Chem Phys 10:6347–6362. doi:10.5194/acp-10-6347-2010
Wang H, Rasch PJ, Feingold G (2011a) Manipulating marine stratocumulus cloud amount and albedo: a process-modelling study of aerosol-cloud-precipitation interactions in response to injection of cloud condensation nuclei. Atmos Chem Phys 11:4237–4249. doi:10.5194/acp-11-4237-2011

Wang PK, Su SH, Charvat Z, St'astka J, Lin HM (2011b) Cross tropopause transport of water by Mid-latitude deep convective storms: a review. Terr Atmos Ocean Sci 22(5):447–462
Wang JS, Seidel DJ, Free M (2012) How well do we know recent climate trends at the tropical tropopause? J Geophys Res 117:D09118
Wilcox EM, Roberts G, Ramanathan V (2006) Influence of aerosols on the shortwave cloud radiative forcing from North Pacific oceanic clouds: results from the Cloud Indirect Forcing Experiment (CIFEX). Geophys Res Lett 33:L21804. doi:10.1029/2006GL027150
Williams ER, Rosenfeld D, Madden N, Gerlach J, Gears N, Atkinson L, Dunnemann N, Frostrom G, Antonio M, Biazon B, Camargo R, Franca H, Gomes A, Lima M, Machado R, Manhaes S, Nachtigall L, Piva H, Quintiliano W, Machado L, Artaxo P, Roberts G, Renno N, Blakeslee R, Bailey J, Boccippio D, Betts A, Wolff D, Roy B, Halverson J, Rickenbach T, Fuentes J, Avelino E (2002) Contrasting convective regimes over the Amazon: implications for cloud electrification. J Geophys Res 107(D20):8082. doi:10.1029/2001JD000380
Wood R (2006) The rate of loss of cloud droplets by coalescence in warm clouds. J Geophys Res 111(D21):205
Wood R (2007) Cancellation of aerosol indirect effects in marine stratocumulus through cloud thinning. J Atmos Sci 64:2657–2669
Wood R (2012) Stratocumulus clouds. Mon Wea Rev 140(8):2373–2423
Wood R, Hartmann DL (2006) Spatial variability of liquid water path in marine boundary layer clouds: the importance of mesoscale cellular convection. J Climate 19:1748–1764
Wood R, Comstock KK, Bretherton CS, Cornish C, Tomlinson J, Collins DR, Fairall C (2008) Open cellular structure in marine stratocumulus sheets. J Geophys Res 113(D12):207
Wood R, Kubar T, Hartmann DL (2009) Understanding the importance of microphysics and macrophysics for warm rain in marine low clouds: Part II. Heuristic models of rain formation. J Atmos Sci 66:2973–2990
Wood R, Bretherton C, Leon D, Clarke A, Zuidema P, Allen G, Coe H (2011) An aircraft case study of the spatial transition from closed to open mesoscale cellular convection over the southeast Pacific. Atmos Chem Phys 11:2341–2370
Wu L, Su H, Jiang JH (2011) Regional simulations of deep convection and biomass burning over South America: 2. Biomass burning aerosol effects on clouds and precipitation. J Geophys Res Atmos 116.D17. doi:10.1029/2011JD016106
Yuan T, Remer LA, Pickering KE, Yu H (2011) Observational evidence of aerosol enhancement of lightning activity and convective invigoration. Geophys Res Lett 38:L04701. doi:10.1029/2010GL046052
Zhang H, McFarquhar GM, Saleeby SM, Cotton WR (2007a) Impacts of Saharan dust as CCN on the evolution of an idealized tropical cyclone. Geophys Res Lett 34:L14812. doi:10.1029/2007GL029876
Zhang R, Li G, Fan J, Wu DL, Molina MJ (2007b) Intensification of Pacific storm track linked to Asian pollution. Proc Natl Acad Sci U S A 104:5295
Zhang H, McFarquhar GM, Cotton WR, Deng Y (2009) Direct and indirect impacts of Saharan dust acting as cloud condensation nuclei on tropical cyclone eyewall development. Geophys Res Lett 36:L06802. doi:10.1029/2009GL037276
Zhang D, Wang Z, Liu D (2010) A global view of midlevel liquid-layer topped stratiform cloud distribution and phase partition from CALIPSO and CloudSat measurements. J Geophys Res 115:D00H13. doi:10.1029/2009JD012143

Improving Understanding of the Global Hydrologic Cycle

Observation and Analysis of the Climate System: The Global Water Cycle

Peter H. Gleick, Heather Cooley, James S. Famiglietti, Dennis P. Lettenmaier, Taikan Oki, Charles J. Vörösmarty, and Eric F. Wood

Abstract Understanding the complexity of the hydrological cycle is central to understanding a wide range of other planetary geological, atmospheric, chemical, and physical processes. Water is also central to other core economic, social, and political issues such as poverty, health, hunger, environmental sustainability, conflict, and economic prosperity. As society seeks to meet demands for goods and services

P.H. Gleick (✉) • H. Cooley
Pacific Institute, 654 13th Street, Preservation Park, Oakland, CA 94612, USA
e-mail: pgleick@pacinst.org; hcooley@pacinst.org

J.S. Famiglietti
Department of Earth System Science, University of California
Center for Hydrologic Modeling (UCCHM), University of California, Irvine,
240K Rowland Hall, Mail Code 4690, Irvine, CA 92697, USA
e-mail: jfamigli@uci.edu

D.P. Lettenmaier
Department of Civil and Environmental Engineering, University of Washington,
164 Wilcox Hall, Seattle, Box 352700, WA 98195-2700, USA
e-mail: dennisl@uw.edu

T. Oki
Institute of Industrial Science, The University of Tokyo,
4-6-1 Komaba, Meguro-ku, Tokyo 153-8505, Japan
e-mail: taikan@iis.u-tokyo.ac.jp

C.J. Vörösmarty
Department of Civil Engineering, City University of New York,
365 Fifth Avenue, New York, NY 10016, USA
e-mail: cvorosmarty@ccny.cuny.edu

E.F. Wood
Department of Civil and Environmental Engineering,
Princeton University, E-208 E-Quad, Princeton, NJ 08544, USA
e-mail: efwood@princeton.edu

G.R. Asrar and J.W. Hurrell (eds.), *Climate Science for Serving Society: Research, Modeling and Prediction Priorities*, DOI 10.1007/978-94-007-6692-1_6,

for a growing population, we must improve our understanding of the fundamental science of the hydrological cycle, its links with related global processes, and the role it plays in ecological and societal well-being. At the same time, human influences on the character and dynamics of the water cycle are growing rapidly. Central to solving these challenges is the need to improve our systems for managing, sharing, and analyzing all kinds of water data, and our ability to model and forecast aspects of both the hydrological cycle and the systems we put in place to manage human demands for water. We need to improve our understanding of each of the components of the hydrological water balance at all scales, and to understand the spatial and temporal variability in the components of the water cycle. This chapter provides a short summary of current WCRP efforts and addresses four primary research challenges:

1. The collection of more comprehensive data and information on all aspects of the hydrologic cycle and human uses of water, at enhanced spatial and temporal resolution and increased precision;
2. Improved management and distribution of these data;
3. Improved representation of the anthropogenic manipulations of the water cycle in the coupled land-atmosphere-ocean models used to forecast climate variations and change at both seasonal to interannual, and decade to century, time scales; and
4. Expanded research at the intersection of hydrological sciences and the technical, social, economic, and political aspects of freshwater management and use.

Keywords Hydrologic cycle • Water • Water systems • Climate • Modeling • Water balance • Data • GEWEX • GRACE • Water-energy nexus

1 Introduction: The Challenge

Water, energy, and climate are physically, spatially, and temporally linked. Energy from the sun and from internal geological processes drives the hydrological cycle. Atmospheric composition, climate system characteristics, and complex feedbacks help determine the planet's energy and water balances and distribution (Oki 1999). Both linear and non-linear dynamics amplify and dampen effects of external forcings. Water on Earth in its three phases is integral to the functioning, dynamics, and variability of the global climatological and biological support systems (Oki et al. 2004). From a purely scientific and academic point of view, understanding the complexity of the hydrological cycle is of paramount interest and central to our understanding of other planetary geological, atmospheric, chemical, and physical processes. But water is more than that: water is key to some of the core economic, social, and political issues of our time such as poverty, health, hunger, environmental sustainability, conflict, and economic prosperity (Gleick 2003).

Perhaps more than any other scientific discipline, hydrological science traces its roots to efforts to tackle challenges of social and economic development, including the provision of safe and reliable drinking water, flood forecasting and protection, wastewater treatment, irrigation development and food production, hydropower generation, and more (Loucks 2007; Wood et al. 2011). As society seeks to meet demands for goods and services for a growing population, the more apparent it becomes that we must improve our understanding of the fundamental science of the hydrological cycle, its links with related global processes, and the role it plays in ecological and societal well-being. At the same time, human influences on the character and dynamics of the water cycle are growing (FC-GWSP 2004a; Vörösmarty et al. 2010; Pokhrel et al. 2011), often faster than our understanding of these influences and their ultimate consequences.

Central to solving these challenges is the need to improve our systems for managing, sharing, and analyzing all kinds of water data, and our ability to model and forecast aspects of both the hydrological cycle and the systems we put in place to manage human demands for water. These improvements would help lead to a far better understanding of the local, regional, and global details of the water balance on timescales from minutes to millennia. In short, we need to improve our understanding of each of the components of the hydrological water balance at all scales, and to understand the spatial and temporal variability in the components of the water cycle. Extensive efforts in some of these areas are ongoing under the auspices of national research centers, universities, and international scientific collaborations, including the World Climate Research Program (WCRP). Recent reviews summarize the current state of understanding and future research priorities in the direct science-related aspects of these problems (for example, Hornberger 2001; FC-GWSP 2004b; Oki et al. 2006; NRC 2007, 2008b; Shapiro et al. 2010; Wood et al. 2011). This assessment expands on those efforts by integrating key scientific research needs with a broader perspective. There is also overlap between the recommendations here and in other reviews of geophysical components of the broad climate system, prepared for the October 2011 WCRP meeting (see, for example, the discussion on satellite observing systems and needs in Trenberth et al. 2011 and Oki et al. 2012).

The hydrological sciences community is faced with a complex moving target in three ways: First, very long-term climatological and hydrological balances are influenced by both cyclical and non-cyclical solar, orbital, and geophysical forcings. Second, climatological and hydrological balances are subject to substantial variability on widely varying timescales of seconds to millennia, and our limited instrumental and paleo observations give us an incomplete understanding of the statistics of extremes and natural variability. Thirdly, humans are now driving changes in atmospheric processes and have also substantially modified the natural hydrological cycle and altered hydrological processes across the land branch of the cycle, with growing evidence of oceanic, continental, and global-scale impacts and resource constraints (Meybeck 2003; FC-GWSP 2004a, b; Oki and Kanae 2006; Vörösmarty et al. 2010; Gleick and Palaniappan 2010).

While our understanding of the role that humans play in altering planetary systems has improved enormously in recent decades, uncertainties in both the science and in our knowledge of future societal factors such as population, economic conditions, technology trends, and energy choices make modeling efforts and future forecasts inherently imperfect. Any effort to summarize future needs must therefore note the important distinctions among the urgent need to improve our basic understanding of the hydrological cycle, the equally urgent need to improve our understanding of how humans are influencing and changing it, and the ultimate consequences of those changes for societal well-being. Perhaps in part as a result of these complexities, few if any of the current generation of land surface models used in coupled land-atmosphere-ocean climate models represent anthropogenic effects on the water cycle, a deficiency that is especially limiting as the demand for climate change information at regional and local scales increases.

This chapter provides a short summary of current WCRP efforts[1] and addresses four primary research challenges:

1. The collection of more comprehensive data and information on all aspects of the hydrologic cycle and human uses of water, at enhanced spatial and temporal resolution and increased precision;
2. Improved management and distribution of these data;
3. Improved representation of the anthropogenic manipulations of the water cycle in the coupled land-atmosphere-ocean models used to forecast climate variations and change at both seasonal to interannual, and decade to century, time scales; and
4. Expanded research at the intersection of hydrological sciences and the technical, social, economic, and political aspects of freshwater management and use.

2 Current WCRP Efforts

WCRP's efforts in the area of hydrology, atmospheric dynamics, thermodynamics, and the interaction between surface-land-ocean-atmosphere processes and the hydrological cycle are addressed mostly by the Global Energy and Water Cycle Experiment (GEWEX), Climate Variability and Predictability (CLIVAR), and Climate and Cryosphere (CLIC) projects.[2] WCRP efforts are linked to the Global Water System Project (GWSP; a partnership with three other global environmental change programs) and the WMO Global Framework for Climate Services (GFCS) efforts. The latter is developing a new working group on climate information and services that is expected to deal with aspects of climate service delivery relative to the water management community. One area that would benefit from better integration of changes in terrestrial systems with oceanic and cryospheric ones is the

[1] Good and more comprehensive summaries of WCRP programs can be found online.

[2] GEWEX was formerly "Global Water and Energy Cycle Experiment" and is now "Global and Regional Energy and Water Exchanges." CLIVAR is the "Climate Variability and Predictability" program.

issue of understanding the dynamics and components of sea-level rise. An example is the effect of reservoir filling globally during the second half of the twentieth century on sea level rise (discussed by Lettenmaier and Milly 2009). Another is the 2010–2011 reduction in the rate of sea-level rise (see http://sealevel.colorado.edu/frontpage). One argument is that this anomaly can be traced to extreme rainfall over several major land areas associated with a strong La Niña event, which had the effect of storing unusually large amounts of water on the global land areas (e.g., Australia and northern South America). GRACE observations have generally confirmed this hypothesis (Behera and Yamagata 2010).

Within each core project there are common themes, including:

1. Making observations and performing analyses
2. Developing, conducting, and evaluating experiments
3. Understanding and evaluating processes
4. Developing applications and services
5. Building technical and management capacity.

A few of the key questions for the future identified by GEWEX (Box 1 below) and CLIVAR include:

- How are the Earth's energy budget and water cycle changing?
- Can we quantify feedback processes in the Earth system and determine how these processes are linked to natural variability?
- Can we accurately model climate variability on the seasonal to interannual timescale?
- What are the impacts of climate variability at different space and time scales on water resources?
- How does and will anthropogenic climate change interact with natural climate variability to alter both the means and extremes of regional water and energy budgets?
- Can we track the flow of energy through the atmospheric and oceanic system and understand the nature of global warming?
- Can we understand the forcings and feedbacks among the different climate system components?[3]

3 Improve Collection of Hydrological and Water System Data

The first recommendation in almost all past reviews of the state of the hydrological sciences is to substantially expand collection of a wide range of geophysical, climatological, and hydrological data. Without adequate data, understanding of existing conditions and dynamic processes will always be constrained. Without adequate

[3]From the WCRP website: http://www.wcrp-climate.org/waterclim.shtml

Box 1 GEWEX Plans for 2013 and Beyond

Mission Statement
To measure and predict global and regional energy and water variations, trends, and extremes (such as heat waves, floods and droughts), through improved observations and modeling of land, atmosphere, and their interactions; thereby providing the scientific underpinnings of climate services.

Vision Statement
Water and energy are fundamental for life on Earth. Fresh water is a major pressure point for society owing to increasing demand and vagaries of climate. Extremes of droughts, heat waves and wild fires as well as floods, heavy rains and intense storms increasingly threaten to cause havoc as the climate changes. Other challenges exist on how clouds affect energy and climate. Better observations and analysis of these phenomena, and improving our ability to model and predict them will contribute to increasing information needed by society and decision makers for future planning.

GEWEX Imperatives
Datasets: Foster development of climate data records of atmosphere, water, land, and energy-related quantities, including metadata and uncertainty estimates.
Analysis: Describe and analyze observed variations, trends and extremes (such as heat waves, floods, and droughts) in water and energy-related quantities.
Processes: Develop approaches to improve process-level understanding of energy and water cycles in support of improved land, ocean, and atmosphere models.
Modeling: Improve global and regional simulations and predictions of ocean evaporation, overall precipitation, clouds, and land hydrology, and thus the entire climate system, through accelerated development of models of the land and atmosphere.
Applications: Attribute causes of variability, trends, and extremes, and determine the predictability of energy and water cycles on global and regional bases in collaboration with the wider WCRP community.
Technology transfer: Develop new observations, models, diagnostic tools, and methods, data management, and other research products for multiple uses and transition to operational applications in partnership with climate and hydro-meteorological service providers.
Capacity building: Promote and foster capacity building through training of scientists and outreach to the user community.
(Source: WCRP 2010)

data, the ability to develop more accurate models for forecasting and planning will be limited. As Hornberger (2001) noted, most major advances in the environmental sciences have resulted from new observations and the acquisition of new, better, or more comprehensive data, not just from the creation of new analytical models. Recent analysis of hydrologic extremes in a changing climate (Trenberth 2011a; NRC 2011a) yet again highlights this issue, in particular the essential need for investments in coherent and long-term observations in light of the "death" of stationarity (Milly et al. 2008) and the growing evidence that changes in the hydrological and climatological cycle due to climate change are already occurring, on land, over the oceans, and in the atmosphere (Meehl et al. 2007, 2009; Zhang et al. 2007; Syed et al. 2010; Trenberth 2011b; Durack et al. 2012).

For example, new analyses of ocean salinity trends and atmospheric water content and fluxes provide evidence for such changes. Syed et al. (2010) used multiple remotely-sensed datasets to analyze the global ocean water balance for changes in water cycle strength. Over the 13-year (1994–2006) study period, they observed significant increases in the rate of oceanic precipitation (240 $km^3/year^2$), oceanic evaporation (768 $km^3/year^2$), and continental discharge (540 $km^3/year^2$), which included ice sheet melting. Durack et al. (2012) noted an increase in ocean salinity, which suggests an accelerating global water cycle. Other studies also support an intensification of the water cycle, including:

1. An increase in atmospheric water content (precipitable water). While not directly indicative of fluxes, these data suggest that the humidity of the atmosphere has been increasing at close to the Clausius–Clapeyron rate, especially over the oceans (Trenberth et al. 2005; Wentz et al. 2007). More work is needed to resolve differences in changes of both absolute and relative humidity.
2. An increase in oceanic evaporation rates. Yu and Weller (2007) find evaporation increasing over the global ocean at 1.3 %/decade since the mid-1970s, due to both warming and intensifying winds. This is above model predictions and close to expectations from Clausius–Clapeyron theory (see also Weimerskirch et al. 2012).
3. Changes in precipitation rates. Wentz et al. (2007) report that global precipitation rates observed from satellites have been increasing with sea-surface temperatures at a rate of about 9 %/°C in the last two decades, though other observations (such as estimates from the Global Precipitation Climatology Project) offer different regional patterns and rates (see, for example, Zhou et al. 2011). At this point, the satellite-based precipitation estimates comprise a short record, and given high natural variability and routine concerns about satellite calibration, additional observations and analysis are warranted.
4. An increase in sea-surface salinity trends. Sea-surface salinity differences have increased by ~8 % over the five decades from 1950 to 2000 (Durack and Wijffels 2010; Durack et al. 2012). The oceanographic data also support these observations, with a consistent pattern found in both the mean salinity and the long term salinity trends (Boyer et al. 2005). Since the oceans have no internal sources or sinks of salinity, the variations are introduced at the surface by changes in

evaporation, precipitation, and runoff. Additional observations and modeling work is needed to improve our understanding of the sensitivity of salinity to temperature and hydrologic changes.

While differences between modeled and observed evaporation and precipitation may be due in part to inadequate data (Allan and Soden 2007) and short observational time series, the trends noted above seem consistent with a strong response of the water cycle to warming. Additional studies should help improve our understanding of these changes.

While many core concepts in hydrological sciences have been largely understood for decades, important basic data on stocks and flows of water, water vapor, and ice are missing for vast regions of the planet – even regions with large populations and highly productive economies. And new opportunities are continuing to emerge, such as understanding the origins and roles of "atmospheric rivers" in long-distance transport of water vapor in the lower atmosphere (Dettinger et al. 2011; Ralph et al. 2011) which now appears to be the driving mechanism for most major floods along the U.S. West Coast and winter-time floods in the U.K. (Lavers et al. 2011). Rapid changes in Arctic sea ice conditions must now also be evaluated because of the likelihood that they will changes the dynamics of important circulation patterns and add a new source of moisture in high latitudes. New data sets focused on water-balance studies are needed because such dynamics and balances are central to the development of useful water models (addressed later). New continental and global hydrometeorological data sets will be required to support these activities. These data sets include observations of streamflow over watershed and continental domains (Fekete et al. 2002), gridded high-resolution precipitation data, and more work to integrate different efforts to improve evapotranspiration estimates at small and continental scales (Jin et al. 2011). Expanded budget studies covering the role of the oceans, snow accumulation, melt, runoff, and evaporation of snow in continental regions are also needed to better understand how snow contributes to the water cycle, and the role of diminishing snowpacks on climate and water availability. Four central data needs include[4]:

- Improvements in precipitation observations sufficient to resolve the diurnal cycle and at a spatial resolution capable of representing variations in precipitation that control runoff generation in small to medium sized watersheds. Precipitation observations should include boundary layer observations, aircraft observations, surface measurements, synoptic-scale information, and coordinated satellite observations.
- Expansion of surface water, ocean surface salinity and moisture flux, and ocean-topography observations are needed to provide data on water storage and flows, including variability, in oceans, rivers, lakes, reservoirs, and wetlands. Efforts should be made to strengthen ocean salinity measurements as an integral measure of water cycle changes through the new salinity satellites Aquarius and SMOS, the ARGO float program, and the Global Drifter program.

[4]Some of these needs will be addressed by planned satellite missions, notably the Surface Water and Ocean Topography mission (SWOT), a joint venture of NASA and CNES, the French space agency.

- Improvements in snow-ice observation networks capable of estimating water storage in snowpacks, especially in mountainous and polar regions, including volumetric measurements of glaciers. Enhanced ice sheet observations are needed, combining satellite remote sensing and deployment of ocean buoys and subsurface floats. Efforts should be made to expand observing systems such as the NRCS SNOTEL of automated snow-water equivalent observations network over the U.S. to provide a global in-situ observational basis for estimating snow water storage in mountainous headwater regions of major river basins.
- Development and deployment of a combination of remote sensing and in situ soil-moisture monitoring systems capable of filling gaps in key elements of the land-surface water balance and land-atmosphere fluxes of heat and water, again of sufficiently high spatial and temporal resolution. In this respect, NASA's planned (2014 launch) SMAP mission, coupled with the COSMOS and other in-situ soil moisture networks over the U.S., should be an important first step.

In addition to these data sets, however, there is a growing need for the collection of far more comprehensive data on human interactions with the hydrologic cycle, including water withdrawals, consumption, and reuse (for example, revising the comprehensive but dated work of Shiklomanov (1997) and the data currently available from UN datasets such as AQUASTAT). Data are also needed on redirection and transfers of water, information on disruptions of nutrient cycles and on contamination by human and industrial wastes (Galloway et al. 2004; He et al. 2011), and on social and economic factors that influence the size and efficiency of water use (Vörösmarty et al. 2005; Gleick et al. 2011). Some work has been done to estimate water withdrawals on a spatially distributed basis, using the distributions of population and irrigation area, for instance, as proxies (Vörösmarty et al. 2000; Oki et al. 2001; Alcamo et al. 2003) but these efforts are limited by data constraints and the strengths and relevance of the proxies chosen.

The global water cycle and related data needs have long been recognized as a top priority for national research programs. In the late 1960s, the International Hydrological Decade pursued studies of world water balances, and pioneering estimates on large-scale hydrologic processes were published in the 1970s (L'vovitch 1973; Korzun 1978; Baumgartner and Reichel 1975). Shiklomanov (1997) assembled country-level statistics on water withdrawals in the past and present and made future projections. These early efforts were expanded with recent advances in information technologies that permit some global water-balance estimates at finer spatial resolution (Alcamo et al. 2007; Shen et al. 2008).

In the U.S., the National Research Council has issued a series of reports addressing research priorities in the areas of global environmental change, the hydrologic sciences, water system management, and climate change (NRC 1998, 1999a, b, 2002a, b, 2005, 2007, 2008a, b, c, 2009, 2010a, b). In 1999, the NRC Committee on Hydrologic Science argued for a comprehensive program of research on the role of the hydrologic cycle in the context of the broader global climate system (NRC 1999a). That same year, the NRC issued another report calling for new strategies for addressing the challenges of watershed science and management (NRC 1999b).

The good news is that we have unprecedented new capabilities in the form of technologies for in-situ and remote sensing and data collection, new approaches for embedded network sensing (ENS), sophisticated computer models for analyzing complex hydrologic processes, techniques for visualization of data, and growing interest and concern on the part of the public and policy makers about a wide range of water challenges.

The bad news is that these tools are not adequately utilized and resources (and sometimes the political will) for collecting even basic data on human uses of water remain limited. For example, the quality of existing remote-sensing data on soil moisture is poor (the recently launched ESA SMOS and upcoming NASA SMAP missions will offer improvements); snow-water equivalent is inadequately monitored at high resolution, especially in mountainous terrain; remote sensing estimates of snow water equivalent are especially problematic in mountain and forested headwaters of major river basins, variations in surface-water levels are not accurately captured by current sensors, and estimates of river discharge remain "an elusive goal" (NRC 2007).

Having better real-time and long-term data on water-balance variables would substantially improve the ability to close the water balances in local and regional watersheds, and the ability to model and understand the global water cycle. Other data of interest include estimates of water vapor transport, wind fields, ocean salinity, cloud structure, extent, and distribution, sea ice, groundwater balances, and a wide range of water-quality conditions.

Improvements are also needed in the resolution and precision of data. These improvements will come about through the development and deployment of new technologies for data collection and observations, expansion of data collection networks, the preservation and broader distribution of existing data sets, and new approaches for identifying unused or underutilized sources of information. The global Earth Science imperative, acknowledged by both international scientific organizations and national academies includes strong recommendations for advances in ground and satellite observational capabilities and implementation of observational data collection and management programs. As stated by the National Research Council (NRC 2007):

> "The scientific challenge posed by the need to observe the global water cycle is to integrate *in situ and space-borne observations* to quantify the key water-cycle state variables and fluxes. The vision to address that challenge is a series of Earth observation missions that will measure the states, stocks, flows, and residence times of water on regional to global scales followed by a series of coordinated missions that will address the processes, on a global scale, that underlie variability and changes in water in all its three phases." (Emphasis added.)

The ultimate goal, not yet realized, is for scientists to be able to track surface, subsurface, and atmospheric water in real-time, over the entire planet, and at sufficiently fine spatial resolution to integrate a complete quantitative picture of the terrestrial water cycle and embed that knowledge into decision support tools for forecasting extreme events for reducing risks and improving the use of water for agriculture and economic development. While such tools will always have limitations because of social, political, and economic factors, it is expected that investments in research and improved models will produce substantial economic benefits. For example, the financial benefits in public-sector weather forecasting and warning systems have been large and positive, estimated at over $30 billion per year on an investment of $5 billion (Lazo et al. 2009).

3.1 Ground-Based, In-Situ Observations

Spatial and temporal observations from surface networks and sensors must be improved and expanded. Regional-scale networks of sites should be developed to record meteorological and surface hydrological variables, soil moisture and dynamics, and groundwater levels and quality. This includes ocean buoys, river gages, snow sampling, new approaches to "embedded network sensing" (ENS), and much more. Such expanded networks should include new inexpensive, linked sensors (e.g., Harmon et al. 2007), establishing monitoring stations near the deltas of major rivers to record water fluxes for dissolved and suspended material, in particular, to improve understanding of carbon and nitrogen cycles, and a wide range of other priorities (NRC 2008b), not the least of which is the protection of deltas from both upstream and ocean-derived threats (Syvitski et al. 2009; Vörösmarty et al. 2009).

Yet even maintaining existing collection networks is difficult. In the U.S., the total number of active streamgages maintained by the U.S. Geological Survey dropped from over 8,250 in 1970 to 6,759 in 1997 due to budget cuts (Fig. 1). Some

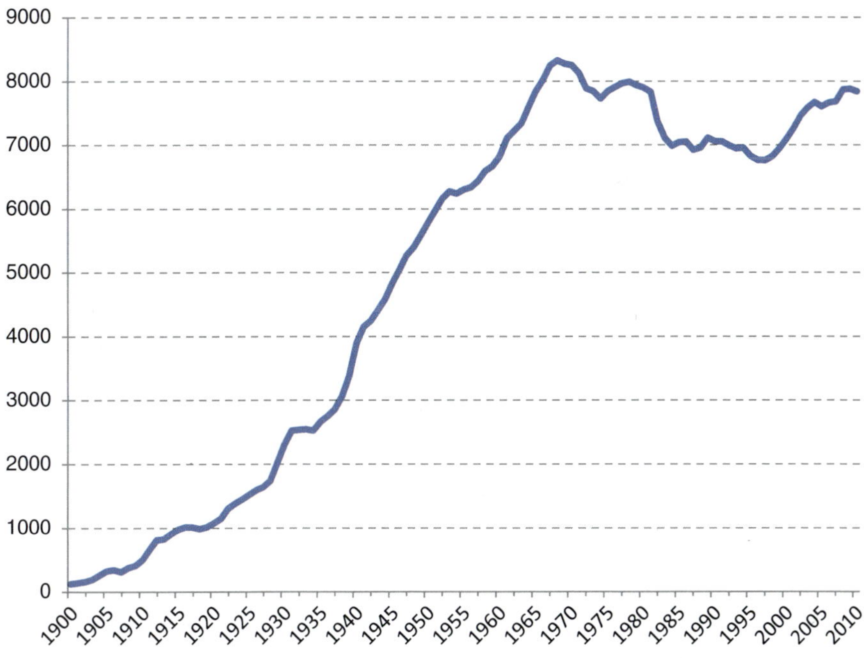

Fig. 1 The number of active USGS streamgages from 1900 to 2010. http://water.usgs.gov/nsip/history1.html

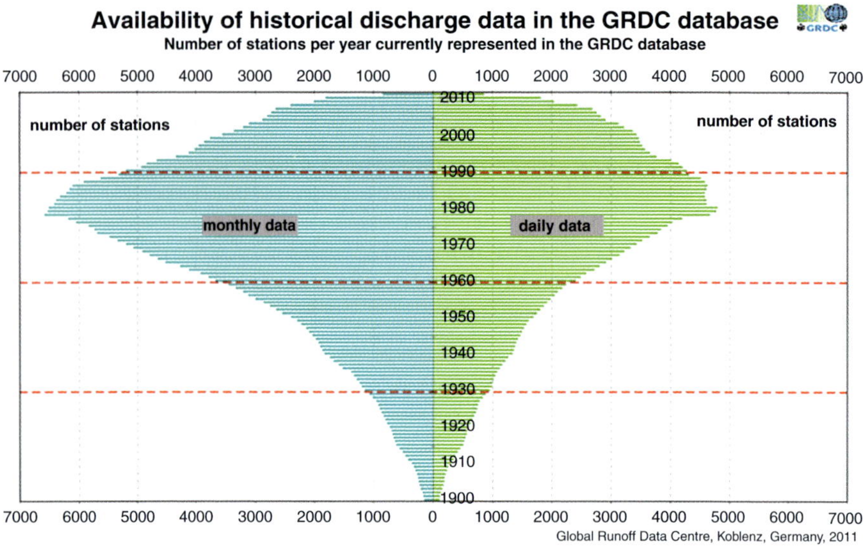

Fig. 2 Availability of historical discharge data in the GRDC database by year (number of stations per year represented in the GRDC database) http://www.bafg.de/cln_031/nn_266918/GRDC/EN/02__Services/services__node.html?__nnn=true

stations have been restored in recent years, but the total number of observing stations is still below the levels of the 1970s and early 1980s. This is a global problem as well, where budget and financing pressures, intellectual property issues, and conflicting policy priorities conspire to discourage monitoring and contribute to the loss of both observational stations and important data sets. Figure 2 shows the declining number of discharge monitoring stations worldwide in the Global Runoff Data Centre (GRDC) archives. Similar trends can be seen at the national level. The number of gages in South Africa dropped from a high of more than 4,000 to around 1,700 by the turn of the twenty-first century. Vast numbers of gages fell into disrepair or were dismantled following the collapse of the former Soviet Union (Stokstad 1999; Shiklomanov et al. 2002). Snow depth in Canada was recorded at over 2,600 stations in 1981 and at fewer than 1,600 in 1999. New methods of data collection and network design may permit more and better data to be collected with fewer stations (Mishra and Coulibaly 2009), but even with these improvements, the current scale of hydrologic data collection is not adequate to satisfy information needs for either science or policy.

3.2 Remotely Sensed Observations

Even with a significant expansion of ground-based monitoring, improved short-term event data collection from aircraft, and additional boundary layer observations, there are concerns that such monitoring is inadequate without increased reliance on satellite systems. There is some limited good news. The synoptic view afforded by satellites is uniquely poised to fill spatial and temporal gaps in ground-based data collection. For instance, improvements in global weather forecasting over the last several decades are largely attributable to better information from satellite retrievals about the distribution of atmospheric water vapor in the Southern Hemisphere. The Tropical Rainfall Monitoring Mission launched in 1997 improved our understanding of mid- and low-latitude precipitation. The GRACE satellites, despite the coarse resolution of their observations, have led to advances in the understanding of water storage changes in ice sheets and groundwater (Box 2).

Unfortunately, few countries and international consortia have the financial and technological resources to commit to comprehensive Earth Observing programs, and growing financial pressures are weakening the budgets allocated to such programs. This results in challenges to agencies, such as NASA in the U.S. and ESA in Europe, that wish to transition research satellites and sensors to other entities. Another aspect of the transition to operations problem (termed the "Valley of Death" by NRC 2000) is resistance by operational agencies to integrating data streams that may have a limited duration. While the authors appreciate this dilemma, we assert that it is in part a matter of culture and motivation. In this respect, a bright spot has been the European Center for Medium-Range Weather Forecasts (ECMWF), which has been working to evaluate the effect of new sensors and data on global weather forecast accuracy. ECMWF is now a global leader in weather forecasting, attributable at least in part to its willingness to evaluate and assimilate new data streams, especially satellite data. We will not review here the diverse and rapidly changing nature of these programs – by the time a final version of this paper is published, details will have changed again. But a general observation is that too little money has been made available to support building and maintaining adequate observing platforms with appropriate instruments, and even those few in development are at high risk of delay or cancellation. One example of a long-term remote observing program is the Global Precipitation Measurement effort, described in Box 3, which began in the late 1990s and is continuing to evolve, with expected launch of the core satellite in early 2014.

New and near-future satellite missions brighten this picture somewhat, but a comprehensive global water cycle platform is desperately needed. The current ESA SMOS (Soil Moisture and Ocean Salinity Mission) and the future NASA SMAP (Soil Moisture Active Passive) missions are positioned to map the water content of the thin veneer of soil near the land surface. The planned joint NASA-CNES SWOT (Surface Water and Ocean Topography) mission will routinely map the heights and inundation extent of inland surface waters. However, current plans for earth observing systems remain inadequate for deliberately moving the science forward in the direction recommended by scientific reviews (Group on Earth Observations 2007; NRC 2007). Worse, the planet is in grave danger of losing a substantial part of the current observing network because replacement systems, including both ground- and ocean-based instruments and satellites, are not being built quickly enough to fill inevitable gaps caused by expected instrument aging and by satellite orbital decay and failure. As one example, the recent budget crisis in the United

Box 2 Gravity Recovery and Climate Experiment (GRACE)

The **G**ravity **R**ecovery **A**nd **C**limate **E**xperiment (GRACE) is a joint mission of NASA and the German Space Agency DLR. Launched in 2002, the twin GRACE satellites are now making extremely accurate measurements of changes in Earth's gravity field caused by mass redistribution over the planet. The major driver of these mass variations on the monthly time scales of GRACE observations is water movement. Hence the gravity maps generated by GRACE provide new detail on how the storage of water is changing in Earth's major land, ocean, and ice reservoirs. When combined with additional ground-based or satellite observations, GRACE data have helped improve the tracking of water flows through river basins, withdrawals of groundwater, rates of ice-sheet melting, and other important hydrologic, oceanographic, and geologic phenomenon (Neumeyer et al. 2006; Ramillien et al. 2004). The data collected by GRACE are also helping to reconcile regional and global terrestrial water budgets (Syed et al. 2008; Sahoo et al. 2011) and allow for water balance estimates of unknown fluxes, including evapotranspiration (Rodell et al. 2004a) and continental discharge (Syed et al. 2009). GRACE-based estimates of groundwater depletion are already influencing the discussion of regional water policies as new data on water withdrawal and storage are made available (Rodell et al. 2010; Famiglietti et al. 2011a).

A follow-on GRACE mission (GRACE-FO) is currently planned for launch in 2017. The GRACE-FO will be essentially identical to the current mission, providing near-continuous measurements of water storage variations from March 2002 through the end of its lifetime. Coupled with the availability of more user-friendly GRACE data projects (Rodell et al. 2010; Landerer and Swenson 2011), the water community will have far-greater access to GRACE data than previously possible. Future, improved versions of the GRACE mission, that would achieve greater spatial and temporal resolution than the current 200,000 km^2, monthly data with 1.5 cm accuracy, are not slated for launch until the next decade (NRC 2007). This so-called GRACE-II (see Table 1) mission will enhance capabilities for monitoring water storage changes at the smaller scales at which water management decisions are made. Moreover, when data from GRACE (or its successor missions) are combined with the remotely-sensed soil moisture and surface water data described here, and integrated into data-assimilating models, an unprecedented picture of global distribution of water, both laterally and vertically, will emerge (Famiglietti 2004).

States and instrument design issues have delayed the Joint Polar Satellite System (JPSS) program and launch to the point where there is now expected to be a major and risky gap in coverage for vital hydrometeorological data (Box 4).

In this context, and while the entries in Table 1 offer hope to estimate a variety of water-cycle variables using remote sensing, a coherent strategy will be necessary to

Box 3 Global Precipitation Measurement (GPM)

The Global Precipitation Measurement (GPM) mission started as an international mission and follow-on and expansion of the Tropical Rainfall Measuring Mission (TRMM) satellite. TRMM, which hosts the first precipitation radar as well as a passive microwave sensor, was launched in November 1997 and continues to make observations almost 16 years later. Its major objective is to measure the global distribution of precipitation accurately with sufficient frequency so that the information provided can improve weather predictions, climate modeling, and understanding of water cycles. An important goal for the GPM mission is the frequent measurement of global precipitation to produce rainfall maps using a TRMM-like core satellite, jointly developed by the U.S. and Japan, and a constellation of multiple satellites that will carry passive microwave radiometers and/or sounders intended to enhance precipitation estimates during the time when the radar is not overhead.

GPM is composed of core system and multiple satellites carrying microwave radiometers and/or sounders (Constellation satellites). The GPM Core Observatory is now schedule to be launched in 2014, and will carry the sensors from multiple countries and agencies designed to collect as much microphysical information as possible for accurate rain estimation, and to provide reference standards for the instruments on the constellation satellites.

Constellation satellites will carry a microwave imager and/or sounder, and are planned to be launched around 2013 by each partner agency for its own purpose. They will contribute to extending coverage and increasing frequency of global rainfall observations. Currently, several satellite missions are planning to contribute to GPM as a part of constellation satellites, including JAXA's Global Change Observation Mission – Water (GCOM-W) series; CNES/ISRO's Megha-Tropiques; EUMETSAT's MetOp series; NOAA's Polar Operational Environmental Satellites (POES) Joint Polar Satellite System (JPSS); and DoD's Defense Meteorological Satellite Program (DMSP) and Defense Weather Satellite System (DWSS).

link these data sources with the dynamics of water-management systems and regional watersheds. One example is need to understand the hydraulics of stream and river systems as well as the statistical time-space domains that different monitoring strategies would have to confront. For example, the technical requirements for developing short-term flood forecast and monitoring are quite different from those needed for long-term water resource assessment, agricultural water efficiency efforts, or integrated management among the energy, water, and food sectors.

A related and often overlooked issue is the need to link remote sensing with in-situ measurements. There is the misperception that satellites measure geophysical parameters. Rather, almost all (GRACE is a notable exception) measure radiation (such as brightness temperatures) and radar backscatter, which are then used to infer

Box 4 Joint Polar Satellite System (JPSS)

NOAA maintains both geostationary weather satellites and polar satellites. Their polar systems provide observations of land, ocean, and atmosphere over the entire Earth. There are only two polar research satellites systems that provide this kind of hydroclimatological data: NOAA's and Europe's EUMETSAT. These two systems provide the primary data for developing National Weather Service (NWS) weather prediction models at high confidence forecasts 2–7 days in advance and they are the backbone of all weather forecasts beyond 48 h. These polar satellites, however, also play other critical roles. They aid in hurricane forecasting and rapid coastal evacuation, provide continuity of the 40+ years of space-based earth observations to monitor and predict climate variability, produce drought forecasts worth $6–8 billion to the farming, transportation, tourism, and energy sectors, support troop deployment operations, and pick up rescue beacon signals. NOAA estimates that satellite observing systems saved 295 lives in the U.S. alone in 2010 and over 28,000 lives worldwide since 1982.

NOAA's current polar satellites are reaching the end of their useful lives. A research satellite known as NPP (NPOESS Preparatory Project) launched in October 2011 to serve as a bridge between from the current polar-orbiting satellites and the next-generation of polar-orbiting satellites, known as the Joint Polar Satellite System (JPSS). NOAA planned to launch the first two JPSS satellites in 2014 but the current budget crisis in the US led Congress to cut NOAA funding forcing a delay in JPSS launch to at least 2017 and possibly beyond. While the President's FY 2012 budget restores full funding, it will not prevent a gap in observation coverage. According to NOAA, it is now a "near-certainty that an unprecedented observational data gap of 15–21 months will occur between the anticipated end of the NPP spacecraft's operational life in 2016 and the date when the first JPSS mission is planned to begin".

Loss of coverage would set back weather observations and forecasting almost a decade to when forecasts were of lesser quality. This problem may reduce forecast accuracy, especially for major weather events such as winter snow storms over the East Coast and hurricane tracks and intensity, by as much as 50 %. Errors in track and intensity forecasts could delay hurricane warnings and evacuations or result in unnecessary evacuations.

geophysical variables. Harmonizing remote sensing data with past ground/in-situ measurements can help to greatly extend spatial and temporal data records. While these harmonization efforts are part of ongoing NASA, NOAA, ESA, JAXA, and EUMETSAT programs, more are needed.

Table 1 Water resources panel candidate missions

Summary of mission focus	Variables	Type of sensors	Coverage	Spatial resolution	Frequency	Synergies with other panels	Related planned or integrated missions
Soil moisture, freeze-thaw state	Surface freeze-thaw state, soil moisture	L-band radar, radiometer	Global	10 km (processed to 1–3 km)	2- to 3-day revisit	Climate Weather	SMAP Aquarius
Surface water and ocean topography	River, lake elevation; ocean circulation	Radar altimeter, nadir SAR interferometer, microwave radiometer, GPS receiver	Global (to ~82° latitude)	Several centimeters (vertical)	3–6 days	Climate Ecosystems Health Weather	SWOT
							SMAP GPM NPP/NPOESS
Snow, cold land processes	Snow-water equivalent, snow depth, snow wetness	SAR, passive microwave radiometry	Global	100 m	3–15 days	Climate Ecosystems Weather	SCLP
Water vapor transport	Water vapor profile; wind speed, direction	Microwave	Global	Vertical		Weather Climate	3D-Winds
							PATH GACM GPSRO

(continued)

Table 1 (continued)

Summary of mission focus	Variables	Type of sensors	Coverage	Spatial resolution	Frequency	Synergies with other panels	Related planned or integrated missions
Sea ice thickness, glacier surface elevation, and glacier velocity	Sea ice thickness, glacier surface elevation; glacier velocity	Lidar, InSAR	Global			Climate Solid Earth	DESDynI ICESat-II
Groundwater storage, ice sheet mass balance, ocean mass	Groundwater storage, glacier mass balance, ocean mass distribution	Laser ranging		100 km		Climate Solid Earth	GRACE-II
Inland, coastal water quality	Inland, coastal water quality; land-use, land-cover change	Hyperspectral imager, multi-spectral thermal sensor	Global or regional	45 m (global), 250–1,500 km (regional)	About days (global), subhourly (regional)	Climate Ecosystems Health	GEO-CAPE

Source: NRC (2007). Because launch dates change so quickly, we have not provided expected dates here

3.3 *Managing Data*

Climate data, including in-situ and satellite observations and model output (such as re-analyses) are not as widely available or readily accessible as they should be. This lack of access is a threat to GEWEX's ability to meet its imperatives (see Box 1) and more broadly, constrains all regional water planning, analysis, and management efforts. The applications goals for GEOSS (Global Earth Observation System of Systems) similarly cannot be met without better access to data. As additional hydrologic data are collected, new systems are needed to manage and distribute those data. Wood et al. (2011) note the additional complications and costs associated with data support systems, but developing such systems is secondary to access and having interoperability across products.

New commitments to establishing and maintaining hydrological data networks are not enough. As articulated by Parson (2011), there is consensus in the science and application communities that open and free access to hydrological and meteorological data is critical for improved utilization of data resources and for transparency in data-based research results and derived data products. Many international bodies such as the World Meteorological Organization, International Science Union and the Group on Earth Observations (ICSU 2004; WMO 1995; Group on Earth Observations 2009) have passed resolutions, advocated for, or created central principles for more open access to data. A global open-access database is highly desirable, and systems must be put in place to ensure access to data and to maintain data in forms that are useful for different research and application needs. These data are crucial for assessing water resources at multiple scales and for verifying hydrological models and evaluating policy solutions.

Many organizations already collect hydrological data using different and often inconsistent platforms for both operational (e.g., National Water Information System (NWIS) of the US Geological Survey (USGS), US EPA, NOAA) and research purposes (NASA, Atmospheric Radiation Measurement (ARM) program, AQUASTAT of the UNFAO, and the Global Runoff Data Center (GRDC)). Data fragmentation and variation makes it difficult for scientists to use data from different sources, to evaluate data accuracy or bias, and to combine mixed data sets without extensive analysis. The lack of data access prevents the development of systems to integrate data from disparate sources like in-situ observations and satellite measurements. Earlier in this article we recognized that satellites are "uniquely poised to fill spatial-temporal gaps in ground-based data," but the development of systems that can integrate and merge such data is seriously hindered by data access barriers. One specific example is the desirability of using TRMM (and in the future GPM), derived precipitation in data sparse regions for flood prediction where both real-time ground observations and satellite-based estimates, when integrated and merged, can lead to improved heavy precipitation monitoring and flood forecasting. Some efforts are now being made to integrate and manage such datasets under the auspices of the Consortium of Universities for the Advancement of Hydrological Science, Inc. (CUAHSI) but these efforts are neither comprehensive nor global (Oki et al. 2006).

The management of global data also remains a challenge. One European-led effort, GRDI2020 (Global Research Data Infrastructures), has been formed to develop "technical recommendations to increase the ability of the research community, industry, and academia to influence the development of a competitive global ICT infrastructure." The International Groundwater Resources Assessment Centre (IGRAC; www.un-igrac.org) offers another example of a new, international hydrological data collection and distribution strategy. Under the IGRAC approach, the continents are discretized into one-degree grid cells. Each one-degree cell has an associated expert, designated by his or her home country, who is responsible for monthly submissions of a short list of key groundwater variables, for example, well levels. The local expert is responsible for determining a representative monthly, one-degree average value for the key variables, and for uploading the averaged and raw data in standardized formats through a user-friendly web-based interface. IGRAC is a new center and as such, its success and the viability of its approach will only become apparent in time. If successful however, the IGRAC approach is one that could conceivably be implemented for other hydrologic variables and organized by UNESCO or the WMO. Other efforts such as CUASHI or GRDI also need to be supported and fostered.

4 Modeling

Models are critical tools for the hydrological sciences. Models are used over a wide range of spatial and temporal scales to forecast future conditions and to reconstruct hydrologic conditions in the instrumental and pre-instrumental past. They are also used to simulate scenarios such as hydrologic stocks and flows or water-quality variations under different observed or hypothetical conditions, and to interpolate observational data, integrate point data over large areas, downscale large-scale data to regional areas, and estimate hydrological variables where no observational data are available. Models help identify water-system risks and test strategies for reducing those risks. Ironically, our ability to develop complex hydrological models has outstripped our ability to provide them with adequate data, hence the need for improving data collection noted above. Despite progress in both model development and data collection and assimilation, Wood et al. (2011) note that the current class of parameterizations used to represent the land surface in numerical weather prediction and climate models is unable to address a wide range of societal needs for water-related information. For example, current weather forecasts are carried out using land surface models with resolutions that are too coarse to represent key local processes (e.g., evapotranspiration along riparian corridors in semi-arid landscapes). Efforts to make the outputs of the current generation of global climate models of use to hydrologists and water resource planners and practitioners, while of growing value, have yet to be completely successful (NRC 2011b). We argue that this should be a priority for WCRP (and GWEX in particular), which previously has placed more emphasis on understanding variations and controls on the global water and energy cycles than on their manifestations over land, and in particular over the most populous parts of the global land area.

Recently, the hydrologic community has begun to call for an acceleration in the development of hydrological models that can be applied to a range of high-priority issues related to food, energy, climate, and economic security, and that represent the land surface hydrology of managed, rather than just natural systems. As one example, the Community Hydrologic Modeling Platform (CHyMP) under development in the U.S., is a broad-based effort that parallels the successful efforts in climate model development (Famiglietti et al. 2009, 2010, 2011a, b). CHyMP will enable fully integrated (snow, ice, surface water, soil moisture, groundwater) modeling of the natural and managed water cycles, across scales, and will provide access to continental-scale models and datasets for a broad swath of research and practicing water scientists and engineers.

Wood et al. (2011) issued a "grand challenge" to the hydrologic community to develop a new generation of "hyperresolution" hydrologic models that can exploit advances in computing power, the internet, and improved access to data. Such models would be capable of representing critical water cycle systems at a high spatial and temporal resolution and would require improved information about existing and projected human modifications such as dams and other artificial storage, groundwater withdrawals and recharge, alterations of nutrient flows, the impacts of urbanization, and much more. This is a core activity required for the development of Earth System models that include human drivers, as described in the next section.

GEWEX's scientific strategy also includes improved prediction. To better represent improved land-surface interactions that include human activities, resolution must increase, and the schemes need to be tested off-line before they are coupled. Steps are being taken in this direction, and the spatial resolution of global hydrological simulations is improving. Current global land-surface hydrologic simulations, such as the Global Land Data Assimilation System (GLDAS) (Rodell et al. 2004b) have grids scales around 25 km and approximately 50 km for offline global simulations (e.g., Dirmeyer et al. 2006; Haddeland et al. 2011); however, a land information system is under development that will have a spatial resolution of 1 km^2 (Oki et al. 2006). This model is designed to be an operational tool that will provide estimates of all major surface hydrological quantities (including evaporation, transpiration, soil moisture, snow depth and melt, and more), using a daily timestep. Forcing data and surface characteristics, including precipitation, radiation, surface winds, and vegetation cover will be provided by both surface (in situ) observations and remote sensing and will be tied into a modeling framework using four dimensional data assimilation (4DDA). The spatial resolution of the resulting model analysis fields eventually will be as high as 100 m globally, which is at least as high as the spatial resolution of most current generation regional hydrological models. If such a system becomes operational on a daily or even hourly timestep, and if observational data of sufficient quality are available to populate and test the model, then it could form an early warning system for hydrological disasters, such as floods and droughts, anywhere in the world (Oki et al. 2006).

Human influences on the hydrological cycles of the Earth are now widespread and often large in magnitude. Yet many current hydrological and land-surface models either exclude or poorly represent human influence on the terrestrial water cycle through activities such as agriculture, forestry, grazing, urbanization, or water diversions. These are critical elements of the contemporary water cycle to understand, with a perhaps more immediate impact than future effects of climate change (which, of course, will be felt in addition to these other anthropogenic influences) (Vörösmarty and Meybeck 2004). Many of these impacts appear to be an inescapable byproduct of economic development, (Vörösmarty et al. 2010), but that does not mean they cannot be mitigated through changes in policies, incentives, behaviors, and technology. An important feature of these influences is that they are, by their nature, interdisciplinary. Another is that they are often local, but with growing regional and even global influence.

Global hydrological models should now consider the effects of human intervention on hydrological cycles. Some efforts in this direction are underway. Several recently developed macro-scale models for water-resources assessment now include reservoir operation schemes (e.g., Haddeland et al. 2006a, b; Hanasaki et al. 2006). Hanasaki et al. (2008a) describe an integrated water-resources model that can simulate the timing and quantity of irrigation requirements and estimate environmental flow requirements. Such an approach can help assess water demand and supply on a daily timescale, and the gaps between water availability and water use on a seasonal basis in the Sahel, the Asian monsoon region, and southern Africa, where conventional water-scarcity indices such as the ratio of annual water withdrawal to water availability and available annual water resources per capita (Falkenmark and Rockström 2004) are not adequate (Hanasaki et al. 2008b). Wisser et al. (2008), Fekete et al. (2010), and Lehner et al. (2011) have worked to assess the implications of large infrastructure projects on water balances. Further improvements in models that couple natural hydrological systems with anthropogenic activities can improve our understanding of key challenges in water management, including the sustainability of water use, ecosystem health, and food production (Hanasaki et al. 2010; Pokhrel et al. 2011).

The effects of anthropogenic alterations in the land surface hydrologic cycle can go far beyond the river basin scale. The scale of human intervention is large enough that we now recognize that the water stored behind reservoirs globally has influenced Earth rotational variations and orbital dynamics, including length of day and polar motion (Chao 1995; Chao and O'Connor 1988). Similarly, Lettenmaier and Milly (2009) estimate that sea level rise, which over the last 50 years has averaged about 3 mm/year, would have been 15–20 % larger in the middle of the last century were it not for the reduction in freshwater flux to the oceans associated with filling of manmade reservoirs (they also note that the rate of filling has since decreased substantially, perhaps to a global net less than zero due to infilling of reservoirs with sediment and slowing of reservoir construction. Recently, Pokhrel et al. (2012) estimated on the basis of an integrated modeling framework that artificial reservoir water impoundment caused a sea level change (SLC) of −0.39 mm/year, while unsustainable groundwater use (groundwater depletion), climate-driven terrestrial

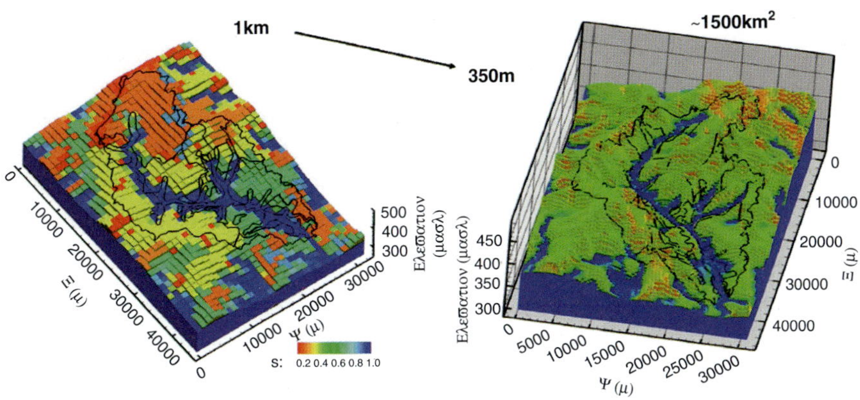

Fig. 3 Higher-resolution models allow better spatial representation of saturated and nonsaturated areas, with implications for runoff generation, biogeochemical cycling, and land-atmosphere interactions. Soil moisture simulations on the Little Washita showing the effect of resolution on its estimation of variables (Kollet and Maxwell 2008)

water storage (TWS) change, and the net loss of water from endorheic basins contributed +1.05, +0.09, and +0.03 mm/year of the SLC, respectively. Therefore, the net TWS contribution to SLC during 1961–2003 is +0.77 mm/year. Their result for the anthropogenic TWS contribution to global SLC partially fills the gap in the global sea level budget reported by the Fourth Assessment Report (AR4) of IPCC (2007) (Fig. 3).

5 Hydrological Sciences Needs for the Twenty-First Century in the Earth System Context

As described above, extensive efforts are underway by the global hydrological sciences community to identify and prioritize needs for data collection, modeling, and analysis. But it is also becoming increasingly apparent that many of the current water-related challenges facing society will not be resolved solely through improvements in scientific understanding. Many of these challenges lie at the intersection between pure and applied science, or require interactions among the sciences, economics, and policy. For example, we must improve our understanding of the societal and economic risks associated with extreme events such as droughts, floods (e.g., Okazawa et al. 2011), and coastal disruptions (NRC 2011a). We must improve our understanding of the role of extreme events and thresholds, the extent to which the water cycle is being modified or intensified (Huntington 2006; Trenberth 2011b), how much of the change is due to human activities, and the social implications of – and possible responses to – such changes. We must improve our understanding of "peak" constraints on water withdrawals from renewable and non-renewable hydrologic systems (Gleick and Palaniappan 2010). IPCC (2012), in the Special Report on Managing the Risks of Extreme Events and Disasters to Advance Climate Change Adaptation (SREX), noted that while risks cannot be fully eliminated, the character and severity of impacts from climate extremes depend not only on the extremes themselves but also on exposure and vulnerability, and emphasized the value of disaster risk management and adaptation strategies that focus on reducing exposure and vulnerability and on enhancing resilience to climate extremes.

As a result, there are new efforts underway to improve our understanding of the complex social, economic, and structural challenges facing water managers and users. These efforts would be greatly enhanced by interdisciplinary research efforts involving the scientific community and a broader range of engineers, economists, utility managers, irrigators, and local communities. Through these efforts, scientists may better understand the data needs of practitioners and some of the constraints they face, thereby helping to ensure that the products produced are actually applied. For example, as one measure of the recognition of these challenges, the Hydrology Section of the American Geophysical Union has just constituted a new Water and Society Technical Committee to heighten the visibility of water policy issues among AGU members and to develop new approaches to addressing a wide range of water-related challenges at the interface of science and policy. While such efforts are not traditionally addressed in the context of efforts by organizations such as the WCRP, it would be worth a serious discussion about the advantages and disadvantages of doing so.

5.1 *Climate, Water, and Social Adaptation*

As large-scale climate models have improved in their parameterizations of hydrologic processes and their spatial resolution, it has become increasingly clear that some of the most likely and unavoidable impacts to society of changes in climate will be changes in water availability, timing, quality, and demand (Kundzewicz et al. 2007; NRC 2011b). For more than a decade, the research community (and sometimes the water management community) has issued increasingly urgent calls for expanded efforts to integrate the findings from climate models with water management and planning efforts at regional levels because of the issues at the intersection of water, food, and energy (increasingly referred to as the "water-energy-agricultural nexus"), and the need to improve our integration of water quality and ecosystem needs into research efforts (AWWA 1997; Gleick 2000; Karl et al. 2009; CDWR 2009; Stakhiv 2011). Each of these topics demands both high-quality science and innovative interdisciplinary thinking. Such integration will require improvements in the quality and detail of information available from global and regional climate models, but will also require new approaches for integrating climate information into water-management institutional planning, improved economic and health risk assessment models, more robust engineering reviews of existing water-related infrastructure, and updated or improved operations rules for water supply, treatment, delivery, and wastewater systems.

5.2 *Water, Energy, Agricultural Nexus*

Connections between water, energy, and food have been recognized for centuries, but most of the focus of attention has been on ensuring the basic availability and reliability of supply of key resources for the production of other goods and services demanded by society. In the past decade or so, there has been new work to expand our understanding of these connections, in part because of adverse consequences caused by ignoring them. For example, changes in the energy policies of some industrialized countries to encourage greatly expanded production of domestic biofuels, such as corn-based ethanol programs, had unanticipated impacts on global food markets and prices and on conflicts over water resources (NRC 2008c, 2010b), with little reflection of biogeophysical realities (Melillo et al. 2009). Similarly, efforts to expand natural gas and oil production from unconventional fields, especially shale oil and gas, has led to unanticipated and poorly studied impacts on water quality, the generation of large volumes of "produced water" with high concentrations of pollutants, and new water demands in some water-scarce regions (Cooley and Donnelly 2012). Growing demands for electricity and for water to cool these systems are also intensifying competition for water in water-short regions and new efforts are underway to pursue alternative water sources and cooling technologies as well as less water-intensive generating systems.

Most current generation land-surface models are not well suited to address these issues. For instance, while climate change will almost certainly affect the availability of cooling water – a key constraint on energy production in many parts of the world – few current models simulate the most critical variable, water temperature. That is beginning to change – recent work by Van Vliet et al. (2012) and Cooley et al. (2011) illustrates the sensitivity of electric power generation to both the hydrological and surface climatic conditions, as well as to assumptions about energy futures and technology choices. This is an area that is deserving of greater attention by both the scientific and applications communities.

5.3 *Water Quality and Ecosystems*

There are serious limitations to our understanding of water quality, including both natural variability and human-induced changes in quality, and the role that water plays in ecosystem dynamics and health. Representations of these complex factors in regional and global models are inadequate and unsophisticated, though some small-scale catchment models have been developed that include physical and biochemical dynamics for some water-quality constituents such as carbon, nitrogen and phosphorus, and sediment (Vörösmarty and Meybeck 2004). Very little work has been done on other chemical components, heavy metals, or new contaminants such as pharmaceuticals (Palaniappan et al. 2010), and the challenge of articulating the additive, and possibly synergistic, interactions of multiple stressors from a variety of sources (broad array of chemicals, thermal pollution, sedimentary impacts) remains (Vörösmarty et al. 2010).

In this context, humans both accelerate and decelerate discharge and biogeochemical (BGC) fluxes through rivers (Meybeck and Vörösmarty 2005). For example, despite huge increases in local erosion from poor land management, around 30 % of global sediment flux is estimated to be trapped upstream behind dams and fails to enter the oceans (Syvitski et al. 2005), placing major coastal landforms like river deltas at risk and altering nutrients available to fisheries. Climate change and its attendant impacts on runoff, carbon and nutrient cycling, and weathering rates will also change these land-to-ocean linkages (Amiotte-Suchet et al. 2003). Frameworks are necessary to handle the component hydrologic, sediment, and biogeochemical dynamics, but notwithstanding ongoing work (e.g., Wollheim et al. 2008) much more needs to be done.

6 A Grand Challenge in Hydrologic and Water-Resources Modeling

Existing vulnerabilities and new threats to water posed by demographic changes, climatic changes, increased exposure to extreme events, and growing economic demands for water and water services are driving urgent needs for improvements in our understanding of the world's water resources and systems (Kundzewicz et al. 2007; Hirschboeck 2009; Shapiro et al. 2010). We will not go back to a time when hydrological sciences could only address pristine, unaltered systems. Humans now not only influence the water-cycle but are integral to it, and we must develop predictive models that represent human interactions with the water cycle at scales useful for water management. This implies that weather and seasonal climate models and land-surface parameterizations must improve in parallel. Without a strong understanding of the dynamics of global and regional water balances and the complex human interactions and influence on water quantity and quality, society risks making incorrect decisions about critical issues around energy, human health, transportation, food production, fisheries, ecosystem protection and management, biodiversity, and national security.

Until recently, anthropogenic effects on the global land water cycle were thought to be small (in part because the global land area is small compared with the oceans and because human populations were small). This is changing: it is now clear that anthropogenic activities such as land use and land cover change, irrigation, groundwater withdrawals, and reservoir storage have influenced sea level (Milly et al. 2010) and even orbital parameters; similarly we have an improved understanding of the role of the oceans in influencing land-surface hydrology. At regional scales, human effects have, in many cases, been large for a longer time – for instance, a number of major global rivers, including the Nile and the Colorado, no longer flow at their mouths as a result of consumptive water use (mostly for agriculture) and trans-basin diversions (Alcamo et al. 2005). In the case of the Colorado, about 1/3 of the river's natural discharge is diverted out of the basin and the rest is used consumptively. Other human influences that substantially affect regional hydrology include groundwater mining, increased soil moisture in irrigated areas, urbanization, and permafrost melt. These effects nonetheless are mostly not represented in regional or global climate models, and regional hydrology models often focus on runoff generation areas far upstream of the parts of the basin that have been most affected by anthropogenic activities. At continental scales, direct anthropogenic effects are probably more modest, but nonetheless can be substantial – especially the effects of land cover change, including irrigation, on moisture recycling and precipitation generation, mostly in the interior of the North America and Eurasia (Haddeland et al. 2007). These effects likewise are rarely represented in land-atmosphere models or their host climate global models.

We therefore argue that the "grand challenge" in the hydrology/water resources/climate arena is to model the role of humans on the water cycle at regional (e.g., large river basin), continental, oceanic, and global scales, including the feedbacks of these effects to the climate system, such as ocean/land interactions. This enterprise will involve the development of new understandings of the complex interactions of humans with the water cycle such as reservoir storage, diversions, and return flows, but even more importantly, of the decision process that will determine the nature of changes in water management as the climate warms. WCRP can serve an important role by fostering activities such as expanded data collection, model development, and intercomparison projects. Furthermore WCRP could and should promote the development of the global data sets that will be required to support the development and testing of these new models. Some of the required data sets have already been developed through activities like the Global Water System Project (GWSP), but effort will be required to assure that they are sufficient for the purposes of land models that ultimately must run within fully coupled Earth System models.

7 Conclusions

Over the last decade there has been a transformation in the way in which we view the continental water cycle. While freshwater systems of the planet are collectively an essential regulator of the non-living dynamics of the Earth System, they also play a central role in human existence and water security. At the same time, we now understand that our contemporary water system is increasingly tightly coupled to economic, social, technological and other factors like climate change. Along with this recognition of a globalized water system has come the awareness that human activities are themselves significantly and increasingly dominating the nature of this major cycle. This dominance takes the form of many "syndromes" that are at once both the causes as well as manifestations of rapid human-induced changes. Although we can increasingly detect and in many cases understand the sources, scope, and mechanisms associated with these changes, we urgently need to improve investments in our observational networks, our basic understanding of the water cycle and the ways it is integrated with energy, climatic, atmospheric, oceanic, and other complex geophysical characteristics of the planet, and our training of the next generation of researchers who increasingly will be called upon to study these larger-scale challenges, which are outside the traditional training perspectives of the hydrologic science community. Otherwise we will be unable to counteract the rising threats to public health, economic progress, and biodiversity caused by a global water system in transition.

Acknowledgements The authors express our gratitude for Ms. Misako Kachi for Box 3.

References

Alcamo J, Döll P, Henrichs T, Kaspar F, Lehner B, Rösch T, Siebert S (2003) Global estimates of water withdrawals and availability under current and future 'business-as-usual' conditions. Hydrol Sci J 48:339–348

Alcamo JH, Grassl H, Hoff P, Kabat F, Lansigan R, Lawford D, Lettenmaier, Lévêque C, Meybeck M, Naiman R, Pahl-Wostl C, Vörösmarty C (2005) Science framework and implementation activities, global water system project, Bonn, 76 pp

Alcamo J, Flörke M, Märker M (2007) Future long-term changes in global water resources driven by socio-economic and climatic changes. Hydrol Sci J 52(2):247–275

Allan RP, Soden BJ (2007) Large discrepancy between observed and simulated precipitation trends in the ascending and descending branches of the tropical circulation. Geophys Res Lett 34:L18705. doi:10.1029/2007GL031460

American Water Works Association (AWWA) (1997) Climate change and water resources. J Am Water Works Assoc 89(11):107–110

Amiotte-Suchet P, Probst J-L, Ludwig W (2003) Worldwide distribution of continental rock lithology: implications for atmospheric/soil CO2 uptake by continental weathering and alkalinity river transport to the oceans. Global Biogeochem Cycles 17:7.1–7.13

Baumgartner F, Reichel E (1975) The world water balance: mean annual global, continental and maritime precipitation, evaporation and runoff. Elsevier, München/Wien

Behera S, Yamagata T (2010) Imprint of the El Niño Modoki on decadal sea level changes. Gepophys Res Lett 37:L23702. doi:10.1029/2010GL045936

Boyer TP, Levitus S, Antonov JI, Locarnini RA, Garcia HE (2005) Linear trends in salinity for the World Ocean, 1955–1998. Geophys Res Lett 32, L01604. doi:10.1029/2004GL021791

California Department of Water Resources (2009) California water plan update 2009. Sacramento. http://www.waterplan.water.ca.gov/cwpu2009/index.cfm

Chao BF (1995) Anthropogenic impact on global geodynamics due to reservoir water impoundment. Geophys Res Lett 22:3529–3532

Chao BF, O'Connor WP (1988) Global surface-water-induced seasonal variations in the Earth's rotation and gravitational field. Geophys J 94(2):263–270. doi:10.1111/j.1365-246X.1988.tb05900.x

Cooley H, Donnelly K (2012) Hydraulic fracturing and water resources: separating the frack from the fiction. Pacific Institute, Oakland, 34 pp

Cooley H, Fulton J, Gleick PH (2011) Water for energy: future water needs for electricity in the intermountain west. Pacific Institute, Oakland, 63 pp

Dettinger MD, Ralph FM, Das T, Neiman PJ, Cayan DR (2011) Atmospheric rivers, floods and the water resources of California. Water 3(2):445–478

Dirmeyer PA, Gao XA, Zhao M, Guo ZC, Oki T, Hanasaki N (2006) GSWP-2 multimodel analysis and implications for our perception of the land surface. Bull Am Meteorol Soc 87:1381–1397

Durack PJ, Wijffels SE (2010) Fifty year trends in global ocean salinities and their relationship to broad-scale warming. J Clim 23:4,342–4,362

Durack PJ, Wijffels SE, Matear RJ (2012) Ocean salinities reveal strong global water cycle intensification during 1950–2000. Science 336(6080):455–458. doi:10.1126/science 1212222

Falkenmark M, Rockström J (2004) Balancing water for humans and nature. Earthscan, London, 247pp

Famiglietti JS (2004) Remote sensing of terrestrial water storage, soil moisture and surface waters. In: Sparks RSJ, Hawkesworth CJ (eds) The state of the planet: frontiers and challenges in geophysics, vol 150, Geophysical monograph series. American Geophysical Union [S.l.]/ International Union of Geodesy and Geophysics, Washington, DC, pp 197–207

Famiglietti J, Murdoch L, Lakshmi V, Hooper R (2009) Rationale and strategy for a community modeling platform in the hydrologic sciences, Community Modeling in Hydrologic Science. Report of the CHyMP scoping workshop held March 26–27, 2008, Washington, DC, CUAHSI technical report #8, April 12. doi:10.4211/techrpts.200911.tr8

Famiglietti J, Murdoch L, Lakshmi V, Hooper R (2010) Towards a framework for community modeling in hydrologic science. Report from the 2nd workshop on a Community Hydrologic Modeling Platform (CHyMP): blueprint for a community hydrologic modeling platform, Memphis, CUAHSI Technical report #9, March 31– April 1, 2009. doi:10.4211/techrpts.20100616.tr9

Famiglietti JS, Lo M, Ho SL, Bethune J, Anderson KJ, Syed TH, Swenson SC, de Linage CR, Rodell M (2011a) Satellites measure recent rates of groundwater depletion in California's Central Valley, Geophys Res Lett 38, L03403. doi:10.1029/2010GL046442

Famiglietti J, Murdoch L, Lakshmi V, Arrigo J (2011b) Establishing a framework for community modeling in hydrologic science. Report from the 3[4d] workshop on a Community Hydrologic Modeling Platform (CHyMP): a strategic and implementation plan, CUAHSI technical report #10, Irvine, March 15–17, 2011

FC-GWSP (Framing Committee of the Global Water System Project) (2004a) Humans transforming the global water system. Eos AGU Trans 85(509):513–514

FC-GWSP (Framing Committee of the Global Water System Project) (2004b) The global water system project: science framework and implementation activities. Earth System Science Partnership Project. Global Water System Project Office, Bonn

Fekete BM, Vörösmarty CJ, Grabs W (2002) High resolution fields of global runoff combining observed river discharge and simulated water balances. Global Biogeochem Cycles 6(3):1. doi :10.1029/1999GB001254

Fekete BM, Wisser D, Kroeze C, Mayorga E, Bouwman L, Wollheim WM, Vorosmarty C (2010) Millennium Ecosystem Assessment scenario drivers (1970–2050): climate and hydrological alterations. Global Biogeochem Cycles 24:GB0A12. doi:10.1029/2009GB003593

Galloway JN, Dentener FJ, Capone DG, Boyer EW, Howarth RW, Seitzinger SP, Asner GP, Cleveland CC, Green PA, Holland EA, Karl DM, Michaels AF, Porter JH, Townsend AR, Vörösmarty CJ (2004) Nitrogen cycle: past, present and future. Biogeochemistry 70:153–226

Gleick PH (2000) Water: the potential consequences of climate variability and change for the water resources of the United States. U.S. Global Change Research Program, Washington, DC, 151 pp

Gleick PH (2003) Global freshwater resources: soft-path solutions for the 21st century. Science 302:1524–1528

Gleick PH, Palaniappan M (2010) Peak water: conceptual and practical limits to freshwater withdrawal and use. Proc Natl Acad Sci (PNAS) 107(25):11155–11162. www.pnas.org/cgi/doi/10.1073/pnas.1004812107

Gleick PH, Christian-Smith J, Cooley H (2011) Water-use efficiency and productivity: rethinking the basin approach. Water Int 36(7):784–798. doi:10.1080/02508060.2011.631873, http://dx.doi.org/

Group on Earth Observations (GEO) (2007) Strategic guidance for current and potential contributors to GEOSS. http://www.earthobservations.org/documents.shtml

Group on Earth Observations (GEO) (2009) Implementation guidelines for the GEOSS data sharing principles. Group on Earth Observations, Geneva

Haddeland I, Lettenmaier DP, Skaugen T (2006a) Effects of irrigation on the water and energy balances of the Colorado and Mekong river basins. J Hydrol 324:210–223

Haddeland I, Skaugen T, Lettenmaier DP (2006b) Anthropogenic impacts on continental surface water fluxes. Geophys Res Lett 33, L08406. doi:10.1029/2006GL026047

Haddeland I, Skaugen T, Lettenmaier DP (2007) Hydrologic effects of land and water management in North America and Asia: 1700–1992. Hydrol Earth Syst Sci 11(2):1035–1045

Haddeland I, Clark D, Franssen W, Ludwig F, Voss F, Arnell NW, Bertrand N, Best M, Folwell S, Gerten D, Gomes S, Gosling SN, Hagemann S, Hanasaki N, Harding R, Heinke J, Kabat P, Koirala S, Oki T, Polcher J, Stacke T, Viterbo P, Weedon GP, Yeh P (2011) Multi-model estimate of the terrestrial global water balance: setup and first results. J Hydrometeorol 12:869–884. doi:10.1175/2011JHM1324.1

Hanasaki N, Kanae S, Oki T (2006) A reservoir operation scheme for global river routing models. J Hydrol 327:22–41

Hanasaki N, Kanae S, Oki T, Masuda K, Motoya K, Shirakawa N, Shen Y, Tanaka K (2008a) An integrated model for the assessment of global water resources – Part 1: model description and input meteorological forcing. Hydrol Earth Syst Sci 12:1007–1025

Hanasaki N, Kanae S, Oki T, Masuda K, Motoya K, Shirakawa N, Shen Y, Tanaka K (2008b) An integrated model for the assessment of global water resources – Part 2: applications and assessments. Hydrol Earth Syst Sci 12:1027–1037

Hanasaki N, Inuzuka T, Kanae S, Oki T (2010) An estimation of global virtual water flow and sources of water withdrawal for major crops and livestock products using a global hydrological model. J Hydrol 384:232–244

Harmon T, Ambrose R, Gilbert R, Fisher J, Stealey M (2007) High-resolution river hydraulic and water quality characteristics using rapidly deployable networked infomechanical systems. Environ Eng Sci 24(2):151–159

He B, Kanae S, Oki T, Hirabayashi Y, Yamashiki Y, Takara K (2011) Assessment of global nitrogen pollution in rivers using an integrated biogeochemical modeling framework. Water Res 45(8):2573–2586

Hirschboeck KK (2009) Future hydroclimatology and the research challenges of a post-stationary world. J Contemp Water Res Educ 142(1):4–9. http://onlinelibrary.wiley.com/doi/10.1111/j.1936-704X.2009.00045.x/pdf

Hornberger C (2001) A plan for a new science initiative on the global water cycle. Chapter 5: An integrated water cycle science plan. Report to the USGCRP from the Water Cycle Study Group. http://www.usgcrp.gov/usgcrp/Library/watercycle/wcsgreport2001/wcsg2001chapter5.htm

Huntington T (2006) Evidence for intensification of the global water cycle: review and synthesis. J Hydrol 319:83–95

ICSU (International Council for Science) (2004) ICSU report of the CSPR assessment panel on scientific data and information. ICSU, Paris

IPCC (2007) Summary for policymakers. In: Solomon S, Qin D, Manning M, Chen Z, Marquis M, Averyt KB, Tignor M, Miller HL (eds) Climate change 2007: the physical science basis. Contribution of working group I to the fourth assessment report of the intergovernmental panel on climate change. Cambridge University Press, Cambridge/New York

IPCC (2012) Summary for policymakers. In: Field CB, Barros V, Stocker TF, Qin D, Dokken DJ, Ebi KL, Mastrandrea MD, Mach KJ, Plattner G-K, Allen SK, Tignor M, Midgley PM (eds) Managing the risks of extreme events and disasters to advance climate change adaptation. A special report of working groups I and II of the intergovernmental panel on climate change, Cambridge University Press, Cambridge/New York, pp 1–19

Jin YF, Randerson JT, Goulden ML (2011) Continental-scale net radiation and evapotranspiration estimated using MODIS satellite observations. Remote Sens Environ 115(9):2302–2319. doi:10.1016/j.rse.2011.04.031, Published: SEP 15 2011

Karl TR, Melillo JM, Peterson TC (eds) (2009) Global change impacts in the United States. Cambridge University Press, Cambridge, 188 pp

Kollet SJ, Maxwell RM (2008) Capturing the influence of groundwater dynamics on land surface processes using an integrated, distributed watershed model. Water Resour Res 44, W02402. doi:10.1029/2007WR006004

Korzun VI (1978) World water balance and water resources of the earth, studies and reports in hydrology, vol 25. UNESCO, Paris

Kundzewicz ZW, Mata LJ, Arnell NW, Döll P, Kabat P, Jiménez B, Miller KA, Oki T, Sen Z, Shiklomanov IA (2007) Freshwater resources and their management. In: Parry ML, Canziani OF, Palutikof JP, van der Linden PJ, Hanson CE (eds) Climate change 2007: impacts, adaptation and vulnerability. Contribution of working group II to the fourth assessment report of the intergovernmental panel on climate change. Cambridge University Press, Cambridge, pp 173–210

L'vovitch MI (1973) The global water balance. Trans Am Geophys Union 54:28–42

Landerer FW, Swenson SC (2011) Accuracy of scaled GRACE terrestrial water storage estimates. Wat Resour Res 48(4). doi:10.1029/2011WR011453

Lavers DA, Allan RP, Wood EF, Villarini G, Brayshaw J, Wade J (2011) Winter floods in Britain are connected to atmospheric rivers. Geophys Res Lett 38:L23803. doi:10.1029/2011GL049783, 2011

Lazo JK, Morss RE, Demuth JL (2009) 300 Billion served: sources, perceptions, uses, and values of weather forecasts. Bull Am Meteorol Soc 90:785–798

Lehner B, Reidy Liermann C, Revenga C, Fekete B, Vörösmarty CJ, Crouzet P, Döll P, Endejan M, Frenken K, Magome J, Nilsson C, Robertson JC, Rödel R, Sindorf N, Wisser D (2011) High resolution mapping of global reservoirs and dams and their downstream river impacts. Front Ecol Environ. doi:10.1890/100125

Lettenmaier DP, Milly PCD (2009) Land waters and sea level. Nat Geosci 2:452–454. doi:10.1038/ngeo567

Loucks D (2007) Water resources and environmental management: issues, challenges, opportunities and options. Water Sci Technol Water Supply 7(2):1–10

Meehl GA, Arblaster JM, Tebaldi C (2007) Contributions of natural and anthropogenic forcing to changes in temperature extremes over the United States. Geophys Res Lett 34, L19709. doi:10.1029/2007GL030948

Meehl GA, Tebaldi C, Walton G, Easterling D, McDaniel L (2009) Relative increase of record high maximum temperatures compared to record low minimum temperatures in the U.S. Geophys Res Lett 36:L23701. doi:10.1029/2009GL040736

Melillo JM, Reilly JM, Kicklighter DW, Gurgel AC, Cronin TW, Paltsev S, Felzer BS, Wang X, Sokolov AP, Schlosser CA (2009) Indirect emissions from biofuels: how important? Science 326:1397–1399. doi:10.1126/science.1180251

Meybeck M (2003) Global analysis of river systems: from earth system controls to Anthropocene syndromes. Philos Trans R Soc Lond Ser B. doi:10.1098/rstb.2003, pp. 1379

Meybeck M, Vörösmarty CJ (2005) Fluvial filtering of land to ocean fluxes: from natural Holocene variations to Anthropocene. Compte Rend 337:107–123

Milly PCD, Betancourt J, Falkenmark M, Hirsch RM, Kundzewicz ZW, Lettenmaier DP, Stouffer RJ (2008) Stationarity is dead: whither water management? Science 319:573–574

Milly PCD, Cazenave A, Famiglietti J, Gornitz V, Laval K, Lettenmaier DP, Sahagian D, Wahr J, Wilson CR (2010) In: Church JA, Woodworth PL, Aarup T, Wilson WS (eds) Terrestrial water-storage contributions to sea-level rise and variability. Wiley/Blackwell, Hoboken, 421 pp

Mishra AK, Coulibaly P (2009) Developments in hydrometric network design: a review. Rev Geophys 47, RG2001. doi:10.1029/2007RG000243

Neumeyer J, Barthelmes F, Dierks O, Flechtner F, Harnisch M, Harnisch G, Hinderer J, Imanishi Y, Kroner C, Meurers B, Petrovic S, Reigber C, Schmidt R, Schwintzer P, Sun HP, Virtanen H (2006) Combination of temporal gravity variations resulting from superconducting gravimeter (SG) recordings, GRACE satellite observations and global hydrology models. J Geod 79(10–11):573–585

NRC (National Research Council) (1998) Global environmental change: research pathways for the next decade. National Academy Press, Washington, DC

NRC (National Research Council) (1999a) Hydrologic science priorities for the U.S. Global change research program. National Academy Press, Washington, DC

NRC (National Research Council) (1999b) New strategies for America's watersheds. National Academy Press, Washington, DC

NRC (National Research Council) (2000) From research to operations in weather satellites and numerical weather prediction: crossing the valley of death. National Academy Press, Washington, DC

NRC (National Research Council) (2002a) Review of USGCRP plan for a New science initiative on the global water cycle. National Academy Press, Washington, DC
NRC (National Research Council) (2002b) Report of a workshop on predictability & limits-to-prediction in hydrologic systems. National Academy Press, Washington, DC
NRC (National Research Council) (2005) The science of instream flows: a review of the Texas instream flow program. National Academy Press, Washington, DC
NRC (National Research Council) (2007) Earth science and applications from space: national imperatives for the next decade and beyond. National Academy Press, Washington, DC
NRC (National Research Council) (2008a) Earth observations from space: the first 50 years of scientific achievements. National Academy Press, Washington, DC
NRC (National Research Council) (2008b) Integrating multiscale observations of U.S. waters. National Academy Press, Washington, DC
NRC (National Research Council) (2008c) Water implications of biofuels production in the United States. National Academy Press, Washington, DC
NRC (National Research Council) (2009) Observing weather and climate from the ground up: a nationwide network of networks. National Academy Press, Washington, DC
NRC (National Research Council) (2010a) Review of the WATERS network science plan. National Academy Press, Washington, DC
NRC (National Research Council) (2010b) Expanding biofuel production: sustainability and the transition to advanced biofuels: summary of a workshop. National Academy Press, Washington, DC
NRC (National Research Council) (2011a) Global change and extreme hydrology: testing conventional wisdom. Committee on hydrologic sciences. National Academy Press, Washington, DC
NRC (National Research Council) (2011b) America's climate choices. National Research Council, committee on America's climate choices. National Academies Press, Washington, DC, 144 pp
Okazawa Y, Yeh PJ-F, Kanae S, Oki T (2011) Development of a global flood risk index based on natural and socio-economic factors. Hydrol Sci J 56(5):789–804
Oki T (1999) The global water cycle. In: Browning K, Gurney R (eds) Global energy and water cycles. Cambridge University Press, Cambridge, pp 10–27
Oki T, Kanae S (2006) Global hydrological cycles and world water resources. Science 313:1068–1072
Oki T, Agata Y, Kanae S, Saruhashi T, Yang D, Musiake K (2001) Global assessment of current water resources using total runoff integrating pathways. Hydrol Sci J 46:983–996
Oki T, Entekhabi D, Harrold T (2004) The global water cycle. In: Sparks R, Hawkesworth C (eds) State of the planet: frontiers and challenges in geophysics, No. 150 in geophysical monograph series. AGU Publication, Washington, DC, p 414
Oki T, Valeo C, Heal K (2006) Hydrology 2020: an integrating science to meet world water challenges. IAHS Publication, Wallingford, p 300
Oki T, Blyth EM, Berbery EH, Alcaraz-Segura D (2012) Land cover and land use changes and their impacts on hydroclimate, ecosystems and society. Plenary paper for the WCRP open science conference, Denver, October 2012. DRAFT, 18 February 2012
Palaniappan M, Gleick PH, Allen L, Cohen MJ, Christian-Smith J, Smith C (2010) Clearing the waters: a focus on water quality solutions. United Nations Environment Programme, Nairobi, 88p
Parson M (2011) Expert report on data policy – open access. GRDI 2020, pp 8. see http://www.grdi2020.eu/StaticPage/About.aspx
Pokhrel Y, Hanasaki N, Koirala S, Cho J, Yeh PJ-F, Kim H, Kanae S, Oki T (2011) Incorporating anthropogenic water regulation modules into a land surface model. J Hydrometeorol 13:255–269
Pokhrel Y, Hanasaki N, Yeh PJ-F, Yamada TJ, Kanae S, Oki T (2012) Model estimates of sea-level change due to anthropogenic impacts on terrestrial water storage. Nat Geosci. doi:10.1038/ngeo1476
Ralph FM, Neiman PJ, Kiladis GN, Weickmann K, Reynolds DW (2011) A multiscale observational case study of a Pacific atmospheric river exhibiting tropical-extratropical connections and a mesoscale frontal wave. Mon Weather Rev 139(4):1169–1189

Ramillien G, Cazenave A, Brunau O (2004) Global time variations of hydrological signals from GRACE satellite gravimetry. Geophys J Int 158(3):813–826

Rodell M, Famiglietti JS, Chen J, Seneviratne S, Viterbo P, Holl SL, Wilson CR (2004a) Basin-scale estimates of evapotranspiration using GRACE and other observations. Geophys Res Lett 31(20):L20504. doi:10.1029/2004GL020873

Rodell M, Houser PR, Jambor U, Gottschalck J, Mitchell K, Meng C-J, Arsenault K, Cosgrove B, Radakovich J, Bosilovich M, Entin JK, Walker JP, Lohmann D, Toll D (2004b) The global land data assimilation system. Bull Am Meteorol Soc 85:381–394. doi:10.1175/BAMS-85-3-381

Rodell M, Famiglietti JS, Scanlon BR (2010) Realizing the potential for satellite gravimetry in hydrology: second GRACE hydrology workshop, August 4, 2009, Austin, TX. EOS Trans AGU 91(10):96

Sahoo AK, Pan M, Troy TJ, Vinukollu RK, Sheffield J, Wood EF (2011) Reconciling the global terrestrial water budget using satellite remote sensing. Remote Sens Environ 115(8):1850–1865

Shapiro M, Shukla J, Brunet G, Nobre C, Béland M, Dole R, Trenberth K, Anthes R, Asrar G, Barrie L, Bougeault P, Brasseur G, Burridge D, Busalacchi A, Caughey J, Chen D, Church J, Enomoto T, Hoskins B, Hov O, Laing A, Le Treut H, Marotzke J, McBean G, Meehl G, Miller M, Mills B, Mitchell J, Moncrieff M, Nakazawa T, Olafsson H, Palmer T, Parsons D, Rogers D, Simmons A, Troccoli A, Toth Z, Uccellini L, Velden C, Wallace M (2010) An earth-system prediction initiative for the 21st century. Bull Am Meteorol Soc 91:1377–1388

Shen Y, Oki T, Utsumi N, Kanae S, Hanasaki N (2008) Projection of future world water resources under SRES scenarios: water withdrawal. Hydrol Sci J 53(1):11–33

Shiklomanov IA (ed) (1997) Assessment of water resources and water availability in the world. Background report for the comprehensive assessment of the freshwater resources of the world, WMO/SEI

Shiklomanov AI, Lammers RB, Vörösmarty CJ (2002) Widespread decline in hydrological monitoring threatens Pan-Arctic research. AGU-Eos Trans 83(13):16–17

Stakhiv EZ (2011) Pragmatic approaches for water management under climate change uncertainty. J Am Water Resour Assoc (JAWRA). doi:10.1111/j.1752-1688.2011.00589

Stokstad E (1999) Scarcity of rain, stream gages threatens forecasts. Science 285(5431):1199–1200

Syed TH, Famiglietti JS, Rodell M, Chen J, Wilson CR (2008) Analysis of terrestrial water storage changes from GRACE and GLDAS. Water Resour Res 44:W02433. doi:10.1029/2006WR005779

Syed TH, Famiglietti JS, Chambers D (2009) GRACE-based estimates of terrestrial freshwater discharge from basin to continental scales. J Hydrometeorol 10(1):22–40. doi:10.1175/2008 JHM993.1

Syed TH, Famiglietti JS, Chambers D, Willis J, Hilburn K (2010) Satellite-based global ocean mass balance estimates of interannual variability and emerging trends in continental freshwater discharge. Proc Natl Acad Sci USA 107(42):17916–17921. doi:10.1073/pnas.1003292107, published ahead of print October 4, 2010

Syvitski JPM, Vörösmarty CJ, Kettner AJ, Green P (2005) Impact of humans on the flux of terrestrial sediment to the global coastal ocean. Science 308:376–380

Syvitski JPM, Kettner AJ, Hannon MT, Hutton EWH, Overeem I, Brakenridge GR, Day J, Vörösmarty C, Saito Y, Giosan L, Nicholls RJ (2009) Sinking deltas due to human activities. Nat Geosci 2:681–686

Trenberth KE (2011a) Attribution of climate variations and trends to human influences and natural variability. Wiley Interdiscip Rev Clim Change 2(6): 925–930

Trenberth KE (2011b) Changes in precipitation with climate change. Clim Res 47:123–138. doi:10.3354/cr00953. [PDF]

Trenberth KE, Belward A, Brown O, Haberman E, Karl TR, Running S, Ryan B, Tanner M, Wielicki B (2011) Challenges of a sustained climate observing system. In: Plenary paper for the WCRP Open Science Conference, Denver, October 2011. DRAFT, 25 July 2011

Trenberth KE, Fasullo J, Smith L (2005) Trends and variability in column-integrated atmospheric water vapor. Clim Dyn 24:741–758

Van Vliet M, Ludwig F, Kabat P, Yearsley JR, Lettenmaier DP (2012) Vulnerability of U.S. and European electricity supply to climate change. Nat Clim Change 2:676–681. doi:10.1038/nclimate1546

Vörösmarty CJ, Leveque C, Revenga C (Convening Lead Authors) (2005) Chapter 7: Fresh water. In: Bos R, Caudill C, Chilton J, Douglas EM, Meybeck M, Prager D, Balvanera P, Barker S, Maas M, Nilsson C, Oki T, Reidy CA (eds) Millennium ecosystem assessment, vol 1 conditions and trends working group report. Island Press, pp 165–207. 966 pp

Vörösmarty CJ, Meybeck M (2004) Responses of continental aquatic systems at the global scale: new paradigms, new methods. In: Kabat P, Claussen M, Dirmeyer PA, Gash JHC, Bravo L, Meybeck M, Pielke RA Sr, Vörösmarty CJ, Hutjes RWA, Lutkemeier S (eds) Vegetation, water, humans and the climate. Springer, Heidelberg, pp 375–413, 566 pp

Vörösmarty CJ, Green P, Salisbury J, Lammers RB (2000) Global water resources: vulnerability from climate change and population growth. Science 289:284–288

Vörösmarty CJ, Day J, DeSherbinen A, Syvitski J (2009) Battling to save the world's river deltas. Bull At Sci 65:31–43

Vörösmarty CJ, McIntyre PB, Gessner MO, Dudgeon D, Prusevich A, Green P, Glidden S, Bunn SE, Sullivan CA, Reidy Liermann C, Davies PM (2010) Global threats to human water security and river biodiversity. Nature 467:555–561

Weimerskirch H, Louzao M, de Grissac S, Delord K (2012) Changes in wind pattern alter albatross distribution and life-history traits. Science 335(6065):211–214. doi:10.1126/science.1210270

Wentz FJL, Ricciardulli KH, Mears C (2007) How much more rain will global warming bring? Science 317:233–235

Wisser D, Frolking S, Douglas EM, Fekete BM, Vörösmarty CJ, Schumann AH (2008) Global irrigation water demand: variability and uncertainties arising from agricultural and climate data sets. Geophys Res Lett 35:L24408. doi:10.1029/2008GL035296

WMO (World Meteorological Organization) (1995) WMO policy and practice for the exchange of meteorological and related data and products including guidelines on relationships in commercial meteorological activities. World Meteorological Organization Congress, Resolution 40 (Cg-XII, 1995), Geneva

Wollheim WM, Vörösmarty CJ, Bouwman AF, Green P, Harrison JA, Meybeck M, Peterson BJ, Seitzinger SP, Syvitski JP (2008) A spatially distributed framework for aquatic modeling of the Earth system (FrAMES). Global Biogeochem Cycles 22:GB2026. doi:10.1029/2007GB002963

Wood EF, Roundy JK, Troy TJ, van Beek LPH, Bierkens MFP, Blyth E, de Roo A, Döll P, Ek M, Famiglietti J, Gochis D, van de Giesen N, Houser P, Jaffé PR, Kollet S, Lehner B, Lettenmaier DP, Peters-Lidard C, Sivapalan M, Sheffield J, Wade A, Whitehead P (2011) Hyperresolution global land surface modeling: Meeting a grand challenge for monitoring Earth's terrestrial water. Water Resour Res 47, W05301. doi:10.1029/2010WR010090

World Climate Research Program (WCRP) (2010) GEWEX Plans for 2013 and Beyond. World Climate Research Programme. http://www.gewex.org/Imperatives.pdf

Yu L, Weller RA (2007) Objectively analyzed air-sea heat fluxes for the global ice-free oceans (1981–2005). Bull Am Meteorol Soc 88:527–539

Zhang X, Zwiers FW, Hegerl GC, Lambert FH, Gillett NP, Solomon S, Stott PA, Nozawa T (2007) Detection of human influence on twentieth-century precipitation trends. Nature 448:461–465. doi:10.1038/nature06025

Zhou YP, Xu KM, Sud YC, Betts AK (2011) Recent trends of the tropical hydrological cycle inferred from Global Precipitation Climatology Project and International Satellite Cloud Climatology Project data. J Geophys Res 116:D09101. doi:10.1029/2010JD015197

Land Use and Land Cover Changes and Their Impacts on Hydroclimate, Ecosystems and Society

Taikan Oki, Eleanor M. Blyth, Ernesto Hugo Berbery, and Domingo Alcaraz-Segura

Abstract This chapter presents recent advances in the understanding of the effect of land cover/land use changes on the hydrologic cycle, and identifies current gaps in the knowledge needed for useful decision-making and water resource management. Research achievements within a framework of Earth System Models (ESM) are introduced, and research needs and future challenges are identified. Land surface provides the lower boundary condition to the atmosphere over continents by controlling the fluxes of momentum, heat, water, and materials such as carbon. In turn, land surface conditions are substantially influenced by atmospheric conditions on various temporal scales. As such, a land-atmosphere coupled system is established through biogeochemical feedbacks. Current land surface models exhibit a wide variety of responses to the same forcings, suggesting the need for increased research at the land-atmosphere interface. Indeed, all Earth System Models require the inclusion and validation of the processes that pertain to the biogeochemical feedbacks. Anthropogenic activities that result in land use and land cover changes

T. Oki (✉)
Institute of Industrial Science, The University of Tokyo,
4-6-1 Komaba, Meguro-ku, Tokyo 153-8505, Japan
e-mail: taikan@iis.u-tokyo.ac.jp

E.M. Blyth
The Centre for Ecology & Hydrology, Maclean Lane,
Crowmarsh Gifford, Wallingford, Oxfordshire OX10 8BB, UK
e-mail: emb@ceh.ac.uk

E.H. Berbery
Earth System Science Interdisciplinary Center, University of Maryland, 5825 University Research Court, Suite 4001, College Park, MD 20740-3823, USA
e-mail: berbery@atmos.umd.edu

D. Alcaraz-Segura
Department of Botany, Faculty of Sciences, University of Granada,
Avenida de Fuentenueva s/n, Granada 18071, Spain
e-mail: dalcaraz@ugr.es

G.R. Asrar and J.W. Hurrell (eds.), *Climate Science for Serving Society: Research, Modeling and Prediction Priorities*, DOI 10.1007/978-94-007-6692-1_7,

affect the land surface characteristics and consequently the land-atmosphere feedbacks and coupling strength. Therefore, human activities play a role in the land-atmosphere coupling system, and thus, in the climate system. Water is essential to societal needs that require the construction of reservoirs, extraction of ground water, irrigation, changes in land use, urbanization among many other influences. The extent and sustainability of those interferences in the natural system remains to be assessed at global scales.

Keywords Land-atmosphere feedback • Vegetation • Ecosystem • Human impacts • Water • Energy • Carbon • Land cover/land use

1 Introduction

The land surface provides lower boundary conditions to the atmosphere: It receives downward short wave and long wave radiation, and emits or reflects upward short wave and long wave radiation. The net radiation is balanced by the fluxes of sensible, latent and ground heat to the atmosphere (Oki 1999). In terms of the water balance, precipitation is balanced by evapotranspiration and runoff (assuming that over long term periods there is no net water storage change on the soil). These exchanges also depend on the atmospheric conditions, including the surface pressure, temperature, humidity and wind. A balance mainly between precipitation, evapotranspiration and surface and deep runoff determines the land surface water cycle. Surface soil moisture, in turn, governs the partitioning of the sensible and latent heat fluxes into the atmosphere, and can affect daily, weekly, intraseasonal, seasonal, and interannual rainfall in various spatial scales through the impacts on development PBL (planetary boundary layers), its longer temporal auto-correlation ("memory" effect), and possibly through interactions with vegetation (see Table 1 of Taylor et al. 2011). Excess water from land discharges into the ocean changing its salinity and temperature, and possibly influences the formation of sea ice and thermohaline circulation at least on local scales (Oki et al. 2004).

The energy, water, and carbon balances determined by land surface processes are characterized by the land surface conditions such as topography, land cover, soil properties, and geological condition. Land cover can be characterized by the vegetation over it, such as forests, shrubs, grass, bare soil, or open water. Since vegetation types are dominantly determined by climatological conditions, land surface interacts with the atmosphere not only on the short time scales but also in longer temporal scales, such as decadal to centennial. Even though storage volumes are not as large as in the ocean, the land stores heat, water, and carbon, and thus, the land surface is one of the key components in the climate system on the Earth.

In many cases, particularly when dealing with extreme events, climatic variations and changes can have significant impacts on human activities; therefore it is critical that climate science includes and develops tools for monitoring and prediction of climatic variations. As climate affects human activities, in turn humans interfere with

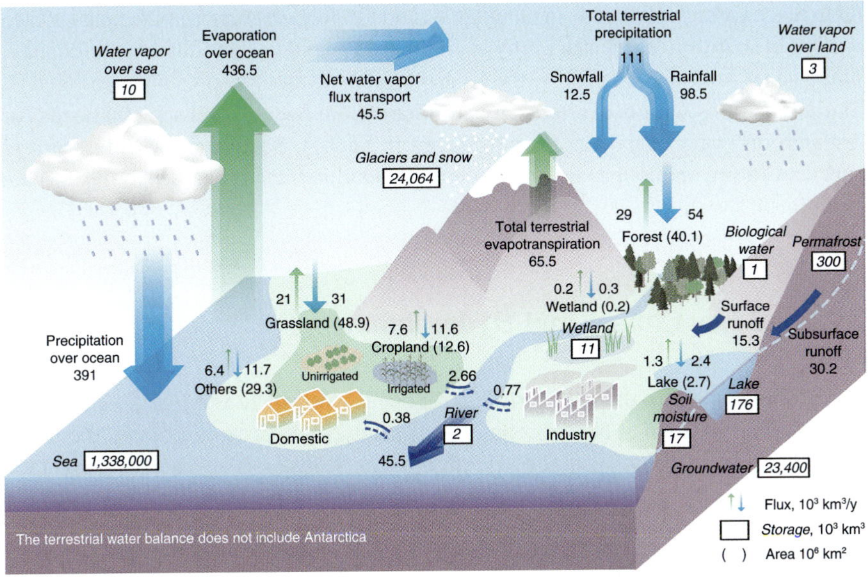

Fig. 1 Global hydrological fluxes (1,000 km³/year) and storages (1,000 km³) with natural and anthropogenic cycles are synthesized from various sources (Dirmeyer et al. 2006; Korzun 1978; Oki et al. 1995; Shiklomanov 1997). *Big vertical arrows* show total annual precipitation and evapotranspiration over land and ocean (1,000 km³/year), which include annual precipitation and evapotranspiration in major landscapes (1,000 km³/year) presented by *small vertical arrows*; *parentheses* indicate area (million km²). The direct groundwater discharge, which is estimated to be about 10 % of total river discharge globally (Church 1996), is included in river discharge

the climate system from local to global scales. Apart from human influences through greenhouse gases (GHGs, not discussed in this chapter), human influences on the ecosystem service of climate regulation occur through changes in land use and land cover (Anderson-Teixeira et al. 2012), as well as through interventions on the water cycle components, for example by irrigation (Rosnay et al. 2003; Guimberteau et al. 2011) and storage in artificial reservoirs (Haddeland et al. 2006; Hanasaki et al. 2006, 2010).

The World Climate Research Programme (WCRP) emphasis on the role of land in the climate system has been mainly conducted through the Global Energy and Water Cycle Experiment (GEWEX). The GEWEX Hydroclimate Panel (GHP) has been promoting and synthesizing field campaigns measuring, estimating, and seeking to close the regional water balances in various climatic zones at continental and sub-continental scales. The Global Land-Atmosphere System Study (GLASS; van den Hurk et al. 2011) has been promoting and organizing numerical studies assessing the coupling between land and atmosphere, and the Global Data and Assessments Panel (GDAP) supports the creation and dissemination of comprehensive datasets of the climatic variables over land. The products from the Second Global Soil Wetness Project (GSWP-2; Dirmeyer et al. 2006) contributed to illustrate the global water cycles as shown in Fig. 1 (Oki and Kanae 2006).

In this chapter, we discuss the feedbacks and interactions between the land surface and the climate system, particularly with regard to land use and land cover change. The role of land use change in the hydro-climate system is presented in Sect. 2. The interactions with ecosystems are summarized in Sect. 3, and societal needs for research on water over land are introduced in Sect. 4. Section 5 identifies current gaps and future challenges for the research on land surface processes in the climate system.

2 Land Use Change and Hydroclimate

Long term changes to the land surface state occur when there is a significant change in the land cover, such as conversions from forest to crops. In cases like this, there will be changes in the biophysical properties of the surface, like its albedo, surface roughness length, and stomatal resistance. In addition, there will be changes to the hydrological functioning of the land surface, with changes in the amount of water available for storage and the runoff, possibly through changes in the soil properties and root uptake.

Many researchers have worked to quantify the impact that such changes have on the atmosphere. For instance, modeling experiments have been carried out to understand the regional climate impact of the wide-scale spread of agriculture that has occurred over the last century (see Pitman et al. 2009). Specifically, the goal was to assess if the current regional climate has been influenced by the anthropogenically altered landscape. The model results showed varying responses of the evaporation and rainfall to the deforestation, as the changes were small and of either sign. Part of the reason is due to difficulties in defining a consistent definition of vegetation characteristics for natural versus anthropogenic land use types and differences in parameterization in the models. However, the models were in better agreement on the changes in the air temperature: removing the forests and replacing them with crops and pasture cools the summer air by about 1° in the last 100 years in the two key regions of largest land use change: the middle of the USA and western Russia. This result is supported by an observational study of evaporation and sensible heat flux observations from a series of paired forest and grass sites across Europe by Teuling et al. (2010), which demonstrated, similar to the models, that the forests generally warm the atmosphere compared to grasses and crops. However, Teuling et al. (2010) also showed how this signal changes during drought conditions, when the grasses dry out and then warm the atmosphere more than the forests. Figure 2 is a schematic summarizing the findings of Teuling et al. (2010) and of Pitman et al. (2009), showing how the forests act to warm the overlying atmosphere under normal climatic periods, while grasses or crops warm the atmosphere during anomalously dry periods. This has important implications for the physical response to land use change and its impact on the regional meteorology, since an increasing cropped area may act to enhance the regional susceptibility to heat waves, while reforestation may act to reduce a heat wave. Clearly, more research and a combined approach to risks and hazards (such as wild fire) are necessary to support this conclusion.

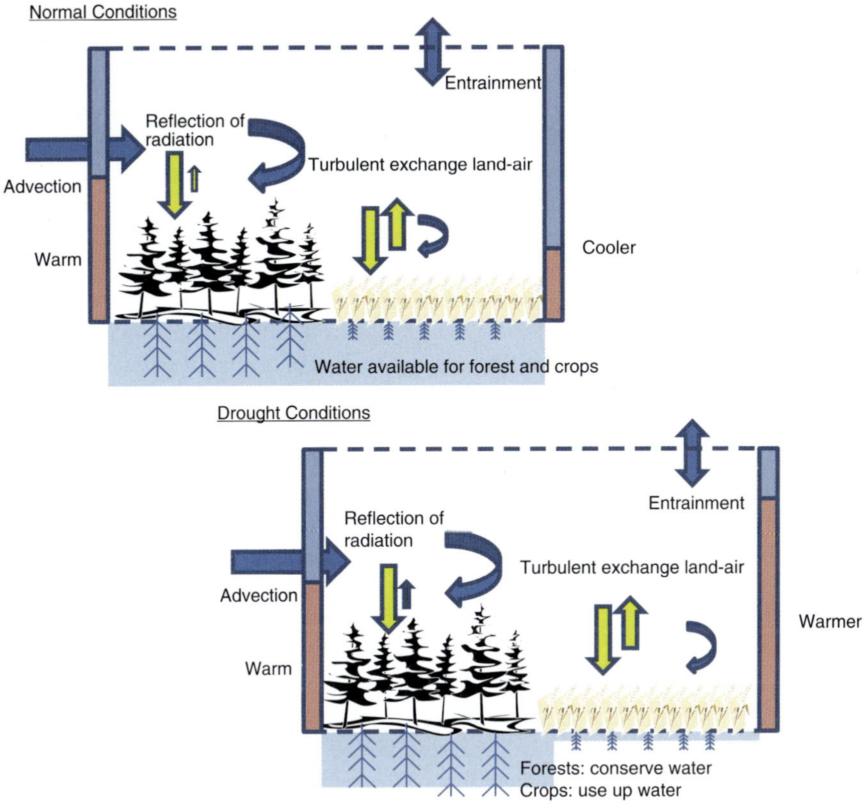

Fig. 2 Summary of impact of land-cover on atmospheric conditions

As well as impacts on the heat and temperature of a region, impacts of land cover change on the hydrological conditions should be expected due to feedbacks in the system. The relationships between the land and the atmosphere are part of the natural interplay that happens all around us: with a long term reduction in rainfall, the land dries out and this warms and dries the atmosphere which leads to further drying out of the land. This positive feedback means that a percentage drop in rainfall leads to a greater percentage drop in runoff and vice versa. Many articles have discussed the mechanisms by which a change in land cover can affect the overlying planetary boundary layer (PBL), its thermodynamic properties and circulation, and consequently the precipitation processes and regional climate (e.g., Pielke and Avissar 1990; Stohlgren et al. 1998; Kanae et al. 2001; Pielke et al. 2007, 2011; Lee and Berbery 2012). This feedback can be important for water resources, for instance, Cai et al. (2009) have demonstrated the role that land-atmosphere feedbacks have had on the recent Australian drought: their model results imply that feedbacks in the system act to exaggerate a drying period and that, during a warm, dry period, the feedbacks in the climate system act to extend the dry period. In contrast, there are areas where the land use change involves extensive moistening of the land through irrigation. This might be the case in India where the strength of the monsoon is determined by the land-sea temperature contrast and decreasing surface temperatures due to irrigation would be expected to reduce the intensity of the monsoon systems (Lee et al. 2009). Tuinenburg et al. (2011)'s study of the observed (from Radiosondes) atmospheric structures in the region show a potential alteration of the timing of the monsoon due to changes in PBL moisture from irrigated land. Douville et al. (2001) conclude that although precipitation does increase as a consequence of increasing evaporation this is somewhat counterbalanced, in the case of the Indian peninsula, by a reduced moisture convergence. Saeed et al. (2011) looked at these influences in more detail using a regional climate model, with and without irrigation. They found increased rainfall over the irrigated areas due to increased local moisture recycling and also an increase of the penetration of rain bearing depressions travelling inland from the Bay of Bengal, caused by a reduction in the westerly flows from the Arabian Sea.

Several researchers have managed to capture this large-scale long-term relationship between climatological precipitation (P), evapotranspiration (E) and potential evapotranspiration (PE) and, by implication, runoff (R), but possibly the most famous empirical equation was derived by Budyko (1974); see also Choudhury (1999):

$$E = \frac{PPE}{(P^{n} + PE^{n})^{\frac{1}{n}}} \tag{1}$$

Where 'n' is a catchment specific dimensionless factor (Roderick and Farquhar 2011). The shape of this curve for various values of 'n' is shown in Fig. 3. Roderick and Farquhar (2011) examined the effect of this relation on freshwater flows at the global scale and how well the climate models are able to represent it. They note that there are different regional responses to the large scale forcing of the water balance: in some regions where 'n' is high, changes in runoff follow closely the changes in precipitation. In other systems or regions where 'n' is low, changes in runoff are always greater than the changes in precipitation. Part of the reason for the differences is associated with different rainfall types (see Porporato et al. 2004) and different topographic and land-cover responses to rainfall. Other influences include atmospheric feedbacks with the atmosphere as outlined in the previous section. In addition, an analysis by Zhang et al. (2004) showed that the land cover is a factor in defining 'n' with forests displaying a higher 'n' compared to data from grass sites (see their Figure 8). This result is confirmed by Yang et al. (2009). The change from forest to grass decreases the 'n' from 2.12 to 1.83. Since it is logical that the value of 'n' is affected by the strength of the land-atmosphere feedbacks, the results from Zhang et al. (2004) suggest that forests have a higher feedback strength than crops, a point that has also been made by Bonan (2008). This is consistent with the result of Teuling et al. (2010) who showed that forests have a conservative approach to the water use, so as precipitation drops and evaporative demand increases, the evaporation decreases quickly. Grasses and crops however do not drop their evaporation so quickly (they have a more linear response to precipitation decrease) and they lose the water, thus leading to hotter drier conditions in drought conditions. The larger feedback strength of forested regions is also consistent with

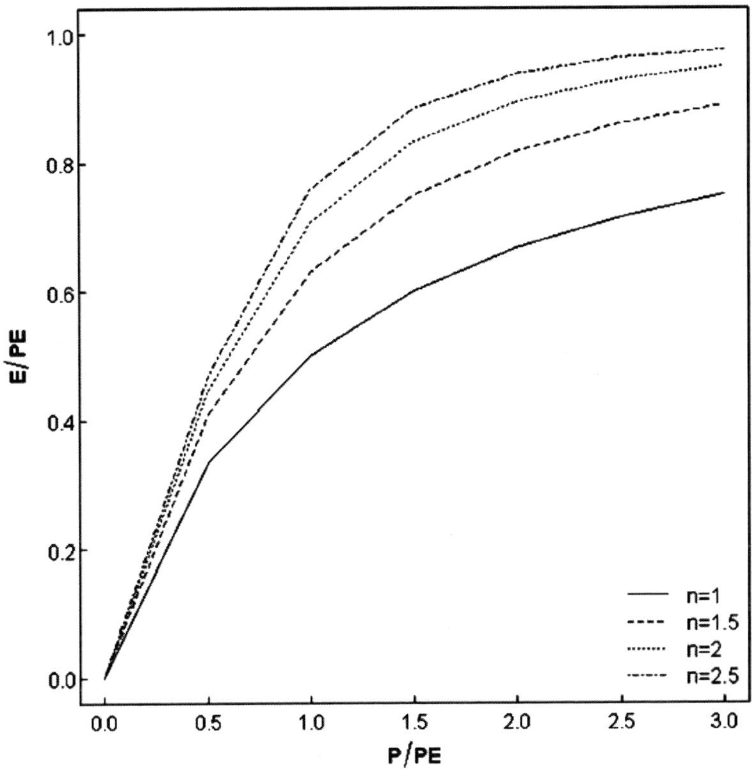

Fig. 3 Ratio of evaporation to potential evaporation as a function of the ratio of precipitation to potential evaporation (aka the Budyko Curve) for different values of 'n'

the finding of McNaughton and Spriggs (1989), who used a PBL model and found that the Priestley-Taylor parameter – which is a measure of the strength of land-atmosphere interactions – should be higher for forests than for grasses.

According to this analysis, the impact of having a decreased level of feedback between the surface and the atmosphere when changing the land cover from forest to crops and pastures is to reduce the sensitivity of the change in runoff to changes in precipitation. This will mean a more linear relationship between changes in precipitation and river flow, with less conservation of water and more drought vulnerability. These conclusions need to be more thoroughly examined with large scale observations and models.

3 Land Use Change and Ecosystems

Climate is the main regional driver of ecosystem structure and functioning through the timing and amount of energy and water that is available in the system (Stephenson 1990). In turn, ecosystems influence climate by determining the energy, momentum, water, and chemical balances between the land-surface and the atmosphere (Chapin et al. 2008). Hence, extensive impacts on ecosystems, both from natural origin and human made (e.g., land use changes), alter one or several pathways of the ecosystem–climate feedbacks, which ends up affecting the regional and global climate.

Indeed, several studies (e.g., Pielke et al. 2002; Kalnay and Cai 2003; Weaver and Avissar 2001; Werth and Avissar 2002) have concluded that the contribution of land-use changes to climate change might be about 10 % of the total global change, but that regionally the relative contribution of land-use change may be notably larger, even larger than that from greenhouse gas emissions. There are conspicuous known cases showing how land-use changes may end up altering the regional climate, such as the aridification of the Mediterranean basin during the Roman Period (Reale and Dirmeyer 2000; Reale and Shukla 2000), or changes in the hydrometeorology of Amazonia after deforestation (Baidya Roy and Avissar 2002; Gedney and Valdes 2000). In South America, inter-annual variability in climate conditions significantly affects vegetation structural and functional properties (Phillips et al. 2009; Brando et al. 2010; Zhao and Running 2010), whose effects may end up influencing the regional climate.

The ecosystem-climate feedbacks are a central problem not only for modeling the land-atmosphere interactions of the climate system (e.g., Mahmood et al. 2010), but also for many other biological and environmental issues. Ecosystem-atmosphere interactions and feedbacks depend on the physical properties of the underlying surface, like surface albedo, surface roughness, and stomatal resistance, among others. These properties affect the radiation balance at the surface as well as the exchange of momentum, heat, moisture, and other gaseous/aerosol materials. Changes in the structure and functioning of the ecosystems will thus have an impact on those exchanges that may end up affecting the climate regulation service that ecosystems provide to societies (Anderson-Teixeira et al. 2012).

Many land surface models do not consider the inter-annual dynamics of ecosystems. Models of intermediate complexity have static vegetation or land-cover classes with look-up tables to identify their corresponding biophysical properties (Chen and Dudhia 2001; Ek et al. 2003). Land cover types are assumed to remain constant but, in reality, they may experience important changes. For instance, the biophysical properties of a typical vegetation type during a wet period should be very different during a drought. The same is true during anomalous periods of intense rain that can create numerous ponds, or flooding. A model that assumes constant surface properties will still be able to represent in general changes in soil moisture content and water stress, but will be unable to represent the different conditions that emerge, e.g., when a field is flooded affecting land-atmosphere interactions, the radiation budget, and the surface water, energy and carbon cycles. Dynamical vegetation models that include the carbon cycle are an attempt to advance in the area of ecosystem-atmosphere interactions, since they allow for changes in vegetation composition and have advanced assumptions regarding surface processes that will feed back into the atmosphere. Yet, direct human-imposed land use change, as deforestation and land cover conversions may have an immediate impact on the atmosphere, as opposed to the slower effects included in a dynamical vegetation model.

Traditionally, land-cover maps are mainly driven by vegetation structure and composition but do not formally include ecosystem functional aspects such as the dynamics of carbon gains. Ecosystems functional attributes (i.e., different aspects of the exchange of matter and energy between the biota and the atmosphere) add some advantages to the traditional use of structural variables. First, variables describing ecosystem functioning have a faster response to disturbances than vegetation structure (Milchunas and Lauenroth 1995). Second, functional attributes allow the quantitative and qualitative characterization of ecosystems services (e.g., carbon sequestration, nutrient and water cycling) (Costanza et al. 1998). Additionally, they can be more easily monitored than structural attributes by using remote sensing at different spatial scales, over large extents, and utilizing a common protocol (Foley et al. 2007). Functional descriptors of ecosystems have been successfully used to define Ecosystem Functional Types (EFTs) (Alcaraz-Segura et al. 2006, 2013; see also Körner 1994; Valentini et al. 1999; Paruelo et al. 2001). In ecology, such classifications into functional units aim to reduce the diversity of biological entities (e.g. ecosystems) on the basis of processes, and allow for the identification of homogeneous groups that show a specific and coordinated response to the environmental factors. EFTs are groups of ecosystems that share functional characteristics in relation to the amount and timing of the exchanges of matter and energy between the biota and the physical environment. In other words, EFTs are homogeneous patches of the land surface that exchange mass and energy with the atmosphere in a common way (Valentini et al. 1999; Paruelo et al. 2001; Alcaraz-Segura et al. 2006, 2013a, 2013b). EFTs are computed from satellite information (e.g., spectral vegetation indices), so they do not identify the functions of a given plant species (as it occurs with plant functional types; see Wright et al. 2006), but instead identify a patch of land that has homogeneous properties in terms of exchanges of energy and mass over a given region. EFTs can thus be considered a top-down functional classification directly based on ecosystem processes.

The definition of EFTs relies in three metrics derived from the NDVI (Normalized Difference Vegetation Index) time series. First, the average of NDVI over 1 year (NDVI-mean) is a linear estimator of the amount of solar energy that is used for photosynthesis, formally called the Fraction of Absorbed Photosynthetically Active Radiation (fAPAR), and is empirically (Paruelo et al. 1997) and conceptually (Monteith 1972) related to net primary production (NPP; Tucker and Sellers 1986). Second, the seasonal coefficient of variation (CV) is a measure of the intra-annual variation of photosynthetic activity, which has been used as an indicator of the seasonality of carbon fluxes or the amplitude of the annual cycle (Oesterheld et al. 1998; Potter and Brooks 1998; Guerschman et al. 2003). Third, the phenology, or date of the absolute maximum of NDVI (DMAX), indicates the intra-annual distribution of the period with maximum photosynthetic activity (Lloyd 1990; Hoare and Frost 2004). These three metrics capture important features of ecosystem functioning for temperate ecosystems (Pettorelli et al. 2005; Lloyd 1990; Paruelo and Lauenroth 1995; Nemani and Running 1997; Paruelo et al. 2001; Virginia et al. 2001) and up to 90 % of the variability of the NDVI temporal dynamics (Paruelo et al. 2001; Alcaraz-Segura et al. 2006, 2009).

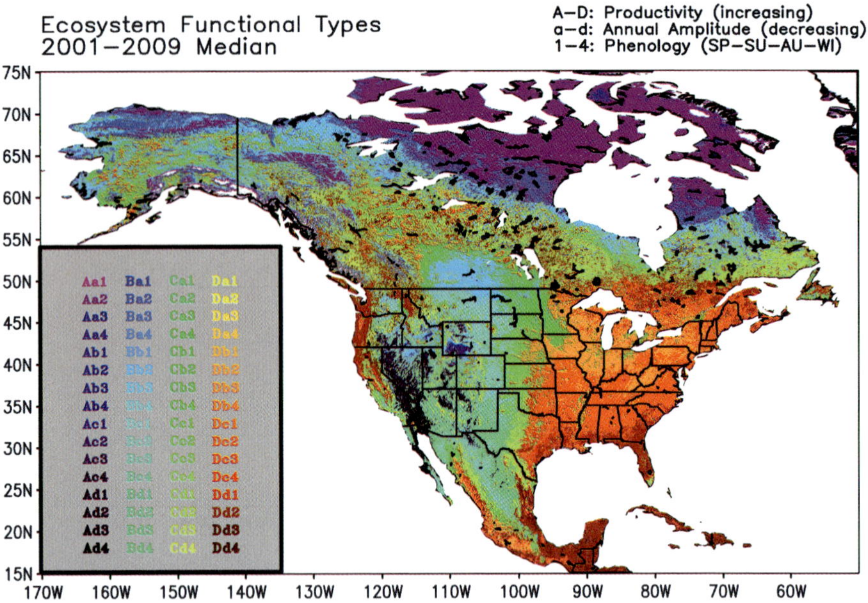

Fig. 4 Ecosystem Functional Types based on three descriptors of the seasonal dynamics of the NDVI estimated from MODIS images for the period 2001–2009

Figure 4 is an example that presents the median of the 64 EFTs for North America as computed from MODIS (or Moderate Resolution Imaging Spectroradiometer) NDVI. The warm colors indicate greatest exchanges of mass and energy between the ecosystems and the atmosphere. As expected, these regions include the coastline of the Gulf of Mexico extending over the Great Plains, subtropical forests surrounding the Gulf of Mexico and the Caribbean, the Pacific coast, the North American Monsoon in northwestern Mexico and the East Coast states. On the other hand, desert regions in Arizona and Nevada, where the net productivity is very low, are depicted with dark colors; tundra is distinctly identified in light purple. The figure depicts the median EFTs for 2001–2009, but since EFTs can be defined on a year-to-year basis, they can give a much better representation of time-varying surface states. Since EFTs are identified from time-series of satellite-derived estimates of the carbon gains dynamics (e.g. spectral vegetation indices such as NDVI and EVI), differences between sensors and datasets may occur due to the corrections applied (Alcaraz-Segura et al. 2010a). Such differences can be used to evaluate the uncertainty of the approach and the sensitivity to different databases (e.g. Alcaraz-Segura et al. 2010b).

Another advantage of Ecosystem Functional Types is that their definition is exclusively based upon the carbon gain dynamics estimated from time-series of satellite images, so EFTs are able to capture differences between natural ecosystems (e.g. native oak forest) and managed ecosystems (e.g. tree plantations) when they differ in their carbon gain dynamics. For instance, Volante et al. (2012) showed how the intrusion of cattle rising and croplands on natural dry forest and shrublands of NW Argentina significantly changed satellite-derived ecosystem functional attributes related to productivity and seasonality and, subsequently, the EFTs composition (Paruelo et al. 2011).

4 Societal Needs for Research on Water Over Land

All organisms, including humans, require water for their survival. Therefore, ensuring that adequate supplies of water are available is essential for human well-being (Millennium Ecosystem Assessment 2005; Oki and Kanae 2006; Vörösmarty et al. 2010). Water issues are related to poverty, and providing access to safe drinking water is one of the key necessities for sustainable development (WHO/UNICEF 2012). However, better information on the hydro-climate system is necessary to understand the issues of supply and demand of water, both in the current climate and the future. Substantial changes to the Earth's climate system, hydrological cycles, and social systems have the potential to increase the frequency and severity of water-related hazards, such as: storm surges, floods, debris flows, and droughts (IPCC 2011). Global population is growing, particularly in the developing world and is accompanied by migration into urban areas, and could be associated with large scale land use/land cover changes. The urbanization threatens to increase the risks of urban flash floods and reduce per-capita water resources. Global economic growth is increasing the demand for food, which further drives demands for irrigation water and drinking water, demands more cropland, and potentially changes land use/land cover. Therefore it is critically important to consider both the social and climate changes in a concerted framework (Kundzewicz et al. 2007) as illustrated in Fig. 5.

In the past, water issues remained local; however, they are becoming a key global issue due to the increased awareness that human induced global warming has large impacts on the water cycle. Further, due to the increase in international trade and mutual interdependence among countries, water issues now often need to

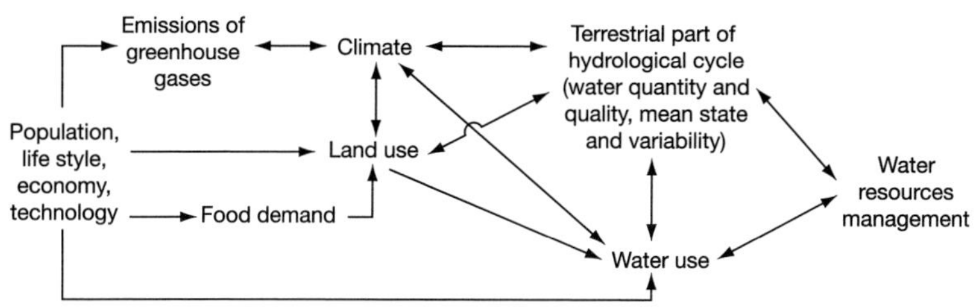

Fig. 5 Impact of human activities on freshwater resources and their management, with climate change being only one of multiple pressures (Modified after Oki 2005)

be dealt on a global scale, and thus require information on global hydrological conditions and their changes associated with climate changes. In trans boundary river basins and shared aquifers, it is necessary to share not only hydrological information but also any development plan that implies modifying LULC to reduce conflicts between relevant parties. In addition, quantitative estimates of recharge amounts or potentially available water resources will assist in implementing sustainable water use.

Global hydrology is not only concerned with global monitoring, modeling, and world water resources assessment. Owing to recent advancements in global earth observation technology and macro-scale modeling capacity, global hydrology can now provide basic information on the regional hydrological cycle which may support the decision making process in the integrated water resources management.

The use of offline land surface models at very fine spatial and temporal scales, e.g., 1-km grid spacing and hourly time intervals, is yet to be fully assessed (Oki et al. 2006; Wood et al. 2011). For such research efforts, observational data from regional studies can provide significant information for validation, and efforts to integrate datasets from various regional studies should be promoted. The recent Coordinated Regional Climate Downscaling Experiment (CORDEX) initiative (Giorgi et al. 2009) from the World Climate Research Program (WCRP) promotes running multiple RCM simulations at higher spatial resolution for multiple regions, and current and future estimates of atmospheric conditions will be provided, although at much lower resolution than that of the offline land surface models.

Certainly another societal need is to assess the impacts of human interferences on the hydrological cycle due to land use changes, such as deforestation and urbanization, reservoir constructions, and water withdrawals for irrigation, industry, and domestic water uses (e.g., Haddeland et al. 2006; Hanasaki et al. 2006, 2010; Pokhrel et al. 2012a).

Withholding water in reservoirs may result in a drop in the sea level. On the other hand, over exploitation of ground water, particularly "fossil water" which has virtually no or very little recharge at present, would have contribution to sea level rise. These effects are studied based on in-situ observations (Gornitz et al. 1997; Konikow 2011), satellite observations (Rodell et al. 2009; Moiwo et al. 2012), and modeling studies (Wada et al. 2010; Pokhrel et al. 2012b). Satellite information like that provided by GRACE (Gravity Recovery and Climate Experiment) serves to monitor the long term changes of these major water storages over land, and provides a powerful tool to assess and validate the global estimates from models.

5 Current Gaps, and Future Challenges

Current global land surface modeling has begun integrating most of the latest achievements in process understanding and regional- or local-scale modeling studies. For example, there are emerging efforts in global simulation of the occurrence, circulation, and balance of solutes and sediments. In addition, improvements to the

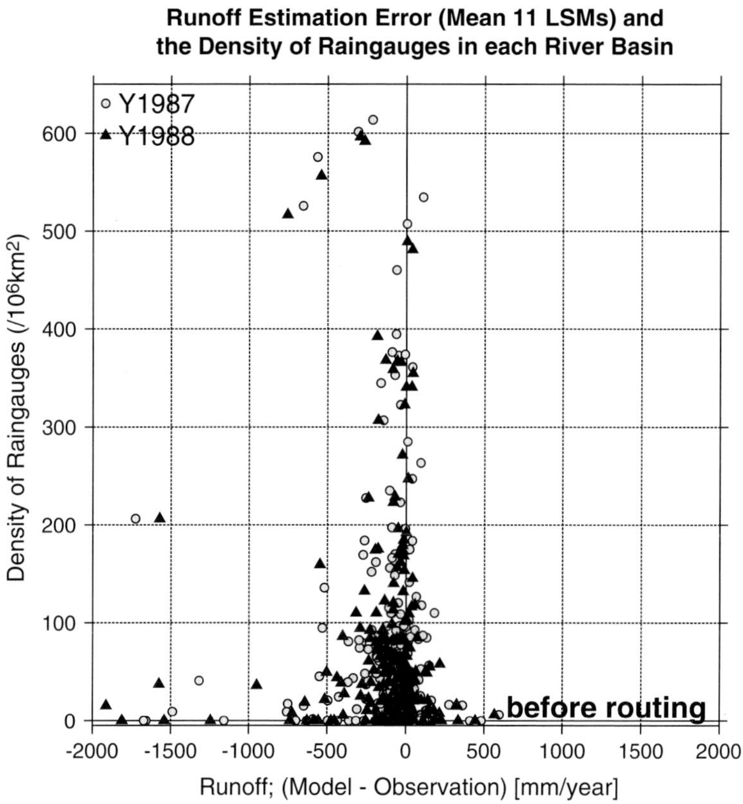

Fig. 6 Comparisons between the density of rain gauge [/10^6 km^2] used in preparing the forcing precipitation and the mean bias error [mm year^{-1}] of 11 LSMs for 150 major river basins in the world in 1987 and 1988 (Oki et al. 1999)

modeling of hydrology and groundwater are being incorporated into the models. Less developed are efforts to consider both natural and anthropogenic sources for nutrients, as well as their coupling to agricultural models that simulate crop growth and yield. Precise information on land use/land cover (LULC) is essential to have better estimates on nutrient, carbon and water cycles. Coupling of the LULC changes with biogeochemical and biogeophysical land surface model would be necessary for better future projections considering both climate and societal changes.

Hydro-meteorological monitoring networks need to be maintained and further expanded to enable the analysis of hydro-climatic trends at the local level and the improvement in the accuracy of predictions, forecasts, and early warnings. As clearly illustrated in Fig. 6 (Oki et al. 1999), global hydrological simulations are relatively poor in areas with little in-situ observations. Basic observational networks on the ground are critically crucial for proper monitoring and modeling of global hydrology; they are also needed to validate remotely sensed information that in turn is needed in order to fill the gaps of in-situ observations. Reliable observational data are essentially necessary not only as the forcing data for global hydrological modeling, but also for the validation of model estimates. River discharge and soil moisture data are critically important for global hydrological studies. Hence the cooperation and coordination of operational agencies in the world need to be prioritized and promoted.

Some key land surface processes, such as hydrology, have been represented in only simple ways in the current global climate models or earth system models due to their relatively minor impacts on the climatic feedbacks from the land surface to the atmosphere on global scales. It has also been pointed out that differences between land surface models is the major source of uncertainty in water balance estimates and multiple impact models are recommended to be used in this type of studies (Haddeland et al. 2011). However, land surface models with higher spatial resolution information are now being developed as impact assessment tools to support decision-making. Integrated land surface models that consider biogeochemical cycles and anthropogenic interventions explicitly (e.g., Hanasaki et al. 2010; Pokhrel et al. 2012a, b) need to be developed and implemented in order to provide more realistic impact assessments and to support the design of practical adaptation measures. In the WCRP conference held in Denver, CO, USA, in October 2011, these research needs and gaps were identified in the Land session. The identified research needs are outlined next:

- The observed and modeled feedbacks between land cover change induced by human activities needs to be assessed. Furthermore, the impact of deforestation on river flow, heat waves and wild fires should be investigated.
- There is a need to check that the earth system models are reproducing the simple signals that have been observed with large scale land use change, such as the cooling effect of deforestation under normal climate conditions, and the opposite warming effect under drought conditions.
- Current earth system models need to include and improve their representation of crop growth in order to better understand the role of land use change on the regional climate and subsequent impacts.
- WCRP, through efforts in GEWEX, has made great advances in understanding the land-atmosphere coupling and its relation to the hydrologic cycle. Yet, there are several areas that currently are poorly covered or not covered at all in the WCRP structure. Two GEWEX panels, GHP and GLASS, are the closest to the themes discussed in this paper, and could either assume or partner with other groups to lead efforts in the following areas: (a) Impacts of irrigation and water management on the hydrologic cycle of large basins; and (b) Effects of LULC on land-atmosphere feedbacks and its subsequent impact on river flows.
- For future states of the climate system, future assessments of the evolution of land use will require an interdisciplinary approach that considers not only the physical science but also societal aspects and economy information.
- A very challenging issue is that of prediction of land use changes based on society's future needs and responses to change. Assessments of future land use are important for climate prediction and climate change scenarios, and in this case

WCRP will have to partner with human dimensions groups (e.g., IHDP) in order to advance our knowledge of future states. Initiatives promoting interdisciplinary research that includes the physical aspects as well as human dynamics will be needed.

6 Concluding Remarks

Land use has had a large impact on water cycles and carbon changes over the twentieth century, and consequently understanding land surface processes is crucial for research of the climate system, and more so in relation with delivering policy relevant knowledge. The choices we make in LULCC will likely influence future climate through the water, carbon and energy balances and cycles.

Major advances in recent Earth System Models (ESMs) include state of art global scale land surface models that include anthropogenic activities such as irrigation, reservoirs and the carbon cycle. They are very promising to assess past, current and future global water crisis and may provide valuable information supporting better policy-making in crop and water management. The relation between biophysical effects of regional LULCC and global GHG is still unclear. For these reasons, LULCC matters at regional scale and so must be included in studies of climate change.

References

Alcaraz-Segura D, Paruelo JM, Cabello J (2006) Identification of current ecosystem functional types in the Iberian Peninsula. Glob Ecol Biogeogr 15:200–212

Alcaraz-Segura D, Cabello J, Paruelo J (2009) Baseline characterization of major Iberian vegetation types based on the NDVI dynamics. Plant Ecol 202(1):13–29. doi:10.1007/s11258-008-9555-2

Alcaraz-Segura D, Chuvieco E, Epstein HE, Kasischke ES, Trishchenko A (2010a) Debating the greening vs. browning of the North American boreal forest: differences between satellite datasets. Glob Chang Biol 16:760–770

Alcaraz-Segura D, Liras E, Tabik S, Paruelo J, Cabello J (2010b) Evaluating the consistency of the 1982–1999 NDVI trends in the Iberian Peninsula across four time-series derived from the AVHRR sensor: LTDR, GIMMS, FASIR, and PAL-II. Sensors 10:1291–1314

Alcaraz-Segura D, Paruelo JM, Epstein HE, Cabello J (2013a) Environmental and human controls of ecosystem functional diversity in temperate South America. Remote Sens 5(1):127–154. doi:10.3390/rs5010127, http://dx.doi.org/10.3390/rs5010127

Alcaraz-Segura D, Berbery HE, Müller O, Paruelo JM (2013b) Characterizing and monitoring climate regulation services. In: Alcaraz-Segura D, Di Bella CM, Straschnoy JV (eds) Earth observation of ecosystem services. CRC Press. ISBN: 9781466505889 http://www.crcpress.com/product/isbn/9781466505889 (in press)

Anderson-Teixeira KJ, Snyder PK, Twine TE, Cuadra SV, Costa MH, DeLucia EH (2012) Climate-regulation services of natural and agricultural ecoregions of the Americas. Nat Climate Change 2(3):177–181. doi:10.1038/nclimate1346

Baidya Roy S, Avissar R (2002) Impact of land use/land cover change on regional hydrometeorology in Amazonia. J Geophys Res 107(D20):LBA 4–1–LBA 4–12

Bonan GB (2008) Forests and climate change: forcings, feedbacks, and the climate benefits of forests. Science 320:1444–1449. doi:10.1126/science.1155121

Brando PM, Goetz SJ, Baccini A, Nepstad DC, Beck PSA, Christman MC (2010) Seasonal and interannual variability of climate and vegetation indices across the Amazon. Proc Natl Acad Sci 107:14685

Budyko MI (1974) Climate and life, English edn. Academic Press, New York

Cai W, Cowan T, Briggs P, Raupach M (2009) Rising temperature depletes soil moisture and exacerbates severe drought conditions across southeast Australia. Geophys Res Lett 36:L21709. doi:10.1029/2009GL040334

Chapin FS III, Randerson JT, McGuire AD, Foley JA, Field JA (2008) Changing feedbacks in the climate-biosphere system. Front Ecol Environ 6:313–320. doi:10.1890/080005

Chen F, Dudhia J (2001) Coupling an advanced land-surface/hydrology model with the Penn State/NCAR MM5 modeling system. Part I: model description and implementation. Mon Weather Rev 129:569–585

Choudhury BJ (1999) Evaluation of an empirical equation for annual evaporation using field observations and results from a biophysical model. J Hydrol 216:99–110. doi:10.1016/S0022-1694(98)00293-5

Church TM (1996) An underground route for the water cycle. Nature 380:579–580

Costanza R, d'Arge R, de Groot R, Farber S, Grasso M, Hannon B, Naeem S, Limburg K, Paruelo J, O'Neill RV, Raskin R, Sutton P, van den Belt M (1998) The value of ecosystem services: putting the issue in perspective. Ecol Econ 25:67–72

de Rosnay P, Polcher J, Laval K, Sabre M (2003) Integrated parameterization of irrigation in the land surface model ORCHIDEE. Validation over Indian Peninsula. Geophys Res Lett 30(19):1986. doi:10.1029/2003GL018024

Dirmeyer PA, Gao XA, Zhao M, Guo ZC, Oki T, Hanasaki N (2006) GSWP-2 multimodel analysis and implications for our perception of the land surface. Bull Am Meteorol Soc 87:1381–1397

Douville H, Chauvin F, Broqua H (2001) Influence of soil moisture on the Asian and African monsoons. Part1: mean monsoon and daily precipitation. J Clim 14:2381–2403

Ek M, Mitchell K, Lin Y, Rogers E, Grunmann P, Koren V, Gayno G, Tarpley D (2003) Implementation of Noah land surface model advances in the National Centers for Environmental Prediction operational mesoscale Eta model. J Geophys Res 108:8851. doi:10.1029/2002JD003296

Foley JA, Asner GP, Costa MH, Coe MT, DeFries R, Gibbs HK, Howard EA et al (2007) Amazonia revealed: forest degradation and loss of ecosystem goods and services in the Amazon basin. Front Ecol Environ 5(1):25–32

Gedney N, Valdes PJ (2000) The effect of Amazonian deforestation on the northern hemisphere circulation and climate. Geophys Res Lett 27:3053–3056

Giorgi F, Jones C, Asrar GR (2009) Addressing climate information needs at the regional level: the CORDEX framework. WMO Bull 58(3):175–183

Gornitz V, Rosenzweig C, Hillel D (1997) Effects of anthropogenic intervention in the land hydrologic cycle on global sea level rise. Glob Planet Change 14:147–161

Guerschman JP, Paruelo JM, Burke IC (2003) Land use impacts on the normalized difference vegetation index in temperate Argentina. Ecol Appl 13:616–628

Guimberteau M, Laval K, Perrier A, Polcher J (2011) Global effect of irrigation and its impact on the onset of the Indian summer monsoon. Clim Dyn. doi:10.1007/s00382-011-1252-5

Haddeland I and coauthors (2011) Multimodel estimate of the global terrestrial water balance: setup and first results. J Hydrometeorol 12:869–884

Haddeland I, Lettenmaier DP, Skaugen T (2006) Effects of irrigation on the water and energy balances of the Colorado and Mekong river basins. J Hydrol 324(1–4):210–223. doi:10.1016/j.jhydrol.2005.09.028

Hanasaki N, Kanae S, Oki T (2006) A reservoir operation scheme for global river routing models. J Hydrol 327(1–2):22–41. doi:10.1016/j.jhydrol.2005.11.011

Hanasaki N, Inuzuka T, Kanae S, Oki T (2010) An estimation of global virtual water flow and sources of water withdrawal for major crops and livestock products using a global hydrological model. J Hydrol 384(3–4):232–244

Hoare D, Frost P (2004) Phenological description of natural vegetation in southern Africa using remotely-sensed vegetation data. Appl Veg Sci 7:19–28

IPCC (2011) Summary for policymakers. In: Field CB et al (eds) Intergovernmental panel on climate change special report on managing the risks of extreme events and disasters to advance climate change adaptation. Cambridge University Press, New York

Kalnay E, Cai M (2003) Impact of urbanization and land-use change on climate. Nature 423:528–531

Kanae S, Oki T, Musiake K (2001) Impact of deforestation on regional precipitation over the Indochina Peninsula. J Hydrometeorol 2:51–70

Konikow LF (2011) Contribution of global groundwater depletion since 1900 to sea-level rise. Geophys Res Lett 38, L17401

Körner C (1994) Scaling from species to vegetation: the usefulness of functional groups. In: Schulze ED, Mooney HA (eds) Biodiversity and ecosystem function. Springer, Berlin, pp 117–139

Korzun VI (1978) World water balance and water resources of the earth, studies and reports in hydrology, vol 25. UNESCO, Paris

Kundzewicz ZW, Mata LJ, Arnell NW, Döll P, Kabat P, Jiménez B, Miller KA, Oki T, Sen Z, Shiklomanov IA (2007) Freshwater resources and their management. In: Parry ML, Canziani OF, Palutikof JP, van der Linden PJ, Hanson CE (eds) Climate change 2007: impacts, adaptation and vulnerability. Contribution of working group II to the fourth assessment report of the intergovernmental panel on climate change. Cambridge University Press, Cambridge, pp 173–210

Lee S-J, Berbery EH (2012) Land cover change effects on the climate of the La Plata basin. J Hydrometeor 13:84–102

Lee E, Chase TN, Rajagopalan B, Barry RG, Biggs TW, Lawrence PJ (2009) Effects of irrigation and vegetation activity on early Indian summer monsoon variability. Int J Climatol 29:573–581

Lloyd D (1990) A phenological classification of terrestrial vegetation cover using shortwave vegetation index imagery. Int J Remote Sens 11:2269–2279

Mahmood R, Pielke RA, Hubbard KG, Niyogi D, Bonan G, Lawrence P, Mcnider R, Mcalpine C, Etter A, Gameda S (2010) Impacts of land use/land cover change on climate and future research priorities. Bull Am Meteorol Soc 91:37–46

McNaughton KG, Spriggs TW (1989) An evaluation of the Priestley Taylor equation. IAHS Press, Wallingford, pp 89–104, IAHS publ 177

Milchunas DG, Lauenroth WK (1995) Inertia in plant community structure: state changes after cessation of nutrient enrichment stress. Ecol Appl 5:452–458

Millennium Ecosystem Assessment (2005) Millennium Ecosystem Assessment Ecosystems and human well-being: synthesis. Island Press, Washington, DC

Moiwo JP, Wenxi L, Tao F (2012) GRACE, GLDAS and measured groundwater data products show water storage loss in Western Jilin, China. Water Sci Technol 65(9):1606–1614. doi:10.2166/wst.2012.053

Monteith JL (1972) Solar radiation and productivity in tropical ecosystems. J Appl Ecol 9:747–766. doi:10.2307/2401901

Nemani RR, Running SW (1997) Land cover characterization using multitemporal red, near-IR, and thermal-IR data from NOAA/AVHRR. Ecol Appl 7:79–90

Oesterheld M, DiBella CM, Kerdiles H (1998) Relation between NOAA-AVHRR satellite data and stocking rate of rangelands. Ecol Appl 8:207–212

Oki T (1999) The global water cycle. In: Browning K, Gurney R (eds) Global energy and water cycles. Cambridge University Press, Cambridge/New York, pp 10–27

Oki T (2005) The hydrologic cycles and global circulation. In: Anderson MG (ed) Encyclopedia of hydrological sciences. Wiley, Chichester

Oki T, Kanae S (2006) Global hydrological cycles and world water resources. Science 313(5790):1068–1072

Oki T, Musiake K, Matsuyama H, Masuda K (1995) Global atmospheric water balance and runoff from large river basins. Hydrol Proc 9:655–678

Oki T, Nishimura T, Dirmeyer P (1999) Assessment of annual runoff from land surface models using Total Runoff Integrating Pathways (TRIP). J Meteorol Soc Jpn 77:235–255

Oki T, Entekhabi D, Harrold T (2004) The global water cycle. In: Sparks R, Hawkesworth C (eds) State of the planet: frontiers and challenges in geophysics, No. 150 in geophysical monograph series. AGU Publication, Washington, DC, p 414

Oki T, Valeo C, Heal K (eds) (2006) Hydrology 2020: an integrating science to meet world water challenges. IAHS Press, Wallingford, IAHS Publication, 300
Paruelo JM, Lauenroth WK (1995) Regional patterns of Normalized Difference Vegetation Index in North American shrublands and grasslands. Ecology 76:1888–1898
Paruelo JM, Epstein HE, Lauenroth WK, Burke IC (1997) ANPP estimates from NDVI for the central grassland region of the United States. Ecology 78:953–958
Paruelo JM, Jobbagy EG, Sala OE (2001) Current distribution of ecosystem functional types in temperate South America. Ecosystems 4:683–698
Paruelo J, Alcaraz-Segura D, Volante JN (2011) El seguimiento del nivel de provisión de los servicios ecosistémicos. In: Laterra P, Jobbágy EG, Paruelo JM (eds) Valoración de servicios ecosistémicos: conceptos, herramientas y aplicaciones para el ordenamiento territorial. Instituto Nacional de Tecnología Agropecuaria, Buenos Aires, pp 141–160
Pettorelli N and coauthors (2005) Trends in ecology & evolution using the satellite-derived NDVI to assess ecological responses to environmental change. http://www.sciencedirect.com/science/article/pii/S016953470500162X
Phillips and coauthors (2009) Drought sensitivity of the Amazon rainforest. Science 323(5919):1344–1347. http://www.sciencemag.org/content/323/5919/1344.short
Pielke RA and coauthors (2007) An overview of regional land-use and land-cover impacts on rainfall. Tellus 59B:587–601
Pielke RA, Avissar R (1990) Influence of structure on local and regional climate. Landsc Ecol 4:133–155
Pielke RA, Marland G, Betts RA, Chase TN, Eastman JL, Niles JO, Niyogi DS, Running SW (2002) The influence of land-use change and landscape dynamics on the climate system: relevance to climate-change policy beyond the radiative effect of greenhouse gases. Philos Trans R Soc Lond A Math Phys Eng Sci 360:1705
Pielke RA Sr, Pitman A, Niyogi D, Mahmood R, McAlpine C, Hossain F, Goldewijk KK, Nair U, Betts R, Fall S, Reichstein M, Kabat P, de Noblet N (2011) Land use/land cover changes and climate: modeling analysis and observational evidence. WIREs Clim Change 2:828–850. doi:10.1002/wcc.144
Pitman AJ and coauthors (2009) Uncertainties in climate responses to past land cover change: first results from the LUCID intercomparison study. Geophys Res Lett 36:L14814. doi:10.1029/2009GL039076
Pokhrel YN, Hanasaki N, Yeh PJ-F, Yamada TJ, Kanae S, Oki T (2012a) Model estimates of sea-level change due to anthropogenic impacts on terrestrial water storage. Nat Geosci. Advance Online Publication. doi:10.1038/Ngeo1476
Pokhrel Y, Hanasaki N, Koirala S, Cho J, Yeh PJ-F, Kim H, Kanae S, Oki T (2012b) Incorporating anthropogenic water regulation modules into a land surface model. J Hydrometeor 13:255–269. doi: 10.1175/JHM-D-11-013.1
Porporato A, Daly E, Rodriguez-Iturbe I (2004) Soil water balance and ecosystem response to climate change. Am Naturalist 164:625–632
Potter CS, Brooks V (1998) Global analysis of empirical relations between annual climate and seasonality of NDVI. Int J Remote Sens 19:2921–2948
Reale O, Dirmeyer P (2000) Modeling the effects of vegetation on Mediterranean climate during the Roman Classical Period: Part I: climate history and model sensitivity. Glob Planet Change 25:163–184
Reale O, Shukla J (2000) Modeling the effects of vegetation on Mediterranean climate during the Roman Classical Period: Part II. Model simulation. Glob Planet Change 25:185–214
Rodell M, Velicogna I, Famiglietti JS (2009) Satellite-based estimates of groundwater depletion in India. Nature 460:999–1002. doi:10.1038/460789a
Roderick ML, Farquhar GD (2011) A simple framework for relating variations in runoff to variations in climatic conditions and catchment properties. Water Resour Res 47:W00G07. doi:10.1029/2010WR009826
Saeed F, Hagemann S, Jacob D (2011) A framework for the evaluation of South Asian summer monsoon in a regional climate model applied to REMO. Int J Climatol. doi:10.1002/joc.2285

Shiklomanov IA (ed) (1997) Assessment of water resources and water availability in the world. Background report for the comprehensive assessment of the freshwater resources of the world, WMO/SEI. Geneva, Switzerland

Stephenson NL (1990) Climatic control of vegetation distribution: the role of the water balance. Am Naturalist 135:649–670

Stohlgren TJ, Chase TN, Pielke RA Sr, Kittel TGF, Baron JS (1998) Evidence that local land use practices influence regional climate, vegetation, and stream flow patterns in adjacent natural areas. Glob Change Policy 4:495–504

Taylor CM and coauthors (2011) New perspectives on land–atmosphere feedbacks from the African Monsoon Multidisciplinary analysis. Atmos Sci Lett 12:38–44. doi:10.1002/asl.336

Teuling AJ and coauthors (2010) Contrasting response of European forest and grassland energy exchange to heatwaves. Nat Geosci 3:722–727. doi:10.1038/ngeo950

Tucker CJ, Sellers PS (1986) Satellite remote-sensing of primary production. Int J Remote Sens 7:1395–1416

Tuinenburg OA, Hutjes RWA, Jacobs CMJ, Kabat P (2011) Diagnosis of local land–Atmosphere feedbacks in India. J Clim 24:251–266

Valentini R, Baldocchi DD, Tenhunen JD, Kabat P (1999) Ecological controls on land-surface atmospheric interactions. In: Integrating hydrology, ecosystem dynamics and biogeochemistry in complex landscapes. Wiley, Berlin, pp 105–116

van den Hurk B, Martin B, Paul D, Andy P, Jan P, Joe S (2011) Acceleration of land surface model development over a decade of glass. Bull Am Meteor Soc 92(12):1593–1600. doi:10.1175/BAMS-D-11-00007.1

Virginia RA, Wall DH, Levin SA (2001) Principles of ecosystem function. Encyclopedia of biodiversity. Academic Press, San Diego, pp 345–352.

Volante JN, Alcaraz-Segura D, Mosciaro MJ, Viglizzo EF, Paruelo JM (2012) Ecosystem functional changes associated with land clearing in NW Argentina. Agric Ecosyst Environ 154:12–22

Vörösmarty CJ, McIntyre PB, Gessner MO, Dudgeon D, Prusevich A, Green P, Glidden S, Bunn SE, Sullivan CA, Reidy Liermann C, Davies PM (2010) Global threats to human water security and river biodiversity. Nature 467(7315):555–561. doi:10.1038/nature09440

Wada Y et al (2010) Global depletion of groundwater resources. Geophys Res Lett 37, L20402

Weaver CP, Avissar R (2001) Atmospheric disturbances caused by human modification of the landscape. Bull Am Meteorol Soc 82:269–281

Werth D, Avissar R (2002) The local and global effects of Amazon deforestation. J Geophys Res 107:8087

WHO/UNICEF (2012) Progress on drinking water and sanitation: 2012 update, WHO/UNICEF Joint Monitoring Programme for Water Supply and Sanitation. http://www.who.int/water_sanitation_health/publications/2012/jmp2012.pdf

Wood EF, Roundy JK, Troy TJ, van Beek LPH, Bierkens MFP, Blyth E, de Roo A, Döll P, Mike E, Famiglietti J, Gochis D, van de Giesen N, Houser P, Jaffé PR, Kollet S, Lehner B, Lettenmaier DP, Peters-Lidard C, Sivapalan M, Sheffield J, Wade A, Whitehead P (2011) Hyperresolution global land surface modeling: meeting a grand challenge for monitoring Earth's terrestrial water. Water Resour Res 47:W05301. doi:10.1029/2010WR010090

Wright JP, Naeem S, Hector A, Lehman C, Reich PB, Schmid B, Tilman D (2006) Conventional functional classification schemes underestimate the relationship with ecosystem functioning. Ecol Lett 9:111–120

Yang D, Shao W, Yeh PJ-F, Yang H, Kanae S, Oki T (2009) Impact of vegetation coverage on regional water balance in the nonhumid regions of China. Water Resour Res 45:W00A14. doi:10.1029/2008WR006948

Zhao M, Running SW (2010) Drought-induced reduction in global terrestrial net primary production from 2000 through 2009. Science 329:940

Zhang L, Hickel K, Dawes WR, Chiew FHS, Western AW, Briggs PR (2004) A rational function approach for estimating mean annual evapotranspiration. Water Resour Res 40:W02502. doi: 10.1029/2003WR002710

Prediction from Weeks to Decades

Ben Kirtman, David Anderson, Gilbert Brunet, In-Sik Kang, Adam A. Scaife, and Doug Smith

Abstract This white paper is a synthesis of several recent workshops, reports and published literature on monthly to decadal climate prediction. The intent is to document: (i) the scientific basis for prediction from weeks to decades; (ii) current capabilities; and (iii) outstanding challenges. In terms of the scientific basis we described the various sources of predictability, e.g., the Madden Jullian Ocillation (MJO); Sudden Stratospheric Warmings; Annular Modes; El Niño and the Southern Oscillation (ENSO); Indian Ocean Dipole (IOD); Atlantic "Niño;" Atlantic gradient pattern; snow cover anomalies, soil moisture anomalies; sea-ice anomalies; Pacific Decadal Variability (PDV); Atlantic Multi-Decadal Variability (AMV); trend among others. Some of the outstanding challenges include how to evaluate and

B. Kirtman (✉)
RSMAS/MPO, University of Miami, 4600 Rickenbacker Causeway, Miami, FL 33149, USA
e-mail: bkirtman@rsmas.miami.edu

D. Anderson
The IDR Group, Lamb Chambers, the Chambers of John Cherry QC,
Lamb Building, Elm Court, Temple, London EC4Y 7AS, UK
e-mail: dlta@btinternet.com

G. Brunet
Meteorological Research Division (MRD)/ASTD, Environment Canada, 2121,
Voie de Service nord, Route, Trans-Canadienne, Dorval, QC, Canada H9P 1J3
e-mail: gilbert.brunet@ec.gc.ca

I.-S. Kang
Atmospheric Science Program, School of Earth and Environmental Sciences,
Soul National University, Gwanak-gu, Seoul 151-742, Korea
e-mail: kang@climate.snu.ac.kr

A.A. Scaife • D. Smith
Met Office, Hadley Center, FitzRoy Road, Exeter, Devon EX1 3PB, UK
e-mail: adam.scaife@metoffice.gov.uk; doug.smith@metoffice.gov.uk

G.R. Asrar and J.W. Hurrell (eds.), *Climate Science for Serving Society: Research, Modeling and Prediction Priorities*, DOI 10.1007/978-94-007-6692-1_8,

validate prediction systems, how to improve models and prediction systems (e.g., observations, data assimilation systems, ensemble strategies), the development of seamless prediction systems.

Keywords Seamless weather and climate prediction • MJO • ENSO • Annular modes • Pacific Decadal Variability • Atlantic Multi-Decadal Variability • Indian Ocean Dipole

1 Introduction

Numerical weather forecasts have seen profound improvements over the last 30-years with the potential now to provide useful forecasts beyond 10 days ahead, especially those based on ensemble, probabilistic systems. Despite this continued progress, it is well accepted that even with a perfect model and nearly perfect initial conditions,[1] the fact that the atmosphere is chaotic causes forecasts to lose predictive information from initial conditions after a finite time (Lorenz 1965), in the absence of forcing from other parts of the Earth's system such as ocean surface temperatures and land surface soil moisture. As a result, for many aspects of weather the "limit of predictability" is about 2 weeks.

So, why is climate prediction[2] (i.e., forecast beyond the limit of weather predictability) possible? While there is a clear limit to our ability to forecast day-to-day weather, there exists a firm scientific basis for the prediction of time averaged climate anomalies. Climate anomalies result from complex interactions among all the components of the Earth system. The atmosphere, which fluctuates very rapidly on a day-to-day basis, interacts with the more slowly evolving components of the Earth system, which are capable of exerting a sustained influence on climate anomalies extending over a season or longer, far beyond the limit of atmospheric predictability from initial conditions alone. The atmosphere, for example, is particularly sensitive to tropical sea surface temperature anomalies such as those that occur in association with El Nino and the Southern Oscillation (ENSO). There is also increasing evidence that external forcings, such as solar variability, greenhouse gas and aerosol concentrations, land use and volcanic eruptions, also 'lend' predictability to the system, which can be exploited on sub-seasonal to decadal timescales.

Consequently, numerical models used for climate prediction have progressed from atmospheric models with a simple representation of the oceans to fully coupled Earth system models complete with fully coupled dynamical oceans, land surface, cryosphere and even chemical and biological processes. In fact, many

[1] Arbitrarily small initial condition errors.

[2] Here we define the prediction of climate anomalies as the prediction of statistics of weather (i.e., mean temperature or precipitation, variance, probability of extremes such as droughts, floods, hurricanes, high winds …).

operational centers around the world now produce sub-seasonal to seasonal predictions using observed initial conditions that include components of the Earth system beyond the atmosphere.

The traditional boundaries between weather forecasting and climate prediction are fast disappearing since progress made in one area can help to accelerate improvements in the other. For example, improvements in the modeling of soil moisture made in climate models can lead to improved weather forecasting of showers over land in summer; and data assimilation, which has been restricted to the realm of weather prediction, is now becoming a requirement of coupled models used for longer term predictions (Brunet et al. 2010).

As the scope of numerical weather forecasting and climate prediction broadens and overlaps, the fact that both involve modeling the same system becomes much more relevant, as many of the processes are common to all time scales. There is much benefit to be gained from a more integrated or "seamless" approach. Unifying modeling across all timescales should lead to efficiencies in model development and improvement by sharing and implementing lessons learned by the different communities. There are many examples of the benefits of this approach (e.g. Brown et al. 2012). These include enabling climate models to benefit from what is learned from data assimilation in weather forecasting, enabling weather forecasting models to learn from the coupling with the oceans in climate models, and sharing the validation and benchmarking of key common processes such as tropical convection. The inclusion of atmospheric chemistry and aerosols, essential components of Earth system models used for projections of climate change, can now be exploited to improve air quality forecasting and the parametrization of cloud microphysics. Predictions of flood events require better representation of hydrological processes at local, regional, continental and global scales, which are important across all time scales. Diagnostic of precipitation model errors show often significant similarity between climate and weather prediction systems hence pointing out to a common solution to the problem. The use of a common core model for various applications is also an opportunity to save human time when porting a system to a new computational platform.

Clearly, there is a growing demand for environmental predictions that include a broad range of space and time scales and that include a complete representation of physical, chemical and biological processes. Meeting this demand could be accelerated through a unified approach that will challenge the traditional boundaries between weather and climate science in terms of the interactions of the bio-geophysical systems. It is also recognized that interactions across time and space scales are fundamental to the climate system itself (Randall et al. 2003; Hurrell et al. 2009; Shukla et al. 2009; Brunet et al. 2010). The large-scale climate, for instance, determines the environment for microscale (order 1 km) and mesoscale (order 10 km) variability which then feedback onto the large-scale climate. In the simplest terms, the statistics of microscale and mesoscale variability significantly impact the simulation of weather and climate and the feedbacks between all the biogeophysical systems. However, these interactions are extremely complex making it difficult to understand and predict the Earth system variability that we observe.

We also note that predictions can be made using purely statistical techniques, or dynamical models, or a combination of both. Statistical and dynamical methods are complementary: improved understanding gained through successful statistical forecasts may lead to better dynamical models, and vice versa. Furthermore, statistical methods provide a baseline level of skill that more complex dynamical models must aim to exceed. Statistical methods are actively used to correct model errors beyond the mean bias so that model output can be used by application models.

Increasingly all forecasts are probabilistic, reflecting the fact that the atmosphere and oceans are chaotic systems and that models do not fully capture all the scales of motion, i.e. the model itself is uncertain (see Slingo and Palmer 2011 for a full discussion of uncertainty). That being the case, skill cannot be judged based on a single case since a probabilistic prediction is neither right nor wrong. Instead an ensemble prediction system produces a range of possible outcomes, only one of which will be realized. Its skill can therefore only be assessed over a wide range of cases where it can be shown that the forecast probability matches the observed probability (e.g., Palmer et al. 2000, 2004; Goddard et al. 2001; Kirtman 2003; DeWitt 2005; Hagedorn et al. 2005; Doblas-Reyes et al. 2005; Saha et al. 2006; Kirtman and Min 2009; Stockdale et al. 2011; Arribas et al. 2011 and others).

Given our current modeling capabilities, a multi-model ensemble strategy may be the best current approach for adequately resolving forecast uncertainty (Derome et al. 2001; Palmer et al. 2004, 2008; Hagedorn et al. 2005; Doblas-Reyes et al. 2005; Wang et al. 2010). The use of multi-model ensembles can give a definite boost to the forecast reliability compared to that obtained by a single model (e.g., Hagedorn et al. 2005; Guilyardi 2006; Jin et al. 2008; Kirtman and Min 2009; Krishnamurti et al. 2000). Although a multi-model ensemble strategy represents the "best current approach" for estimating uncertainty, it does not remove the need to improve models and our understanding.

Another factor in climate prediction is that, unlike weather forecasting, model-specific biases grow strongly in a fully coupled ocean–atmosphere system, to the extent that the distribution of probable outcomes in seasonal to decadal forecasts may not reflect the observed distribution, and thus the forecasts may not be reliable. It is essential, therefore, that forecast reliability is assessed using large sets of model hindcasts. These enable the forecast probabilities to be calibrated based on past performance and the model bias to be corrected. However, these empirical correction methods are essentially linear and yet we know that the real system is highly nonlinear. As Turner et al. (2005) have demonstrated, there is inherently much more predictive skill if improvements in model formulation could be made that reduce these biases, rather than correcting them after the fact.

2 Sub-seasonal Prediction

Forecasting the day-to-day weather is primarily an atmospheric initial condition problem, although there can be an influence from land and sea-ice (Pellerin et al. 2004; Smith et al. 2012) conditions and ocean temperatures. Forecasting at the seasonal-to-interannual range depends strongly on the slowly evolving components of the Earth system, such as the ocean surface, but all the components can influence the evolution of the system. In between these two time-scales is sub-seasonal variability.

2.1 Madden Julian Oscillation

Perhaps the best known source of predictability on sub-seasonal timescales is the Madden-Jullian Oscillation (MJO, Madden and Julian 1971). This has a natural timescale in the range 30–70 days. It is associated with regions of enhanced or reduced precipitation, and propagates eastwards, with speeds of ~5 m/s, depending on its longitude. The MJO clearly influences precipitation in the tropics. It influences tropical cyclone activity in the western and eastern north Pacific, the Gulf of Mexico, southern Indian Ocean and Australia (See Vitart 2009 for references). It also influences the Asian and Australian monsoon onset and breaks and is associated with northward moving events in the Bay of Bengal (Lawrence and Webster 2002). Recent estimates of the potential predictability associated with the MJO suggest that it may be as much as 40 days (Rashid et al. 2011).

Interaction with the ocean may play some role in the development and propagation of the MJO, but does not appear to be crucial to its existence (Woolnough et al. 2007; Takaya et al. 2010). The way convection is represented in numerical models does influence the characteristics of the MJO quite strongly, however. Until recently the MJO was quite poorly represented in most models. There are now some models that have something resembling an MJO (Pegion and Kirtman 2008; Vitart and Molteni 2010; Waliser et al. 2009; Wang et al. 2010; Gottschalck et al. 2010; Lin et al. 2010a, b; Lin and Brunet 2011) but more remains to be done.

Not only is the MJO important in the tropics, there is growing evidence that it has an important influence on northern hemisphere weather in the PNA (Pacific North American pattern) and even in the Atlantic and European sectors. Cassou (2008) and Lin et al. (2009) have studied the link from the MJO to modes of the northern hemisphere including the North Atlantic Oscillation. In Lin et al. (2009) time-lagged composites and probability analysis of the NAO index for different phases of the MJO reveal a statistically significant two-way relationship between the NAO and the tropical convection of the MJO (see Table 1). A significant increase of the NAO amplitude happens about 1–2 weeks after the MJO-related convection anomaly reaches the tropical Indian Ocean and western Pacific region. The development of the NAO is associated with a Rossby wave train in the upstream Pacific and North American region. In the Atlantic and African sector, there is an extratropical influence on the tropical intraseasonal variability. Certain phases of the MJO are preceded by 2–4 weeks by the occurrence of strong NAOs. A significant change of upper zonal wind in the tropical Atlantic is caused by a modulated transient westerly momentum flux convergence associated with the NAO.

The MJO has also been found to influence the extra-tropical weather in various locations. For example, Higgins et al. (2000) and Mo and Higgins (1998) investigated

Table 1 Lagged probability composites of the NAO index with respect to each MJO phase

MJO phase	1	2	3	4	5	6	7	8
NAO Lag –5		–35	–40			+49	+49	
Lag –4						+52	+46	
Lag –3		–40					+46	
Lag –2						+50		
Lag –1								
Lag 0				+45				–42
Lag 1			+47	+45				–46
Lag 2		+47	+50	+42		–41	–41	–42
Lag 3		+48				–41	–48	
Lag 4						–39	–48	
Lag 5				–41				

From Lin et al. (2009)
Lag n means that the NAO lags the MJO of the specific phase by n pentads, while Lag –n indicates that the NAO leads the MJO by n pentads. Positive values are for the upper tercile, while negative values are for the lower tercile. Values shown are only for those having a 0.05 significance level according to a Monte Carlo test

the relationships between tropical convection associated with the MJO and U.S. West Coast precipitation. Vecchi and Bond (2004) found that the phase of the MJO has a substantial systematic and spatially coherent effect on sub-seasonal variability in wintertime surface air temperature in the Arctic region. Wheeler et al. (2009) documented the MJO impact on Australian rainfall and circulation. Lin and Brunet (2009) and Lin et al. (2010b) found significant lag connection between the MJO and the intra-seasonal variability of temperature and precipitation in Canada. It is also observed that with a lead time of 2–3 weeks, the MJO forecast skill is significantly influenced by the NAO initial amplitude (Lin and Brunet 2011) (Fig. 1).

The importance of the tropics in extra-tropical weather forecasting has been illustrated by several authors. Early results from Ferranti et al. (1990) indicated that better representation of the MJO led to better mid-latitude forecasts in the northern hemisphere, and the benefit of the connection of the MJO and NAO in intra-seasonal forecasting has been demonstrated in Lin et al. (2010a). With a lead time up to about 1 month the NAO forecast skill is significantly influenced by the existence of the MJO signal in the initial condition. A strong MJO leads to a better NAO forecast skill than a weak MJO. These results indicate that it is possible to increase the predictability of the NAO and the extra-tropical surface air temperature with an improved tropical initialization, a better prediction of the tropical MJO and a better representation of the tropical-extra-tropical interaction in dynamical models.

2.2 *Other Sources of Sub-seasonal Predictability*

An important source of potential predictability comes from the relatively persistent variations in the lower stratosphere following sudden stratospheric warmings and other stratospheric flow changes, which have been shown to precede anomalous

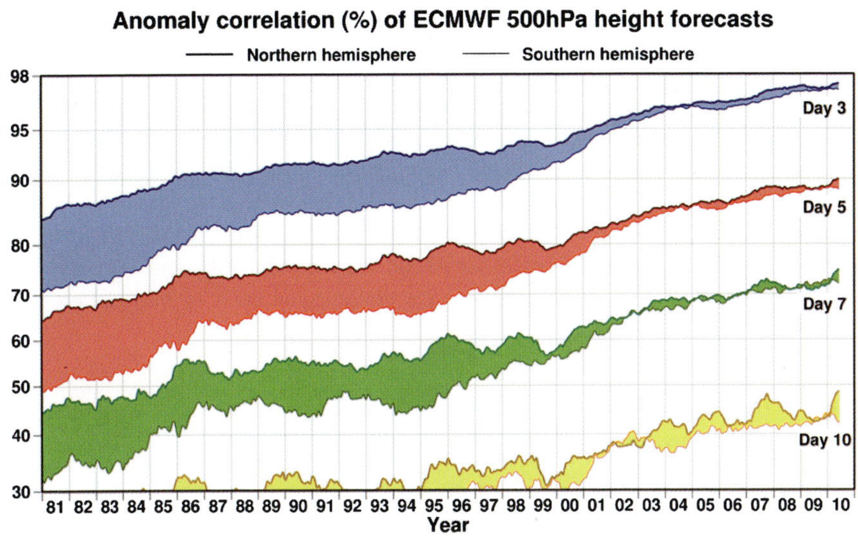

Fig. 1 Evolution of ECMWF forecast skill for varying lead times (3 days in *blue*; 5 days in *red*; 7 days in *green*; 10 days in *yellow*) as measured by 500-hPa height anomaly correlation. *Top line* corresponds to the Northern Hemisphere; *bottom line* corresponds to the Southern hemisphere. Large improvements have been made, including a reduction in the gap in accuracy between the hemispheres (Source: Courtesy of ECMWF. Adapted from Simmons and Holligsworth (2002))

circulation conditions in the troposphere (Kuroda and Kodera 1999; Baldwin and Dunkerton 2001). The long radiative timescale and wave-mean flow interactions in the stratosphere can lead to persistent anomalies in the polar circulation. These can then influence the troposphere, particularly in the mid-latitudes to produce persistent anomalies in the storm track regions and highly populated areas around the Atlantic and Pacific basins (Thompson and Wallace 2000). Once they occur, stratospheric sudden warmings provide further predictability during winter and spring, although the extent to which they are themselves predictable is generally limited to 1–2 weeks (Marshall and Scaife 2010a).

Soil moisture memory spans intraseasonal time scales depending on the season. Memory in soil moisture is translated to the atmosphere through the impact of soil moisture on the surface energy budget, mainly through its impact on evaporation. Soil moisture initialization in forecast systems is known to affect the evolution of forecast precipitation and air temperature in certain areas during certain times of the year on intraseasonal time scales (e.g., Koster et al. 2010). Model studies (Fischer et al. 2007) suggest that the European heat wave of summer 2003 was exacerbated by dry soil moisture anomalies in the previous spring.

Hudson et al. (2011a, b) and Hamilton et al. (2012) have shown that modes of climate variability, such as ENSO, the Indian Ocean Dipole (IOD) and the Southern Annular Mode (SAM), are sources of intra-seasonal predictability; if ENSO/IOD/SAM are in extreme phases, intra-seasonal prediction is extended. These studies argue that it is not predicting intra-seasonal variations in the tropics per se that matters, but that these slow variations shift the seasonal probabilities of daily weather one way or the other and this shift can be detected as short as 2 weeks into the forecast.

Although the field is still in its infancy, early results concerning the extent of polar predictability also show promise (e.g., Blanchard-Wrigglesworth et al. 2011). Most of these efforts have taken place in Europe or North America and have therefore focused on the Arctic and North Atlantic. Operational seasonal prediction systems for the Arctic show the impact of summertime sea-ice and fall Eurasian snow-cover anomalies, and September Arctic sea-ice extent appears to be predictable given knowledge of the springtime ice thickness or early to mid summer sea ice extent.

3 Seasonal-to-Interannual Prediction

In many respects seasonal prediction is the most mature of the three timescales under consideration in this paper. Statistical methods have been used for many decades, especially for the Indian Summer Monsoon, and the seasonal timescale has been the primary focus of the early development of ensemble prediction systems. The seasonal timescale is also one in which the low frequency forcing from the ocean, especially El Nino/La Nina, really begins to dominate and provide significant levels of predictability.

3.1 El Nino Southern Oscillation (ENSO)

The largest source of seasonal-to-interannual prediction is ENSO. ENSO is a coupled mode of variability of the tropical Pacific that grows through positive feedbacks between sea surface temperature (SST) and winds – a weakening of the easterly trade winds produces a positive SST anomaly in the eastern tropical Pacific which in turn alters the atmospheric zonal (Walker) circulation to further reduce the easterly winds. The time between El Niño events is typically about 2–7 years, but the mechanisms controlling the reversal to the opposite La Niña phase are not understood completely, nor are those that lead to sustained La Nina events extending beyond 1 year.

ENSO influences seasonal climate almost everywhere (see Fig. 2 taken from Smith et al. 2012), either by directly altering the tropical Walker circulation (Walker and Bliss 1932), or through Rossby wave trains that propagate to mid and high latitudes (Hoskins and Karoly 1981), substantially modifying weather patterns over North America. There is also a notable influence on the North Atlantic Oscillation (NAO), especially in late winter (Brönimann et al. 2007). It has also been shown that ENSO governs much of the year-to-year variability of global mean temperature (Scaife et al. 2008). However, the strongest impacts of ENSO

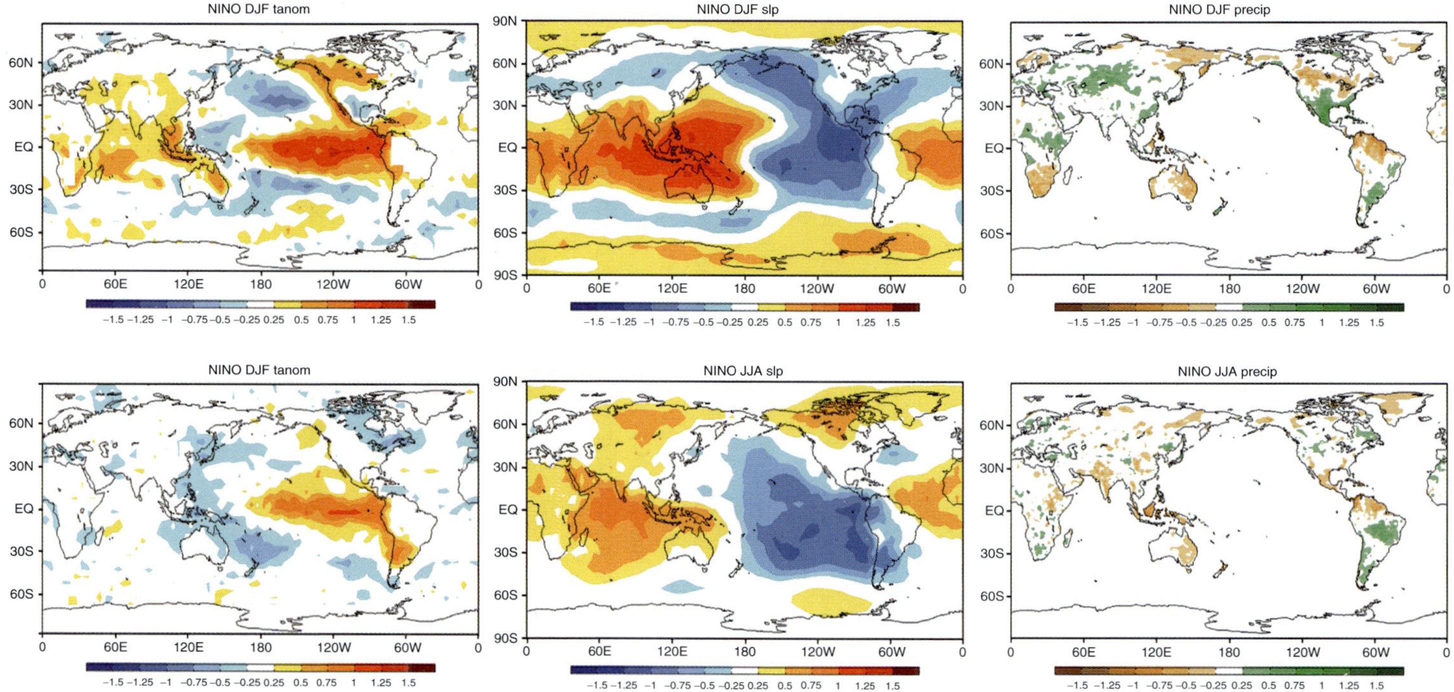

Fig. 2 Observed ENSO teleconnections. Composite differences between positive and negative phases of ENSO, for boreal winter (DJF, *top row*) and summer (JJA, *bottom*). Composite differences are divided by 2 to show the amplitude of the variability. The contour interval is 0.25 (standard deviations), with values greater than 0.2 in magnitude significant at the 95 % level based on a one-sided t test. SSTs are taken from HadISST (Rayner et al. 2003), surface temperatures are taken from HadCRUT3 (Brohan et al. 2006), sea level pressures from HadSLP2 (Allan and Ansell 2006), and precipitation from GPCC (Rudolf et al. 2005). Positive ENSO years are 1902, 1911, 1913, 1918, 1925, 1930, 1939, 1940, 1957, 1965, 1972, 1982, 1986, 1991, 1997 and 2009. Negative ENSO years are 1916, 1917, 1942, 1949, 1955, 1967, 1970, 1973, 1975, 1984, 1988, 1999 and 2007 (Figure redrawn following Smith et al. (2012))

occur in Indonesia, North and South America, East and South Africa, India and Australia. A notable recent example was the intense rainfall and flooding in Northeast Australia during 2010/2011 during a pronounced La Nina event – the strongest since 1973/1974.

The ability to predict the seasonal variations of the tropical climate dramatically improved from the early 1980s to the late 1990s. This period was bracketed by two of the largest El Niño events on record: the 1982–1983 event and the 1997–1998 event. In the case of the former, there was considerable confusion as to what was happening in the tropical Pacific (see Anderson et al. 2011). As a result the NOAA Tropical Atmosphere Ocean (TAO) array of tethered buoys was implemented across the equatorial Pacific, providing essential observations of the ocean's sub-surface behavior. By contrast the development of the 1997–1998 El Nino was monitored very carefully and considerably better forecast. This improvement was due to the convergence of many factors. These included: (i) a concerted international program, called TOGA (Tropical Oceans Global Atmosphere), with the remit to observe, understand and predict tropical climate variability; (ii) the application of theoretical understanding of coupled ocean-atmosphere dynamics, and (iii) the development and application of models that simulate the observed variability with some fidelity. The improvement led to considerable optimism regarding our ability to predict seasonal climate variations in general and El Niño/Southern Oscillation (ENSO) events in particular.

Despite these successes, basic questions regarding our ability to model the physical processes in the tropical Pacific remain open challenges in the forecast community. For instance, it is unclear how the MJO, Westerly Wind Bursts (WWBs), intraseasonal variability or atmospheric weather noise influence the predictability of ENSO (e.g., Thompson and Battisti 2001; Kleeman et al. 2003; Flugel et al. 2004; Kirtman et al. 2005) or how to represent these processes in current models. It has been suggested that enhanced MJO and WWB activity was related to the rapid onset and the large amplitude of the 1997–1998 event (e.g., Slingo et al. 1999; Vecchi and Harrison 2000; Eisenman et al. 2005). However, more research is needed to fully understand the scale interactions between ENSO and the MJO and the degree that MJO/WWB representation is needed in ENSO prediction models to better resolve the range of possibilities for the evolution of ENSO (Lengaigne et al. 2004; Wittenberg et al. 2006).

After the late 1990s, however, the ability of some models to predict tropical climate fluctuations reached a plateau with only modest subsequent improvement in skill; but see for example Stockdale et al. (2011) who document progress with one coupled system over more than a decade of development. Arguably, there were substantial qualitative forecasting successes – almost all the models predicted a warm event during the boreal winter of 1997/1998, one to two seasons in advance. Despite these successes, there have also been some striking quantitative failures. For example, according to Barnston et al. (1999) and Landsea and Knaff (2000) none of the models predicted the early onset or the amplitude of that event, and many of the dynamical forecast systems (i.e., coupled ocean–atmosphere models) had difficulty capturing the demise of the warm event and the development of cold anomalies that persisted through 2001. In subsequent forecasts, many models failed to predict the three consecutive years (1999–2001) of relatively cold conditions and the development of warm anomalies in the central Pacific during the boreal summer of 2002. Accurate forecasts can still sometimes be a challenge even at relatively modest lead-times (Barnston 2007, Personal communication) although the recent 2009/2010 El Nino and 2010/2011, 2011/2012 La Nina events were well predicted at least 6 months in advance by most operational centers.

Typically, prediction systems do not adequately capture the differences between different ENSO events such as the recently identified different types of ENSO event (Ashok et al. 2007). In essence, the prediction systems do not have a sufficient number of degrees of freedom for ENSO as compared to nature. There are also apparent decadal variations in ENSO forecast quality (Balmaseda et al. 1995; Ji et al. 1996; Kirtman and Schopf 1998), and the sources of these variations are the subject of some debate. It is unclear whether these variations are just sampling issues or are due to some lower frequency changes in the background state (see Kirtman et al. 2005 for a detailed discussion).

Chronic biases in the mean state of climate models and their intrinsic ENSO modes remain, and it is suspected that these biases have a deleterious effect on El Nino/La Nina forecast quality and the associated teleconnections. Some of these errors are extremely well known throughout the coupled modeling community. Three classic examples, which are likely interdependent, are (1) the so-called double ITCZ problem, (2) the excessively strong equatorial cold tongue typical to most models, and (3) the sub-tropical eastern Pacific and Atlantic warm biases endemic to all models. Such biases may limit our ability to predict seasonal-to-interannual climate fluctuations, and could be indicative of errors in the model formulations. Resolution may be one cause of some of these errors (e.g. Luo et al. 2005). Studies with models that employ higher resolution in both the atmosphere and ocean have demonstrated significant improvements in the mean state of the tropical Pacific and the simulation of El Nino and its teleconnections (e.g. Shaffrey et al. 2008).

3.2 Tropical Atlantic Variability

On seasonal-to-interannual time scales, tropical Atlantic SST variability is typically separated into two patterns of variability – the gradient pattern and the equatorial pattern (Kushnir et al. 2006). The gradient pattern is characterized as a north–south dipole centered at the equator with the largest signals in the sub-tropics, and is typically associated with variability in the southern-most position of the inter-tropical convergence zone (ITCZ). The equatorial pattern is sometimes referred to as the zonal mode (e.g., Chang et al. 2006), or the "Atlantic Nino" because of its structural similarities to the ENSO pattern in the Pacific, although the phase locking with the annual cycle is quite different and the air-sea feedbacks are weaker leading to a more clearly damped mode of variability (e.g., Nobre et al. 2003).

The gradient pattern is linked to large rainfall variability over South America and the northeast region (Nordeste) of Brazil in particular during the boreal spring (Moura and Shukla 1981; Nobre and Shukla 1996). The positive gradient pattern (i.e., warm SSTA to the north of the equator) is associated with a failure of the ITCZ to shift its southern most location during boreal spring. This leads to large-scale drought in much of Brazil and coastal equatorial Africa. The equatorial pattern in the positive phase is linked to increased maritime rainfall just south of the climatological position of the boreal summer ITCZ. The associated terrestrial rainfall anomalies are typically relatively small.

Early predictability studies (Penland and Matrosova 1998) suggest that the north tropical Atlantic component of the gradient pattern (and variability in the Caribbean) can be predicted one to two seasons in advance largely due to the "disruptive" or excitation influence from the Indo-Pacific SSTA, but this does not suggest that local coupled processes in the region are unimportant (e.g., Nobre et al. 2003). The NAO can also be an external excitation mechanism, but again local processes remain important for the life cycle of the variability. The predictability of the southern subtropical Atlantic component of the gradient mode has not been well established, and is largely viewed as independent from ENSO (Huang et al. 2002). There has been little success in predicting the zonal mode.

3.3 *Tropical Indian Ocean Variability*

There are three dominant patterns of variability in the tropical Indian Ocean that affect remote seasonal-to-interannual rainfall variability over land: (i) a basin- wide pattern that is remotely forced by ENSO (e.g., Krishnamurthy and Kirtman 2003); (ii) the so-called Indian Ocean Dipole/Zonal Mode (IOD for simplicity) that can be excited by ENSO, but also can also develop independently of ENSO (e.g., Saji et al. 1999; Webster et al. 1999; Huang and Kinter 2002); and (iii) a gradient pattern similar to the Atlantic that is prevalent during boreal spring (Wu et al. 2008). The basin wide pattern is slave to ENSO and thus its predictability is largely determined by the predictability of ENSO. The IOD plays an important role in the Indian Ocean sector response to ENSO and contributes to regional rainfall anomalies that are independent of ENSO. Idealized predictability studies suggest that the IOD should be predictable up to about 6-months (Wajsowicz 2007; Zhao and Hendon 2009), but prediction experiments are less optimistic (e.g., Zhao and Hendon 2009). Shi et al. (2012) compare the skill of several operational seasonal forecast models, and consider whether larger amplitude events are more skillfully predicted. The predictability of the Indian Ocean meridional mode has not been investigated to date.

Mechanistically, the basin wide mode is captured in thermodynamic slab mixed layer models suggesting that ocean dynamics is of secondary importance and that the pattern is due to an "atmospheric bridge" associated with ENSO (e.g., Lau and Nath 1996; Klein et al. 1999). The IOD, on the other hand, depends on coupled air-sea interactions and ocean dynamics. For example, Saji et al. (1999) noted that the IOD was associated with east-west shifts in rainfall and substantial wind anomalies. Huang and Kinter (2002) argued for well defined (although not as well defined as for ENSO) interannual oscillations where thermocline variations due to asymmetric equatorial Rossby waves play an integral role in the evolution of the IOD. The importance of thermocline variations are a potential source of ocean memory and hence predictability. The development and decay of the meridional mode is largely driven by local thermodynamic cloud and wind feedbacks induced by either ENSO or the IOD, whereas thermocline variations do not seem to be important (Wu et al. 2008).

3.4 *Other Sources of Seasonal to Interannual Predictability*

3.4.1 Upper Ocean Heat Content

On seasonal-to-interannual time scales upper ocean heat content is a known source of predictability. The ocean can store a tremendous amount of heat. The heat capacity of 1 m^3 of seawater is around 3,500 times that of air. Sunlight penetrates the upper ocean, and much of the energy associated with sunlight can be absorbed directly by the top few meters of the ocean. Mixing processes further distribute heat through the surface mixed layer, which can be tens to hundreds of meters thick. With the difference in heat capacity, the energy required to cool the upper 2.5 m of the ocean by 1 °C could heat the entire column of air above it by the same 1 °C. The ocean can also transport warm water from one location to another, so that warm tropical water is carried by the Gulf Stream off New England, where in winter during a cold-air outbreak, the ocean can heat the atmosphere at a rate of many hundreds of W/m^2, similar to the heating rate from solar irradiation.

Ocean heat can also be sequestered below the surface to re-emerge months later and provide a source of predictability (e.g., Alexander and Deser 1994). This occurs in the North Pacific and has been well documented in the North Atlantic where Spring atmospheric circulation patterns associated with a strong (weak) Atlantic jet drive positive (negative) tripolar anomalies in Atlantic ocean heat content (Hurrell et al. 2003). A positive tripole here indicates cold anomalies in the Labrador and subtropical Atlantic and warm anomalies just south of Newfoundland. The shoaling of the thermocline in summer then preserves these heat content anomalies in the subsurface until late Autumn or early winter when the more vigorous storm track deepens the mixed layer and the original heat content anomalies can "re-emerge" at the surface (Timlin et al. 2002) to influence the atmosphere again. This has been the basis of some statistical methods of seasonal forecasting (Folland et al. 2011) and it appears to have played a role in some recent extreme events (Taws et al. 2011). However it is still the case that models produce only a weak response to Atlantic ocean heat content anomalies, and higher resolution (e.g. Minobe et al. 2008; Nakamura et al. 2005) or other atmosphere–ocean interactions may need to be represented if the levels of predictability suggested in some studies from this coupling are to be fully realized.

3.4.2 Snow Cover

Snow acts to raise surface albedo and decouple the atmosphere from warmer underlying soil. Large snowpack anomalies during winter also imply large surface runoff and soil moisture anomalies during and following the snowmelt season, anomalies that are of direct relevance to water resources management and that in turn could feed back on the atmosphere, potentially providing some predictability at the seasonal time scale.

The impact of October Eurasian snow cover on atmospheric dynamics may improve the prediction quality of northern hemisphere wintertime temperature forecasts (Cohen and Fletcher 2007), and winter snow cover can affect predictive skill of spring temperatures (Shongwe et al. 2007). The autumn Siberian snow cover anomalies have also been used for prediction of the East Asian winter monsoon strength (Jhun and Lee 2004; Wang et al. 2009) and spring-time Himalayan snow anomalies may affect the Indian monsoon onset (Turner and Slingo 2011). Becker et al. (2001) demonstrated that Eurasian spring-time snow anomalies may also affect Indian summer monsoon strength through the influence of soil moisture anomalies on Asian circulation patterns.

3.4.3 Stratosphere

Recent investigations suggest that variations in the stratospheric circulation may precede and affect tropospheric anomalies (e.g. Baldwin and Dunkerton 2001; Ineson and Scaife 2009; Cagnazzo and Manzini 2009). The long timescales of the stratospheric QBO could also have an effect under some circumstances (e.g. Boer and Hamilton 2008; Marshall and Scaife 2009). All of these influences act on the surface climate via the northern and southern annular modes (or their regional equivalents such as the NAO). Currently skill is very limited in these patterns of variability and given their key role in extratropical seasonal anomalies this could be an important area for future development. A key factor in this is the vertical resolution of the models used for seasonal prediction, which typically do not include an adequately resolved stratosphere, but should.

3.4.4 Vegetation and Land Use

Vegetation structure and health respond slowly to climate anomalies, and anomalous vegetation properties may persist for some time (months to perhaps years) after the long-term climate anomaly that spawned them subsides. Vegetation properties such as species type, fractional cover, and leaf area index help control evaporation, radiation exchange, and momentum exchange at the land surface; thus, long-term memory in vegetation anomalies could be translated into the larger Earth system (e.g. Zeng et al. 1999). Furthermore a significant portion of the Earth's land surface is cultivated and hence the seasonality of vegetation cover may be different from natural vegetation. Early work with coupled crop-climate models suggests that this may also contribute to seasonal variations that may be predictable (e.g. Osborne et al. 2009).

3.4.5 Polar Sea Ice

Sea ice is an active component of the climate system and is coupled with the atmosphere and ocean at time scales ranging from weeks to decadal. When large anomalies are established in sea ice, they tend to persist due to inertial memory and feedback in the atmosphere–ocean-sea ice system. These characteristics suggest that some aspects of sea ice may be predictable on seasonal time scales. In the Southern Hemisphere, sea ice concentration anomalies can be predicted statistically by a linear Markov model on seasonal time scales (Chen and Yuan 2004). The best cross-validated skill is at the large climate action centers in the southeast Pacific and Weddell Sea, reaching 0.5 correlation with observed estimates even at 12-month lead time, which is comparable to or even better than that for ENSO prediction.

On the other hand we have less understanding of how well sea ice impacts the predictability of the overlying atmosphere although some studies now suggest a negative AO response to declining Arctic Sea Ice (e.g. Wu and Zhang 2010).

4 Decadal Prediction

4.1 Potential Sources of Decadal Predictability

4.1.1 External Forcing

Anthropogenic forcing effects from greenhouse gases and aerosols are a key source of skill in decadal predictions, and are incorporated through the initial conditions and boundary forcings (e.g. Smith et al. 2007). The forcing from greenhouse gases and aerosols are included in the initial condition in that they affect the current state of the climate system. A first order estimate of the likely effects of anthropogenic forcings is provided by the trend since 1900 (Fig. 3 from Smith et al. 2012). This is over-simplified because not this entire trend is attributable to human activities. The response to greenhouse gases is non-linear so that future human-induced changes could be different, and other sources of anthropogenic forcing such as aerosols and ozone could produce responses very different to the trend. Nevertheless, in many regions the trend is comparable to the natural climate variability, suggesting that anthropogenic climate change is a potentially important source of decadal prediction skill.[3]

Solar variations have also been recurring themes historically in discussions of decadal prediction. Variations in solar forcing are, however, generally comparatively small and tend to operate on long timescales with the most notable being the

[3] In some of the literature a "prediction" corresponds to an initial value problem and the "projection" corresponds to a boundary forced problem. Here we recognize that decadal prediction and even seasonal prediction is a both an initial value and a boundary value problem. Throughout the text we refer to the combined initial value and boundary value problem as prediction problem.

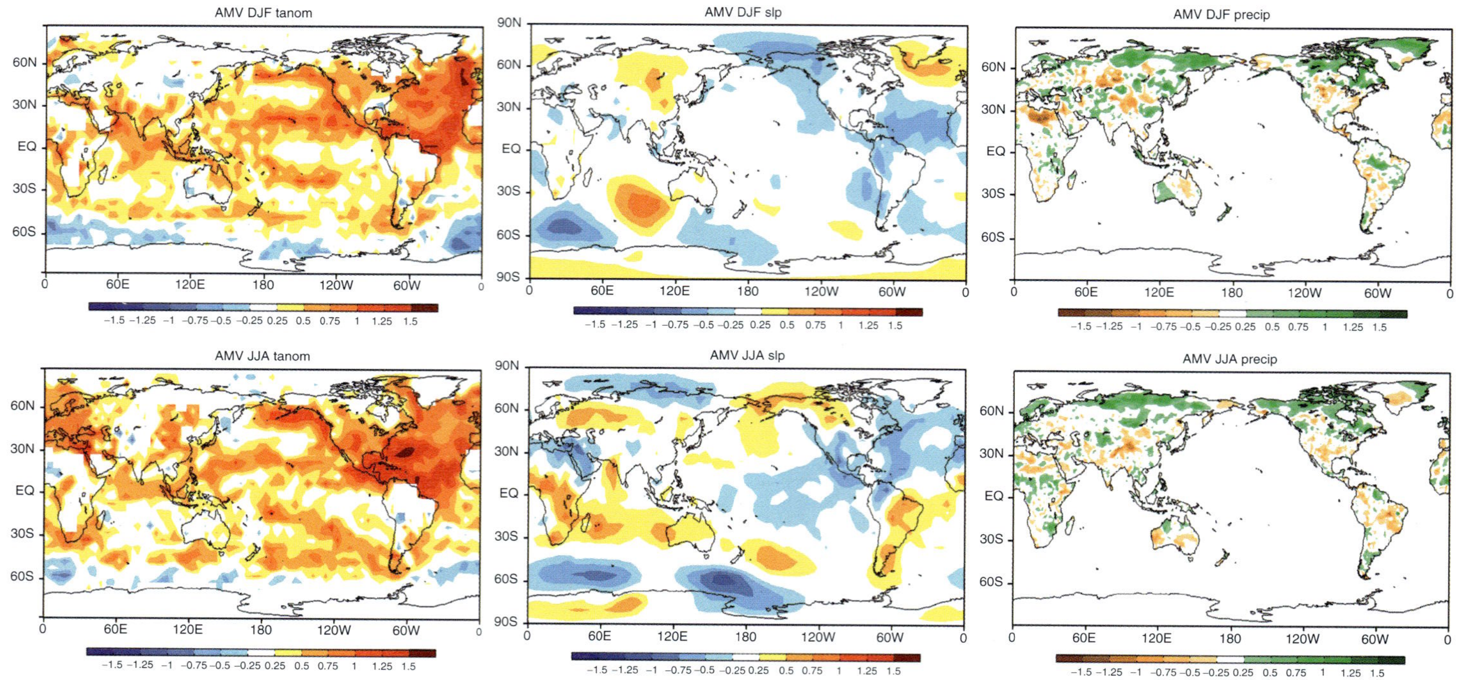

Fig. 3 As Fig. 2 but for Atlantic multi-decadal variability (*AMV*). All were smoothed with a 9-year running mean. Positive AMV years are 1934–1942, 1948, 1952–1957, 1999–2005. Negative AMV years are 1906–1922, 1971–1978. Assuming 4° of freedom, the contour values ±0.25 and ±0.5 are statistically significant at the 87 and 95 % levels, respectively (Figure redrawn following Smith et al. (2012))

11-year solar cycle. Van Loon et al. (2007) review some aspects of solar forcing, and Ineson et al. (2011) have recently shown that the 11-year solar cycle could be an important component of extra-tropical decadal predictability on regional scales, especially in the Euro-Atlantic sector, provided models contain an adequate representation of the stratosphere.

Explosive volcanic eruptions, although relatively rare (typically less than one per decade) also have a significant impact on climate (Robock 2000) and can 'lend' predictability on timescales from seasons to several years ahead. Aerosol injected into the stratosphere during an eruption cools temperatures globally for a couple of years. The hydrological cycle and atmospheric circulation are also affected, globally. Precipitation rates generally decline due to the reduced water carrying capacity of a cooler atmosphere, but winters in northern Europe and central Asia tend to be milder and wetter due to additional changes in the NAO.

Volcanic eruptions are not predictable in advance, but once they have occurred they are a potential source of forecast skill (e.g. Marshall et al. 2009). A similar approach has been considered for seasonal forecasting; once the atmospheric loading has been estimated based on the severity and type of explosion, this could be used in the forecast model. Furthermore, volcanoes impact ocean heat and circulation for many years, even decades (Stenchikov et al. 2009). In particular, the Atlantic meridional overturning circulation (AMOC) tends to be strengthened by volcanic eruptions. Volcanoes could therefore be a crucial source of decadal prediction skill (Otterå et al. 2010), although further research is needed to establish robust atmospheric signals on these timescales. Moreover, there is also evidence that volcanism can reduce the AMOC and may have been a contributor to the Little Ice Age onset (e.g., Miller et al. 2012).

4.1.2 Atlantic Multi-decadal Variability

Atlantic multi-decadal variability (AMV) is likely to be a major source of decadal predictability (Fig. 4 from Smith et al. 2012). Observations and models indicate that north Atlantic SSTs fluctuate with a period of about 30–80 years, linked to variations of the AMOC (Delworth et al. 2007; Knight et al. 2005). The AMOC and AMV can vary naturally (Vellinga and Wu 2004; Jungclaus et al. 2005) or through external influences including volcanoes (Stenchikov et al. 2009; Otterå et al. 2010), anthropogenic aerosols and greenhouse gases (IPCC 2007).

Idealized model experiments suggest that natural fluctuations of the AMOC and AMV are potentially predictable at least a few years ahead (Griffies and Bryan 1997; Pohlmann et al. 2004; Collins et al. 2006; Dunstone and Smith 2010; Matei et al. 2012). If skilful AMV predictions can be achieved in reality, observational and modeling studies suggest that important climate impacts, including rainfall over the African Sahel, India and Brazil, Atlantic hurricanes and summer climate over Europe and America, might also be predictable (Sutton and Hodson 2005; Zhang and Delworth 2006; Knight et al. 2006; Dunstone et al. 2011).

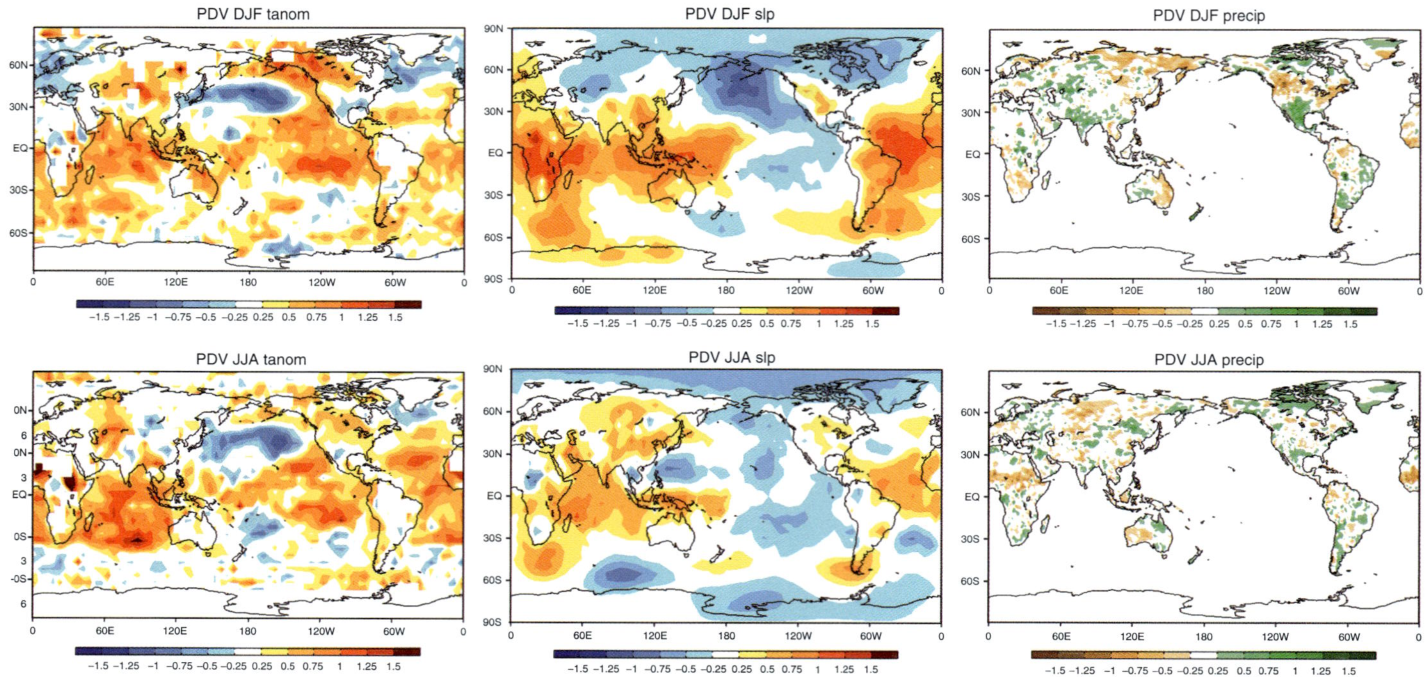

Fig. 4 As Fig. 3 but for Pacific decadal variability (PDV). Positive PDV years are 1937–1941, 1981–1991. Negative PDV years are 1948–1961, 1964–1975 (Figure redrawn following Smith et al. (2012))

4.1.3 Pacific Decadal Variability

Pacific decadal variability (PDV; Fig. 5 from Smith et al. 2012) is also associated with potentially important climate impacts, including rainfall over America, Asia, Africa and Australia (Power et al. 1999; Deser et al. 2004). The combination of PDV, AMV and climate change appears to explain nearly all of the multi-decadal US droughts (McCabe et al. 2004) including key events like the American dustbowl of the 1930s (Schubert et al. 2004). However, mechanisms underlying PDV are less clearly understood than for AMV. Furthermore, predictability studies show much less potential skill for PDV than AMV (Collins 2002; Boer 2004; Pohlmann et al. 2004).

4.1.4 Other Sources of Decadal Predictability

As mentioned above, another potential source of interannual predictability is the Quasi-Biennial Oscillation (QBO) in the stratosphere. The QBO is a wave-driven reversal of tropical stratospheric winds between easterly and westerly with a mean period of about 28 months. The QBO influences the stratospheric polar vortex and hence the winter NAO and Atlantic-European climate. Because the QBO is predictable a couple of years ahead, this may provide some additional predictability of Atlantic winter climate (Boer and Hamilton 2008; Marshall and Scaife 2009).

The ongoing decline in Arctic sea ice volume (e.g. Schweiger et al. 2011) as a result of global warming may also provide another element that influences decadal prediction. As already discussed, there is emerging evidence that reduced Arctic sea ice favors negative AO circulation patterns in winter; as yet there is no evidence for how an increasingly ice-free summer Arctic may affect the summer circulation but much more research needs to be done.

4.2 *Achievements So Far*

Decadal prediction is much less mature than seasonal prediction and does not benefit from a dominant mode of variability, ENSO, as is the case for seasonal to interannual prediction. Skilful statistical predictions of temperature have been demonstrated, both for externally forced signals (Lean and Rind 2009) and for idealized model internal variability (Hawkins et al. 2011). Lee et al. (2006) found evidence for skilful temperature predictions using dynamical models forced only by external changes. Furthermore, several studies show improved skill through initialization, although whether this represents skilful predictions of internal variability or a correction of errors in the response to external forcing cannot be determined. In addition to demonstrating useful predictions of global temperature (Smith et al. 2007), initialization also improves regional predictions of surface temperature, mainly in the north Atlantic and Pacific Ocean (Pohlmann et al. 2009; Mochizuki et al. 2009; Smith et al. 2010). Evidence for improved predictions over land is less convincing.

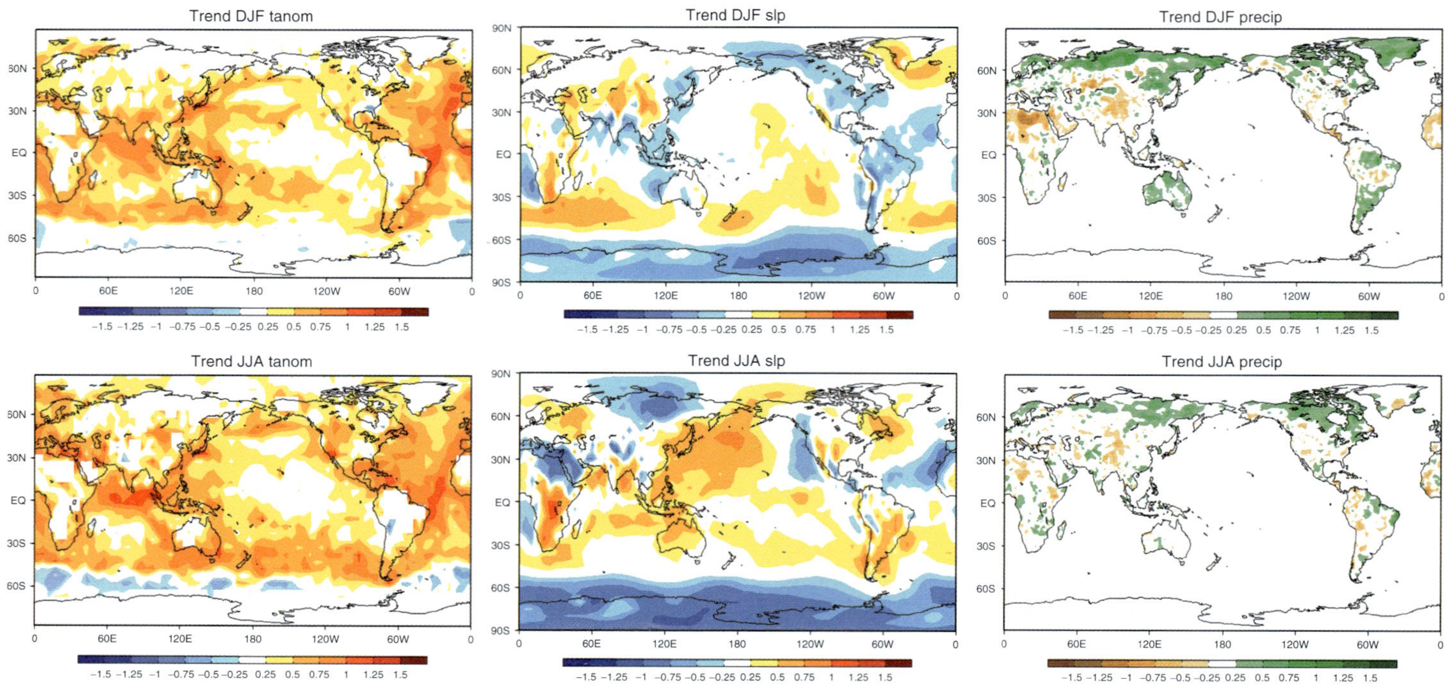

Fig. 5 As Fig. 3 but for trends. For comparison with AMV and PDV, which show a transition from neutral to peak conditions over about 15 years, we show 15-year normalized differences (Figure redrawn following Smith et al. (2012))

Skillful retrospective predictions of Atlantic hurricane frequency out to years ahead have been achieved (Smith et al. 2010). As discussed earlier, some of this skill is attributable to external forcing from a combination of greenhouse gases, aerosols, volcanoes and solar variations, but their relative importance has not yet been established. Initialization improves the skill mainly through atmospheric teleconnections from improved surface temperature predictions in the north Atlantic and tropical Pacific.

On longer timescales, studies of potential predictability within a "perfect model" framework suggest multi-year predictability of the internal variability over the high-latitude oceans in both hemispheres. The first attempts at decadal prediction have identified the Atlantic subpolar gyre as a key source of predictability, with a teleconnection to tropical Atlantic SSTs (Smith et al. 2010).

Based on model predictability experiments, improved skill in north Atlantic SST is expected to be related to skilful predictions of the Atlantic meridional overturning circulation (AMOC), but this cannot be verified directly because of a lack of observations. However, recent multi-model ocean analyses (Pohlmann et al. 2013) provide a consistent signal that the AMOC at 45°N increased from the 1960s to the mid-1990s, and decreased thereafter. This is in agreement with related observations of the NAO, Labrador Sea convection and north Atlantic sub-polar gyre strength. Furthermore, the multi-model AMOC is skilfully predicted up to 5 years ahead. However, models forced only by external factors showed no skill, highlighting the importance of initialization.

5 Summary

The societal requirement for climate information is changing. Across many sectors, the need to be better prepared for and more resilient to adverse weather and climate events is increasingly evident and that is placing new demands on the climate science community. Even without global warming, society is becoming more vulnerable to natural climate variability through increasing exposure of populations and infrastructure, so the need for reliable monthly to interannual predictions is growing, especially in the Tropics. Also, it is now generally accepted that the global climate is warming and the requirement to adapt to current and unavoidable future climate change is becoming more urgent. The emphasis is moving quite rapidly from end-of-the-century climate scenarios towards more regional and impacts-based predictions, with a focus on monthly to decadal timescales.

Various physical mechanisms exist to support long-range predictability beyond the influence of atmospheric initial conditions. These come from slowly varying components of the Earth system, such as the ocean, and boundary conditions such as increasing greenhouse gases or solar variability. While there have been important developments in representing these processes to provide skill in monthly to decadal prediction, there are likely to be other sources of predictability that are currently not exploited due to lack of scientific understanding and/or the ability to capture them in models.

Major areas of research include.

5.1 *Improving the Fidelity of the Climate Models at the Heart of Forecast Systems*

Model biases remain one of the most serious limitations in the delivery of more reliable and skillful predictions. The current practice of bias correction is unphysical and neglects entirely the non-linear relationship between the climate mean state and modes of weather and climate variability. Reducing model bias is arguably the most fundamental requirement going forward. **A key activity must be the evaluation of model performance with a greater focus on processes and phenomena that are fundamental to reducing model bias and for delivering improved confidence in the predictions.** Likewise, the potential predictability in the climate system for monthly to decadal timescales is probably underestimated because of model shortcomings.

Recent research has already shown that higher horizontal and vertical resolution has the potential to increase significantly the predictability in parts of the world where it is currently low, such as western Europe, and **a coordinated effort to assess the value of model resolution to improved predictability is needed.**

5.2 *Developing More Sophisticated Measures of Defining and Verifying Forecast Reliability and Skill for the Different Lead Times*

The development of probabilistic systems for weather forecasting and climate prediction means that the concept of skill has to be viewed differently from the traditional approaches used in deterministic systems. The skill and reliability of probabilistic forecasts have to be assessed against performance across a large number of past events, the hindcast set, so that the prediction system can be calibrated.

The process of forecast calibration using hindcasts presents some serious challenges, however, when the lead time of the predictions extends beyond days to months, seasons and decades. That is because to have a high enough number of cases in the hindcast set means testing the system over many realizations, which can extend to many decades in the case of decadal prediction. The observational base has improved substantially over the last few decades, especially for the oceans, and so the skill of the forecasts may also improve just because of better-defined initial conditions. The fact that the observing system is changing can introduce spurious variability making calibration and validation difficult. Additionally, the process of calibration assumes that the current climate is stationary, but there is clear evidence that the climate is changing (see the Fourth Assessment Report of the Intergovernmental Panel on Climate Change (IPCC 2007)), especially in temperature. The potentially increasing numbers of unprecedented extreme events challenges our current approach to calibrating monthly to decadal predictions and interpreting their results.

Although both the limited nature of the observational base and a changing climate pose some problems for seasonal prediction, for decadal prediction, they are extremely challenging. As already discussed, there is decadal predictability in the climate system through phenomena such as the Atlantic multi-decadal oscillation and the Pacific decadal oscillation, but our understanding of these phenomena is still limited largely owing to the paucity of ocean observations.

A review of the current methods of quantifying forecast skill and reliability in a changing climate is needed and an assessment of their fit for purpose going forward.

5.3 *Design of Ensemble Prediction Systems*

Ensemble prediction systems (EPS) are now established in extended range weather and climate prediction, but the techniques to represent forecast uncertainty and to sample adequately the phase space of the climate system are quite diverse. One of the challenges in the past has been ensuring that the spread of the probabilistic system is sufficient to capture the range of possible outcomes. One of the implications of model bias is a restriction in the spread of the ensemble, and a response to this was to develop multi-model ensembles. **There is still more research to be done on how to best combine multiple forecasting tool as well as how to measure progress.**

The techniques used to sample forecast uncertainty range from initial condition uncertainty (including optimal perturbations and ensemble data assimilation), through stochastic physics to represent the influence of unresolved processes, to the use of perturbed parameters in the parametrizations to represent model uncertainty, and on longer timescales uncertainties in the boundary forcing (e.g. anthropogenic GHG and aerosol emissions). **New activities in coupled data assimilation and in defining more physically-based approaches to representing stochastic, unresolved processes in models are recommended.**

The methods outlined above essentially address different aspects of forecast and model uncertainty, but there is currently little understanding of the relative importance of each for forecasts on different lead times. **A new research activity is proposed that will bring together the various techniques used in weather forecasting and climate prediction to develop a seamless EPS.**

5.4 *Utility of Monthly to Decadal Predictions*

There is a growing appreciation of the importance of hazardous weather in driving some of the most profound impacts of climate variability and change, and a clear message from users that current products, such as 3-month mean temperatures and

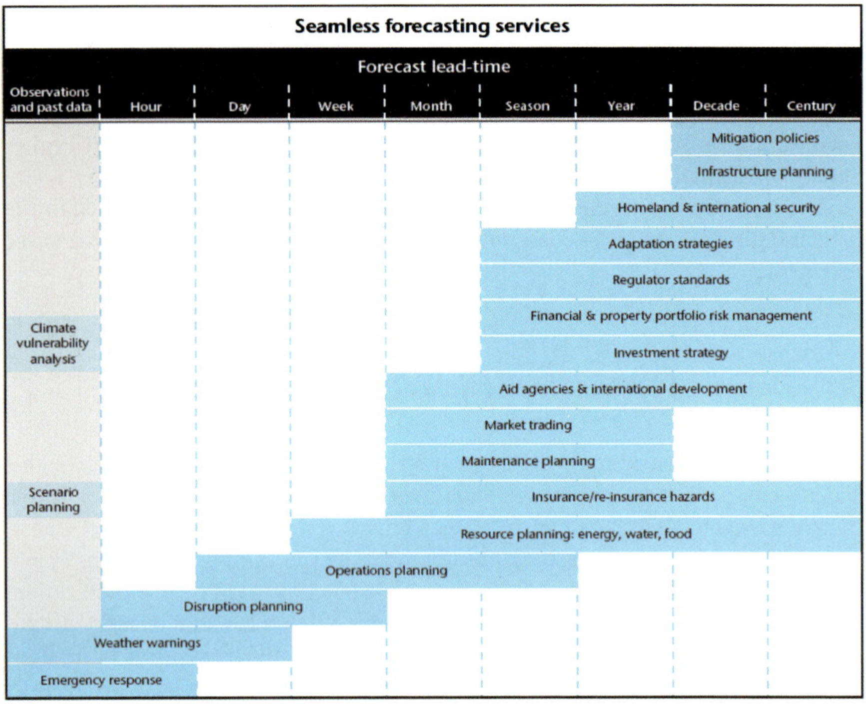

Fig. 6 Seamless forecasting services and potential users of monthly to decadal predictions (From Met Office Science Strategy: http://www.metoffice.gov.uk/media/pdf/a/t/Science_strategy-1.pdf)

precipitation, are not very helpful. Instead, **information on weather and climate variables that directly feed into decision-making (such as the onset of the rainy season, the likelihood of days exceeding critical temperature thresholds, the number of land-falling tropical cyclones) is needed** (see Fig. 6).

Increased computational power has meant that it is now possible to perform simulations that represent synoptic weather systems more accurately (~50 km) and are closer to the global resolutions used in weather forecasting. This raises the questions of how best to exploit the wealth of weather information in monthly to decadal prediction systems; how to understand more fully the weather and climate regimes in which hazardous weather forms; and how to derive products and services that address levels of risk that relate to customer needs. **Stronger links must be established between the science and the service provision.**

Acknowledgements This manuscript was greatly improved by the comments and suggestions made by Julia Slingo. The authors also thank the anonymous reviewers for helpful comment on improving the manuscript. Ben Kirtman was supported by NOAA grants NA10OAR4320143 and NA10OAR4310203. Adam Scaife and Doug Smith were supported by the Joint DECC/Defra MetOffice Hadley Centre Climate Programme (GA01101).

References

Alexander MA, Deser C (1994) A mechanism for the recurrence of wintertime midlatitude SST anomalies. J Phys Oceanogr 25:122–137

Allan R, Ansell T (2006) A new globally complete monthly historical gridded mean sea level pressure dataset (HadSLP2): 1850–2004. J Climate 19:5816–5842

Anderson DLT et al (2011) Current capabilities in sub-seasonal to seasonal prediction. http://www.wcrp-climate.org/documents/CAPABILITIES-IN-SUB-SEASONAL-TO-SEASONALPREDICTION-FINAL.pdf

Arribas A, Glover M, Maidens A, Peterson K, Gordon M, MacLachlan C, Graham R, Fereday D, Camp J, Scaife AA, Xavier P, McLean P, Colman A, Cusack S (2011) The GloSea4 ensemble prediction system for seasonal forecasting. Mon Weather Rev 139(6):1891–1910. doi:10.1175/2010MWR3615.1

Ashok K, Behera SK, Rao SA, Weng H, Yamagata T (2007) El Niño Modoki and its possible teleconnection. J Geophys Res 112:C11007. doi:10.1029/2006JC003798

Baldwin MP, Dunkerton TJ (2001) Stratospheric harbingers of anomalous weather regimes. Science 244:581–584

Balmaseda MA, Davey MK, Anderson DLT (1995) Decadal and seasonal dependence of ENSO prediction skill. J Clim 8:2705–2715

Barnston AG, Glantz M, He Y (1999) Predictive skill of statistical and dynamical climate models in SST forecasts during the 1997–98 El Nino and the 1998 La Nina onset. Bull Am Meteorol Soc 80:217–243

Becker BD, Slingo JM, Ferranti L, Molteni F (2001) Seasonal predictability of the Indian Summer Monsoon: what role do land surface conditions play? Mausam 52:175–190

Blanchard-Wrigglesworth E, Armour KC, Bitz CM, DeWeaver E (2011) Persistence and inherent predictability of Arctic sea ice in a GCM ensemble and observations. J Clim 24:231–250. http://dx.doi.org/10.1175/2010JCLI3775.1

Boer GJ (2004) Long time-scale potential predictability in an ensemble of coupled climate models. Clim Dyn 23:29–44. doi:10.1007/s00382-004-0419-8

Boer GJ, Hamilton K (2008) QBO influence on extratropical predictive skill. Clim Dyn 31:987–1000

Brohan P, Kennedy J, Harris I, Tett SFB, Jones PD (2006) Uncertainty estimates in regional and global observed temperature changes: a new dataset from 1850. J Geophys Res 111:D12106

Brönimann S, Xoplaki E, Casty C, Pauling A, Luterbach J (2007) ENSO influence on Europe during the last centuries. Clim Dyn 28:181–197

Brown A, Sean M, Mike C, Brian G, John M, Ann S (2012) Unified modeling and prediction of weather and climate: a 25-year journey. Bull Am Meteorol Soc 93:1865–1877. doi:10.1175/BAMS-D-12-00018.1

Brunet G, and 13 others (2010) Collaboration of the weather and climate communities to advance sub-seasonal to seasonal prediction. Bull Am Meteor Soc 91(10):1397–1406. doi: 10.1175/2010BAMS3013.1

Cagnazzo C, Manzini E (2009) Impact of the stratosphere on the winter tropospheric teleconnections between ENSO and the North Atlantic and European Region. J Clim 22:1223–1238. doi: 10.1175/2008JCLI2549.1

Cassou C (2008) Intraseasonal interaction between the Madden–Julian Oscillation and the North Atlantic Oscillation. Nature 455:523–527. doi:10.1038/nature07286

Chang P, and Coauthors (2006) Climate fluctuations of tropical coupled systems – the role of ocean dynamics. J Clim 19:5122–5174

Chen D, Xiaojun Yuan (2004) A Markov model for seasonal forecast of Antarctic sea ice. J Clim 17:3156–3168. doi:10.1175/1520-0442(2004)

Cohen J, Fletcher C (2007) Improved skill of Northern Hemisphere winter surface temperature predictions based on land–atmosphere fall anomalies. J Clim 20:4118–4132

Collins M (2002) Climate predictability on interannual to decadal time scales: the initial value problem. Clim Dyn 19:671–692. doi:10.1007/s00382-002-0254-8
Collins M et al (2006) Interannual to decadal climate predictability in the North Atlantic: a multi-model ensemble study. J Clim 19:1195–1203
Delworth TL, Zhang R, Mann ME (2007) Decadal to centennial variability of the Atlantic from observations and models In Ocean Circulation: mechanisms and impacts, Geophysical Monograph Series 173. American Geophysical Union, Washington, DC, pp 131–148
Derome J, Brunet G, Plante A, Gagnon N, Boer GJ, Zwiers FW, Lambert SJ, Sheng J, Ritchie H (2001) Seasonal predictions based on two dynamical models. Atmos–Ocean 39:485–501
Deser C, Phillips AS, Hurrell JW (2004) Pacific interdecadal climate variability: linkages between the tropics and the North Pacific during Boreal Winter since 1900. J Clim 17:3109–3124
DeWitt DG (2005) Retrospective forecasts of interannual sea surface temperature anomalies from 1982 to present using a directly coupled atmosphere–ocean general circulation model. Mon Weather Rev 133:2972–2995
Doblas-Reyes FJ, Hagedorn R, Palmer TN (2005) The rationale behind the success of multi-model ensembles in seasonal forecasting – II. Calibration and combination. Tellus A 57:234–252. doi:10.1111/j.1600-0870.2005.00104.x
Dunstone NJ, Smith DM (2010) Impact of atmosphere and sub-surface ocean data on decadal climate prediction. Geophys Res Lett 37, L02709. doi:10.1029/2009GL041609
Dunstone NJ, Smith DM, Eade R (2011) Multi-year predictability of the tropical Atlantic atmosphere driven by the high latitude north Atlantic ocean. Geophys Res Lett 38:L14701. doi:10.1029/2011GL047949
Eisenman I, Yu L, Tziperman E (2005) Westerly wind bursts: ENSO's tail rather than the dog? J Clim 18:5224–5238
Ferranti L, Palmer TN, Molteni F, Klinker E (1990) Tropical-extratropical interaction associated with the 30–60 day oscillation and its impact on medium and extended range prediction. J Atmos Sci 47(18):2177–2199
Fischer EM, Seneviratne SI, Vidale PL, Lüthi D, Schär C (2007) Soil moisture–atmosphere interactions during the 2003 European summer heat wave. J Clim 20:5081–5099
Flugel M, Chang P, Penland C (2004) The role of stochastic forcing in modulating ENSO predictability. J Clim 17:3125–3140
Folland CK, Scaife AA, Lindesay J, Stephenson D (2011) How predictable is European winter climate a season ahead? Int J Clim. doi:10.1002/joc.2314
Goddard L, Mason SJ, Zebiak SE, Ropelewski CF, Basher R, Cane MA (2001) Current approaches to seasonal-to-interannual climate predictions. Int J Climatol 21:1111–1152
Gottschalck J and 13 others (2010) A framework for assessing operational MJO forecasts: a project of the Clivar MJO working group. Bull Am Meteorol Soc. doi:10.1175/2010BAMS2816.1
Griffies SM, Bryan K (1997) Predictability of North Atlantic multidecadal climate variability. Science 275:181. doi:10.1126/science.275.5297.181
Guilyardi E (2006) El Nino-mean state-seasonal cycle interactions in a multi-model ensemble. Clim Dyn 26:329–348. doi:10.1007/s00382-005-0084-6
Hagedorn R, Doblas-Reyes FJ, Palmer TN (2005) The rationale behind the success of multi-model ensembles in seasonal forecasting – I. Basic concept. Tellus A 57:219–233. doi:10.1111/j.1600-0870.2005.00103.x
Hamilton E, Eade R, Graham RJ, Scaife AA, Smith DM, Maidens A, MacLachlan C (2012) Forecasting the number of extreme daily events on seasonal timescales. J Geophys Res 117:D03114. doi:10.1029/2011JD016541
Hawkins E, Robson JI, Sutton R, Smith D, Keenlyside N (2011) Evaluating the potential for statistical decadal predictions of sea surface temperatures with a perfect model approach. Clim Dyn 37:2459–2509. doi:10.1029/2011JD016541
Higgins RW, Schemm J-KE, Shi W, Leetmaa A (2000) Extreme precipitation events in the western United States related to tropical forcing. J Clim 13:793–820
Hoskins BJ, Karoly DJ (1981) The steady linear response of a spherical atmosphere to thermal and orographic forcing. J Atmos Sci 38:1179–1196

Huang BH, Kinter JL (2002) Interannual variability in the tropical Indian Ocean. J Geophys Res-Oceans 107:20-1–20-26

Huang BH, Schopf PS, Pan ZQ (2002) The ENSO effect on the tropical Atlantic variability: a regionally coupled model study. Geophys Res Lett 29

Hudson D, Alves O, Hendon HH, Marshall AG (2011a) Bridging the gap between weather and seasonal forecasting: intraseasonal forecasting for Australia. Q J R Meteorol Soc 137(656):673–689

Hudson D, Alves O, Hendon HH, Wang G (2011b) The impact of atmospheric initialisation on seasonal prediction of tropical Pacific SST. Clim Dyn 36:1155. doi:10.1007/s00382-010-0763-9

Hurrell JW, Kushnir Y, Ottersen G, Visbeck M (eds) (2003) The North Atlantic oscillation: climatic significance and environmental impact, vol 134, Geophysical Monograph Series. AGU, Washington, DC, 279 pp, doi:10.1029/GM134

Hurrell J, Meehl GA, Bader D, Delworth T, Kirtman B, Wielicki B (2009) Climate system prediction. Bull Am Meteorol Soc 90(12):1819–1832. doi:10.1175/2009BAMS2752.1

Ineson S, Scaife AA (2009) The role of the stratosphere in the European climate response to El Nino. Nat Geosci 2:32–36

Ineson S, Scaife AA, Knight JR, Manners JC, Dunstone NJ, Gray LJ, Haigh JD (2011) Solar forcing of winter climate variability in the Northern Hemisphere. Nat Geosci 4:753–757. doi:10.1038/ngeo1282

IPCC (2007) Climate change 2007: the physical science basis. In: Solomon S et al (eds) Contribution of Working Group I to the fourth assessment report of the Intergovernmental Panel on Climate Change. Cambridge University Press, Cambridge, UK

Jhun J-G, Lee E-J (2004) A new East Asian winter monsoon index and associated characteristics of the winter monsoon. J Clim 17:711–726

Ji M, Leetmaa A, Kousky VE (1996) Coupled model forecasts of ENSO during the 1980s and 1990s at the National Meteorological Center. J Clim 9:3105–3120

Jin EK, and Coauthors (2008) Current status of ENSO prediction skill in coupled models. Clim Dyn 31(6):647–664

Jungclaus JH, Haak H, Latif M, Mikolajewicz U (2005) Arctic-North Atlantic interactions and multidecadal variability of the Meridional Overturning Circulation. J Clim 18:4013–4031

Kirtman BP (2003) The COLA anomaly coupled model: ensemble ENSO prediction. Mon Weather Rev 131:2324–2341

Kirtman BP, Dughong M (2009) Multimodel ensemble ENSO prediction with CCSM and CFS. Mon Weather Rev 137:2908–2930. doi:10.1175/2009MWR2672.1

Kirtman BP, Schopf PS (1998) Decadal variability in ENSO predictability and prediction. J Clim 11:2804–2822

Kirtman BP, Pegion K, Kinter S (2005) Internal atmospheric dynamics and climate variability. J Atmos Sci 62:2220–2233

Kleeman R, Tang Y, Moore AM (2003) The calculation of climatically relevant singular vectors in the presence of weather noise as applied to the ENSO problem. J Atmos Sci 60:2856–2868

Klein SA, Soden BJ, Lau NC (1999) Remote sea surface temperature variations during ENSO: evidence for a tropical atmospheric bridge. J Clim 12:917–932

Knight JR, Allan RJ, Folland CK, Vellinga M, Mann ME (2005) A signature of persistent natural thermohaline circulation cycles in observed climate. Geophys Res Lett 32:L20708. doi:10.1029/2005GL024233

Knight JR, Folland CK, Scaife AA (2006) Climatic impacts of the Atlantic Multidecadal Oscillation. Geophys Res Lett 33:L17706. doi:10.1029/2006GL026242

Koster RD, Mahanama S, Yamada TJ, Balsamo G, Boisserie M, Dirmeyer P, Doblas-Reyes F, Gordon CT, Guo Z, Jeong J-H, Lawrence D, Li Z, Luo L et al (2010) The contribution of land surface initialization to subseasonal forecast skill: first results from the GLACE-2 project. Geophys Res Lett 37:L02402. doi:10.1029/2009GL04167

Krishnamurthy V, Kirtman BP (2003) Variability of the Indian Ocean: relation to monsoon and ENSO. Q J R Meteorol Soc 129:1623–1646

Krishnamurti TN, Kishtawal CM, Zhan Zhang, LaRow T, Bachiochi D, Williford E, Gadgil S, Surendran S (2000) Multimodel ensemble forecasts for weather and seasonal climate. J Clim 13:4196–4216. doi:10.1175/1520-0442(2000)

Kuroda Y, Kodera K (1999) Role of planetary waves in the stratosphere-troposphere coupled variability in the northern hemisphere winter. Geophys Res Lett 26(15):2375–2378
Kushnir Y, Robinson WA, Chang P, Robertson AW (2006) The physical basis for predicting Atlantic sector seasonal-to-interannual climate variability. J Clim 19:5949–5970
Landsea CW, Knaff JA (2000) How much skill was there in forecasting the very strong 1997–98 El Nino? Bull Am Meteorol Soc 81:2107–2120
Lau NC, Nath MJ (1996) The role of the "atmospheric bridge" in linking tropical Pacific ENSO events to extratropical SST anomalies. J Clim 9:2036–2057
Lawrence D, Webster PJ (2002) The boreal summer intraseasonal oscillation and the South Asian monsoon. J Atmos Sci 59:1593–1606
Lean JL, Rind DH (2009) How will Earth's surface temperature change in future decades? Geophys Res Lett 36:L15708. doi:10.1029/2009GL038932
Lee TCK, Zwiers FW, Zhang X, Tsao M (2006) Evidence of decadal climate prediction skill resulting from changes in anthropogenic forcing. J Clim 19:5305–5318
Lengaigne ME, Guilyardi E, Boulanger J-P, Menkes C, Inness PM, Delecluse P, Cole J, Slingo JM (2004) Triggering of El Nino by westerly wind events in a coupled general circulation model. Clim Dyn 23:6. doi:10.1007/s00382-004-0457-2
Lin H, Brunet G (2009) The influence of the Madden-Julian Oscillation on Canadian wintertime surface air temperature. Mon Weather Rev 137:2250–2262
Lin H, Brunet G (2011) Impact of the North Atlantic Oscillation on the forecast skill of the Madden-Julian Oscillation. Geophys Res Lett 38:L02802. doi:10.1029/2010GL046131
Lin H, Brunet G, Derome J (2009) An observed connection between the North Atlantic Oscillation and the Madden-Julian Oscillation. J Clim 22:364–380
Lin H, Brunet G, Fontecilla J (2010a) Impact of the Madden-Julian Oscillation on the intraseasonal forecast skill of the North Atlantic Oscillation. Geophys Res Lett 37:L19803. doi:10.1029/2010GL044315
Lin H, Brunet G, Mo R (2010b) Impact of the Madden-Julian Oscillation on wintertime precipitation in Canada. Mon Weather Rev 138:3822–3839
Lin H, Brunet G, Fontecilla JS (2010c) Impact of the Madden Julian Oscillation on the intraseasonal forecast skill of the North Atlantic Oscillation. Geophys Res Lett 37:L19803
Lorenz EN (1965) A study of the predictability of a 28-variable atmospheric model. Tellus 17:321–333
Luo J-J, Yamagata T, Roeckner E, Madec G, Yamagata T (2005) Reducing climatology bias in an ocean–atmosphere CGCM with improved coupling physics. J Clim 18:2344–2360
Madden RA, Julian PR (1971) Detection of a 40–50 day oscillation in the zonal wind in the tropical Pacific. J Atmos Sci 28:702–708
Marshall AG, Scaife AA (2009) Impact of the QBO on surface winter climate. J Geophys Res 114:D18110. doi:10.1029/2009JD011737
Marshall AG, Scaife AA, Ineson S (2009) Enhanced seasonal prediction of European winter warming following volcanic eruptions. J Climate 22:6168–6180
Matei D, Baehr J, Jungclaus JH, Haak H, Müller WA, Marotzke J (2012) Multiyear prediction of monthly mean Atlantic meridional overturning circulation at 26.5°N. Science 335:76–79
McCabe GJ, Palecki MA, Betancourt JL (2004) Pacific and Atlantic Ocean influences on multidecadal drought frequency in the United States. Proc Nat Acad Sci 101:4136–4141. doi:10.1073/pnas.0306738101
Miller GH et al (2012) Abrupt onset of the little ice age triggered by volcanism and sustained by sea-ice/ocean feedbacks. Geophys Res Lett 39:L02708. doi:10.1029/2011GL050168
Minobe S, Kuwano-Yoshida A, Komori N, Xie S-P, Small RJ (2008) Influence of the Gulf stream on the troposphere. Nature 452:206–210
Mo KC, Higgins RW (1998) Tropical convection and precipitation regimes in the western United States. J Clim 11:2404–2423
Mochizuki T et al (2009) Pacific decadal oscillation hindcasts relevant to near-term climate prediction. Proc Natl Acad Sci 107:1833–1837

Moura AD, Shukla J (1981) On the dynamics of droughts in Northeast Brazil – observations, theory and numerical experiments with a general-circulation model. J Atmos Sci 38:2653–2675
Nakamura M, Enomoto T, Yamane S (2005) A simulation study of the 2003 heatwave in Europe. J Earth Simul 2:55–69
Nobre P, Shukla J (1996) Variations of sea surface temperature, wind stress, and rainfall over the tropical Atlantic and South America. J Clim 9:2464–2479
Nobre P, Zebiak SE, Kirtman BP (2003) Local and remote sources of tropical Atlantic variability as inferred from the results of a hybrid ocean–atmosphere coupled model. Geophys Res Lett 30(5):8008. doi:10.1029/2002GL015785
Osborne TM, Slingo JM, Lawrence D, Wheeler TR (2009) Examining the influence of growing crops on climate using a coupled crop-climate model. J Clim 22:1393–1411. doi:10.1175/2008JCLI2494.1
Otterå OH, Bentsen M, Drange H, Suo L (2010) External forcing as a metronome for Atlantic multidecadal variability. Nat Geosci. doi:10.1038/NGEO955
Palmer TN, Brankovic C, Richardson DS (2000) A probability and decision-model analysis of PROVOST seasonal multimodel ensemble integrations. Q J R Meteorol Soc 126:2013–2034
Palmer TN, and Coauthors (2004) Development of a European multi-model ensemble system for seasonal-to-interannual prediction (DEMETER). Bull Am Meteorol Soc 85:853–872
Palmer TN, Doblas-Reyes F, Weisheimer A, Rodwell M (2008) Toward seamless prediction. Calibration of climate change projections using seasonal forecasts. Bull Am Meteorol Soc 459–470. See also reply to Scaife et al. 2009 in Bull Am Meteorol Soc Oct 2009, 1551–1554. doi: 10.1175/2009BAMS2916.1
Pegion K, Kirtman BP (2008) The impact of air-sea interactions on the predictability of the Tropical Intra-Seasonal Oscillation. J Clim 22:5870–5886
Pellerin PH, Ritchie FJ, Saucier F, Roy S, Desjardins MV, Lee V (2004) Impact of a two-way coupling between an atmospheric and an ocean-ice model over the Gulf of St. Lawrence. Mon Weather Rev 132(6):1379–1398
Penland C, Matrosova L (1998) Prediction of tropical Atlantic sea surface temperatures using linear inverse modeling. J Clim 11:483–496
Pohlmann H, Botzet M, Latif M, Roesch A, Wild M, Tschuck P (2004) Estimating the decadal predictability of a coupled AOGCM. J Clim 17:4463–4472
Pohlmann H, Jungclaus J, Köhl A, Stammer D, Marotzke J (2009) Initializing decadal climate predictions with the GECCO oceanic synthesis: effects on the North Atlantic. J Clim 22:3926–3938
Pohlmann H, Smith DM, Balmaseda MA, da Costa ED, Keenlyside NS, Masina S, Matei D, Muller WA, Rogel P (2013) Skillful predictions of the mid-latitude Atlantic meridional overturning circulation in a multi-model system. Climate Dyn doi:10.1007/s00382-013-1663-6
Power S, Casey T, Folland C, Colman A, Mehta V (1999) Interdecadal modulation of the impact of ENSO on Australia. Clim Dyn 15:319–324
Randall D, Khairoutdinov M, Arakawa A, Grabowski W (2003) Breaking the cloud parameterization deadlock. Bull Am Meteorol Soc 84:1547–1564
Rashid HA, Hendon HH, Wheeler MC, Alves O (2011) Prediction of the Madden-Julian Oscillation with the POAMA dynamical prediction system. Clim Dyn 36:649–661. doi:10.1007/s00382-010-0754-x
Rayner NA, Parker DE, Horton EB, Folland CK, Alexander LV, Rowell DP, Kent EC, Kaplan A (2003) Global analyses of sea surface temperature, sea ice, and night marine air temperature since the late nineteenth century. J Geophys Res 108(D14):4407. doi:10.1029/2002JD002670
Robock A (2000) Volcanic eruptions and climate. Rev Geophys 38:191–219
Rudolf B, Beck C, Grieser J, Schneider U (2005) Global precipitation analysis products. Global Precipitation Climatology Centre (GPCC), DWD, Internet publication, pp 1–8
Saha S, and Coauthors (2006) The NCEP climate forecast system. J Climate 19:3483–3517. doi:10.1175/JCLI3812.1
Saji NH, Goswami BN, Vinayachandran PN, Yamagata T (1999) A dipole mode in the tropical Indian Ocean. Nature 401:360–363
Scaife AA, Folland CK, Alexander LV, Moberg A, Knight JR (2008) European climate extremes and the North Atlantic Oscillation. J Clim 21:72–83

Schubert SD, Suarez M, Pegion PJ, Koster RD, Bacmeister JT (2004) On the cause of the 1930s dustbowl. Science 33:1855–1859
Schweiger A, Lindsay R, Zhang J, Steele M, Stern H (2011) Uncertainty in modeled arctic sea ice volume. J Geophys Res 117:C00D06. doi:10.1029/2011JC007084
Shukla J, Hagedorn R, Miller M, Palmer TN, Hoskins B, Kinter J, Marotzke J, Slingo J (2009) Strategies: revolution in climate prediction is both necessary and possible: a declaration at the world modelling summit for climate prediction. Bull Am Meteor Soc 90:175–178. doi:10.1175/2008BAMS2759.1
Shaffrey L, Stevens I, Norton W, Roberts M, Vidale PL, Harle J, Jrrar A, Stevens D, Woodage M, Demory M-E, Donners J, Clark D, Clayton A, Cole J, Wilson S, Connolley W, Davies T, Iwi A, Johns T, King J, New A, Slingo JM, Slingo A, Steenman-Clark L, Martin G (2008) UK-HiGEM: the new UK high resolution global environment model. Model description and basic evaluation. J Clim 22:1861–1896
Shi Li, Hendon HH, Alves O, Jing-Jia Luo, Balmaseda M, Anderson D (2012) How predictable is the Indian ocean dipole? Mon Weather Rev 140(12):3867–3884
Shongwe ME, Ferro CAT, Coelho CAS, van Oldenborgh GJ (2007) Predictability of cold spring seasons in Europe. Mon Weather Rev 135:4185–4201. doi:10.1175/2007MWR2094.1
Slingo JM, Palmer TN (2011) Uncertainty in weather and climate prediction. Philos Trans R Soc A 369(1956):4751–4767. doi:10.1098/rsta.2011.0161
Slingo JM, Rowell DP, Sperber KR, Nortley F (1999) On the predictability of the interannual behaviour of the Madden-Julian Oscillation and its relationship with El Nino. Q J R Meteorol Soc 125:583–609
Smith DM, Cusack S, Colman AW, Folland CK, Harris GR, Murphy JM (2007) Improved surface temperature prediction for the coming decade from a global climate model. Science 317:796–799
Smith DM, Eade R, Dunstone NJ, Fereday D, Murphy JM, Pohlmann H, Scaife AA (2010) Skilful multi-year predictions of Atlantic hurricane frequency. Nat Geosci 3:846–849. doi:10.1038/NGEO1004
Smith DM, Scaife AA, Kirtman B (2012) What is the current state of scientific knowledge with regard to seasonal and decadal forecasting? Environ Res Lett 7:015602. doi:10.1088/1748-9326/7/1/015602
Simmons AJ, Hollingsworth A (2002) Some aspects of the improvement in skill of numerical weather prediction. Q J R Meteorol Soc 128:647–677. doi:10.1256/003590002321042135
Stenchikov G, Delworth TL, Ramaswamy V, Stouffer RJ, Wittenberg A, Zeng F (2009) Volcanic signals in the oceans. J Geophys Res 114:D16104. doi:10.1029/2008JD011673
Stockdale T, Anderson D, Balmaseda M, Doblas-Reyes F, Ferranti L, Mogensen K, Palmer TN, Molteni F, Vitart F (2011) ECMWF seasonal forecast system 3 and its prediction of sea surface temperature. Clim Dyn 37(3–4):455–471
Sutton RT, Hodson DLR (2005) Atlantic Ocean forcing of North American and European summer climate. Science 309:115–118
Takaya Y, Vitart F, Balsamo G, Balmaseda M, Leutbecher M, Molteni F (2010) Implementation of an ocean mixed layer model in IFS. ECMWF Tech Memo 622 (available from ECMWF Shinfield Park Reading RG2 9AX UNITED KINGDOM)
Taws SL, Marsh R, Wells NC, Hirschi J (2011) Re-emerging ocean temperature anomalies in late-2010 associated with a repeat negative NAO. Geophys Res Lett 38:L20601. doi:10.1029/2011GL048978
Thompson CJ, Battisti DS (2001) A linear stochastic dynamical model of ENSO. Part II: analysis. J Clim 14:445–466
Thompson DWJ, Wallace JM (2000) Annular modes in the extratropical circulation. Part I: month-to-month variability. J Clim 13:1000–1016. doi:10.1175/1520-0442(2000)
Timlin MS, Alexander MA, Deser C (2002) On the reemergence of North Atlantic SST anomalies. J Clim 15:2707–2712
Turner AG, Slingo JM (2011) Using idealized snow forcing to test teleconnections with the Indian summer monsoon in the Hadley Centre GCM. Clim Dyn 36:1717–1735. doi:10.1007/s00382-010-0805-3

Turner AG, Inness PM, Slingo JM (2005) The role of the basic state in the ENSO-Monsoon relationship and implications for predictability. Q J R Meteorol Soc 131:781–804
van Loon H, Meehl GA, Shea DJ (2007) Coupled air-sea response to solar forcing in the Pacific region during northern winter. J Geophys Res 112:D02108. doi:10.1029/2006JD007378
Vecchi GA, Bond NA (2004) The Madden-Julian Oscillation (MJO) and northern high latitude wintertime surface air temperatures. Geophys Res Lett 31:L04104. doi:10.1029/2003GL018645
Vecchi GA, Harrison DE (2000) Tropical Pacific sea surface temperature anomalies, El Niño, and equatorial westerly wind events. J Clim 13:1814–1830
Vellinga M, Wu P (2004) Low-Latitude freshwater influence on centennial variability of the Atlantic Thermohaline Circulation. J Clim 17:4498–4511
Vitart F (2009) Impact of the Madden Julian Oscillation on tropical storms and risk of landfall in the ECMWF forecast system. Geophys Res Lett 36:L15802. doi:10.1029/2009GL039089
Vitart F, Molteni F (2010) Simulation of the Madden-Julian Oscillation and its teleconnections in the ECMWF forecast system. Q J R Meteorol Soc 136:842–855. doi:10.1002/qj.623
Wajsowicz RC (2007) Seasonal-to-interannual forecasting of tropical Indian Ocean sea surface temperature anomalies: potential predictability and barriers. J Clim 20:3320–3343
Waliser D, and others (2009) MJO simulation diagnostics. J Climate 22:3006–3030
Walker GT, Bliss EW (1932) World weather V. Mem R Meteorol Soc 4(36):53–84
Wang Bin, and 27 others (2009) Advance and prospectus of seasonal prediction: assessment of the APCC/CliPAS 14-model ensemble retrospective seasonal prediction (1980–2004). Climate Dyn 33:93–117. doi:10.1007/s00382-008-0460-0
Wang Wanqiu, Mingyue Chen, Kumar A (2010) An assessment of the CFS real-time seasonal forecasts. Weather Forecast 25:950–969. doi:10.1175/2010WAF2222345.1
Webster PJ, Moore AM, Loschnigg JP, Leben RR (1999) Coupled ocean–atmosphere dynamics in the Indian Ocean during 1997–98. Nature 401:356–360
Wheeler M, Hendon HH, Cleland S, Meinke H, Donald A (2009) Impacts of the Madden-Julian oscillation on Australian Rainfall and circulation. J Clim 22:1482–1498
Wittenberg AT, Rosati A, Ngar-Cheung Lau, Ploshay JJ (2006) GFDL's CM2 global coupled climate models. Part III: tropical Pacific climate and ENSO. J Clim 19:698–722. doi:10.1175/JCLI3631.1
Woolnough SJ, Vitart F, Balmaseda MA (2007) The role of the ocean in the Madden-Julian Oscillation: implications for MJO prediction. Q J R Meteorol Soc 133:117–128
Wu Q, Zhang X (2010) Observed forcing–feedback processes between Northern Hemisphere atmospheric circulation and Arctic sea ice coverage. J Geophys Res 115, D14119
Wu R, Kirtman BP, Krishnamurthy V (2008) An asymmetric mode of tropical Indian Ocean rainfall variability in boreal spring. J Geophys Res-Atmos 113
Zeng N, Neelin JD, Lau K-M, Tucker CJ (1999) Enhancement of interdecadal climate variability in the Sahel by vegetation interaction. Science 286(5444):1537–1540. doi:10.1126/science.286.5444.1537
Zhang R, Delworth TL (2006) Impact of Atlantic multidecadal oscillations on India/Sahel rainfall and Atlantic hurricanes. Geophys Res Lett 33:L17712. doi:10.1029/2006GL026267
Zhao M, Hendon HH (2009) Representation and prediction of the Indian Ocean dipole in the POAMA seasonal forecast model. Q J R Meteorol Soc 135:337–352

Assessing the Reliability of Climate Models, CMIP5

Bart van den Hurk, Pascale Braconnot, Veronika Eyring, Pierre Friedlingstein, Peter Gleckler, Reto Knutti, and Joao Teixeira

Abstract In spite of the yet incomplete subsample of the 5th phase of the Coupled Model Intercomparison Project (CMIP5) model ensemble to date, evaluation of these models is underway. Novel diagnostics and analysis methods are being utilized in order to explore the skill of particular processes, the degree to which models have improved since CMIP3, and particular features of the hindcasts, decadal and centennial projections. These assessments strongly benefit from the increasing availability of state-of-the-art data sets and model output processing techniques. Also paleo-climate analysis proves to be useful for demonstrating the ability of models to simulate climate conditions that are different from present day. The existence of an increasingly wide ensemble of model simulations re-emphasizes the need to carefully consider the implications of model spread. Disparity between projected results does imply that model uncertainty exists, but not necessarily reflects a true estimate of this uncertainty.

B. van den Hurk (✉)
Royal Netherlands Meteorological Institute (KNMI), AE De Bilt,
Post Box 201 NL-3730, The Netherlands
e-mail: hurkvd@knmi.nl

P. Braconnot
Insitut Pierre Simon Laplace/Laboratoire des Sciences du Climat et de l'Environnement, unité mixte de recherches CEA-CNRS-UVSQ, Bât.712, Orme des Merisiers CE-Saclay, Gif sur Yvette Cedex F-91191, France
e-mail: pascale.braconnot@lsce.ipsl.fr

V. Eyring
Deutsches Zentrum fuer Luft- und Raumfahrt (DLR), Institut für Physik der Atmosphäre (IPA), Oberpfaffenhofen, Wessling 82234, Germany
e-mail: Veronika.Eyring@dlr.de

P. Friedlingstein
College of Engineering, Mathematics and Physical Sciences, University of Exeter, Harrison Building, Streatham Campus, North Park Road, Exeter EX4 4QF, UK
e-mail: P.Friedlingstein@exeter.ac.uk

G.R. Asrar and J.W. Hurrell (eds.), *Climate Science for Serving Society: Research, Modeling and Prediction Priorities*, DOI 10.1007/978-94-007-6692-1_9,

Projections generated by models with a similar origin or utilizing parameter perturbation techniques generally show more mutual agreement than models with different development histories. Weighting results from different models is a potentially useful technique to improve projections, if the purpose of the weighting is clearly identified. However, there is yet no consensus in the community on how to best achieve this.

These findings, discussed at the session "Assessing the reliability of climate models: CMIP5" of the World Climate Research Program (WCRP) Open Science Conference (OSC), illustrate the need for comprehensive and coordinated model evaluation and data collection. The role that WCRP can play in this coordination is summarized at the end of this chapter.

Keywords Climate model assessment • Evaluation • Model ensembles • Process verification • CMIP5 • WCRP coordinations

List of Acronymns

AMIP	Atmospheric Model Intercomparison Project
CMIP3	CMIP5 3rd, 5th Coupled Model Intercomparison Project
ENSO	El Nino Southern Oscillation
ESM	Earth System Model
GCM	General Circulation Model
IGBP	International Geosphere-Biosphere Program
IHDP	International Human Dimensions Program
IPCC	Intergovernmental Panel on Climate Change
ISCCP	International Satellite Cloud Climatology Project
MME	Multi Model Ensemble
OSC	Open Science Conference
PCMDI	Program for Climate Model Diagnosis and Intercomparison
PPE	Perturbed Physics Ensemble
WCRP	World Climate Research Program

P. Gleckler
Program for Climate Model Diagnosis and Intercomparison, Lawrence Livermore National Laboratory, 7000 East Avenue, Livermore, CA 94550, USA
e-mail: gleckler1@llnl.gov

R. Knutti
Institute for Atmospheric and Climate Science, CHN N 12.1, Universitätstrasse 16, ETH Zurich, CH-8092, Switzerland
e-mail: reto.knutti@env.ethz.ch

J. Teixeira
Jet Propulsion Laboratory, M/S 169-237, 4800 Oak Grove Drive, Pasadena, CA 91109, USA
e-mail: Joao.Teixeira@jpl.nasa.gov

1 Introduction

The assessment of the reliability of climate models is needed to have confidence in the information about future development of the climate system generated with these models. It is generally applied by confronting climate model output with observations over a past period, and interpreting the performance of the model to replicate observed trends, spatial and temporal variability patterns, mean seasonal cycles, responses to perturbation and mutual relationships between relevant quantities. However, this assessment is subject to a large number of aspects:

- the quality and representativity of the reference observational data set
- the knowledge on the initial and time-varying boundary conditions needed to force the climate model
- the comparability between the observed and modeled quantities
- interpretation of discrepancies in terms of model or observational deficiencies, etc.

Yet, this assessment is rapidly evolving and improving. In the context of the 5th Coupled Model Intercomparison Project (CMIP5) an increasing number of climate model simulations is becoming available, and a wide range of analyses currently based on a sub-set of the anticipated model ensemble is being undertaken. During the WCRP Open Science Conference (OSC) held in Denver, October 2011, a selection of studies dedicated to the assessment of the reliability of these climate models was presented in the parallel session B7. Many studies referred back to results from the earlier CMIP3 project, which likewise benefited from the public availability of a large set of model results, leading to a revolution of model evaluation tools, observations and diagnostics. This revolution is ongoing as CMIP5 is running ahead, but important new findings can already be noted. Here we provide an overview of the main topics that emerged during the OSC, which reflect the current state-of-the-art assessments for the reliability of climate models. In particular we summarize what has been implied from the spread in model results, provide examples of novel observations and diagnostics, and give a set of examples of ongoing process evaluation studies that have been discussed as part of the session. Recommendations for WCRP to the governance of this important activity are given at the end of this document.

2 The Implications (and Usefulness) of Model Spread

CMIP5 is clearly more ambitious than its predecessors (in particular CMIP3): although it is still under development, more experiments and associated research questions, more participating models, more model fields, a better documentation of models, and more data storage are becoming available (CLIVAR 2011; Taylor et al. 2012). As model data are submitted to the Program for Climate Model Diagnosis and Intercomparison (PCMDI) and the several storage nodes that are linked together, it becomes evident that model spread will still be substantial. Part of the difference between model results can be attributed to unforced variability, originating from the nonlinear nature of the variable climate system. The impact of this unforced variability reduces as the projection horizon, spatial scale or averaging period increases (Hawkins and Sutton 2009), although natural variability may still be pronounced at the small spatial scales at the end of the twenty-first century. At short time scales (i.e., less than 5 years) unforced uncertainty may potentially be reduced by a realistic initialization of the forecasts, a procedure at the heart of the decadal projections contained in CMIP5.

The spread between equally forced models at longer projection time scales or averaging intervals can be considered to be related to the total model uncertainty. However, this spread does not reflect systematic biases in the models, it assumes that the model sample is representative across the model space, and may be limited due to model formulations that are mutually similar. In general, model spread does imply that model uncertainty exists, but it may not reflect what we think the "true" model uncertainty is, because models are related by taking some observations as reference for the model tuning (Masson and Knutti 2011) and may not sample all possible uncertainties.

Although an increased model ensemble does not capture model deficiencies common to all models (like missing small-scale processes), it is informative about our ability to reliably simulate the climate and its response to external forcings. The design of CMIP5 allows assessment of the importance of many processes in the climate system (see examples of process analyses below), and can help set research priorities in order to reduce this aspect of uncertainty (Dufresne and Bony 2008). Some of these process uncertainties can potentially be reduced by making use of observations showing variability due to comparable forcings at seasonal or interannual time scales. Examples include the evaluation of the snow-albedo feedback (Hall and Qu 2006) and the evaluation of the terrestrial biosphere in the carbon-climate feedback (see below). A paleo-modeling experiment is explicitly included in CMIP5, focusing on the mid Holocene, the Last Glacial Maximum (LGM, where large ice sheets and low greenhouse gas levels were present) and the Past Millennium are considered specifically. Also paleo-observations of SST during the LGM from the MARGO synthesis (Margo Project Members 2009) allow "out-of-sample" evaluation of climate models, that is, evaluation of models under climate conditions different from present day. In spite of a fair amount of uncertainty of these observations, a reliable model ensemble should encompass the observed observation range, showing as a preferably uniform rank histogram of model-observation differences. Comparisons between MARGO data and a Perturbed Physics Ensemble (PPE) generated by perturbing physical parameters clearly showed this PPE to be incapable of capturing the large SST-responses of the LGM relative to the current climate. The available CMIP5 Multi-Model Ensemble (MME) was shown to be able to encompass the LGM observed global mean SST-response, although spatial patterns of this response and individual model results were not as reliable (Hargreaves et al. 2011).

Model spread can also be utilized to diagnose the inherent predictability of the climate at decadal time scales. Climate states can be considered to be predictable if their probability of occurrence conditional to the initial state is significantly different from the climatological probability of occurrence. An ensemble of initialized model simulations will at some stage diverge to their climatological probability distribution, preferably at the same rate as nature. The predictability of the natural climate system given an initial state cannot be assessed, as we have only a single realization of the near future. But long integrations of climate models can be used to infer their inherent predictability, for instance by calculating the rate of divergence from an ensemble of episodes that have analogous start conditions (Branstator and Teng 2010). This procedure does assume the absence of model error and thus maps the inherent predictability in the modeled climate, which can be considered as an upper limit of predictability under the assumption that models are free of systematic errors. Analysis of the predictability of the 5 year low-pass filtered ocean heat content in the upper 300 m from six climate models for which a long unforced integration was available revealed that the time range in which useful predictions could be made varied between 5 and 20 year for the North Atlantic basin and somewhat shorter for the Pacific. An evaluation with 10 CMIP5 models shows comparable results, albeit that the results varied widely across the ensemble. Assessment of the inherent predictability should be carried out for every model participating in the decadal prediction simulations (e.g. Matei et al. 2012).

The existence of an MME and their varying degree of consistence with a wide range of observations raises the question as whether a probability distribution of future climate conditions could be constructed by weighting the models using performance metrics. Model quality metrics obtained by combining multi-variable performance metrics such as those presented by Reichler and Kim (2008) demonstrate an increase in skill of climate models over time, but their interpretations are not clear, and not very useful for any particular purpose. For example, climate change assessments in the Arctic regions will be tempted to give a stronger weight to metrics that represent sea ice conditions, whereas climate change assessments for the Sahel will need other variables to be represented well (Knutti 2008). The increased skill of climate models does not imply that the spread in future projections is reducing. On the contrary, preliminary analysis of a small subset of CMIP5 (10 models) shows that the spread in twenty-first century global mean temperature is similar to the CMIP3 ensemble, despite considerable model development. New observational analyses put extra constraints on the range in modeled climate sensitivity (e.g. the analysis of land-ocean contrasts in longwave radiation by Huber et al. 2011), but the probability distribution of this climate sensitivity is still wide.

The Intergovernmental Panel on Climate Change (IPCC) Expert Meeting on "Assessing and Combining Multi Model Climate Projections[1]" held in Boulder in January 2010 (Knutti et al. 2010) gave a list of properties of performance metrics to be useful. In particular:

- they should be simple to interpret
- they should be related to the prediction purpose
- they should reflect known processes

[1]https://www.ipcc-wg1.unibe.ch/publications/supportingmaterial/IPCC_EM_MultiModel Evaluation_MeetingReport.pdf

- relevant observations with sufficiently low uncertainty should be available
- they should be robust against their exact definition and aggregation procedure.

When metrics comply with these criteria, they could be used to generate a model weighting or selection procedure, provided that the rationale and implementation of the weighting are clearly defined and documented.

The existence of a range of models is a prerequisite of generating useful climate change assessment. No model is perfect, and even a model based on "the best available knowledge" cannot easily be defined since the "best knowledge" concerning a particular process or regime cannot easily be defined. The variability in the multi-model ensemble should be utilized and tailored to the application at hand.

3 New Observations and Diagnostics

Since the CMIP3 era a wealth of new observations, diagnostics and analysis methods have evolved, tailored to the evaluation of physical processes, their interactions, prediction skill and reliability for describing climate change. A full review of these developments is out of the scope of this report, but a few noticeable developments were discussed at the session, which are summarized here.

Use of observations is often implicit in the development and tuning of climate models: no model can be constructed without them. Model output evaluation using observations implies an explicit use of these. This evaluation supports the further development of the models, and gives credibility to the projections produced. In practice model evaluation and generation of "operational" climate model projections are parallel processes, where the model versions that usually have some inertia between upgrades. It preferably should be designed to highlight concrete model components that should be changed or replaced in order to improve the model's skill. An assessment of the model uncertainty in quantities that are subject to many processes (such as near surface temperature or precipitation) is in itself useful (see previous section), but often does not reveal the necessary adjustments to models with limited skill. In the context of model evaluation, observations should be as much as possible analogous to the model variables that are generated (see the CMIP5 protocol of Taylor et al. 2012), and readily be available to the research community. For this a strong collaboration between model developers and data collectors (including satellite mission teams) is mandatory, not only concerning the technical infrastructure that allows model-to-observation comparisons, but also in the area of defining comparison metrics and skill thresholds.

A promising initiative in this respect is the presence of a "Obs4MIPS" tab on the PCMDI website (Teixeira et al. 2011) that discloses a number of satellite products designed to evaluate model cloud, precipitation and radiation characteristics. For instance, Jiang et al. (2012) compared A-train ice/water cloud and integrated water vapor observations to a range of CMIP3 and CMIP5 models, generally showing an improvement of the modeled cloud characteristics over the recent past. A traditional approach to compare model output to satellite data is to transfer the observed radiances into physical fields using retrieval algorithms. However, using this approach error propagation and aggregation are difficult to assess. Therefore, significant progress has been made in CMIP5 by building in satellite simulators into the GCMs, allowing evaluation of radiances instead (e.g. Bodas-Salcedo et al. 2011). However, the translation of radiance errors back to model improvement is often concealed by the many processes considered in the forward radiance modeling. A new stage in model-to-observation comparison is to map radiances back to model fields that takes the observational postprocessing and aggregation into account, but yet allows a model evaluation in geophysical units.

A powerful analysis method is to conditionally sample observations and model data in order to obtain quantities that are representative for a certain climate regime. A Cloud Regime Error Metric is derived by Williams and Webb (2009) by decomposing cloud regimes from the International Satellite Cloud Climatology Project (ISCCP) data archive discriminating classes of cloud top temperature and optical thickness. Utilizing this conditional sampling approach model biases can be specified for particular cloud regimes, enabling discerning errors due to a misrepresentation of cloud radiative forcing for particular classes and signals attributable to changes in the relative frequency of occurrence of particular cloud regimes. Using transpose-AMIP simulations from a single CMIP5 model for which all necessary data were available it was shown that biases in cloud radiative properties develop very fast in the forecast (already present 1 day after the initialization). Also a persistent problem of undersampled frequency of mid-level clouds is evident from this analysis.

Advanced statistical techniques are also being utilized to detect the scale dependence of skill of GCMs (Sakaguchi et al. 2012). The skill of GCM-generated surface temperature trends over the past decades obviously varies over temporal and spatial scales: global mean and long term trends are more easily reproduced than similar trends at smaller scales. The detection of the spatial and temporal scale of model skill is strongly relevant for the confidence in model climate projections. Global mean temperature trends from a sub-set of CMIP5 climate models are shown to be accurate: the uncertainty is smaller than the observational uncertainty. At smaller spatial scales the CMIP5 subset outperforms the earlier CMIP3 ensemble, at least for longer time scales.

Also for evaluation of land surface processes more data sets have become available. Jung et al. (2009) used a regression tree analysis to extrapolate Fluxnet site observations of surface evaporation and gross primary production (GPP) to all land areas by means of a set of climate data and vegetation indices satellite products. This data set is useful for evaluation of global patterns of mean GPP and evaporation, but by nature of its construction trend analysis cannot be applied. Leaf area index (LAI) data from MODIS show that simulations by the CMIP5 Earth System models show a fair correlation for the northern hemisphere (Anav et al. 2013).

For paleo-studies an increasing number of observations becomes available. Mutually independent data sets exist that reveal information on vegetation (pollen and other tracers), fires (charcoal deposition), regional hydrology (lake level marks) and aerosol level (dust deposition). Schmittner et al. (2011) used the Univ. of Victoria climate model to constrain the likelihood range of the climate sensitivity using LGM temperature reconstructions. Similar exercises are currently being undertaken using CMIP5 model output. In general, climate models seem to reproduce first order responses to different climate conditions fairly well, but do not reconstruct the regional signature of these responses and the various feedbacks at the millennium time scale (such as vegetation feedback) very well.

4 Examples of Process Evaluations Currently in Progress

Many process evaluations are currently underway, exploring the (yet limited) CMIP5 data archive with sophisticated analysis methods. Here we give a small number of examples of such studies that were presented during the session, not attempting to give a complete overview of the ongoing analyses.

An important source of uncertainty is the degree to which terrestrial and oceanic fluxes of CO_2 respond to future climate change. In the C4MIP experiment, Friedlingstein et al. (2006) showed that uncertainty in this response, represented in an ensemble of Earth System Models (ESMs) representing the carbon cycle and its interactions with the climate system, has a strong impact on the projected global temperature, due to pronounced feedbacks between the climate and the carbon cycle. Increased ecosystem release of carbon under warmer climate conditions may imply a strong positive carbon-climate feedback. Determination of the strength of this feedback is one of the outstanding problems in climate research.

Hall and Qu (2006) used the pronounced seasonal cycle in observed snow cover to determine the strength of the snow-albedo feedback (where reduced snow cover leads to higher radiative absorption which in turn promotes snow melt), and compared this to climate change projections from a range of GCMs. The physical mechanism of the snow-albedo feedback is fairly well understood and operates similarly at the seasonal and centennial time scale, thus allowing to determine the optimal feedback strength that should be present in the model simulations. Cox et al. (2012) similarly utilize observations collected at time scales covered in the current data record to infer an estimate of the carbon-climate feedback strength. Notifying that the observed interannual variability of atmospheric CO_2 concentration is primarily due to terrestrial biosphere responses to (ENSO-modulated) temperature fluctuations, the terrestrial carbon loss per degree warming can be derived from the observational record. During ENSO years the carbon uptake by vegetation is much weaker than during non-ENSO years. CMIP5 models can similarly be evaluated on such interannual time scales and tested against the observed CO2-climate sensitivity.

Using a similar approach, Mahlstein and Knutti (JGR submitted) use the relation between Arctic temperature and sea ice extend to estimate future ice-cover area as a function of regional temperature projections from CMIP5. According to this simple extrapolation the Arctic will be free of ice during summer when global mean temperature increases by 2K above present, whereas the uncalibrated CMIP3 models suggest that this does not occur until a 3K global mean warming.

Analysis of feedbacks between processes is the key in evaluating the ocean component of GCMs. This feedback analysis requires advanced processing of available observations and the use of informative conceptual frameworks. The CMIP5 archive and individual models participating to this experiment are currently explored intensively to diagnose the complex physical feedbacks between the ocean and atmosphere. Guilyardi et al. (2011) revisit the classical ENSO theory of the interplay between the dynamical Bjerkness feedback and the heat flux feedback, and conclude that the disparity between a subset of 6 CMIP5 models is largest in the strength and sign of the Bjerkness feedback. Spatial patterns of the shortwave radiative forcing play a key role. Evaluation of the CCSM4 hindcasts by Bates et al. (2012) reveal the existence of compensating errors between solar and evaporative fluxes. Using the Common Ocean-ice Reference Experiments data set the atmospheric feedbacks are further disentangled in a heat flux equation decomposition, and errors in the surface wind fields explain a significant portion of the model disparity. The wind driven forcing also tends to play a role in explaining model errors in Atlantic meridional heat transport at 26°N, where a trade off between overturning and wind driven gyre transport takes place. Msadek (2011) found that the slope of the wind driven gyre transport as function of the Atlantic Meriodinal Overturning Circulation strength (another example of an advanced feedback diagnostic) has the wrong sign in the GFDL model. These diagnostic studies are crucial to gain confidence in decadal predictions which critically depend on a right initialization and propagation of anomalies in the ocean component of coupled AOGCMs, and on centennial time scale in which the ocean mixing properties play a crucial role in determining the time scale of the transient climate response to the changed radiative forcing.

The recent trends in surface solar radiation, attributed to global dimming and brightening, provide a useful testbed for evaluation of the representation of the clear sky direct aerosol radiative forcing. CMIP3 models underestimate the amplitude of the dimming/brightening signals particularly over China, Europe and India (Dwyer et al. 2010), which was partly attributed to incorrect aerosol emission scenarios. Allen et al. (2012) revisit this analysis for CMIP5, where in contrast to CMIP3 only a single aerosol emission inventory was used. 14 CMIP5 models with a total of 54 ensemble members were available, and compared to observations from the Global Energy Balance Archive, ISCCP, and surface data sets. Clear sky radiation was calculated by removing radiation variability explained by cloud cover. In spite of the increased consistency in the aerosol fields, the dimming trend was not well captured by CMIP5: the timing of the reversion from dimming to brightening in Europe was about right, but the amplitude both in Europe and China is still too small. The conclusions are robust after correcting for cloud cover in the observations and models. It is of interest to explore the ability of detailed radiation process models in simulating the dimming and brightening features.

5 Summary and Recommendations for WCRP

Since CMIP3 significant progress has been made in the design of multi-model experiments, the interpretation of model spread, the availability and usage of observations, and the diagnostics of complex processes and their interactions. As the

CMIP5 archive is filling up these analyses will further develop. Many other studies, not reported here, are underway along the lines sketched above. The coordinating role of WCRP in these developments has been very beneficial. But where can WCRP play a further role?

The evident increase in the level of sophistication in the practice of model evaluation, data collection and development of diagnostics and experimental design is clearly reflected in a number of recent WCRP coordination meetings and documents, such as the world modeling summit (2008), and the WCRP modeling coordination meeting (2010). Many recommendations documented in the workshop reports call for enhanced collaboration, particular focus areas and promotion of development of e.g. better observational data bases. Here we will review these recommendations in the light of the discussions and developments reported previously.

The current overall organizational design of WCRP is well targeted to establish the required improved collaboration of experts with a different disciplinary background. Specifically, improved links should be encouraged between observationalists and modelers, NWP and climate model developers, and physical and statistical experts. Links between observation and model experts should be organized around the development of agreed observable model evaluation diagnostics and performance metrics (see e.g. the "Good practice paper" by Knutti et al. 2010). Frequents meetings between model developers and application experts can benefit from a cross-fertilization of common practices in these communities. Frequent evaluation of climate models using data assimilation and routine observations, as commonly applied in NWP, can help target the most important biases and their causes in climate models. Long climate integrations can highlight systematic shortcomings in NWP systems normally masked by routine application and model state adjustments. The concept of seamless prediction is a fruitful research area where climate and NWP applications are joined. And finally, the involvement of statisticians is important to improve the detection and evaluation of extreme events in the model suite.

Another step forward is the identification of a number of key model deficiencies. An inventory over >100 experts, carried out in 2010, revealed a number of persistent shortcomings in model performance, that urgently need improvement. The issues mentioned most frequently were:

- tropical variability and biases
- moist processes (clouds, convection, precipitation)
- carbon cycle and land/ocean–atmosphere coupling
- troposphere-stratosphere interaction
- formulation of physics in high resolution models.

The first three topics are well covered in the studies reported previously in this paper, while the agreed need to improve the predictability of the atmospheric circulation, and the representation of extremes are reflected by the last two topics. Improvements in these areas require an increased investment in model development capacity (Jacob 2011), but also require improved experimental design (e.g. initialized forecasts, specific feedback experiments, experiments aimed at describing specific (extreme) events) and diagnostics. The WCRP working group structure is well capable for designing these focused studies.

Finally, WCRP can continue to play its role as ambassador aiming at targeting funding resources, improving interdisciplinary links and engaging experts and students. It can do so by organizing targeted conferences and sessions, and provide input to circuits where decisions are being made. Important overarching targets are for instance:

- the continued need to close the gaps between observations and models
- the continuation of the collection and storage of high-quality homogenized observation records
- the design of focused field observation studies
- the involvement of the NASA and ESA climate initiatives
- the call for focused modeling studies
- the promotion of development of comprehensive Earth System Models including components of e.g. the biosphere, cryosphere, and human dimensions, requiring strengthened links with IGBP and IHDP.

Most recommendations require efforts from the researchers in the fields: submit targeted research proposal, commit to coordinated activities, maintain or improve the interdisciplinary network. By its organizational design with its working groups and conference sessions, WCRP can synchronize the activities of the wide range of involved researchers, and as such help in improving the important understanding of our environment.

References

Allen RJ, Norris JR, Wild M (2012) Evaluation of multidecadal variability in CMIP5 surface solar radiation and inferred underestimation of aerosol direct effects. Submitted to J Geophys Res

Anav A, Friedlingstein P, Kidston M, Bopp L, Ciais P, Cox P, Jones C, Jung M, Myneni R, Zhu Z (2013) Evaluating the land and ocean components of the carbon cycle in the CMIP5 Earth System Models. J Climate. doi:10.1175/JCLI-D-12-00417.1 (in press)

Bates SC, Fox-Kemper B, Jayne SR, Large WG, Stevenson S, Yeager SG (2012) Mean biases, variability, and trends in air-sea fluxes and SST in the CCSM4. J Climate 25:7781–7801. doi:10.1175/JCLI-D-11-00442.1

Bodas-Salcedo A and Coauthors (2011) COSP: satellite simulation software for model assessment. Bull Am Meteorol Soc 92:1023–1043. doi:10.1175/2011BAMS2856.1

Branstator G, Teng H (2010) Two limits of initial-value decadal predictability in a CGCM. J Clim 23(23):6292–6311. doi:10.1175/2010JCLI3678.1

CLIVAR (2011) WCRP Coupled Model Intercomparison Project – Phase 5 – CMIP5 –, CLIVAR exchanges, Special issue no 56, vol 16(2), May 2011

Cox PM, Pearson D, Booth BB, Friedlingstein P, Huntingford C, Jones CD, Luke CM (2012) Sensitivity of tropical carbon to climate change constrained by carbon dioxide variability. Nature 494:341–344. doi:10.1038/nature11882

Dufresne J-L, Bony S (2008) An assessment of the primary sources of spread of global warming estimates from coupled atmosphere–ocean models. J Clim 21:5135–5144

Dwyer JG, Norris JR, Ruckstuhl C (2010) Do climate models reproduce observed solar dimming and brightening over China and Japan? J Geophys Res 115:D00K08. doi:10.1029/2009JD012945

Friedlingstein P et al (2006) Climate–carbon cycle feedback analysis: results from the C4MIP model intercomparison'. J Clim 19(15):3337–3353

Guilyardi E, Cai W, Collins M, Fedorov A, Jin F-F, Kumar A, Sun D-Z, Wittenberg A (2011) New strategies for evaluating ENSO processes in climate models. BAMS. doi:10.1175/BAMS-D-11-00106.1

Hall A, Qu X (2006) Using the current seasonal cycle to constrain snow albedo feedback in future climate change. Geophys Res Lett 33:L03502. doi:10.1029/2005GL025127

Hargreaves HC, Paul A, Ohgait R, Abe-Ouchi A, Annan JD (2011) Are paleoclimate model ensembles consistent with the MARGO data synthesis? Clim Past Discuss 7:775–807. doi:10.5194/cpd-7-775-2011

Hawkins E, Sutton RT (2009) The potential to narrow uncertainty in regional climate predictions. BAMS 90:1095. doi:10.1175/2009BAMS2607.1

Huber M, Mahlstein I, Wild M, Fasullo J, Knutti R (2011) Constraints on climate sensitivity from radiation patterns in climate models. J Clim 24:1034–1052. doi:10.1175/2010JCLI3403.1

Jacob C (2011) From regional weather to global climate; oral presentation at OSC. http://conference2011.wcrp-climate.org/abstracts/jackob_A4.pdf

Jiang JH, Su H, Zhai C, Perun VS et al (2012) Evaluation of cloud and water vapor simulations in CMIP5 climate models using NASA A-train satellite observations. J Geophys Res 117(D1410):24 pp. doi:10.1029/2011JD017237

Jung M, Reichstein M, Bondeau A (2009) Towards global empirical upscaling of FLUXNET eddy covariance observations: validation of a model tree ensemble approach using a biosphere model. Biogeosciences 6:2001–2013

Knutti R (2008) Should we believe model predictions of future climate change? Trienn Issue Earth Sci Philos Trans R Soc A 366:4647–4664. doi:10.1098/rsta.2008.0169

Knutti R et al (2010) Good practice guidance paper on assessing and combining multi model climate projections. In: Stocker TF, Qin D, Plattner G.-K, Tignor M, Midgley PM (eds) Meeting report of the Intergovernmental Panel on Climate Change expert meeting on assessing and combining multi model climate Projections, IPCC Working Group I Technical Support Unit, University of Bern, Bern, Switzerland

MARGO Project Members (2009) Constraints on the magnitude and patterns of ocean cooling at the Last Glacial Maximum. Nat Geosci 2:127–132. doi:10.1038/ngeo411

Masson D, Knutti R (2011) Climate model genealogy. Geophys Res Lett 38:L08703. doi:10.1029/2011GL046864

Matei D, Baehr J, Jungclaus JH, Haak H, Müller WA, Marotzke J (2012) Multiyear prediction of monthly mean atlantic meridional overturning circulation at 26.5°N. Science 335:76–79. doi:10.1126/science.1210299

Msadek R (2011) Comparing the meridional heat transport at 26.5ºN and its relationship with the MOC in two CMIP5 coupled models and in RAPID-array observations (oral presentation WCRP OSC Denver, Oct 2011)

Reichler T, Kim J (2008) How well do coupled models simulate today's climate? Bull Am Meteorol Soc 89:303–311

Sakaguchi K, Xubin Z, Brunke MA (2012) Temporal- and spatial-scale dependence of three CMIP3 climate models in simulating the surface temperature trend in the twentieth century. J Clim 25:2456–2470. doi:10.1175/JCLI-D-11-00106.1, http://dx.doi.org/

Schmittner A, Urban NM, Shakun JD, Mahowald NM, Clark PU, Bartlein PJ, Mix AC, Rosell-Melé A (2011) Climate ensitivity estimated from temperature reconstructions of the last glacial maximum. Science 334(6061):1385–1388. doi:10.1126/science.1203513

Taylor KE, Stouffer RJ, Meehl GA (2012) An overview of CMIP5 and the experiment design. Bull Am Meteorol Soc 93:485–498

Teixeira J, Waliser D, Ferraro R, Gleckler P, Potter G (2011) Satellite observations for CMIP5 simulations. CLIVAR Exchanges No. 56, 16(2) May 2011

Williams KD, Webb MJ (2009) A quantitative performance assessment of cloud regimes in climate models. Clim Dyn 33:141–157. doi:10.1007/s00382-008-0443-1

Changes in Variability Associated with Climate Change

Karen H. Rosenlof, Laurent Terray, Clara Deser, Amy Clement, Hugues Goosse, and Sean Davis

Abstract In this paper, we briefly discuss changes in large-scale oscillations such as the El Nino/Southern Oscillation (ENSO), the North Atlantic Oscillation (NAO), and the northern and southern annular modes (NAM and SAM), changes in the polar and tropical troposphere, and interactions between the stratosphere and troposphere in a changing climate. We consider both changes in variability as well as trends in the

K.H. Rosenlof (✉)
Chemical Sciences Division, NOAA Earth Systems Research Laboratory,
325 Broadway, R/CSD8, Boulder, CO 80305, USA
e-mail: Karen.H.Rosenlof@noaa.gov

L. Terray
Sciences de l'Univers au CERFACS URA1875, CERFACS/CNRS,
42 Avenue Gaspard Coriolis, 31057, Cedex 01 Toulouse, France
e-mail: terray@cerfacs.fr

C. Deser
Climate and Global Dynamics, National Center for Atmospheric Research,
PO Box 3000, Boulder, CO 80307-300, USA
e-mail: cdeser@ucar.edu

A. Clement
RSMAS/MPO, University of Miami, 4600 Rickenbacker Causeway,
Miami, FL 33149, USA
e-mail: aclement@rsmas.miami.edu

H. Goosse
Earth and Life Institute, Centre de recherches sur la terre et le climat Georges Lemaître, SST/ELI/ELIC, Université catholique de Louvain, Place Louis Pasteur, 3, boite L4.03.08, Louvain-la-Neuve B-1348, Belgium
e-mail: hugues.goosse@uclouvain.be

S. Davis
NOAA ESRL Chemical Sciences Division, Cooperative Institute for Research in Environmental Sciences (CIRES), University of Colorado at Boulder,
325 Broadway, Boulder, CO 80305, USA
e-mail: sean.m.davis@noaa.gov

G.R. Asrar and J.W. Hurrell (eds.), *Climate Science for Serving Society: Research, Modeling and Prediction Priorities*, DOI 10.1007/978-94-007-6692-1_10,

mean state. We conclude, that to fully understand how modes of variability will change in a changing climate, we need additional analysis of observations, both paleo and present day, and a solid fundamental understanding of mechanisms. Understanding of mechanisms necessarily requires use of models, ranging from simple to complex. Such models need to be fully coupled, between atmosphere and ocean, and need to include a fully resolved middle atmosphere as well.

Keywords Climate variability • Climate change • Annular modes • El Niño Southern Oscillation • Sea ice • Greenhouse gases • Ozone • Stratosphere

1 Introduction

Climate change involves changing statistical properties in the climate system over an extended period of time. Such changes may be induced through long-term changes in solar or orbital parameters, long periods of enhanced volcanic activity or through long-term changes in radiatively significant gases. Whatever the actual forcing is, the end result will likely be long-term changes in the mean state or in the variability of the system or both. There are multiple ways to assess whether such changes may be occurring. Extended model simulations, where appropriate forcing parameters are varied are one means of assessing changes in the mean state or variability of the climate system. These simulations can then be used to provide estimates of the climate response to changing forcings as well as assess the internal variability, both of which are needed for detection and attribution studies. Past changes in climate variability can be addressed via analysis of historical data, using both recent measurements as well and geologic records or ice core records. As we are currently in the midst of a large scale climate change experiment, with changes in radiative gases and surface conditions induced by anthropogenic activity, analysis of existing climate data over the industrial era is another means of assessing impacts on variability due to changes in forcings of the climate system. We are interested in changes in the mean circulation and variability of that circulation ultimately because it impacts surface temperature and precipitation.

In this paper, we briefly cover an extremely broad topic: "How climate change impacts climate variability" with focus on the identification and mechanisms for modes and regimes of large-scale variability in different climates. We are basing the content of this paper largely on work discussed at the WCRP Open Science Conference held in Denver in 2011. Although there are many modes of variability that can be addressed in a review paper such as this, we will concentrate our efforts on just a few topics. In particular, we will briefly discuss changes in large-scale oscillations such as the El Nino/Southern Oscillation (ENSO), the North Atlantic Oscillation (NAO), and the northern and southern annular modes (NAM and SAM), changes in the polar and tropical troposphere, and interactions between the stratosphere and troposphere in a changing climate. We consider both changes in variability as well as trends in the mean state in our discussion. We will then make recommendations as to key issues that require additional research.

2 How Do Changes in Greenhouse Gases and Solar and Orbital Parameters Impact the Tropics?

2.1 *ENSO*

Decades of observational, theoretical, and numerical modeling research has shown that El Nino/Southern Oscillation (ENSO) is the result of dynamical coupling between the ocean and atmosphere which results in growth of perturbations to the tropical Pacific climate on seasonal to interannual timescales, generally referred to as the 'Bjerknes feedback.' However, while the fundamental mechanism of ENSO is fairly well understood, there are still some important open questions, particularly with regards to how ENSO will change in the future in response to anthropogenic forcing.

One way to address this is to look for detectable trends in the behavior of ENSO over the twentieth century that might be attributable to external forcing. For example, in the 1990s, it was argued that the persistent, weak El Nino that occurred from 1990 to 1995 was highly unlikely given the character of the record prior to that time (Trenberth and Hoar 1996, 1997), and that this may be an indication of 'El Nino and Climate Change' (though no formal attribution statement was made in those studies). What we have learned since then is that the details of ENSO continue to present new puzzles with practically every realization of the phenomenon (Stevenson et al. 2012). For example, there seems to be an increasing number of 'central Pacific' events, which, in contrast to the classic eastern Pacific event, have their maximum temperature response confined to the central basin (Yeh et al. 2009; Newman et al. 2011). There also appears to be variations in the predictability of ENSO, which depend on the mean state (e.g. Kirtman and Schopf 1998). In short, the 'natural' behavior of ENSO is so varied that detecting anthropogenic trends in likely to take an extremely long record (Wittenberg 2009).

One might look back further using paleoclimate records, and then the basic story is actually fairly straightforward: A simple, first-order answer which is supported both by paleoclimate records and by climate models is that the Pacific is characterized by large seasonal and interannual variability with global impacts no matter what the state of the mean climate is. Seasonally-resolved tropical Pacific paleoclimate records from periods in the Earth's history that were both warmer and colder than today show that ENSO-like interannual variability was present. Available Pliocene records, for example, when the Earth was several degrees warmer than present and ice sheets were minimal in extent, show that ENSO frequency and amplitude were not significantly different from today (Watanabe et al. 2011; Scroxton et al. 2011). The same goes for the glacial climates: Koutavas and Joanidis (2009) have shown using isotope measurements on individual forams that there is large variability at the Last Glacial Maximum, and coral records from prior glacial stages also suggest considerable interannual variability (Tudhope et al. 2001). Climate models have thus far not been able to rid the tropical Pacific of ENSO variability by either warming (Huber and Caballero 2003; Galeotti et al. 2010; von der Heydt et al. 2011) or cooling the climate (Zheng et al. 2008). Nor does there appear to be an obvious relationship between radiative forcing and ENSO behavior over the last millennium, when solar and volcanic forcing as well as the mean climate state all have varied (Cobb et al. 2003; Emile-Geay et al. 2012).

The only external climate forcing that, thus far, has been shown to influence ENSO in a systematic way that is consistent in observations and models is precessional forcing. It appears that when perihelion occurs in Northern Hemisphere summer (every 21 kyr), ENSO variance is reduced. The mechanism varies from model to model, but fundamentally it is the altered annual cycle of the large-scale atmosphere–ocean circulation that appears to influence ENSO. Models underestimate the influence of this effect compared with observations (c.f. Brown et al. 2008).

Looking forward using climate models, there is also considerable uncertainty. First, there are large biases in climate model simulations of the mean tropical Pacific climate, which may impact their ability to simulate ENSO (Roberts and Battisti 2011). That said, if models are run into the future with greenhouse gas forcing, they robustly simulate an enhanced equatorial warming (IPCC 2007; Liu et al. 2005), but this should not be thought of as a change in ENSO. Rather, models do not simulate a consistent response in ENSO, and the changes are generally small even with the large 4×CO_2 forcing (Fig. 1, from Guilyardi et al. 2012). An analysis of CMIP3 models by DiNezio et al. (2012) suggests an explanation: They argue that ENSO is fairly insensitive to greenhouse gas forcing because the winds and thermocline actually have opposing effects on ENSO. As the climate warms, the Walker circulation weakens, which on its own would weaken ENSO variability. However, the ocean response to the weaker trade winds is a less tilted, but *shallower* thermocline, and this effect would strengthen ENSO variability. DiNezio et al. (2012) argue that these competing mechanisms can explain why climate models simulate overall little change in ENSO in response to greenhouse gas forcing, and the same arguments can be made for interpreting past climate changes as well.

These prior studies point to some important, outstanding questions about the response of ENSO to external forcing:

1. Can we develop a more complete characterization of the ENSO system with paleoclimate proxies including the internal variability as well as the radiatively-forced changes?
2. What are the mechanisms by which changes in the mean state can influence interannual variability of ENSO? Can models represent these mechanisms, and can observational networks support their characterization in the real system?
3. What are the mechanisms that contribute to the different 'flavors' of ENSO, and how may these be altered by a changing mean state?
4. What are the prospects for improving the predictability of ENSO on seasonal, interannual and even decadal timescales? Is predictability altered by an externally forced change in mean state?

2.2 *Width of the Tropics*

A geographic definition of the "tropical belt" is the region between the Tropics of Cancer and Capricorn, which are currently at 23°27′S and 23°27′N, respectively. This geographic definition of the tropical edge latitudes is a consequence of the

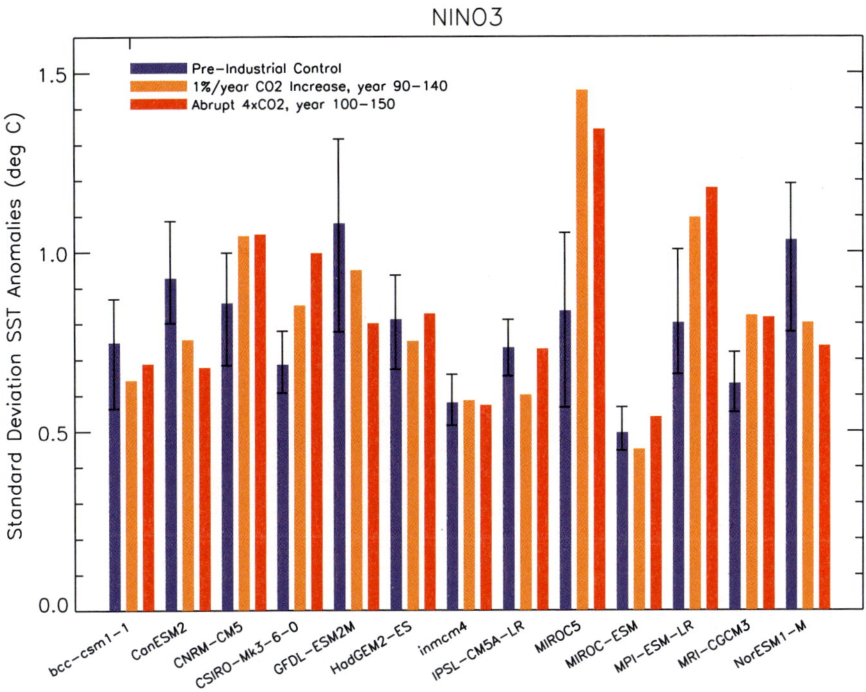

Fig. 1 Standard deviation of Niño3 SST anomalies for CMIP5 model experiments. *Blue bars*, preindustrial control experiments, *orange bars*, years 90–140 from the 1 %/year CO_2 increase experiments, *red bars* years 50–150 from the abrupt 4×CO_2. Calculations are performed for the models indicated on the x axis. The *black 'error bar'* indicates the minimum and maximum of 50 year windowed standard deviation of Niño3 anomalies computed from the multi-century control experiments. Thus, when the Niño3 standard deviation in one of the CO_2 runs falls *below* or *above* the error bar, the changes are deemed to be significant. If significant changes are seen in both experiments that indicates a more robust response in that model (After Guilyardi et al. 2012)

axial tilt of the earth, which varies with a periodicity of approximately 41,000 years. From an atmospheric perspective, there is no similarly simple definition of the latitudinal extent of the tropical belt, but most definitions refer approximately to the region equatorward of the Hadley cell edges. These atmospheric tropical edge latitudes have been quantitatively diagnosed from observations and models by identifying arbitrary thresholds or local extremes in meteorological properties (e.g., winds, tropopause height) as they change with latitude from their tropical to extratropical values. Figure 2 shows examples of tropical edge diagnostics considered in the literature. From a surface climate perspective, the tropical edges are significant because they are essentially the poleward boundaries of the subtropical deserts. Potential changes in the latitudinal extent of the tropics are thus related to changes in precipitation patterns, and could lead to significant regional impacts in the ecosystem health, water resources, and agriculture.

Additionally, there are more subtle implications regarding transport of mass and species into the stratosphere. The tropical upper troposphere is the primary location

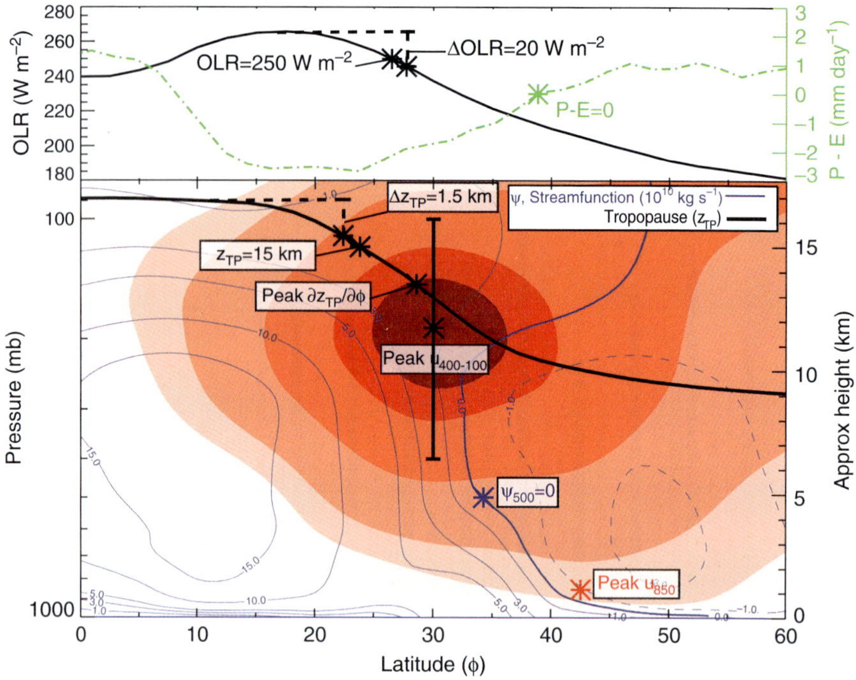

Fig. 2 An example monthly-mean zonal-mean cross section of atmospheric properties from the NCEP/NCAR reanalysis and NCEP OLR dataset for January 2008. The zonal-mean zonal wind is *shaded* (5 m s^{-1} contours), and the mean meridional streamfunction (ψ) based on the zonal-mean meridional wind is contoured in *blue*. Tropical edge diagnostics are denoted by *asterisks*. *Dashed vertical bars* illustrate the relative threshold metrics, and the solid vertical bar illustrates the vertical range of averaging for the peak wind metric (From Davis and Rosenlof 2012)

through which air enters the stratosphere; it is the gateway for stratospheric entry of tropospheric trace gasses, some of which are potentially ozone depleting or radiatively active. If the tropical upwelling region widens, it can conceivably alter the amount of such species entering the stratosphere, especially if the upwelling region encompasses a greater concentration of populated regions emitting anthropogenic pollutants. Hence, how the tropical belt responds to natural and anthropogenic forcings is not only significant for regional surface climate in the subtropics, it also has a potentially global-scale impact via changes in stratospheric composition and radiative forcing.

Multiple independent analyses using chemical constituent measurements, meteorological observations, and meteorological fields from reanalyses have identified changes in the latitudinal extent and character of the tropical belt during the past 40–50 years. Specifically, studies have noted an intensification and poleward expansion of the Hadley cell as defined by OLR and the meridional mass streamfunction (Hu and Fu 2007; Johanson and Fu 2009; Mitas and Clement 2005), the region of high-altitude tropical tropopause (Lu et al. 2009; Seidel and Randel 2007; Seidel et al. 2008), and the region of "tropical"-like low column ozone (Hudson et al. 2006).

Other studies based on reanalyses have suggested changes in both the strength and position of the subtropical and polar jet streams (Archer and Caldeira 2008; Strong and Davis 2007), and a poleward shift in storm tracks (Fyfe 2003; McCabe et al. 2001). A lack of trend has also been noted (Swart and Fyfe 2012). Although changes in the eddy-driven jets are discussed in Sect. 3.1, and are not strictly related to the other tropical edge diagnostics discussed here, it is worth noting that the jet latitudes and Hadley cell edge latitudes are correlated during Austral summer in the SH (Kang and Polvani 2011).

Tropical widening and poleward expansion of the jets has been detected in climate model simulations with anthropogenic forcings, and pre-industrial control runs indicate that the magnitude of the late-twentieth century widening cannot be explained by natural variability alone (Johanson and Fu 2009; Lu et al. 2009; Yin 2005). However, the rate of widening is greater in observations than in models for the few diagnostics that have been tested (Johanson and Fu 2009). For example, the late-twentieth century poleward expansion rates from several Hadley cell diagnostics span a range of ~ 0.6–1.8° decade^{-1}, whereas comparable model estimates are 0.1–0.2° decade^{-1} (Hu and Fu 2007; Johanson and Fu 2009).

A better understanding of the dynamical mechanisms for tropical belt expansion is very important for assessing the relative importance of ozone depletion and anthropogenic greenhouse gas (GHG) forcing of tropical widening, and may help in reconciling the discrepancy between observations and models, thus allowing for better predictions of future widening. Several dynamical mechanisms have been proposed for explaining the poleward expansion of the tropics and jets, and in general these mechanisms involve interactions between the atmospheric thermal structure/gradients, winds, and wave breaking. More specifically, tropical belt changes have been proposed to occur due to polar stratospheric cooling (Polvani and Kushner 2002; Polvani et al. 2011a, b; Tandon et al. 2011), increases in upper tropospheric static stability associated with global warming (Frierson et al. 2007; Lu et al. 2007), warming in the tropical Indo-Pacific ocean (Johanson and Fu 2009; Lau et al. 2008), and changes in wave propagation and breaking associated with changes in the near-tropopause meridional temperature gradient (Chen and Held 2007; Lorenz and Deweaver 2007; Simpson et al. 2009).

A comparison of a variety of estimates of changes in the width of the tropics over the past 30 years is given in Davis and Rosenlof (2012). This study demonstrated that there is a large spread among tropical width trends calculated from different edge definitions, as well as from different reanalyses. The study also shows that the use of objective definitions gives trends that are smaller than previous subjective estimates (i.e., $z_{TP} = 15$ km and OLR = 250 W m^{-2} in Fig. 2). For one metric (the Hadley cell streamfunction, ψ_{500}), the reanalysis trends are large (> 1° decade^{-1}), statistically significant, and significantly larger than those derived from climate models. Other than the Hadley cell metric, reanalysis trends over the past 30 years from objective definitions are mostly positive but not statistically significant.

To date, much work has focused on trends, and relatively little work has been done comparing the co-variability of various metrics on seasonal, interannual, hemispheric, and longitudinally-resolved scales. Such comparisons, for models, observations, and between models and observations, would help give a clearer picture of what aspects of tropical widening are robust. Clearly, additional studies are needed to ascertain the reasons for the differences in model- and observation-based trends, and mechanisms for the changes in tropical width need to be further explored. Key questions that need answering include:

1. Are historical (i.e., satellite- and reanalysis-based) trends in tropical width accurate? How well can the observational- and model-based trends of the past several decades be reconciled?
2. What are the predominant drivers of historical and future tropical width trends in models (e.g., natural variability, greenhouse gasses, stratospheric ozone depletion)?
3. What are the dynamical mechanisms by which these drivers affect the tropical width? To what extent can trend variations (e.g., as a function of season, hemisphere, definition) be used to test these proposed mechanisms?
4. How do tropical width trends relate to changes in other modes of climate variability?
5. Are there feedback processes operating whereby tropical width changes impact stratospheric composition, leading to a radiative impact on surface temperatures or further tropical width changes?

3 How Does Climate Change Impact Middle and High Latitudes?

3.1 The Northern Annular Mode and Related Latitudinal Shifts of the Eddy-Driven Jet

The Northern Annular Mode (NAM), also called the Arctic Oscillation, is the main atmospheric mode of variability in the northern extratropics (Thompson and Wallace 2000). It is usually defined as the first empirical orthogonal function (EOF) of Northern Hemisphere (20°–90°N) winter sea level pressure but other definitions exist. While the NAM structure is very similar to the North Atlantic Oscillation (NAO) pattern in the Atlantic, it exhibits stronger anomalies over the North Pacific, leading to a more zonally symmetric structure. It has been argued that the two patterns, NAM and NAO, may in fact represent two different paradigms of the Northern Hemisphere variability: the sectoral paradigm (NAO) and the annular one (NAM) (Deser 2000; Ambaum and Hoskins 2002). While the debate of which of them is more appropriate remains unresolved, here we take a simple and pragmatic viewpoint: it likely depends on the asked scientific question and context. Consequently, we will be alternatively using both paradigms. In the Atlantic, the NAM/NAO is also strongly related to latitudinal displacements of the eddy-driven jet although other modes of variability (such as the East Atlantic pattern) are also needed to fully account for the jet variability (Woollings and Blackburn 2012). NAM/NAO positive phases are characterized by a strong subpolar jet (the eddy-driven jet) that is well separated from the subtropical jet. During negative phase periods, in contrast, the two jets merge and lead to a more zonal circulation across the Atlantic. The NAM/NAO is an intrinsic mode of atmospheric variability as it always appears in long atmospheric simulations with climatological SST forcing. NAM and related jet stream variations are due to mean flow forcing associated with the breaking of transient, synoptic-scale Rossby wave (Benedict et al. 2004; Franzke et al. 2004; Riviere and Orlanski 2007). The NAM/NAO can also be viewed as a stochastic process with a characteristic e-folding time around 10 days (Feldstein 2000). On longer interannual time scales, it exhibits long-range dependence with more power than a simple red-noise process (Stephenson et al. 2000).

The observed interannual persistence of positive NAO phase winter events in the 1990s following the mostly negative phases in the 1960s has led to a strong NAM/NAO trend and many related climate impacts in the Northern Hemisphere (Hurrell et al. 2003; Hurrell and Deser 2009). This remarkable phenomenon has spurred a strong interest in the research community on the possible influence of low-frequency external forcings, such as interannual-to-decadal SST variability or the increasing GHG concentrations, onto the NAM/NAO. A couple of studies then suggested detection of an anthropogenic influence on sea level pressure (SLP) with a response pattern projecting strongly on the NAM in the northern extratropics (Gillett et al. 2003, 2005). However, they also pointed out that the climate models used in these studies strongly underestimated the amplitude of the response. A more recent study (Gillett and Stott 2009) carried out a global seasonal SLP detection and attribution analysis suggesting detection of an anthropogenic influence for the tropics but not for the southern and northern extratropics. This indicates that the NAM pattern did not dominate the anthropogenic fingerprint identified in previous SLP detection results. Note however that this last study uses a single climate model and needs to be extended using a multimodel approach to assess the robustness of the findings. Furthermore, the recent winter NAO/NAM trend has considerably weakened when updated to 2011 due to the dominance of negative phase years since 2000, to the extent that it is no longer significant at the 5 % level.

The study of the influence of external forcing upon the extratropical atmospheric circulation has often been based on the following paradigm (often termed the nonlinear paradigm): the forced response and the leading mode(s) of the unperturbed climate have similar structure implying also that the dominant patterns of intrinsic variability remain unchanged. Among the various arguments which have been proposed to sustain this paradigm (Branstator and Selten 2009), the following explanation is the most often invoked. The atmospheric variability exhibits preferred flow states or regimes such as blocked and zonal flows (Vautard 1990; Cheng and Wallace 1993; Kimoto and Ghil 1993; Hannachi 2007). In this framework, the response to anthropogenic forcing may be a change in the residence frequency of the most dominant regimes such as the phases of the NAM/NAO (Palmer 1999; Corti et al. 1999; Terray et al. 2004). Recent work suggests that this paradigm might not be adapted

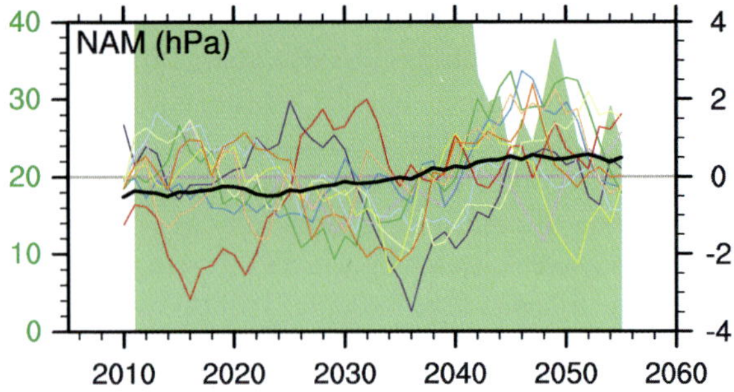

Fig. 3 Projected changes in the 10 year running mean wintertime Northern Annular Mode (NAM) from CCSM3 forced by the SRES A1B GHG scenario. The *thick black curve* shows the forced response, estimated from the average of the 40 realizations. The *thin colored curves* show results for ten individual realizations in which internal variability is superimposed upon the forced response. The *green shaded curve* shows the minimum number of model realizations needed to detect a 95 % significant change relative to the decade centered on 2010 (After Deser et al. 2012)

to fully capture the atmospheric response to anthropogenic forcing. First, while the structure of the NAM is mainly barotropic, it has been suggested from CMIP3 model studies that the response to anthropogenic forcing has a strong baroclinic component in the Arctic due to strong surface warming induced by ice melting (Woollings 2008). Second, the horizontal pattern of the mean response is never exactly the NAM (and even less so the NAO) but rather projects onto the NAM/NAO with different amplitudes depending on the models and periods used, the size of the ensembles, and other parameters. Third, in the context of the anthropogenic influence on the Northern Hemisphere extratropical circulation, the signal-to-noise ratio is likely to be low as shown by a study of a very large ensemble (40 members) of twenty-first century climate scenarios performed with one climate model (Deser et al. 2012). For example, more than 25 members are needed to detect the forced response in the NAM as estimated by the ensemble mean (Fig. 3).

Many studies have documented the possible impacts of increased amounts of greenhouse gases (GHG) upon the mid-to-high latitude atmospheric circulation changes in the Northern hemisphere. Among them, one can cite the rise in the height of the tropopause (Lorenz and DeWeaver 2007), the increase in dry static stability (Frierson 2006) and a NAM-related poleward shift of the tropospheric jet streams and storm tracks (Yin 2005). The latter is well marked in the Southern Hemisphere and less so in the Northern one, where it can actually be missing in some models. When present, this change is related to changes in baroclinicity with different effects between low and upper-level baroclinicity in the context of the twenty-first century GHG increase. Stronger warming in the tropical upper troposphere leads to an increase in upper-level horizontal temperature gradients in mid-latitudes while the increased warming of the polar regions due to sea ice melting leads to a decrease of low-level baroclinicity. Several mechanisms have been proposed to support the dynamical interpretation of the jet stream poleward shift due to enhanced upper-level baroclinicity. They usually invoke the increase and poleward shift of eddy kinetic energy as well as an increase in eddy length scale and its effects on the nature of wave breaking (Kidston et al. 2010; Riviere 2011). Other changes such as the tropopause height increase (Lorenz and DeWeaver 2007) or changes in subtropical stability (Lu et al. 2010) could also lead to similar effects.

While much has been learned about the impact of radiative forcing associated with changes in GHG and other constituents upon the NAM/NAO and related changes in the jet streams and other characteristics of the extratropical circulation, some key questions remain, including:

1. What are the relative impacts of the known mechanisms of the NAM and jet streams response? To what extent do competing changes in low and upper-level baroclinicity explain the model spread in the poleward shift of the Jet streams? If yes, what are the relative roles and spread of the surface and upper-level temperature response?
2. Does the NAM affect subtropical and tropical atmospheric circulations, and if so, by what processes? Is the potential interaction between tropical and extratropical modes going to change with increasing GHGs?
3. Do stratospheric dynamics play a role in the tropospheric NAM response? Is there a two-way interaction in the response to GHGs?
4. Is the non-linear paradigm still useful? Should we think instead in terms of two-way interaction between the response and variability?

To answer these, both observational and modeling approaches (including coupled ocean–atmosphere-land-sea ice models, those with a well resolved stratosphere and with and without interactive atmospheric chemistry) are needed. Simpler models such as the three-level quasi-geostrophic (QG) model or dry GCMs with simple setups must be widely used to provide dynamically coherent mechanisms and support complex GCM analysis.

3.2 The Southern Annular Mode

The Southern Annular Mode (SAM) refers to a seesaw of atmospheric mass between the middle and high latitudes of the Southern Hemisphere (SH; e.g. Thompson et al. 2000; Thompson and Wallace 2000; Marshall 2003; Fogt and Bromwich 2006). It is the leading pattern of tropospheric circulation variability over the extra-tropical SH, accounting for the largest fraction of variance on time scales longer than a few weeks (e.g. Thompson et al. 2000). The positive phase of the SAM is associated with reduced Sea Level Pressure (SLP) at polar latitudes and increased SLP at middle latitudes, evident as a strengthening of the westerly winds along their poleward flank; the negative phase shows opposite-sign changes (Thompson et al. 2000). The SAM is an intrinsic mode of atmospheric variability resulting from unstable dynamical feedbacks between the climatological zonal flow and high-frequency transient eddies along the storm track (e.g. Limpasuvan and Hartmann 2000). Although it is an intrinsic property of the atmosphere, it is also sensitive to a variety of external forcing factors including changes in radiative forcing associated with the build-up of greenhouse gas (GHG), depletion of stratospheric ozone concentrations, and alterations in earth's orbital parameters (e.g., Arblaster and Meehl 2006; Arblaster et al. 2011; Son et al. 2009, 2010; Polvani et al. 2011a, b; Hall and Visbeck 2002; McLandress et al. 2011). The SAM is also sensitive to changes in Sea Surface Temperatures (SSTs) both in the extra-tropics (e.g., Sen Gupta and England 2007) and in the tropics in association with the El Nino/Southern Oscillation (ENSO) phenomenon (e.g., L'Heureux and Thompson 2006; Seager et al. 2005, 2010; Fogt et al. 2011; Schneider et al. 2011). While present year-round, the SAM is most prominent in austral summer (December-February) and autumn (March-May).

Assessing the response of the SAM to each of the forcing agents listed above is a complex task due to: (1) the limited duration of the observational record; (2) the high level of unforced (internal) variability in the SAM; and (3) the covariability of the forcings (e.g., the build-up of GHGs and the depletion of stratospheric ozone have occurred in tandem over the past 50 years or so). Reliable instrumental records of barometric pressure suitable for documenting the SAM extend back to approximately 1957 (Marshall 2003). Attempts have been made to extend this record further back in time, but the degree of reliability of such efforts is not clear. The recent positive trend in the SAM since the late 1950s has been argued to be in part a response to both the increase in GHG concentrations and the decrease in polar stratospheric ozone amounts, based on a variety of atmospheric general circulation model experiments (e.g., Kushner et al. 2001; Arblaster and Meehl 2006; Deser and Phillips 2009; Son et al. 2009; Gillett and Thompson 2003). Arblaster and Meehl (2006) show further that the impact of ozone depletion is mainly limited to austral summer while the effect of increased GHGs is evident year-round. The case for the impact of ozone depletion upon the SAM has also been made in observational analyses, relying on the time-lagged response of the lower troposphere to radiative changes in the stratosphere to argue cause-and-effect (e.g., Thompson and Solomon 2002). A positive trend in tropical Pacific SSTs associated with ENSO since the late 1950s has also been shown to contribute to the upward trend in the SAM in austral summer (e.g., Schneider et al. 2011).

Lower-stratospheric ozone levels are expected to return to near-normal conditions in the next 30–50 years as a result of the Montreal Protocol to reduce ozone-depleting chemicals (Waugh et al. 2009). The increase in ozone levels is expected to drive a negative trend in the SAM which will counteract the tendency associated with increased GHG forcing (Perlwitz et al. 2008; Arblaster et al. 2011; Son et al. 2010). The net result may be a near-cancellation of radiative forcing and a lack of trend in the SAM, at least in austral summer (e.g., Polvani et al. 2011b).

It should also be noted, that while many of the same issues pertain to the NAM and SAM responses to anthropogenic forcing, there are some differences in the factors affecting the two annular modes. A primary consideration is that polar stratospheric ozone depletion has been stronger in the Southern Hemisphere (SH) than in the Northern Hemisphere (NH) over the past few decades. Given that ozone depletion and GHG increases both act to strengthen the SAM and to shift it poleward, one may expect the annular mode response to anthropogenic forcing to be stronger in the

SH compared to the NH in the late twentieth century and weaker in the twenty-first century due to SH ozone hole recovery.

While much has been learned about the impact of radiative forcing associated with changes in GHG and stratospheric ozone concentrations upon the SAM, some key questions remain, including:

1. What are the mechanisms of the SAM response, and to what extent are changes in SSTs (in both the tropics and extra-tropics) and sea ice involved?
2. To what extent does the SAM affect subtropical and tropical atmospheric circulations, and by what processes?
3. To what extent do changes in the stratospheric Brewer-Dobson circulation impact the SAM, and by what mechanisms?
4. To what extent do known modes of multi-decadal climate variability such as the "Pacific Decadal Oscillation" and the "Atlantic Multi-decadal Oscillation" affect the SAM?

To answer these, both observational and modeling approaches (including coupled ocean–atmosphere-land-sea ice models with and without interactive atmospheric chemistry) are needed. In particular, paleo-climate proxy records with demonstrated sensitivity to the SAM are needed to provide a longer-term perspective on past variations in the SAM and associations with fluctuations in CO_2 and SSTs, both tropical and extra-tropical. It is important that these proxy records provide information on austral summer and autumn conditions when the SAM is most prominent and distinguishable from another important mode of atmospheric circulation variability, the "Pacific-South American" pattern, which also affects middle and high latitude SH climate. While progress has been made towards answering these questions, for example Kang et al. (2011) address #2 and Li et al. (2010) address #3, additional studies are still needed.

3.3 Sea Ice and Associated Atmospheric and Oceanic Circulations

Sea ice responds directly to the changes in wind stress and heat fluxes associated to standard modes of atmospheric variability. For instance, when the NAO is in its positive phase, the enhanced south-westerly atmospheric flow in the Barents Sea induces a reduction of the sea ice cover while the more northerly winds in the Labrador Sea favors a higher sea ice extent there (Deser et al. 2000; Rigor et al. 2002). Anomalous circulation over the North Pacific also has a potential impact on sea ice in the Bering Sea up to the central Arctic (Overland and Wang 2005). In the Southern Ocean, SAM is associated with a decreased ice extent in the Bellingshausen Sea and an increase in the Ross Sea (Lefebvre et al. 2004). ENSO also has a clear impact in the Bellingshausen Sea, adding or subtracting its effect to the one of SAM there, depending on the polarity of the two modes (Stammerjohn et al. 2008; Pezza et al. 2011). Any change in atmospheric variability, as discussed in the other sections of this paper, thus have a clear impact on the sea ice cover. In a similar way, changes in the state of the ocean, bringing more or less heat to the sea ice, have an imprint on the sea ice in both hemispheres (e.g. Polyakov et al. 2010). However, the role of the ocean in explaining sea ice variability has been much less studied than the one of the atmosphere.

In turn, variations in the ice concentration or thickness modify the surface albedo as well as the heat and freshwater transfers between the ocean and the atmosphere, inducing temperature and circulation changes in those two media (e.g., Deser et al. 2007; Raphael et al. 2011) Those changes are present both locally, close to the anomalies, and at the large-scale. In this framework, the effect on the atmospheric circulation of the reduced ice cover during the recent summers has received particular attention because of the expected further reduction in the coming decades. In particular, it has been suggested that a low summer ice extent could be associated with stronger easterly winds (or reduced westerlies) in the following seasons, leading to colder conditions in some regions of Eurasia in Autumn and winter (e.g., Honda et al. 2009; Overland and Wang 2010; Petoukhov and Semenov 2010). However, additional work is still required to confirm and refine this hypothesis.

When analyzing changes in sea ice variability as a function of the mean conditions, we must take into account that sea ice displays a fundamental difference compared to the ocean and atmosphere as the surface it covers depends directly on the state of the system. By comparing various model results in different set up, it has been shown that the standard deviation of the summer ice extent in the Southern Ocean is roughly proportional to the root square of the mean extent (Goosse et al. 2009). The proposed explanation simply states that the largest fraction of the variability occurs nears the mean ice edge. A larger ice extent corresponds thus to a longer ice edge and thus to a larger variability. The variability of the ice extent is also strongly dependent on the mean state in the Arctic. However, its geometry, with a central basin surrounded by continents compared to the roughly annular Southern Ocean, induces a maximum standard deviation of the ice extent in summer for a sea ice cover of about 3×10^6 km^2, i.e. when enough ice remains to sustain large variability but the mean limit of the ice edge is still far away from the continent to allow strong variability of its position both southward and northward during cold and warm years (Goosse et al. 2009; Eisenman 2010).

Consequently, sea ice obviously plays a larger role in the setting up the mean conditions and the variability of the climate during cold periods such as the Last Glacial Maximum than in much warmer ones where it was absent of the surface of the Earth (Polyak et al. 2010). The few paleorecords of sea ice concentration are generally related to the presence or absence of sea ice during some periods covering centuries to millennia, with not much information on interannual to decadal variability. However, information from the early Holocene suggest low frequency variations of the sea ice transport, likely related to changes in atmospheric circulation and possibly to the forcing changes, as well as periods with larger multi-decadal variability of the ice cover compared to more quiet ones (e.g. Funder et al. 2011). The large-scale structure of those suggested changes and the mechanisms potentially responsible for them are still largely unknown and thus deserve attention.

In the future, models suggest an increase of the variability of the summer ice extent in the Arctic as sea ice extent is reduced (Holland et al. 2008; Goosse et al. 2009). This is consistent with the geographic arguments mentioned above but could also be related to a thinning of the ice pack (Notz 2009). Such a higher variability, combined with the reduction caused by anthropogenic forcings, can explain the very low ice extent observed during some recent years and the abrupt reductions of the sea ice extent simulated in response to the warming (e.g., Holland et al. 2006, 2008). An alternative explanation is that those large reductions would occur when the system is crossing a threshold (or tipping point) but this hypothesis appears unlikely on the basis of our present knowledge (Holland et al. 2008; Notz 2009; Eisenman and Wettlaufer 2009).

This brief overview illustrates that, although we have learned much about sea ice variability over the past decades, many questions remain. Some keys ones include

1. What is the response of atmospheric and oceanic circulation to anomalies in the sea ice cover?
2. Will the knowledge of the sea ice concentration and thickness bring predictability at the seasonal to decadal time scale?
3. Are warm states in polar regions (as for instance in the Arctic during the mid-twentieth century, the so-called Medieval Climate Anomaly roughly 1,000 year ago and the early Holocene) characterized by different amplitude/modes of variability than colder periods?
4. What is the role of ocean in setting up sea-ice variability at multidecadal time-scales?

Answering the first two questions will help us to understand and predict the impact of changes of the sea ice cover (mean and variability). Answering the third one will provide essential information on the behavior of the system that will help us to refine our projections of future changes. However, this will require additional high-resolution proxy records and simulations devoted to the target periods. Additional long time series will also be required in order to address question 4 but this will allow better estimates of the mechanisms responsible for the multi-decadal variability of the ice extent. A striking example is the occurrence of the big Weddell Polynya, covering about 250 10^3 km^2 during the entire austral winters of 1974, 1975, and 1976 (Carsey 1980). We still do not know if this major event is extremely rare one and may still occur in the future or if it was a recurring feature of the Southern Ocean that is not observed anymore because of some shifts in the Southern Ocean.

4 How Do Greenhouse Gas Induced Climate Changes Interact with Ozone Depletion?

Just as the GHG induced climate changes impact stratospheric ozone, changes to the ozone layer will also affect climate (Forster et al. 2011; Eyring et al. 2010). Changes in both long-lived greenhouse gases and stratospheric ozone influence surface climate directly via radiative effects and indirectly by forcing circulation and temperature changes. There have been a number of studies looking at the impact of Antarctic stratospheric ozone depletion on climate. In particular, lower-stratospheric cooling associated with the Antarctic ozone hole during austral spring and early summer strengthens the SH polar stratospheric vortex compared with pre-ozone hole periods (see Thompson and Solomon 2002; Baldwin et al., 2007; Forster et al. 2011). There may also be impacts on rainfall patterns (Kang et al. 2011) and the latitudinal extent of the tropics (Lu et al. 2009; Polvani et al. 2011a). Additionally there has been work considering the climate interactions between greenhouse gas increases and stratospheric ozone recovery. Key issues under current research include assessing how climate change may impact ozone recovery, how the current levels of stratospheric ozone depletion have affected surface climate and how changes in ozone expected with the decreases in concentrations of ozone depleting substances will impact the troposphere.

It has been found that changes in stratospheric ozone, water vapor and aerosols all radiatively affect surface climate. Observations and model simulations show that the Antarctic ozone hole is the major contributor to SH circulation changes over the past 50 or so years (Polvani et al. 2011a), and these changes extend all the way to the SH tropics. Additionally, ozone increases expected with the reduction in ozone depleting substances in the stratosphere may act to counteract some SH circulation changes expected from CO_2 increases (Polvani et al. 2011b). Recent literature (Scaife et al. 2012) has shown that there is a significant impact of stratospheric changes on tropospheric climate. It is quite likely that stratospheric ozone changes will alter the temperature and wind structure of the stratosphere. This will ultimately impact surface climate regimes.

The horizontal structure, seasonality and amplitude of the observed trends in the SH tropospheric jet are only reproducible in climate models forced with Antarctic ozone depletion. The southward shift of the SH tropospheric jet due to the ozone hole has been linked to a range of observed climate trends over SH mid and high latitudes during summer. Because of this shift, the ozone hole has contributed to summertime changes in surface winds, warming over the Antarctic Peninsula, and cooling over the high plateau. Other impacts of the ozone hole on surface climate have been investigated but have yet to be fully quantified. These include a potential impact in sea ice area averaged around Antarctica (e.g., Sigmond and Fyfe 2010), a southward shift of the SH storm track and associated precipitation, warming of the sub-surface Southern Ocean at depths up to several hundred meters; and decreases of carbon uptake over the Southern Ocean (see Forster et al. 2011 and references therein). Robust connections between NH ozone depletion and surface climate have not yet been established with observational data, possibly due to the fact that NH ozone losses are much smaller than those observed in the SH.

In addition to ozone changes impacting surface climate as noted in the previous discussion here on the SAM, GHG changes also can alter stratospheric ozone chemistry. This is through the GHG contribution to stratospheric temperature change, which then impacts ozone concentrations through changes in reaction rates for ozone controlling chemical reactions that are temperature-dependent. GHG forced changes in the stratospheric circulation can in turn alter the ozone distribution in the stratosphere and the flux of ozone into the troposphere. As noted above, ozone depletion/NH climate connections are not robust, however it remains to be seen whether NH ozone losses will increase with expected increases in GHGs and associated stratospheric cooling, thereby potentially altering NH surface climate as well.

There may very well be coincident changes in ozone, water vapor (a key GHG) and circulation. Randel et al. (2006) demonstrated a strengthening in tropical upwelling led to decreases in stratospheric water vapor as well decreases in ozone in a narrow layer near the tropical tropopause. They note that part of the temperature changes may also be explained as a radiative response to the observed ozone decreases. The changes in water vapor were subsequently used in a model study that demonstrated that the water vapor change may have induced a surface temperature response (Solomon et al. 2010). There are clearly feedback processes operating here, but they are not fully understood and warrant additional study.

Important questions remain that require further research.

1. When will stratospheric ozone recover to values seen prior to the discovery of the Antarctic ozone hole?
2. How will stratospheric ozone recovery interact with changing greenhouse gas concentrations?
3. How will changes in ozone impact surface climate? (primarily see discussion on the SAM)
4. What are the feedbacks between ozone changes, other radiatively active gases, and circulation changes?

To answer these questions, observations of stratospheric ozone and ozone depleting constituents need to continue. Key to furthering our understanding of stratospheric ozone/climate relationships are development and analysis of climate models that fully represent stratospheric processes.

5 Summary

There are strong indications that some aspects of climate variability either will change or have already changed with variations in GHGs. In this paper, we have discussed climate changes related to ENSO, tropical width, the NAM and SAM, sea ice and variations in stratospheric ozone. There is solid fundamental knowledge in regards to control mechanisms. However, there are many open questions in regards to all of these as well.

In regards to ENSO, there is more work that can be done using paleoclimate proxies. There are also questions as to how mean state changes impact interannual variability and predictability. More work is needed both in regards to modeling and observations.

In analysis of the latitudinal extent of the tropics there remain many unknowns. First off, there are questions on the accuracy of historical trends and details on mechanisms. More observations are needed, and further analysis of models that include all potentially relevant processes, both tropospheric and stratospheric. Similar conclusions can be drawn in regards to the state of knowledge for the NAM and the SAM, stratospheric ozone and sea ice variability in a changing climate.

The bottom line is that to fully understand how modes of variability will change we need additional analysis of observations, both paleo and present day, and solid fundamental understanding of mechanisms. Understanding of mechanisms necessarily requires use of models, ranging from simple to complex. Because coupling, between ocean and atmosphere, and between different segments of the atmosphere is important for many of the phenomena discussed, there should be an emphasis on fully coupled general circulation models. Stratospheric processes are also likely to be important, so models also need to include ozone chemistry. These topics involve all of the core WCRP projects, and answering the key questions will involve cooperative research between all the WCRP communities.

References

Ambaum MHP, Hoskins BJ (2002) The NAO troposphere-stratosphere connection. J Clim 15:1969–1978

Arblaster JM, Meehl GA (2006) Contributions of external forcings to southern annular mode trends. J Clim 19:2896–2905

Arblaster JM, Meehl GA, Karoly DJ (2011) Future climate change in the southern hemisphere: competing effects of ozone and greenhouse gases. Geophys Res Lett 38, L02701. doi:10.1029/2010GL045384

Archer CL, Caldeira K (2008) Historical trends in the jet streams. Geophys Res Lett 35:L08803. doi:10.1029/2008GL033614

Baldwin M, Dameris M (Lead Authors) (2007) Climate-ozone connections, Chapter 5 in Scientific assessment of ozone depletion: 2006, Global Ozone Research and Monitoring Project–Report No. 50. World Meteorological Organization, Geneva, 572 pp

Benedict JJ, Lee S, Feldstein SB (2004) Synoptic view of the North Atlantic oscillation. J Atmos Sci 61:121–144. doi:10.1175/1520-0469(2004)

Branstator G, Selten FM (2009) Modes of variability and climate change. J Clim 22:2639–2658. doi:10.1175/2008JCLI2517.1

Brown J, Collins M, Toniazzo T (2008) Modelling mid-Holocene tropical climate and ENSO variability: towards constraining predictions of future change with paleo-data. Clim Dyn 30:19–36

Carsey FD (1980) Microwave observations of the Weddell Polynya. Mon Weather Rev 108:2032–2044

Chen G, Held IM (2007) Phase speed spectra and the recent poleward shift of Southern Hemisphere surface westerlies. Geophys Res Lett 34(21), L21805. doi:10.1029/2007GL031200

Cheng X, Wallace JM (1993) Cluster analysis of the northern hemisphere wintertime 500 hPa height field: spatial patterns. J Atmos Sci 50:2674–2696

Cobb KM, Charles CD, Cheng H, Edwards RL (2003) El Nino/Southern oscillation and tropical Pacific climate during the last millennium. Nature 424:271–276

Corti S, Molteni F, Palmer TN (1999) Signature of recent climate change in frequencies of natural atmospheric circulation regimes. Nature 398:799–802

Davis SM, Rosenlof KH (2012) A multidiagnostic intercomparison of tropical-width time series using reanalyses and satellite observations. J Clim 25(4):1061–1078. doi:10.1175/JCLI-D-11-00127.1

Deser C (2000) On the teleconnectivity of the "Arctic oscillation". Geophys Res Lett 27:779–782
Deser C, Phillips AS (2009) Atmospheric circulation trends, 1950–2000: the relative roles of sea surface temperature forcing and direct atmospheric radiative forcing. J Clim 22:396–413. doi: 10.1175/2008JCLI2453.1
Deser C, Walsh JE, Timlin MS (2000) Arctic sea ice variability in the context of recent atmospheric circulation trends. J Clim 13:617–633
Deser C, Tomas RA, Peng S (2007) The transient atmospheric circulation response to North Atlantic SST and sea ice anomalies. J Clim 4751–4767
Deser C, Phillips AS, Bourdette V, Teng H (2012) Uncertainty in climate change projections: the role of internal variability. Clim Dyn 38:527–546. doi:10.1007/s00382-010-0977-x
DiNezio PN, Kirtman B, Clement AC, Lee S-K, Vecchi GA, Wittenberg A (2012) Mean climate controls on the simulated response of ENSO to increasing greenhouse gases. J Climate 25(21):7399–7420. doi:10.1175/JCLI-D-11-00494.1
Eisenman I (2010) Geographic muting of changes in the Arctic sea ice cover. Geophys Res Lett 37, L16501
Eisenman I, Wettlaufer JS (2009) Nonlinear threshold behavior during the loss of Arctic sea ice. Proc Natl Acad Sci USA 106(1):28–32
Emile-Geay J, Cobb KM, Mann ME, Wittenberg A (2012) Estimating tropical Pacific SST variability over the last millennium; Part 2: Reconstructions and uncertainties. 10.1175/JCLI-D-11-00511.1
Eyring V et al (2010) Sensitivity of 21st century stratospheric ozone to greenhouse gas scenarios. Geophys Res Lett 37, L16807. doi:10.1029/2010GL044443
Feldstein SB (2000) Is interannual zonal mean flow variability simply climate noise? J Clim 13:2356–2362
Fogt RL, Bromwich DH (2006) Decadal variability of the ENSO teleconnection to the high-latitude South Pacific governed by coupling with the Southern Annular Mode. J Clim 19:979–997
Fogt R, Bromwich D, Hines K (2011) Understanding the SAM influence on the South Pacific ENSO teleconnection. Clim Dyn 36:1555–1576. doi:10.1007/s00382-010-0905-0
Forster P, Thompson D (Lead Authors) (2011) Stratospheric changes and climate, Chapter 4 in Scientific assessment of ozone depletion: 2010, Global Ozone Research and Monitoring Project–Report no. 52. World Meteorological Organization, Geneva
Franzke C, Lee S, Feldstein SB (2004) Is the North Atlantic oscillation a breaking wave? J Atmos Sci 61:145–160
Frierson DMW (2006) Robust increases in midlatitude static stability in simulations of global warming. Geophys Res Lett 33:L24816. doi:10.1029/2006GL027504
Frierson DMW, Lu J, Chen G (2007) Width of the Hadley cell in simple and comprehensive general circulation models. Geophys Res Lett 34(18), L18804. doi:10.1029/2007GL031115
Funder S, Goosse H, Jepsen H, Kaas E, Kjær KH, Korsgaard NJ, Larsen NK, Linderson H, Lyså A, Möller P, Olsen J, Willerslev E (2011) A 10,000 year record of Arctic Ocean sea ice variability – view from the beach. Science 333:747–750
Fyfe JC (2003) Extratropical southern hemisphere cyclones: harbingers of climate change? J Clim 16:2802–2805
Galeotti S, von der Heydt A, Huber M, Bice D, Dijkstra HA, Jilbert T, Lanci L, Reichart G (2010) Evidence for active ENSO in the late Miocene greenhouse climate. Geology 1017
Gillett NP, Thompson DWJ (2003) Simulation of recent Southern Hemisphere climate change. Science 302:273–275
Gillett NP, Stott PA (2009) Attribution of anthropogenic influence on seasonal sea level pressure. Geophys Res Lett 36:L23709. doi:10.1029/2009GL041269
Gillett NP, Zwiers FW, Weaver AJ, Stott PA (2003) Detection of human influence on sea level pressure. Nature 422:292–294
Gillett NP, Allan RJ, Ansell TJ (2005) Detection of external influence on sea level pressure with a multi-model ensemble. Geophys Res Lett 32(19), L19714. doi:10.1029/2005GL023640

Goosse H, Arzel O, Bitz CM, de Montety A, Vancoppenolle M (2009) Increased variability of the Arctic summer ice extent in a warmer climate. Geophys Res Lett 36, L23702. doi:10.1029/2009GL040546

Guilyardi E, Bellenger H, Collins M, Ferrett S, Cai W, Wittenberg A (2012) A first look at ENSO in CMIP5 Submitted to CLIVAR Exchanges, 58, 17(1):29–32

Hall A, Visbeck M (2002) Synchronous variability in the Southern Hemisphere atmosphere, sea ice, and ocean resulting from the annular mode. J Clim 15:3043–3057

Hannachi A (2007) Tropospheric planetary wave dynamics and mixture modeling: two preferred regimes and a regime shift. J Atmos Sci 64:3521–3541

Holland MM, Bitz CM, Tremblay B (2006) Future abrupt reductions in the summer Arctic sea ice. Geophys Res Lett 33, L23503. doi:10.1029/2006GL028024

Holland MM, Bitz CM, Tremblay LB, Bailey DA (2008) The role of natural versus forced change in future rapid summer Arctic ice loss, in Arctic sea ice decline: observations, projections, mechanisms, and implications. Geophys Monogr Ser 180:133–150

Honda M, Inoue J, Yamane S (2009) Influence of low Arctic sea-ice minima on anomalously cold Eurasian winters. Geophys Res Lett 36, L08707. doi:10.1029/2008GL037079

Hu Y, Fu Q (2007) Observed poleward expansion of the Hadley circulation since 1979. Atmos Chem Phys 7(19):5229–5236

Huber M, Caballero R (2003) Eocene El Nino: evidence for robust tropical dynamics in the "hothouse". Science 299:877–881

Hudson RD, Andrade MF, Follette MB, Frolov AD (2006) The total ozone field separated into meteorological regimes – part II: Northern Hemisphere mid-latitude total ozone trends. Atmos Chem Phys 6:5183–5191

Hurrell JW, Deser C (2009) North Atlantic climate variability: the role of the North Atlantic oscillation. J Mar Syst 78(1):28–41. doi:10.1016/j.jmarsys.2008.11.026

Hurrell JW, Kushnir Y, Visbeck M, Ottersen G (2003) An overview of the North Atlantic oscillation. The North Atlantic oscillation: climate significance and environmental impact. In: Hurrell JW, Kushnir Y, Ottersen G, Visbeck M (eds) Geophysical Monograph Series 134, pp 1–35

IPCC (2007) Climate change 2007: the physical science basis. In: Solomon S, Qin D, Manning M, Chen Z, Marquis M, Averyt KB, Tignor M, Miller HL (eds) Contribution of working group I to the fourth assessment report of the intergovernmental panel on climate change. Cambridge University Press, Cambridge/New York, 996 pp

Johanson CM, Fu Q (2009) Hadley cell widening: model simulations versus observations. J Clim 22:2713–2725. doi:10.1175/2008JCLI2620.1

Kang SM, Polvani LM (2011) The interannual relationship between the latitude of the eddy-driven jet and the edge of the Hadley cell. http://dx.doi.org/10.1175/2010JCLI4077.1. 24:563–568. doi:10.1175/2010JCLI4077.1

Kang S, Polvani LM, Fyfe JC, Sigmond M (2011) Impact of polar ozone depletion on subtropical precipitation. Science 332:951–954

Kidston J, Frierson DMW, Renwick JA, Vallis GK (2010) Observations, simulations, and dynamics of jet stream variability and annular modes. J Clim 23:6186–6199

Kimoto M, Ghil M (1993) Multiple flow regimes in the Northern Hemisphere winter. Part II: sectorial regimes and preferred transitions. J Atmos Sci 50:2645–2673

Kirtman BP, Schopf PS (1998) Decadal variability in ENSO predictability and prediction. J Clim 11:2804–2822

Koutavas A, Joanidis S (2009) El Nino during the last glacial maximum. G Eochimica et Cosmochimica Acta 73(13):A690–A690

Kushner PJ, Held IM, Delworth TL (2001) Southern-hemisphere atmospheric circulation response to global warming. J Clim 14:2238–2249

L'Heureux ML, Thompson DWJ (2006) Observed relationships between the El-Nino/Southern oscillation and the extratropical zonal-mean circulation. J Clim 19:276–287

Lau N-C, Leetmaa A, Nath MJ (2008) Interactions between the responses of North American climate to El Nino-La Nina and to the secular warming trend in the Indian-Western Pacific Oceans. J Clim 21(3):476–494. doi:10.1175/2007JCLI1899.1

Lefebvre W, Goosse H, Timmermann R, Fichefet T (2004) Influence of the Southern Annular Mode on the sea-ice-ocean system. J Geophys Res 109, C090005. doi:10.1029/2004JC002403
Li F, Newman PA, Stolarski RS (2010) Relationships between the Brewer-Dobson circulation and the southern annular mode during austral summer in coupled chemistry-climate model simulations. J Geophys Res 115, D15106. doi:10.1029/2009JD012876
Limpasuvan V, Hartmann DL (2000) Wave-maintained annular modes of climate variability. J Clim 13:4414–4429
Liu Z, Vavrus S, He F, Wen N, Zhong Y (2005) Rethinking tropical ocean response to global warming: the enhanced equatorial warming. J Clim 18(22):4684–4700. doi:10.1175/JCLI3579.1
Lorenz DJ, Deweaver ET (2007) Tropopause height and zonal wind response to global warming in the IPCC scenario integrations. J Geophys Res 112, D10119. doi:10.1029/2006JD008087
Lu J, Vecchi GA, Reichler T (2007) Expansion of the Hadley cell under global warming. Geophys Res Lett 34, L06805. doi:10.1029/2006GL028443
Lu J, Deser C, Reichler T (2009) Cause of the widening of the tropical belt since 1958. Geophys Res Lett 36, L03803. doi:10.1029/2008GL036076
Lu J, Chen G, Frierson D (2010) The position of the midlatitude storm track and eddy-driven westerlies in aquaplanet AGCMs. J Atmos Sci 17:3984–4000
Marshall GJ (2003) Trends in the Southern Annular Mode from observations and reanalyses. J Clim 16:4134–4143
McCabe GJ, Clark MP, Serreze MC (2001) Trends in Northern Hemisphere surface cyclone frequency and intensity. J Clim 14:2763–2768
McLandress C, Shepherd TG, Scinocca JF, Plummer DA, Sigmond M, Jonsson AI, Reader MC (2011) Separating the dynamical effects of climate change and ozone depletion. Part II: Southern Hemisphere troposphere. J Clim 24:1850–1868. doi:10.1175/2010JCLI3958.1
Mitas CM, Clement A (2005) Has the Hadley cell been strengthening in recent decades? Geophys Res Lett 32:L03809. doi:10.1029/2004GL021765
Newman M, Shin S-I, Alexander MA (2011) Natural variation in ENSO flavors. Geophys Res Lett 38:L14705. doi:10.1029/2011GL047658
Notz D (2009) The future of ice sheets and sea ice: between reversible retreat and unstoppable loss. Proc Natl Acad Sci 106(49):20,590–20,595
Overland JE, Wang M (2005) The third Arctic climate pattern: 1930s and early 2000s. Geophys Res Lett 32, L23808. doi:10.1029/2005GL024254
Overland JE, Wang M (2010) Large-scale atmospheric circulation changes are associated with the recent loss of Arctic sea ice. Tellus 62A:1–9
Palmer TN (1999) A nonlinear dynamical perspective on climate change. J Clim 12:575–591
Perlwitz J, Pawson S, Fogt RL, Nielsen JE, Neff WD (2008) Impact of stratospheric ozone hole recovery on Antarctic climate. Geophys Res Lett 35, L08714. doi:10.1029/2008GL033317
Petoukhov V, Semenov VA (2010) A link between reduced Barents-Kara sea ice and cold winter extremes over northern continents. J Geophys Res 115, D21111. doi:10.1029/2009JD013568
Pezza AB, Rashid HA, Simmonds I (2011) Climate links and recent extremes in Antarctic sea ice, high-latitude cyclones, Southern Annular Mode and ENSO. Clim Dyn. doi:10.1007/s00382-011-1044-y
Polvani LM, Kushner PJ (2002) Tropospheric response to stratospheric perturbations in a relatively simple general circulation model. Geophys Res Lett 29:1114. doi:10.1029/2001GL014284
Polvani LM, Waugh DW, Correa GJP, Son S-W (2011a) Stratospheric ozone depletion: the main driver of twentieth-century atmospheric circulation changes in the Southern Hemisphere. J Clim 24(3):795–812. doi:10.1175/2010JCLI3772.1
Polvani LM, Previdi M, Deser C (2011b) Large cancellation, due to ozone recovery, of future Southern Hemisphere atmospheric circulation trends. Geophys Res Lett 38(4):L04707. doi:10.1029/2011GL046712
Polyak L, Alley RB, Andrews JT, Brigham-Gretted J, Cronine TM, Darbyf DA, Dykeg AS, Fitzpatrich JJ, Funderi S, Holland M, Jennings AE, Millerc GH, O'Regan M, Savelle J, Serreze M, St K, John JWCW, Wolff E (2010) History of sea ice in the Arctic. Q Sci Rev 29:1757–1778

Polyakov IV, Timokhov LA, Alexeev VA, Bacon S, Dmitrenko IA, Fortier L, Frolov IE, Gascard J-C, Hansen E, Ivanov VV, Laxon S, Mauritzen C, Perovich D, Shimada K, Simmons HL, Sokolov VT, Steele M, Toole J (2010) Arctic Ocean warming contributes to reduced polar ice cap. J Phys Ocean 40:2743–2756

Randel WJ, Wu F, Vömel H, Nedoluha GE, Forster P (2006) Decreases in stratospheric water vapor after 2001: links to changes in the tropical tropopause and the brewer-Dobson circulation. J Geophys Res 111, D12312. doi:10.1029/2005JD006744

Raphael MN, Hobbs W, Wainer I (2011) The effect of Antarctic sea ice on the Southern Hemisphere atmosphere during the southern summer. Clim Dyn 36(7–8):1403–1417

Rigor IG, Wallace JM, Colony RL (2002) Response of sea ice to the Arctic oscillation. J Clim 15:2648–2663

Rivière G (2011) A dynamical interpretation of the poleward shift of the jet streams in global warming scenarios. J Atmos Sci 68:1253–1272

Rivière G, Orlanski I (2007) Characteristics of the Atlantic storm-track eddy activity and its relation with the North Atlantic oscillation. J Atmos Sci 64:241–266. doi:10.1175/JAS3850.1

Roberts WGH, Battisti DS (2011) A new tool for evaluating the physics of coupled atmosphere–ocean variability in nature and in general circulation models. Clim Dyn 36:907–923. doi:10.1007/s00382-010-0762-x

Scaife A et al (2012) Climate change projections and stratosphere–troposphere interaction. Clim Dyn. doi:10.1007/s00382-011-1080-7

Schneider DP, Deser C, Okumura Y (2011) An assessment and interpretation of the observed warming of West Antarctica in the austral spring. Clim Dyn 38:323–347. doi:10.1007/s00382-010-0985-x

Scroxton N, Bonham SG, Rickaby REM, Lawrence SHF, Hermoso M, Haywood AM (2011) Persistent El Niño–Southern oscillation variation during the Pliocene Epoch. Paleoceanography 26, PA2215. doi:10.1029/2010PA002097

Seager R, Harnik N, Robinson WA, Kushnir Y, Ting MF, Huang HP, Velez (Nakamura) J (2005) Mechanisms of ENSO-forcing of hemispherically symmetric precipitation variability. Q J R Meteorol Soc 131(608):1501–1527

Seager R, Naik N, Ting M, Cane MA, Harnik N, Kushnir Y (2010) Adjustment of the atmospheric circulation to tropical pacific SST anomalies: variability of transient eddy propagation in the Pacific-North America sector. Q J R Meteorol Soc 136:277–296. doi:10.1002/qj.588

Seidel DJ, Randel WJ (2007) Recent widening of the tropical belt: evidence from tropopause observations. J Geophys Res 112(D20):D20113. doi:10.1029/2007JD008861

Seidel DJ, Fu Q, Randel WJ, Reichler TJ (2008) Widening of the tropical belt in a changing climate. Nat Geosci 1:21–24. doi:10.1038/ngeo.2007.38

Sen Gupta A, England MH (2007) Coupled ocean–atmosphere feedback in the Southern Annular Mode. J Clim 20:3677–3692

Sigmond M, Fyfe JC (2010) Has the ozone hole contributed to increased Antarctic sea ice extent? Geophys Res Lett 37, L18502. doi:10.1029/2010GL044301

Simpson IR, Blackburn M, Haigh JD (2009) The role of eddies in driving the tropospheric response to stratospheric heating perturbations. J Atmos Sci 66:1347–1365. doi:10.1175/2008JAS2758.1

Solomon S, Rosenlof KH, Portmann R, Daniel J, Davis S, Sanford T, Plattner G-K (2010) Contributions of stratospheric water vapor changes to decadal variations in the rate of global warming. Science 327:1219–1223

Son S-W, Tandon NF, Polvani LM, Waugh DW (2009) Ozone hole and Southern Hemisphere climate change. Geophys Res Lett 36, L15705. doi:10.1029/2009GL038671

Son S-W et al (2010) Impact of stratospheric ozone on Southern Hemisphere circulation change: a multimodel assessment. J Geophys Res 115:D00M07. doi:10.1029/2010JD014271

Stammerjohn SE, Martinson DG, Smith RC, Yuan X, Rind D (2008) Trends in Antarctic annual sea ice retreat and advance and their relation to El Niño-Southern oscillation and Southern Annual Mode variability. J Geophys Res 113:C03S90. doi:10;1029/2007JC004269

Stephenson DB, Pavan V, Bojariu R (2000) Is the North Atlantic oscillation a random walk? Int J Climatol 20:1–18. doi:10.1002/(SICI)1097-0088(200001)20:1<1::AID-JOC456>3.0.CO;2-P

Stevenson S, Fox-Kemper B, Jochum M, Neale R, Deser C, Meehl G (2012) Will there be a significant change to El Niño in the twenty-first century? J Climate 25:2129–2145. doi:10.1175/JCLI-D-11-00252.1
Strong C, Davis RE (2007) Winter jet stream trends over the Northern Hemisphere. Q J R Meteorol Soc 133(629):2109–2115. doi:10.1002/qj.171
Swart NC, Fyfe JC (2012) Observed and simulated changes in the Southern Hemisphere surface westerly wind-stress. Geophys Res Lett 39:L16711. doi:10.1029/2012GL052810
Tandon NF, Polvani LM, Davis SM (2011) The response of the tropospheric circulation to water vapor-like forcings in the stratosphere. J Climate, 110701092225006, doi:10.1175/JCLI-D-11-00069.1
Terray L, Demory ME, Déqué M, de Coetlogon G, Maisonnave E (2004) Simulation of late-twenty-first-century changes in wintertime atmospheric circulation over Europe due to anthropogenic causes. J Climate 17:4630–4635. doi: http://dx.doi.org/10.1175/JCLI-3244.1
Thompson DWJ, Solomon S (2002) Interpretation of recent Southern Hemisphere climate change. Science 296:895–899
Thompson DWJ, Wallace JM (2000) Annular modes in the extratropical circulation. Part I: month-to-month variability. J Clim 13:1000–1016
Thompson DWJ, Wallace JM, Hegerl GC (2000) Annular modes in the extratropical circulation. Part II: trends. J Clim 13:1018–1036
Trenberth KE, Hoar TJ (1996) The 1990–1995 El Niño-Southern oscillation event: longest on record. Geophys Res Lett 23:57–60
Trenberth KE, Hoar TJ (1997) El Niño and climate change. Geophys Res Lett 24:3057–3060
Tudhope AW, Chilcott CP, McCulloch MT, Cook ER, Chappell J, Ellam RM, Lea DW, Lough JM, Shimmield GB (2001) Variability in the El Nino-Southern oscillation through a glacial-interglacial cycle. Science 291:1511–1517
Vautard R (1990) Multiple weather regimes over the North Atlantic analysis of precursors and successors. Mon Weather Rev 118(10):2056–2081
Von der Heydt AS, Nnafie A, Dijkstra HA (2011) Cold tongue/Warm pool and ENSO dynamics in the Pliocene. Clim Past Discuss 7:997–1027. doi:10.5194/cpd-7-997-2011
Watanabe T et al (2011) Permanent El Niño during the Pliocene warm period not supported by coral evidence. Nature 471:209–211. doi:10.1038/nature09777
Waugh DW et al (2009) Impact of climate change on ozone recovery. Geophys Res Lett 36:L03805. doi:10.1029/2008GL036223
Wittenberg AT (2009) Are historical records sufficient to constrain ENSO simulations? Geophys Res Lett 36:L12702. doi:10.1029/2009GL038710
Woollings T (2008) The vertical structure of anthropogenic zonal-mean atmospheric circulation change. Geophys Res Lett 35:L19702. doi:10.1029/2008GL034883
Woollings T, Blackburn M (2012) The North Atlantic jet stream under climate change, and its relation to the NAO and EA patterns. J Clim 25(3):886–902
Yeh SW et al (2009) El Nino in a changing climate. Nature 461:511–514. doi:10.1038/nature08316
Yin J (2005) A consistent poleward shift of the storm tracks in simulations of 21st century climate. Geophys Res Lett 32(18):L18701. doi:10.1029/2005GL023684
Zheng W, Braconnot P, Guilyardi E, Merkel U, Yu Y (2008) ENSO at 6ka and 21ka from ocean–atmosphere coupled model simulations. Clim Dyn 30(7–8):745–762

Understanding and Predicting Climate Variability and Change at Monsoon Regions

Carolina Vera, William Gutowski, Carlos R. Mechoso, B.N. Goswami, Chris C. Reason, Chris D. Thorncroft, Jose Antonio Marengo, Bruce Hewitson, Harry Hendon, Colin Jones, and Piero Lionello

Abstract The chapter highlights selected scientific advances made under WCRP leadership in understanding climate variability and predictability at regional scales with emphasis on the monsoon regions. They are mainly related to a better understanding of the physical processes related to the ocean-land-atmosphere interaction that characterize the monsoon variability as well as to a better knowledge of the sources of climate predictability. The chapter also highlights a number of challenges that are considered crucial to improving the ability to simulate and thereby predict regional climate variability. The representation of multi-scale convection and its interaction with coupled modes of tropical variability (where coupling refers both to ocean-atmosphere and/or land-atmosphere coupling) remains the leading problem to be addressed in all aspects of monsoon simulations (intraseasonal to decadal prediction, and to climate change).

C. Vera (✉)
Centro de Investigaciones del Mar y la Atmosfera (CIMA/CONICET-UBA), DCAO/FCEN, UMI IFAECI/CNRS, Intendente Guiraldes 2160 – Ciudad Universitaria, Pabellón II – 2do. piso, C1428EGA Buenos Aires, Argentina
e-mail: carolina@cima.fcen.uba.ar

W. Gutowski
Department of Geological & Atmospheric Sciences, Department of Agronomy, Iowa State University, 3021 Agronomy, Ames, IA 50011, USA
e-mail: gutowski@iastate.edu

C.R. Mechoso
Department of Atmospheric and Oceanic Sciences, University of California, Los Angeles (UCLA), 7127 Math Sciences Building, 405 Hilgard Avenue, Los Angeles, CA 90095-1565, USA
e-mail: mechoso@atmos.ucla.edu

B.N. Goswami
Indian Institute of Tropical Meteorology, Dr. Homi Bhabha Rd., Pashan, Pune 411008, India
e-mail: goswami@tropmet.res.in

G.R. Asrar and J.W. Hurrell (eds.), *Climate Science for Serving Society: Research, Modeling and Prediction Priorities*, DOI 10.1007/978-94-007-6692-1_11,

Systematic errors in the simulation of the mean annual and diurnal cycles continue to be critical issues that reflect fundamental deficiencies in the representation of moist physics and atmosphere/land/ocean coupling. These errors do not appear to be remedied by simple model resolution increases, and they are likely a major impediment to improving the skill of monsoon forecasts at all time scales. Other processes, however, can also play an important role in climate simulation at regional levels. The influence of land cover change requires better quantification. Likewise, aerosol loading resulting from biomass burning, urban activities and land use changes due to agriculture are potentially important climate forcings requiring better understanding and representation in models. More work is also required to elucidate mechanisms that give rise to intraseasonal variability. On longer timescales an improved understanding of interannual to decadal monsoon variability and predictability is required to better understand, attribute and simulate near-term climate change and to assess the potential for interannual and longer monsoon prediction.

C.C. Reason
Department of Oceanography, University of Cape Town, Private Bag X3, Rondebosch, Cape Town 7701, South Africa
e-mail: chris.reason@uct.ac.za

C.D. Thorncroft
Department of Atmospheric and Environmental Sciences, University at Albany, SUNY, DAES – ES351, Albany, NY 12222, USA
e-mail: chris@atmos.albany.edu

J.A. Marengo
Centro de Ciencia do Sistema Terrestre (CCST), INPE, São Paulo, Brazil
e-mail: jose.marengo@inpe.br

B. Hewitson
Department of Environmental & Geographical Science, Climate Systems Analysis Group (CSAG), University of Cape Town, Rondebosch 7701, RSA, South Africa
e-mail: hewitson@csag.uct.ac.za

H. Hendon
Centre for Australian Weather and Climate Research, Bureau of Meteorology, GPO Box 1289, Melbourne, VIC 3001, Australia
e-mail: h.hendon@bom.gov.au

C. Jones
Rossby Center, Swedish Meteorological and Hydrological Institute, SE-601 76 Norrköping, Sweden
e-mail: colin.jones@smhi.se

P. Lionello
Università del Salento – Dipartimento di Scienze e Tecnolog, University of Salento and CMCC, Piazza Tancredi N7-73100, Lecce, Italy
e-mail: piero.lionello@pd.infn.it

A need is found to strengthen the links between model evaluation at the applications level and process-oriented refinement of model formulation. Further work is required to develop and sustain effective communication among the observation, model user, and model development communities, as well as between the academic and "operational" model development communities. More research and investment is needed to translate climate data into actionable information at the regional and local scales required for decisions.

Keywords Monsoons • Climate variability • Climate change • Regional climate modeling • Predictability

1 Introduction

The better understanding, simulation and prediction of climate and its variability at regional and local scales have challenged the scientific community for many decades. These are complex subjects due to the physical processes and interactions that occur on space and time scales among the different elements of the climate system. In addition, climate variability is of great importance for society. Particularly difficult are the world's monsoon regions where more than two thirds of the Earth's populations live.

Understanding, simulating and predicting monsoons involves multiple aspects of the physical climate system (i.e., atmosphere, ocean, land, and cryosphere), as well as the impact of human activities. During the last decades World Climate Research Programme (WCRP) has promoted international research programs that have implemented modeling activities and field experiments aimed to the fundamental processes that shape the monsoons. The WCRP/CLIVAR panel on the Variability of American Monsoon Systems (VAMOS, Mechoso 2000) contributed to the organization of multinational research on the American Monsoons. VAMOS encouraged the realization of the South American Low Level Jet experiment (SALLJEX, Vera et al. 2006a), the North American Monsoon Experiment (NAME, Higgins et al. 2006), La Plata Basin (LPB) Regional Hydroclimate Project (Berbery et al. 2005), VAMOS Ocean-Cloud-Atmosphere-Land Study (VOCALS, Wood and Mechoso 2008) in the south-eastern Pacific, and more recently the Intra-Americas Study of Climate Processes (IASCLIP) Program (Enfield et al. 2009). The West African Monsoon (WAM) has also received considerable attention through the international African Monsoon Multidisciplinary Analysis (AMMA) program (Redelsperger et al. 2010). Several observational campaigns, such as the GEWEX/CEOP (Coordinated Enhanced Observing Period) and the YOTC (Year of Tropical Convection) have archived both in-situ and satellite observation data, providing a continuous record of observations for studies on processes and interactions affecting monsoon variability. WCRP also recently sponsored the Asian Monsoon Years (2007–2012).

The chapter presents outstanding scientific advances made under WCRP leadership in understanding, simulating and predicting climate variability and change at regional scales with an emphasis in the monsoon regions. The chapter also discusses important related challenges to be addressed by the WCRP community in references to the monsoons

2 Regional Perspectives

This section is a brief review of progress in understanding the different monsoon systems. The focus is on monsoon variability and predictability on time scales of great societal value, such as intraseasonal, interannual, decadal and longer including climate change. Rather than being comprehensive, the review highlights major advances made mainly during the last decade.

2.1 Asian-Australian Monsoons

2.1.1 Regional Variability and Predictability

A prominent feature of the Asia-Australian (AA) monsoon is its intraseasonal variation (ISV). This consists of a series of active and break cycles, which typically originate over the western equatorial Indian Ocean. Enhanced and suppressed convective activity associated with boreal summer intraseasonal oscillations propagate both poleward over land and eastward over the ocean during the summer monsoon, exhibiting both 10–20 and 30–50 day modes (Goswami 2005). During the Australian summer monsoon, ISV is dominated by the Madden Julian Oscillation (MJO) with a periodicity between 30 and 50 days and propagation primarily west-east with only limited poleward influence over subtropical Australia (Wheeler et al. 2009). The AA monsoon ISV with far greater amplitude than the interannual variation can have a dramatic impact on the region. For example, the intraseasonal break in the monsoon over India in July 2002 resulted in only 50 % of normal rainfall that month, causing enormous loss of crops and livestock. The ISV influences predictability of the seasonal mean climate (Goswami and Ajaya Mohan 2001) and shortest time scales through modulating the frequency of occurrence of synoptic events such as lows, depressions and tropical cyclones (Maloney and Hartmann 2000; Goswami et al. 2003; Bessafi and Wheeler 2006).

There is some evidence that models that are more successful in simulating the seasonal mean climate of the AA monsoon region tend to make better predictions of intraseasonal activity (e.g. Kim et al. 2009). Important westward propagating variations also occur on the 10–20 day time scale during the boreal summer Asian monsoon (Annamalai and Slingo 2001), but the ability to predict these features has yet to be demonstrated.

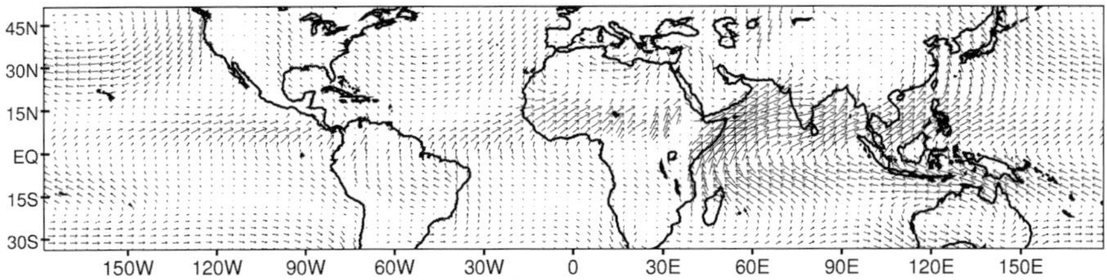

Fig. 1 Seasonal change in observed lower tropospheric wind (925 hPa) over the tropical monsoon regions (JJA minus DJF). Note the obvious reversal from north-easterly to south-westerly winds near West Africa and India and from anticlockwise to clockwise circulation in tropical South America, from northern hemisphere winter to summer (Courtesy of A. Turner)

Rainfall in the greater AA monsoon is surprisingly consistent from year to year, reflecting the robust forcing arising from the seasonal land-surface heating (Fig. 1). However, even relatively small percentage variations, when set against large seasonal rainfall totals, can have dramatic impacts on society, particularly where agriculture remains the main source of living (Gadgil and Kumar 2006). Floods are also common disasters in monsoon Asia. Due to the recent growth of Asian economies, flood damage is increasing, particularly in larger cities.

El Niño Southern Oscillation (ENSO) is the dominant forcing of AA monsoon interannual variability (IAV). ENSO's warm phase (El Niño) tends to be associated with reduced summer monsoon rainfall, although in the case of the Australian monsoon, the impact of El Niño is stronger in the pre-monsoon season (e.g., Hendon et al. 2012). In addition, antecedent Eurasian snow cover has been reported to contribute to monsoon IAV (e.g. Goswami 2006) while tropical Atlantic Ocean temperatures have also been associated with variations of the Indian summer monsoon (Rajeevan and Sridhar 2008; Kucharski et al. 2008). The Southern Annular mode (SAM) also influences the Australian summer monsoon through a poleward shift of the Australian anticyclone during SAM positive phase, resulting in stronger easterly winds impinging on eastern Australia, enhancing summer rainfall (Hendon et al. 2007). A large fraction of the AA monsoon IAV is unexplained by known, slowly varying forcing and may be considered 'internal' IAV arising from interactions with extra-tropics (e.g. Krishnan et al. 2009) or scale-interactions within the tropics (e.g. Neena et al. 2011).

Seasonal prediction of land-based seasonal rainfall in the AA monsoon region with the most modern dynamical coupled models such as those that contributed to Asian-Pacific Economic Cooperation Climate Center (APCC)/Climate Prediction and its Application to Society (CliPAS) Project (Wang et al. 2009) and in DEMETER Project (Kang and Shukla 2006) remains too low to be of practical use, even at the shortest lead times. Poor seasonal predictions of the AA monsoon seems to be related to model difficulties with the representation of land surface processes and uncertainty of initial conditions over land, but it also stems from local air-sea interaction in the surrounding oceans that tends to damp ocean-atmosphere variability in regions of monsoonal westerlies (Hendon et al. 2012). However, an encouraging trend of improvement in prediction skill of the Asian monsoon has emerged in recent models such as in ENSEMBLES Project (e.g. Rajeevan et al. 2011; Delsol and Shukla 2012). The high level of unpredictable intraseasonal variability during the AA monsoon is another contributing factor, which on top of poor MJO simulation and other monsoon ISV further limits the ability to predict and simulate monsoon variability.

The finding that the multi-decadal variability of the south Asian monsoon (Goswami 2006) and the Atlantic Multi-decadal Oscillation (AMO) are strongly linked (Goswami et al. 2006), raised hope of decadal predictability of the monsoon. However, the recent long decreasing trend of Indian monsoon rainfall since 1960 and decoupling with AMO indicates a changing character of multi-decadal variability of south Asian monsoon, also supported by reconstruction of rainfall over the past 500 years (Borgaonkar et al. 2010). Character and robustness of decadal variability of all monsoon systems need to be established from the instrumental records supplemented by multi-proxy reconstructions. In order to exploit predictability of such decadal variability, the ability to simulate observed decadal monsoon variability by the current coupled ocean-atmosphere models need to be established.

2.1.2 Long-Term Trends and Projections

Lack of an increasing trend of South Asian monsoon rainfall in the backdrop of a clear increasing trend of surface temperature (Kothawale and Rupa 2005) has been reconciled as due to contribution from a increasing trend of extreme rainfall events being compensated by contributions from a decreasing trend of low and moderate events (Goswami et al. 2006). It is also suggested that an increased intensity of short-lived extreme rain events may lead to a decreasing predictability of monsoon weather (Mani et al. 2009).

Future projections based on the Coupled Model Intercomparison Project-3 (CMIP3, Meehl et al. 2007b) show monsoon precipitation increasing in South and East Asia during June-August and over the equatorial regions and parts of eastern Australia in December-February, though model consistency is not high locally, especially for Australia (Fig. 2). The projected increase in the monsoon precipitation comes with large uncertainty (Krishna Kumar et al. 2011) making it difficult to influence policy decisions. Unfortunately, even the CMIP5 models (Taylor et al. 2012) show similar uncertainty in regional projection of precipitation (Kitoh 2012). Both CMIP3 and CMIP5 models indicate that while projected monsoon precipitation is likely to increase, the monsoon circulation strength is likely to decrease in warmer climate (Kitoh 2012). Projected changes in the atmospheric circulation impact those on regional precipitation (Kitoh 2011). For example, in the East Asian summer monsoon, a projected intensification of the Pacific subtropical high, defining the Meiyu-Changma-Baiu frontal zone and the associated moisture flux, may bring about increase rainfall (Kitoh 2011). Most models project an increase in the interannual variability of monthly mean precipitation (Krishna Kumar et al. 2011). The intensity of precipitation events is also projected to increase, with a shift

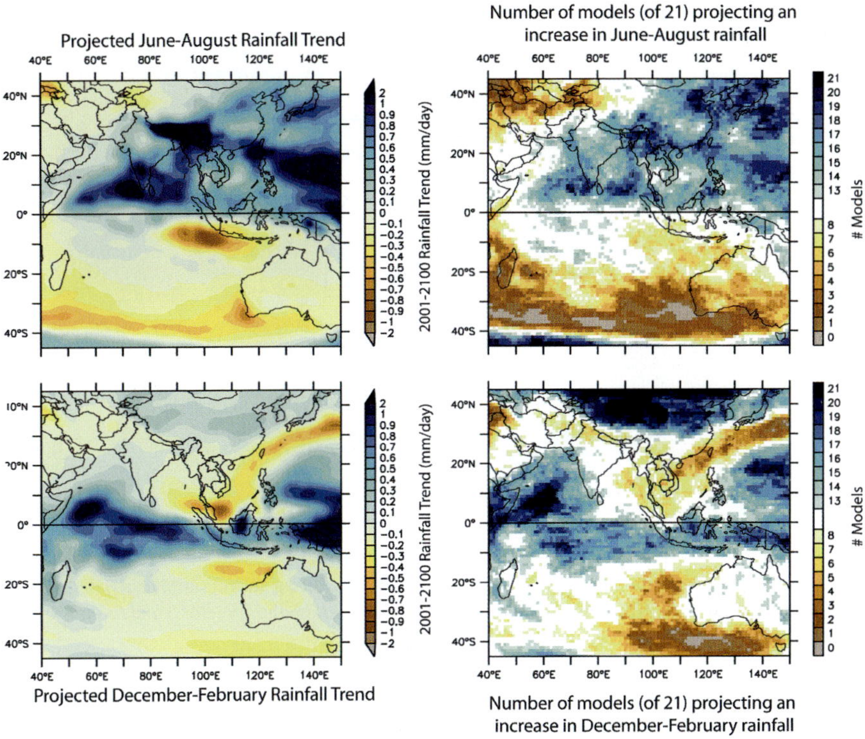

Fig. 2 Projected change in precipitation amount over the Asian-Australian monsoon region in June-August (*top row*) and December-February (*bottom row*) due to anthropogenic climate change using the CMIP-3 models. The *left panels* show the 2001–2100 trend in mm/day (21-model average), and the *right panels* show the number of models (of 21) that have an increasing trend (Figure provided by G. Vecchi (GFDL), following analysis method of Christensen et al. (2007) using the CMIP3 model archive described in Meehl et al. (2007b))

towards an increased frequency of heavy precipitation events (e.g. >50 mm day-1). Changes in extreme precipitation follow the Clausius–Clapeyron constraint and are largely determined by changes in surface temperature and water vapor content (e.g. Turner and Slingo 2009).

2.2 *American Monsoon Systems*

2.2.1 Regional Variability and Predictability

During the warm season, the MJO modulates a number of weather phenomena affecting the North American monsoon system (NAMS) and the inter-American seas (IAS) region, like tropical cyclones, tropical easterly waves, and Gulf of

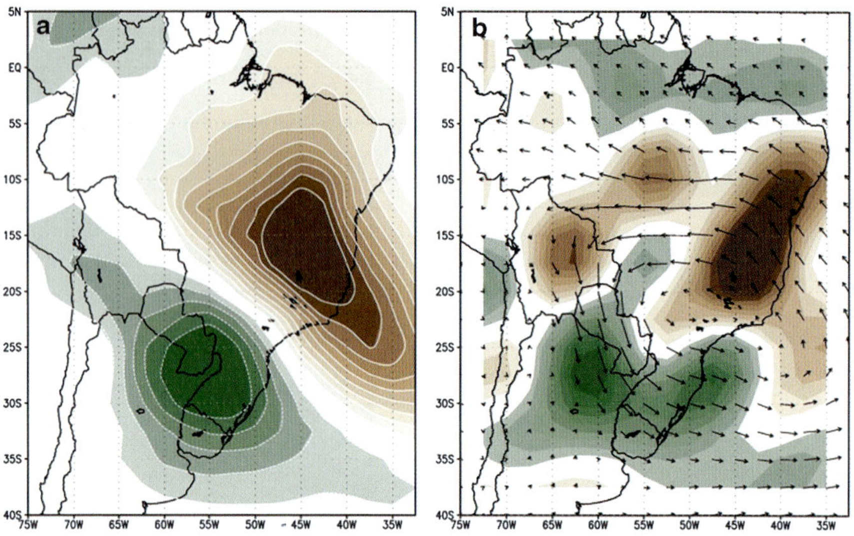

Fig. 3 (**a**) EOF1 pattern for 10–90 day filtered OLR anomalies during austral summer. (**b**) Regression map between EOF1 principal component and 850-hPa wind anomalies (vectors) and the associated divergence (*shading*) (Courtesy of Paula Gonzalez, IRI)

California surges (Barlow and Salstein 2006; Yu et al. 2011). Intraseasonal (and even interannual and interdecadal) variations of South American Monsoon System (SAMS) appear to be dominated by a continental-scale eddy centered over eastern subtropical South America (e.g. Robertson and Mechoso 2000; Zamboni et al. 2012). In the cyclonic phase of this eddy, the South Atlantic Convergence Zone (SACZ) intensifies and precipitation weakens to the south, resembling a dipole-like structure in the precipitation anomalies; the anticyclonic phase (Fig. 3b) shows opposite characteristics (e.g. Nogues-Paegle and Mo 1997, 2002; Ma and Mechoso 2007). Such an anomaly dipole pattern seems to have a strong component due to internal variability of the atmosphere, but it is also is influenced on intraseasonal timescales (Fig. 3a) by the MJO (e.g. Liebmann et al. 2004) and, on interannual timescales, by both ENSO (Nogues-Paegle and Mo 2002) and surface conditions in the southwestern Atlantic (Doyle and Barros 2002).

El Niño and La Niña tend to be associated with anomalously dry and wet events, respectively, in the equatorial belt of both NAMS and SAMS. ENSO influences NAMS and SAMS activity through changes in the Walker/Hadley circulations of the eastern Pacific and through extratropical teleconnections extended across both the North and South Pacific Oceans (PNA and PSA, respectively). During austral spring, climate variability in southeastern South America is influenced by combined activity of ENSO (Vera et al. 2006b) and SAM (Silvestri and Vera 2003). Influences of ENSO on rainfall in the IAS region is complicated by concurrent influences from sea surface temperature (SST) anomalies in the tropical Atlantic Ocean; the Pacific and Atlantic rainfall responses are comparable in magnitude but opposite in sign

(Enfield 1996). An additional complication is the reported change in Atlantic-Pacific Niños since the late 1960s, according to which summer Atlantic Niños (Niñas) alter the tropical circulation favoring the development of Pacific Niñas (Niños) in the following winter (Rodríguez-Fonseca et al. 2009).

Contemporary GCMs are able to capture large-scale circulation features of the American Monsoon Systems. Moreover, the models can reasonably predict early-season rainfall anomalies in NAMS, but they have difficulty in maintaining useful forecast skill throughout the monsoon season (Gochis 2011). In general, models still have difficulty in producing realistic simulations of the statistics of American monsoon precipitation and their modulation by the large-scale circulation (Wang et al. 2005; Marengo et al. 2011, 2012). Model limitations are more evident with the intensity of the mid summer drought and the SACZ, the timing of monsoon onset and withdrawal, diurnal cycle, and in regions of complex terrain (e.g. Gutzler et al. 2003; Ma and Mechoso 2007). Assessment of simulated behavior is also limited by uncertainties in spatially averaged observations (Gutzler et al. 2003), which undermines model improvement.

Accurate MJO activity forecasts could be expected to lead to significant improvements in the skill of warm season precipitation forecasts in the tropical Americas (e.g., Jones and Schemm 2000). On the other hand, CGCM skill in predicting seasonal mean precipitation in both NAMS and SAMS core domains are low and consistent with a weak ENSO impact. In contrast, north and south of the SAMS core region, higher predictability can be attributed to stronger ENSO impacts (Marengo et al. 2003).

Land surface processes and land use changes can significantly impact both NAMS and SAMS (e. g. Vera et al. 2006b). The continental-scale pattern of NAMS IAV shows anomalously wet (dry) summers in the southwest U.S. are accompanied by dry (wet) summers in the Great Plains of North America. Stronger and weaker NAMS episodes often follow northern winters characterized by dry (wet) conditions in the southwest U.S. Moreover, land-atmosphere interactions have to be considered to reproduce correctly the temperature and rainfall anomalies over all South America during El Niño events (Grimm et al. 2007; Barreiro and Diaz 2011). Moreover SAMS precipitation seems to be more responsive to reductions of soil moisture than to increases (Collini et al. 2008; Saulo et al. 2010). Recently Lee and Berbery (2012) examined through idealized numerical experiments potential changes in the regional climate of LPB due to land cover changes. They found that replacement of forest and savanna by crops in the northern part of the basin, leads to overall increase in albedo which in turns leads to reduction of sensible heat flux and surface temperature. Moreover, a reduction of surface roughness length favors a reduction of moisture flux convergence and thus precipitation. They found opposite changes in the southern part of the basin where crops replace grasslands.

On decadal and multidecadal time scales, the influence of the Pacific Decadal Oscillation (PDO) on precipitation has been described in both NAMS (Brito-Castillo et al. 2003; Englehart and Douglas 2006 and SAMS (e.g. Robertson and Mechoso 2000; Zhou and Lau 2001; Marengo et al. 2009) regions. The warm PDO phase tends to have dry (wet) El Niño and wet (dry) La Niña summers in North

America (southern South America) (Englehart and Douglas 2006; Kayano and Andreoli 2007). The North Atlantic Oscillation (NAO) and, the AMO can also influence the American Monsoons (Hu and Feng 2008; Chiessi et al. 2009) and the IAS region (e.g. Giannini et al. 2001), while decadal changes in the SAM influence on precipitation anomalies in southeastern South America have also been recorded (Silvestri and Vera 2009).

2.2.2 Long-Term Trends and Projections

Between 1943 and 2002, NAMS onset has become increasingly later and NAMS rainfall more erratic, though the absolute intensity of rainfall has been increasing (Englehart and Douglas 2006). In the NAMS core region, daily precipitation extremes have shown significant positive trends during the second half of the twentieth century (e.g. Arriaga-Ramirez and Cavazos 2010), while consecutive dry days with periods longer than 1 month have significantly increased in the U.S. southwest (Groisman and Knight 2008). The SAMS has shown a climate shift in the mid 1970s, starting earlier and finishing later after that date (Carvalho et al. 2010). Positive trends in warm season mean and extreme rainfall have been documented in southeastern South America during the twentieth century (e.g. Marengo et al. 2009; Re and Barros 2009).

Climate change scenarios for the twenty-first century show a weakening of the NAMS, through a weakening and poleward expansion of the Hadley cell (Lu et al. 2007). Projected changes in ENSO have, however, substantial uncertainty with regard to the hydrological cycle of the NAMS (Meehl et al. 2007a). Changes in daily precipitation extremes in the NAMS have inconsistent or no signal of future change (e.g. Tebaldi et al. 2006). CMIP3 models do not indicate significant changes in SAMS onset and demise under the A1B scenario (Carvalho et al. 2010). On the other hand, the majority of CMIP3 models project positive trends in summer precipitation for the twenty-first century over southeastern South America (e.g. Vera et al. 2006c). That trend has been recently related to changes in the activity of the dipolar leading pattern of precipitation IAV (Junquas et al. 2012). In addition, a weak positive trend in the frequency of daily rainfall extremes has been projected in southeastern part South America by the end of the twenty-first century, associated with more frequent/intense SALLJ events (e.g. Soares and Marengo 2009).

2.3 Sub-Saharan Africa

2.3.1 Regional Variability and Predictability

WAM (Fig. 4) is characterized by rainfall ISV dominating in two distinct periods: 10–25 and 25–90 days (Sultan and Janicot 2003; Matthews 2004; Lavender and Matthews 2009; and Janicot et al. 2010). In the 10–25 day range, rainfall variability

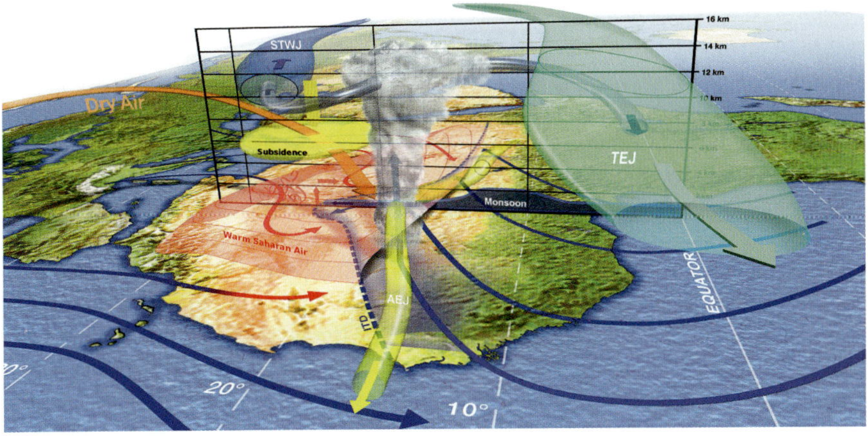

Fig. 4 Three-dimensional schematic view of the West African monsoon (see text for details) (Adapted from Lafore (2007), illustration: François Poulain)

has been associated with a "quasi-biweekly-zonal-dipole mode" that includes a notable eastward propagating signal between Central America and West Africa (Mounier et al. 2008), and a "Sahelian mode" that includes a westward propagating signal in the Sahelian region (Mounier and Janicot 2004). In the 25–90 day range, rainfall variability appears to have a significant MJO contribution but the mechanisms for impact are not straightforward, possibly arising in association with a westward propagating Rossby wave signal that can be equatorial or sub-tropical (e.g. Janicot et al. 2010; Ventrice et al. 2011) as well as eastward propagating Kelvin waves (e.g. Matthews 2004).

Recent studies confirm the importance of SST IAV in the Atlantic, Pacific – Indian and the Mediterranean basins on the WAM (e.g. Losada et al. 2009; Mohino et al. 2010, 2011; Rodriguez-Fonseca et al. 2011). It has also been suggested that vegetation IAV (affected by the previous year's rainy season) influences the early stages of the following rainy season (Philippon et al. 2005). Abiodun et al. (2008) examined the impacts on the WAM of large-scale deforestation or desertification in West Africa. Either change yielded strengthened moisture transport by easterly flow, which led to reduced moisture for precipitation. Short rains over equatorial East Africa are strongly sensitive to ENSO (e. g. Ogallo 1988; Hastenrath et al. 1993) and to the Indian Ocean Dipole (e.g. Saji et al. 1999; Webster et al. 1999). One of the strongest SST-rainfall correlations anywhere on the African continent exists between East African rainfall and tropical Indian Ocean SST in October-November-December. Teleconnections between the NAO and austral autumn Congo River discharge and regional rainfall have also been documented (Todd and Washington 2004).

In general, warm (cool) SST anomalies east of South Africa are associated with above (below) average summer rainfall over southeastern Africa (Reason and Mulenga 1999). ENSO also exerts a strong influence on summer rainfall over southern Africa. The South Indian Ocean SST dipole, which influences summer rainfall over southern Africa (Behera and Yamagata 2001; Reason 2001, 2002), has its southwestern pole in the greater Agulhas Current region. In addition, warm (cold) events in the Angola-Benguela Frontal Zone (ABFZ) region during summer/autumn, not only disrupt fisheries but also often produce large positive (negative) rainfall anomalies along the Angolan and Namibian coasts and inland (Rouault et al. 2003). A teleconnection between the SAM, tropical southeast Atlantic SST and central / southern African rainfall has also been identified (Grimm and Reason 2011). On the other hand, local re-circulation of moisture (e.g., Cook et al. 2004), and land surface feedbacks (e.g., Mackellar et al. 2010) can also contribute to climate variability in southern Africa. Strong relationships between the frequency of dry spells during the summer rainy season and Nino 3.4 SST have been found for areas in northern South Africa/southern Zimbabwe, and Zambia (Reason et al. 2005; Hachigonta and Reason 2006). A weaker relationship exists between dry spell frequency and the Indian Ocean Dipole.

Predictability of the seasonal and intra-seasonal regional climate over Africa depends strongly on location, season, and state of global modes of variability that couple to a given region. In most regions demonstrable statistical is readily shown. For example, Ndiaye et al. (2011) examine the performance of eight AGCMs and eight coupled atmosphere-ocean GCMs (CGCMs) over the Sahel and find skill levels of correlation between predicted and observed Sahel rainfall at up to 6 month lead time. The same study explores the relative merits of AGCMs versus CGCMs, and while there are indications that AOGCMs have the advantage, how beneficial this is to skill enhancement remains an open question. Comparable results are found over southern Africa, for example Landman and Beraki (2012), who also highlight the added value of multi-model approaches for improving skill. Generally the seasonal forecasting studies collectively show forecast skill strongly variable in time, especially when the equatorial Pacific Ocean is in a neutral state (Landman et al. 2012). Nonetheless, the value to society, when translating this measure of predictive skill into the realms of decision maker, remains a point of debate. While some positive experiences with using forecasts have led to valuable lessons (e.g. Tall et al. 2012), the interface with decision makers in the context of a variable skill forecast product remains a significant challenge.

Multi-decadal SST variability in both the Atlantic and Pacific has been shown to be important for the WAM (e.g. Rodriguez-Fonseca et al. 2011) and southern Africa (Reason et al. 2006). The partial recovery in West African rainfall over the past decade has received substantial debate over the respective roles of the Atlantic and Indian basins (e.g. Giannini et al. 2003; Knight et al. 2006; Hagos and Cook 2008; Mohino et al. 2011). It has been argued that at interannual time scales the relationship between West African rainfall and tropical SST is non-stationary (Losada et al. 2012). That is, the impact on West African rainfall of SST anomalies in a tropical ocean basin differs before and after the 1070s because in the more recent period those basin anomalies tend to develop simultaneously with others in the global tropics. Such findings emphasize the need for proper initial conditions in the forecasts. Central to the decadal predictability, however, is the challenge of how to initialize the models (Meehl et al. 2009), and whether the initial state can adequately capture the mechanisms central to regional predictive skill (for example, the AMO, PDO, etc.). For Africa this is particularly important, especially southern Africa where the regional response is linked to a broad range of hemispheric-scale processes. Liu et al. (2012) explore this initialization issue, and show that while initialization leads to improvements in hindcast simulations over the oceans, the improvement with initialization of the land areas was detectable, but limited. Chikamoto et al. (2012) likewise examine predictability with a hindcast ensemble experiment, and note the value of ocean subsurface temperatures for decadal signals, but find this most notable for the north Pacific and Atlantic and the corresponding connection to North and West Africa. The skill for southern Africa remains more complicated. Perhaps especially important for Africa in general, is that it remains unclear what level of skill is required to support stakeholder decisions on a decadal scale (Meehl et al. 2009).

2.3.2 Long-Term Trends and Projections

The spatial patterns and seasonality of African rainfall trends since 1950 seem to be related to the atmosphere's response to SST variations (e.g. Hoerling et al. 2006). While drying over the Sahel during boreal summer seems to be a response to warming of the South Atlantic relative to North Atlantic SST, Southern African drying during austral summer seems to be a response to Indian Ocean warming (e.g. Hoerling et al. 2006). A reduction in precipitation over eastern and southern Africa has also been detected in relation with Indian Ocean warming (Funk et al. 2008). In general, an increasing delay in wet season onset has been detected over Africa during the last part of the twentieth century (Kniveton et al. 2009).

Climate change projections from WCRP/CMIP3 models fail in showing agreement on changes of West African rainfall for the twenty-first century (Biasutti and Giannini 2006; Christensen et al. 2007; Joly et al. 2007). However, precipitation changes derived from empirical downscaling applied to GCM projection ensemble, show larger agreement in projecting an increased precipitation along the southern Africa coast, widespread increase in late summer precipitation across south-east Africa, reduced precipitation in the interior, and a less spatially coherent early summer decrease (Hewitson and Crane 2006; Tadross et al. 2009). In general across southern Africa there are indications of future drying in the west and wetter condition in the east (Hewitson and Crane 2006; Giannini et al. 2008; Batisani and Yarnal 2010). Hewitson and Crane (2006) further note that the interplay between change in derivative aspects of rainfall (such as increasing intensity but reducing frequency) can be masked in the more common representation of seasonal averages.

3 Regional Climate Simulation

3.1 Regionalization Needs

Access to quality-controlled high-resolution, regional climate data is key for assessing regional climate vulnerability, impacts and the subsequent development of informed adaptation strategies. Currently, CGCMs used for seasonal to decadal prediction and climate change projection typically employ horizontal resolutions of ~1–2°. This limits their ability to represent important effects of complex topography, surface heterogeneity, and coastal and regional water bodies, all of which modulate the large-scale climate on local scales. Coarse resolution also limits the ability of CGCMs to simulate extreme weather events that contribute non-linearly to the societal impact of regional climate variability. To increase the utility of CGCM simulations some form of downscaling or regionalization is usually applied to increase the spatial detail of the simulated data.

Regionalization techniques currently include (i) Dynamical downscaling (DD), where a Regional Climate Model (RCM) is run at increased resolution over a limited area, forced at the boundaries by GCM data (Giorgi and Mearns 1999), (ii) Global Variable Resolution Models (GVAR), that employ a telescoping procedure to locally increase model resolution over a limited area within a continuous AGCM (Deque and Piedelievre 1995) and (iii) Empirical-Statistical Downscaling (ESD), where statistical relationships, developed between observed large-scale predictors and local scale predictands, are applied to GCM output (Hewitson and Crane 1996). Most of these techniques aim to add regional detail without changing the large-scale climate derived from the GCM. All regionalization methods are, to a first-order, dependent on the quality of the large-scale climate simulated by the driving GCM.

3.2 Coordinated Downscaling Exercises

Several large-scale efforts have been pursued to assess regional climate change based on the development of ensembles of RCMs in an attempt to sample a fraction of the uncertainty space associated with projecting regional climate change. Efforts over North America have occurred in NARCCAP (Mearns et al. 2009, 2012) and over South America in CLARIS (Menendez et al. 2010) and other regional projects (Marengo et al. 2009, 2011). A number of coordinated RCM projects have focused on specific regional phenomena, such as PIRCS for North American summer season precipitation (Takle et al. 1999; Anderson et al. 2003), WAMME for the west African monsoon (Druyan et al. 2010), R-MIP for East Asia (Fu et al. 2005) and the Mediterranean region (Gualdi et al. 2011). The GEWEX-sponsored ICTS project (Takle et al. 2007) investigated the transferability of RCMs across a range of different regions using unmodified model formulations. Over the past 15 years such activities, many sponsored by WCRP, have provided detailed knowledge of the RCMs' ability to simulate important regional climate processes and climate change.

In 2008 the WCRP initiated the Coordinated Regional Downscaling Experiment (CORDEX), with the intention to (i) provide a coordinated framework for the development, and evaluation of accepted downscaling methodologies; (ii) generate an ensemble of high-resolution, regional climate projections for all land-regions, through downscaling of CMIP5 projections; (iii) make these projections available to climate researchers and the impact-adaptation-vulnerability community and support the use of such data in their activities; and (iv) foster international collaboration in regional climate science, with an emphasis on increasing the capacity of developing nations to generate and utilize climate data local to their region. CORDEX is an unprecedented opportunity for scientists to collaborate in order to evaluate and improve downscaling methods for different regions of the world and to engage more closely with users of this data (Giorgi et al. 2009; Jones et al. 2011).

CORDEX has defined a set of target domains along with a standard resolution for regional data of 50 km. The evaluation phase of CORDEX entails downscaling global reanalysis data for the past 20 years over all regions for which a group plans to generate downscaled future projections (e.g. Africa, South America, Europe, etc.). For each CORDEX area, evaluation teams have been established to define key climate processes and metrics of performance pertinent to that region, in order to make a detailed evaluation of downscaling methods for the recent past. Subsequent to this, DD and ESD methods will be applied to CMIP5 projections for the same regions. 1950–2010 will used be available for evaluation while 2010–2100 constitutes the time period over which regional projections will be made. While each of the CORDEX regions will be targeted by groups local to the region, the international downscaling community has agreed to target Africa as a common domain for the coming few years, with an aim of generating an ensemble of climate projections for Africa to support the Intergovernmental Panel on Climate Change (IPCC) fifth Assessment process.

4 Challenges in Monsoon Simulation and Prediction

Although there has been substantial progress in understanding and simulating regional climate as a result of research promoted by WCRP, the successful prediction and simulation of the monsoon and surrounding subtropical regions remains elusive. Limiting factors to improving simulation of the Earth's monsoon systems include the inability to adequately resolve multi-scale interactions that contribute to the maintenance of those systems (Sperber and Yasunari 2006). A discussion of some selected processes requiring improved simulation and prediction in the different monsoon systems is presented in the following subsections.

4.1 Large to Regional Scale Processes Influencing Monsoon Variability and Predictability

The identification and understanding of phenomena offering some degree of intra-seasonal to inter-annual predictability, is necessary to skillfully predict climate fluctuations in those scales (CLIVAR 2010). In that sense, a prerequisite for a successful simulation of regional-scale monsoon variability is an accurate representation of large-scale modes of variability (e.g. MJO, ENSO). While CGCMs are improving in their ability to simulate such modes, capturing their remote impact on monsoon variability requires also simulating atmospheric and oceanic teleconnections from the mode source regions into the monsoon regions as well as simulating related regional features (e.g. Alexander et al. 2002).

A number of phenomena that directly impact the quality of simulated monsoon climates include both large-scale features as well as monsoon features. An example is the MJO that can propagate into, and out of the monsoon region while locally influencing monsoon ISV. While dynamical prediction of the MJO has improved in recent years (e.g. Rashid et al. 2010; Kang and Kim 2010; Gottschalck et al. 2010), climate models of the sort used for seasonal prediction have difficulties with the simulation and prediction of monsoon intraseasonal variability, which compounds the problem of trying to predict relatively low interannual variability together with the modest relationship with El Niño (CLIVAR 2010). The inadequate simulation of MJO and monsoon ISV and in general, the inadequate simulation of the interaction of organized tropical convection with large-scale circulation has limited extensively the studies of predictability of monsoon ISV (CLIVAR 2010). Some of the model limitations in predicting monsoon on intraseasonal and seasonal time scales are related to the fact that predictable variations of the monsoons associated with El Niño are typically confined to pre-monsoon and post monsoon, while most of the variability of the main monsoon appears to be associated with internally generated (i.e. independent of slow boundary forcing) intraseasonal variations (CLIVAR 2010).

Regarding longer time scales, the recent increased capacity of global decadal prediction brings up the issue of how to address regional decadal prediction. On decadal timescales, natural variability overlaps with trends and signals associated to anthropogenic climate change, which might induce different regional climate responses (as it was discussed in Sect. 2). Moreover, the magnitude of decadal variability exceeds in many regions of the world those associated with the trends resulting from anthropogenic changes. The provision of present and future climate information on decadal time scales is important considering the need of climate information on those timescales for decision making of many different socio-economic sectors (Vera et al. 2010).

The better understanding of how energy and water cycles of the monsoon systems change as the climate warms is a critical problem (GEWEX 2011). Hydrological responses to changes in precipitation and evaporation are complex and vary between regions. For example, it has been shown that a direct influence of global warming leads to increase water vapor in the atmosphere and more precipitation. However, with more precipitation and, thus, more latent heat release per unit of upward motion in the atmosphere, the atmospheric circulation weakens, causing monsoons to falter. Therefore, sorting out the role of natural variability from climate change signals and from effects due to land-use change is a key challenge for monsoon related research (GEWEX 2011). In addition, a warming climate is expected to alter the occurrence and magnitude of extreme events, especially, droughts, heavy precipitation and heat waves. How both, natural variability combined with anthropogenic climate change signal, affect the nature of climate extremes at regional scales is also a grand challenge for future research (GEWEX 2011).

Progress in understanding and quantifying predictability of regional decadal climate variations require climate model simulations that resolve and capture regional processes accurately. Moreover, developing skillful decadal predictions at regional scales relies on better understanding of the associated mechanisms and in particular of the identification of the climate patterns that offer some degree of decadal predictability (e.g. PDO, AMO, CLIVAR 2010). Doblas-Reyes et al. (2011) have evaluated the skill of decadal predictions made with the European Centre for Medium-Range Weather Forecasts coupled forecast system using an initialization common in seasonal prediction with realistic initial conditions. Despite model drift and model limitation in reproducing several climate processes, positive correlations between decadal predictions with observations are found for tropospheric air temperature for many regions of the world with increasing skill with forecast time. On the other hand, precipitation does not show significantly positive skill beyond the first year. The recent availability of the CMIP5 prediction experiments (Meehl et al. 2009; Taylor et al. 2012) should help to expand research on monsoon decadal variability and predictability. Evaluation of decadal predictions over monsoon regions is a challenge by itself in view of the limited availability of enough long and spatially dense records. Paleo-climate proxy records might provide useful information for the validation task (CLIVAR 2010).

Besides the influence of large-scale climate variability on monsoon systems, regional phenomena may also impact the monsoon circulation simulation depending on how they are locally reproduced. Examples include; Tibetan plateau snow cover and its impact on large-scale thermal gradients and thereby the monsoon circulation (Shen et al. 1998; Becker et al. 2001) or the Saharan heat low and its impact on the West African monsoon (Fig. 4, Lavaysse et al. 2009). Interactions between regional orography and monsoon circulations have been documented for South America (Lenters and Cook 1999), South Asia (Wu et al. 2007) and East Africa (Slingo et al. 2005). Over Asia, regionally aerosol emissions can modify both surface and atmospheric solar heating, altering thermal gradients and the monsoon-scale circulation (Meywerk and Ramanathan 1999, Meehl et al. 2008). Similar effects were found by Konare et al. (2008), related to radiative cooling due to Saharan dust and the West African monsoon. Such processes are particularly important to represent when estimating potential changes in monsoon circulations in response to future GHG and aerosol emissions (Ramanathan et al. 2001).

4.2 Key Local to Regional Processes Influencing Monsoon Variability and Predictability

A number of local- to regional-scale processes strongly influence the accuracy and utility of simulated monsoon data. These processes are all highly regional, involving complex interactions across a range of spatial and temporal scales, but are often fundamental to the specific development of each monsoon. Improvement in the understanding and simulation of such processes is crucial for progress in predicting monsoon variability and change.

4.2.1 Surface Heterogeneity

Land surface processes and land use change play an important role in regional monsoon variability. CLIVAR (2010) concludes that during a monsoon early stages, when the surface is not sufficiently wet, soil moisture anomalies may modulate the onset and development of precipitation. Furthermore, when the soil is not too dry or not too wet, the soil conditions can control the amount of water being evaporated, and also can produce fundamental changes in the planetary boundary layer (PBL) structure that affects the development of convection and precipitation.

Koster et al. (2004) identified a number of "hot spots" of land–atmosphere coupling, where sub-seasonal precipitation variability is modulated by regional soil water characteristics. Strong coupling was identified over the Great Plains of North America, northern India and West Africa–Sahel. In these regions accurate estimates of soil water, either in initial conditions or during model integration, will likely impact simulated intra-seasonal monsoon variability. Dirmeyer et al. (2009), showed that regions and seasons that have large soil moisture memory predominate in both summer and winter monsoon regions in the period after the rainy season wanes, excepting the Great Plains of the North America and the Pampas/Pantanal of South America, where there are signs of land-atmosphere feedback throughout most of the year. Soil moisture anomalies seem to have a significantly larger impact on rain rates in the African monsoon than over South Asia, likely due to a weaker oceanic moisture contribution to Africa and to the South Asian monsoon (Douville et al. 2001). Taylor et al. (2005) further showed that a more responsive and heterogeneous surface vegetation scheme impact both the simulated diurnal cycle of convection, as well as the frequency and intensity of convective events over West Africa. Xue et al. (2006) showed that, during the austral summer, consideration of explicit vegetation processes in a GCM does not alter the monthly mean precipitation at the planetary scales, but produces a more successful simulation of the South American monsoon system at continental scales. The improvement is particularly clear in reference to the seasonal southward displacement of precipitation during the monsoon onset and its northward merging with the intertropical convergence zone during its mature stage, as well as better monthly mean precipitation over the South American continent. Kelly and Mapes (2010) showed that biases in land surface fluxes reduce the accuracy of seasonal precipitation in the North American monsoon.

Adequate representation of the land surface conditions should be then carefully included in monsoon climate predictions (CLIVAR 2010). For example, recently Guo et al. (2012) showed using forecast experiments from the second phase of the Global Land- Atmosphere Coupling Experiment (GLACE-2, e.g. Koster et al. 2011) that predictability of air temperature and precipitation in climate models over North America rebounds during late spring to summer because of information stored in the land surface. Coupling becomes established in late spring, enabling the effects of

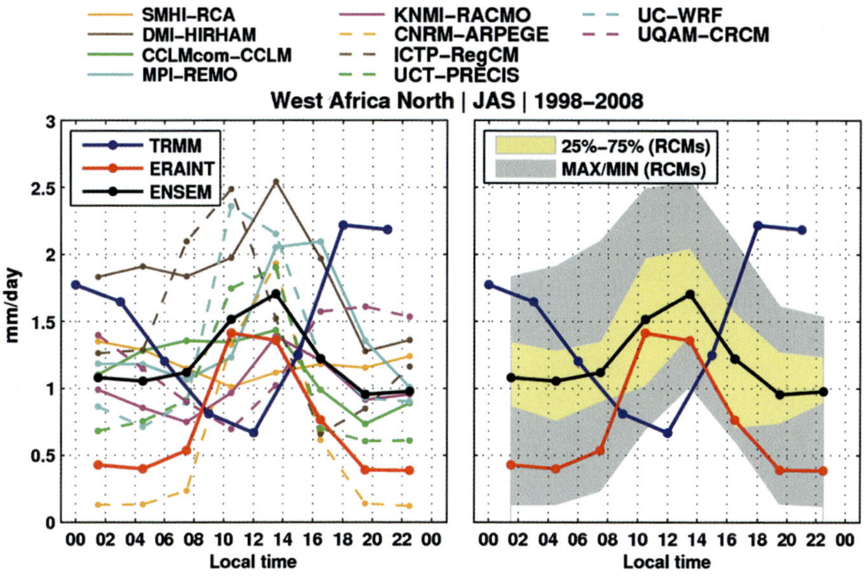

Fig. 5 Mean diurnal cycle of precipitation averaged over West Africa and for the period 1998–2008. *Left panel* shows TRMM342B, ERA-interim, the 10 RCM ensemble mean and results from each RCM. *Right panel* plots in *yellow shading* the spread of the 50 % most accurate RCMs and the full spread of the RCM results

soil moisture anomalies to increase atmospheric predictability in 2-month forecasts. The latter indicates that climate prediction in that particular region could be significantly improved with soil moisture observations during spring.

4.2.2 Diurnal Cycle

An accurate representation of the diurnal cycle of convection over tropical lands remains an unresolved problem in climate models employing convection parameterizations, with convection systematically triggered too early in the day and precipitation maxima often phased with local noon, some 6–8 h earlier than observed (Yang and Slingo 2001; Guichard et al. 2004). Figure 5 presents the mean diurnal cycle of rainfall for July-August-September, averaged over of West Africa from 10 RCMs that downscaled ERA-interim using the CORDEX-Africa domain (see for details Nikulin et al. 2012). TRMM is used as an observational reference, with a clear peak in precipitation from ~18.00 local time to 03.00 in the night and a minimum at local noon. ERA-interim 24-h forecast precipitation is completely out of phase with TRMM, exhibiting a maximum at local noon and minimum from early evening to early morning. Most RCMs show the same out of phase shape for the diurnal cycle. Two models exhibit an evening/nocturnal precipitation maximum (UQAM-CRCM and SMHI-RCA). These models employ variants of the Kain-Fritsch convection scheme (Kain and Fritsch 1990; Bechtold et al. 2001) with relatively advanced convective trigger functions and entrainment/detrainment schemes that are responsive to large-scale conditions (Kain and Fritsch 1990). Although parameter adjustments in convective schemes can reduce diurnal cycle errors, much deeper physical insight is needed in boundary-layer/convection coupling, triggering processes (e.g., Lee et al. 2007) and the multi-scale behavior of convective systems (Tao and Moncrieff 2009; Stechmann and Stevens 2010). Clearly, despite concerted efforts, the problem remains challenging and the need for better physical understanding of convective processes implies that simply increasing model resolution will not resolve the problem. Thus, much work remains to fully simulate all components of the precipitation diurnal cycle over tropical land regions.

Excessive triggering of convection over land contributes to models precipitating too frequently and at too low intensities (Dai 2006), while an incorrect phase to the diurnal cycle of convection and associated precipitation and clouds can induce systematic biases in the diurnal cycle of surface temperature and surface evaporation (Betts and Jakob 2002). Such errors may have a cumulative impact on soil moisture through the rainy season. Recent studies have thrown new light on the diurnal cycle of convection (Grabowski et al. 2006; Khairoutdinov and Randall 2006; Hohenegger et al. 2008) and suggest a number of extensions to convection parameterizations that may improve the diurnal cycle. These include; advanced convective trigger functions that account for heterogeneous surface and atmospheric forcing (Rio et al. 2009; Rogers and Fritsch 1996), super-parameterizations that embed cloud-permitting models in each grid box (Xing et al. 2009), convective entrainment that is sensitive both to the size of developing convective systems and the surrounding environment (Grabowski et al. 2006), the inclusion of evaporatively driven downdrafts and the impact of cold pools on vertical stability (Khairoutdinov and Randall 2006; Rio et al. 2009), and updraft mass-flux detrainment that impacts the convergence of convective outflows with low-level jets (Anderson et al. 2003).

4.2.3 Low Level Jets

As discussed above, LLJs are integral part of many monsoon systems. Statistically significant relationships have been found between nocturnally-peaking LLJs and nocturnal precipitation extremes in numerous disparate regions of the world (Monaghan et al. 2010). Widespread changes in the amplitude of near-surface diurnal heating cycles have been recorded as an important component of LLJ maintenance and that careful assessment of the impact of these changes on future LLJ activity is required. The complicated interactions involved in LLJ formation and maintenance provides an excellent testbed for understanding interactions of a multitude of physical parameterizations. Improvement in the simulation of LLJs should lead to a better representation of the phase and amplitude of the diurnal cycle of precipitation and thus warm season rain, though appropriate coupling of LLJs and convection is required (Anderson et al. 2003). This is a severe test for models given the unique land-sea distributions, surface types, and orographic influences of the disparate monsoon regions (Sperber and Yasunari 2006).

4.2.4 Regional Ocean-Atmosphere Coupling

The primary source of water for monsoon rainfall is evaporation from the ocean. Processes influencing SSTs and ocean thermocline depth are therefore likely important for a good representation of monsoon precipitation. There are indications that detailed representation of coastal ocean processes may lead to improvements in model simulations of monsoon ISV in some regions (Annamalai et al. 2005). Furthermore, it is well established that cyclone variability in the Bay of Bengal seems sensitive to a detailed representation of ocean mixed layer processes (e.g. Pasquero and Emanuel 2008). On seasonal time scales, coupled ocean-atmosphere models are required to simulate the observed negative correlation between precipitation and SST over the warm waters of the AA monsoon region (Wang et al. 2005). Furthermore, Xie et al. (2007) using a regional coupled model of the tropical East Pacific, highlight its ability to simulate tropical ocean instability waves, Central American gap winds, and their impact on coastal SSTs.

The better monitoring and understanding of air-sea interaction processes in subtropical anticyclones/subtropical and tropical gyres in the South Pacific, South Atlantic and South Indian Oceans will likely lead to improvements in the understanding and modeling of climate variability in Africa, South America and Oceania. The VOCALS program (Wood and Mechoso 2008), which grew out within the VAMOS panel, focuses on the South East Pacific climate and emphasizes the interactions among major climate components: atmosphere, ocean, clouds, and the aerosol. The program has a field component (Wood et al. 2011), and a model assessment of cloud and PBL which compared the regional performance of a number of different models (Wyant et al. 2010). The comparison of model outputs with VOCALS observations showed a good representation of large-scale dynamics, but a poor representation of clouds in general, with too shallow coastal model boundary layers. Moreover, the model assessment analyses has clarified quantitatively the erroneous way in which models reproduce the SST underneath the stratocumulus decks in the region (de Szoeke et al. 2010). Model improvements under VOCALS, nevertheless, have had more impact on the simulated SSTs in the Pacific than for the Atlantic. The latter could be due either to a more complex nature of the bias problem in addition to a lack of focused attention from the research community (Zuidema et al. 2011).

5 Challenges in Generating Actionable Regional Climate Information

The importance of climate information systems that provide products and services relevant to climate-related risk management and decision-making has risen dramatically in the last few years, a trend that is likely to continue. However, science and scientific capacity-building on climate variability and change has been so far insufficiently translated into policy relevant discourse and action. The lessons learned strongly suggest that the way forward needs a cultural change in the interaction of the climate science community and the users (Goddard et al. 2010; Vera et al. 2010;

and references therein). This change should consider the demand side as the starting point and the main focus of this interaction, as opposed to using a supply-oriented approach (e.g. Lemos et al. 2002; Ziervogel 2004). In addition, it is also essential to enhance natural–social science coupling as well as to improve dialogue with decision-makers. Such coupling needs to be built into climate modeling institutions and programs (e.g., SDWG 2012). Building effective partnerships between the providers and users of climate information are multi-faceted and often not straightforward, but it is crucial if the investments in climate science and their potential benefits to society are to be made (e.g. Barsugli et al. 2009; Vera et al. 2010, and references therein).

A key need for any climate service is the provision of timely and reliable predictions of the likelihood of hazardous weather and climate events. Defining what hazardous means, for whom and where, requires detailed understanding of the vulnerability of society and key systems (e.g. food and water) to changes in the patterns and characteristics of weather and climate. It also needs to consider how interactions with other components of the earth system act to mediate the impacts of hazardous weather and climate (e.g. soil moisture in intensifying heat waves, atmospheric chemistry in linking blocking to poor air quality, oceans and the cryosphere in determining sea level rise), along the underpinning research required to represent those processes. These multi-scale, interdisciplinary challenges require the WCRP to work closely with WWRP, IGBP and IHDP.

The development of climate services needs to be made in parallel to improving model capability. Besides the overall tasks that WCRP will do in the future to build better climate models, the effort must include regional-to-local scale verification of climate predictions pursued together with a dynamical understanding of the processes behind the predictability, and a determination of the quality of experimental predictions (including initialization issues) to provide guidance for climate model improvement (Vera et al. 2010; Goddard et al. 2010).

A fundamental component of climate services must be the provision of historical climate data and assessments of the current climate. Improved reanalyses drawing on the latest developments in models and data assimilation should be promoted as fundamental to climate services. In particular, ways to assemble, quality-check, reprocess and reanalyze datasets relevant to climate prediction at regional and local scales are needed. Also development of quantitative climate information for a wide range of variables in addition to surface temperature and precipitation is required at regional and local scales. Efforts should also be made for a better determination and availability of agreed and reliable datasets and variables required addressing specific socio-economic sector vulnerability, and identification of the specific regions where society is most vulnerable to changes in the near-future climate (Vera et al. 2010 and references therein).

Climate services need to provide probabilistic predictions at regional and local scales which allow users to manage their own risks in an objective way. Characterization of the uncertainties associated with climate predictions are needed including properly accounting for those aspects that are and are not predictable. Ensemble prediction systems are now well established in climate prediction, but the techniques to represent prediction uncertainty are quite diverse. Future research should consider how these diverse approaches can be brought together and the relative value of each assessed (Goddard et al. 2010).

6 Concluding Remarks

This chapter highlighted a number of advances made in monsoon research, mainly related to a better understanding of the physical processes related to the ocean-land-atmosphere interaction that characterize the monsoon variability as well as to a better knowledge of the sources of climate predictability. Considerable challenges need to be overcome, however, before predictions of regional monsoon variability can be achieved at a level of accuracy required by society applications. These challenges relate both to our basic understanding of physical processes, as well as to their successful representation in numerical models and the ability to translate that knowledge in climate information actionable for decision makers. This chapter presented challenges that we consider crucial to improve our ability to simulate and predict regional climate variability, particularly in monsoon regions. Central to many of these is the representation of moist convection and its interaction with regional dynamics and surface processes. For all aspects of monsoon simulations (intraseasonal to decadal prediction and to climate change) the representation of multi-scale convection and its interaction with coupled modes of tropical variability (where coupling refers both to ocean-atmosphere and/or land-atmosphere coupling) remains the leading problem to be addressed.

Systematic errors in the simulation of the mean annual and diurnal cycles continue to be critical issues that reflect fundamental deficiencies in the representation of moist physics and atmosphere/land/ocean coupling (they do not appear to be remedied by simple model resolution increases), and are likely a major impediment to improving the skill of monsoon forecasts at all time scales. Other processes, however, can also play an important role in climate simulation at regional levels. The influence of land cover change requires better quantification. Likewise, aerosols loading resulting from biomass burning, urban activities or, land use changes due to agriculture are potentially important climate forcings requiring better understanding and representation in models.

Besides the progress already made, more work is required to elucidate mechanisms that give rise to intraseasonal variability. This timescale is key for users of climate forecasts and so there is a high societal need to exploit any potential predictability present using current dynamical and/or statistical models. It is expected that new observational and modeling campaigns, such as DYNAMO (Dynamics of Madden-Julian Oscillation) and YOTC (Year of Tropical Convection) will contribute to improving the understanding and numerical representation of active and break monsoon cycles. Alongside this, it is important to consider how the time-varying, large-scale environment interacts with variability in regional weather systems including MCS, easterly waves and tropical cyclones.

On decadal and multi-decadal timescales an improved understanding of monsoon variability and predictability is required to better understand, simulate and attribute near-term climate change and to assess the potential for monsoon prediction. CMIP5 simulations (Taylor et al. 2012) provide improved regional-scale information compared to earlier GCM intercomparison projects, through the use of higher resolution models. Careful analysis of these simulations will provide new indications of how climate change may affect monsoon systems particularly in the coming decades. Community analysis of simulated monsoon processes in these runs are expected, with some activities having already started (e.g. by CLIVAR-AAMP). CORDEX will also downscale CMIP5 runs over monsoon land regions, allowing the benefits of increased model resolution in simulating e.g. intraseasonal variability of the various monsoons to be assessed.

Intense work is currently dedicated in many WCRP programs and projects to improve models, data-assimilation and data-gathering components of numerical climate prediction systems in order to increase forecast skill. However, further advances are needed to accelerate the improvement of overall model performance, and strengthening the links between model evaluation at the level of the application and the process-oriented refinement of the model formulation. There is a very large amount of information generated by numerous process studies and multi-model analyses that potentially could be used by projects aimed to improve climate models. The community needs to develop and sustain effective communication and implementation of this new knowledge. In order to facilitate the access to it by modeling groups, such synthesis efforts require closer collaboration among the observational data, model user, and model development communities, and also between the academic and "operational" model development communities (Jakob 2010). An important area of common research is the design of metrics to quantify the ability of models to simulate key features of regional climate systems. It should also be noted that models largely developed and tuned in the Northern Hemisphere may not perform optimally over parts of South America or sub-Saharan Africa and attention needs to be given to regionally sensitive parameterizations.

The community must also exploit the rapidly evolving computing opportunities afforded by advances in computer hardware and software engineering. Priority must be given to developing multi-model, multi-member prediction systems, running models with sufficient resolution to resolve key topographic features and mesoscale factors that mold regional climate. Complementing this effort is the need to expand climate models into earth system models that more thoroughly represent the climate system (Shukla et al. 2009).

A challenge will still remain to connect predictions of regional climate variability and projections of change to practical outcomes. More research and investment is needed to translate climate data into actionable information at the regional and local scales required for decisions (Vera et al. 2010). The expansion of activities must include: (i) Better determination and availability of agreed and reliable sets of data/variables required to address specific socio-economic sector vulnerability; (ii) ways of securing climate observing systems, particularly in less developed regions; (iii) ways to assemble, quality-check, reprocess and reanalyze datasets relevant to climate prediction at regional and local scales; (iv) characterization of the uncertainties associated with climate predictions including properly accounting for those aspects that are and are not predictable; (v) tailoring climate information to local scales and sector needs, and (vi) supporting long-term training of climate scientists in developing nations, coupled with an effort to ensure suitable infrastructures by which scientists in these regions can access, analyze, and ultimately develop prediction data and subsequently distribute this data to users in the region.

Acknowledgements The coauthors take great pleasure in acknowledging the WCRP community, in particular that of CLIVAR, GEWEX, WGSIP, and WGCM, for the tremendous scientific contributions to our understanding of the monsoon systems as summarized here. We thank the reviewers Filippo Giorgi and E. Hugo Berbery, for valuable comments that substantially improved the manuscript. The editing assistance of Hannah Maxwell is highly appreciated.

References

Abiodun BJ, Pal J, Afiesimama EA, Gutowski WJ, Adedoyin A (2008) Simulation of West Africa monsoon using the RegCM3. Part II: impact of deforestation and desertification. Theor Appl Clim 93:245–261. doi:10.1007/s00704-007-0333-1

Alexander M, Bladé I, Newman M, Lanzante JR, Lau N-C, Scott JD (2002) The atmospheric bridge: the influence of ENSO teleconnections on air–sea interaction over the global oceans. J Clim 15:2205–2231. doi:10.1175/1520-0442

Anderson CJ, Arritt RW, Takle ES, Pan Z, Gutowski WJ, da Silva R, PIRCS modelers (2003) Hydrologic processes in regional climate model simulations of the central United States flood of June-July 1993. J Hydrometeorol 4:584–598

Annamalai H, Slingo JM (2001) Active/break cycles: diagnosis of the intraseasonal variability of the Asian summer monsoon. Clim Dyn 18:85–102

Annamalai H, Liu P, Xie S-P (2005) Southwest Indian ocean SST variability: its local effect and remote influence on Asian monsoons. J Clim 18:4150–4167

Arriaga-Ramírez S, Cavazos T (2010) Regional trends of daily precipitation indices in Northwest Mexico and the Southwest United States. J Geophys Res Atmos 115(D14111)

Barlow M, Salstein D (2006) Summertime influence of the Madden-Julian oscillation on daily rainfall over Mexico and central America. Geophys Res Lett 33:6

Barreiro M, Diaz N (2011) Land-atmosphere coupling in El Niño influence over South America. Atmos Sci Lett. doi:10.1002/asl.348

Barsugli J, Anderson C, Smith JB, Vogel JM (2009) Options for improving climate modeling to assist water utility planning for climate change. Water Utility Climate Alliance, 146 pp. www.wucaonline.org/assets/pdf/actions_whitepaper_120909.pdf

Batisani N, Yarnal B (2010) Rainfall variability and trends in semi-arid Botswana: implications for climate change adaptation policy. Appl Geogr 30:483–489

Bechtold P, Bazile E, Guichard F, Mascart P, Richard E (2001) A mass flux convection scheme for regional and global models. Q J R Meteorol Soc 127:869–886

Becker BD, Slingo JM, Ferranti L, Molteni F (2001) Seasonal predictability of the Indian summer monsoon: what role do land surface conditions play? Mausam 52:175–190

Behera SK, Yamagata T (2001) Subtropical SST dipole events in the southern Indian ocean. Geophys Res Lett 28:327–330

Berbery EH, Silva Dias MA, Baethgen W, Barros V, Collischonn W, Failache N, Lettenmaier D, Marengo J, Menendez A, Monte Domecq R, Moraes O, Morales C, Paruelo J, Seluchi M, Silva Dias P, Terra R, Ereño C, Meitín J (2005) La Plata Basin (LPB) Continental Scale Experiment – Implementation plan. Available from http://www.eol.ucar.edu/projects/lpb/

Bessafi M, Wheeler MC (2006) Modulation of south Indian ocean tropical cyclones by the Madden-Julian oscillation and convectively-coupled equatorial waves. Mon Weather Rev 134:638–656

Betts AK, Jakob C (2002) Study of diurnal cycle of convective precipitation over Amazonia using a single column model. J Geophys Res 107(D23):4732. doi:10.1029/2002JD002264

Biasutti M, Giannini A (2006) Robust Sahel drying in response to late 20th century forcings. Geophys Res Lett 33:L11706. doi:10.1029/2006GL026067

Borgaonkar HP, Sikder AB, Ram S, Pant GB (2010) El Nino and related monsoon drought signals in 523-year-long ring width records of teak (Tectona grandis L.F.) trees from south India. Palaeogeogr Palaeoclimatol Palaeoecol 285:74–84. doi:10.1016/j.palaeo.2009.10.026

Brito-Castillo L, Diaz-Castro S, Salinas-Zavala CA, Douglas AV (2003) Reconstruction of long-term winter streamflow in the Gulf of California continental watershed. J Hydrol 278:39–50

Carvalho C, Jones LMV, Silva AE, Jones C, Liebmann B, Silva Dias PL (2010) The south American monsoon system and the 1970s climate transition. Int J Climatol. doi:10.1002/joc.2147

Chiessi CM, Mulitza S, Patzold J, Wefer G, Marengo JA (2009) Possible impact of the Atlantic multidecadal oscillation on the South American summer monsoon. Geophys Res Lett 36

Chikamoto Y, Kimoto M, Ishii M, Mochizuki T, Sakamoto TT, Tatebe H, Komuro Y, Watanabe M, Nozawa T, Shiogama H, Mori M, Yasunaka S, Imada Y (2012) An overview of decadal climate predictability in a multi-model ensemble by climate model MIROC. Clim Dyn. doi:10.1007/s00382-012-1351-y

Christensen et al (2007) Regional climate projections. In: Climate change (2007) The physical science basis. In: Solomon S, Qin D, Manning M, Chen Z, Marquis M, Averyt KB, Tignor M, Miller HL (eds) Contribution of working group I to the fourth assessment report of the intergovernmental panel on climate change. Cambridge University Press, Cambridge/New York

CLIVAR (2010) CLIVAR Scientific frontiers and imperatives. Available at www.clivar.org

Collini EA, Berbery EH, Barros V, Pyle M (2008) How does soil moisture influence the early stages of the South American monsoon? J Clim 21:195–213

Cook C, Reason CJC, Hewitson BC (2004) Wet and dry spells within particular wet and dry summers in the south African summer rainfall region. Clim Res 26:17–31

Dai A (2006) Precipitation characteristics in eighteen coupled climate models. J Clim 19:4605–4630

de Szoeke SP, Fairall CW, Wolfe DE, Bariteau L, Zuidema P (2010) Surface flux observations on the Southeastern tropical Pacific Ocean and attribution of SST errors in coupled ocean–atmosphere models. J Clim 23:4152–4174

Delsol T, Shukla J (2012) Climate models produce skilful predictions of summer monsoon rainfall. Geophys Res Lett 39:L09703. doi:10.1029/2012GL051279

Déqué M, Piedelievre JP (1995) High resolution climate simulation over Europe. Clim Dyn 11(6):321–339. doi:10.1007/BF00215735

Dirmeyer PA, Schlosser CA, Brubaker KL (2009) Precipitation, recycling, and land memory: an integrated analysis. J Hydrometeorol 10:279–288

Doblas-Reyes FJ, Balmaseda MA, Weisheimer A, Palmer TN (2011) Decadal climate prediction with the European centre for medium-range weather forecasts coupled forecast system: impact of ocean observations. J Geophys Res 116, D19111

Douville H, Chauvin F, Broqua H (2001) Influence of soil moisture on the Asian and African monsoons. Part I: mean monsoon and daily precipitation. J Clim 14:2381–2403

Doyle ME, Barros VR (2002) Midsummer low-level circulation and precipitation in subtropical South America and related sea surface temperature anomalies in the South Atlantic. J Clim 15:3394–3410

Druyan LM, Feng J, Cook KH, Xue Y, Fulakeza M, Hagos S, Konaré A, Moufouma-Okia W, Rowell DP, Vizy EK (2010) The WAMME regional model intercomparison study. Clim Dyn 35(1):175–192. doi:10.1007/s00382-009-0676-7

Enfield DB (1996) Relationships of inter-American rainfall to tropical Atlantic and Pacific SST variability. Geophys Res Lett 23:3305–3308

Enfield D, Wang C, Lee S-K (2009) Cross-hemispheric interactions between variability of the American monsoon systems regions. VAMOS! Newsletter 5:11–14. Available at http://www.clivar.org/organization/vamos

Englehart PJ, Douglas AV (2006) Defining intraseasonal rainfall variability within the North American monsoon. J Clim 19:4243–4253

Fu CB, Wang S, Xiong Z, Gutowski WJ, Lee D-K, McGregor JL, Sato Y, Kato H, Kim J-W, Su M-S (2005) Regional climate model intercomparison project for Asia. Bull Am Meteorol Soc 86:257–266

Funk C, Dettinger MD, Michaelsen JC, Verdin JP, Brown ME, Barlow M, Hoell A (2008) Warming of the Indian ocean threatens Eastern and Southern African food security but could be mitigated by agricultural development. Proc Natl Acad Sci USA 105:11081–11086

Gadgil S, Kumar KR (2006) The Asian monsoon – agriculture and economy. In: Wang B (ed) The Asian monsoon, 844 pp. Springer, Berlin, pp 651–683

GEWEX (2011) GEWEX Imperatives. Available at www.gewex.org

Giannini A, Kushnir Y, Cane MA (2001) Interannual variability of Caribbean rainfall, ENSO, and the Atlantic Ocean. J Clim 13:297–311

Giannini A, Saravanan R, Chang P (2003) Oceanic forcing of Sahel rainfall on interannual to interdecadal time scales. Science 302:1027–1030

Giannini A, Biasutti M, Held IM, Sobel AH (2008) A global perspective on African climate. Clim Chang 90:359–383

Giorgi F, Mearns LO (1999) Introduction to special section: regional climate modeling revisited. J Geophys Res 104(D6):6335–6352. doi:10.1029/98JD02072

Giorgi F, Jones C, Asrar G (2009) Addressing climate information needs at the regional level: the CORDEX framework. WMO Bull 58(3):75–179

Gochis D (2011) A view forward for North American Monsoon research. CLIVAR Exchanges 16, 1(55):18–22

Goddard L, Aitchellouche Y, Baethgen W, Dettinger M, Graham R, Hayman P, Kadi M, Martínez R, Meinke H (2010) Providing seasonal-to-interannual climate information for risk management and decision-making. Procedia Environ Sci 1:81–101

Goswami BN (2005) South Asian monsoon (Chap 2). In: Lau WKM, Waliser DE (eds). Intraseasonal variability in the atmosphere-ocean climate system. Springer-Praxis, Berlin/Heidelberg, pp 19–61

Goswami BN (2006) The Asian monsoon: interdecadal variability. In: Chang CP, Wang B, Lau NC (eds) The global monsoon system: research and forecast. World Meteorological Organization (WMO)/TD No 1266. www.wmo.int

Goswami BN, Ajaya Mohan RS (2001) Intra-seasonal oscillations and inter-annual variability of the Indian summer monsoon. J Clim 14:1180–1198

Goswami BN, Xavier PK, Sengupta D (2003) Clustering of low pressure systems during the Indian summer monsoon by intraseasonal oscillations. Geophys Res Lett 30:1431. doi:10.1029/2002GL016734

Goswami BN, Venugopal V, Sengupta D, Madhusoodanan MS, Xavier PK (2006) Increasing trend of extreme rain events over India in a warming environment. Science 314:1442–1445

Gottschalck JM, Wheeler K, Weickmann F, Vitart N, Savage H, Lin H, Hendon D, Waliser K, Sperber M, Nakagawa C, Prestrelo M, Flatau WH (2010) A framework for assessing operational model MJO forecasts: a project of the CLIVAR madden-Julian oscillation working group. Bull Am Meteorol Soc 91:1247–1258

Grabowski WW, Bechtold P, Cheng A, Forbes R, Halliwell C, Khairoutdinov M, Lang S, Nasuno T, Petch J, Tao W-K, Wong R, Wu X, Xu K-M (2006) Daytime convective development over land: a model intercomparison based on LBA observations. Q J R Meteorol Soc 132:317–344

Grimm AM, Reason CJC (2011) Does the South American monsoon affect African rainfall? J Clim 24:1226–1238

Grimm A, Pal J, Giorgi F (2007) Connection between spring conditions and peak summer monsoon rainfall in South America: role of soil moisture, surface temperature, and topography in Eastern Brazil. J Clim 20:5929–5945

Groisman PY, Knight RW (2008) Prolonged dry episodes over the conterminous United States: New tendencies emerging during the last 40 years. J Clim 21(9):1850–1862
Gualdi S et al (2011) Future climate projections. In: Navarra A, Tubiana L (eds) Regional assessment of climate change in the Mediterranean. Springer, Dordrecht
Guichard F, Petch JC, Redelsperger J-L, Bechtold P, Chaboureau J-P, Cheinet S, Grabowski W, Grenier H, Jones CG, Köhler M, Piriou J-M, Tailleux R, Tomasini M (2004) Modeling the diurnal cycle of deep precipitating convection over land with cloud-resolving models and single-column models. Q J R Meteorol Soc 130:3139–3172
Guo Z, Dirmeyer P, DelSole T, Koster R (2012) Rebound in atmospheric predictability and the role of the land surface. J Clim 25:4744–4749
Gutzler D, Kim H-K, Higgins W, Juang H, Kanamitsu M, Mitchell K, Ritchie L, Schemm J-K, Schubert S, Yang R, Mo K, Song Y, Pegion P (2003) The North American monsoon Model Assessment Project (NAMAP). NCEP/Climate prediction center ATLAS no. 11, 32pp
Hachigonta S, Reason CJC (2006) Interannual variability in dry and wet spell characteristics over Zambia. Clim Res 32:49–62
Hagos SM, Cook KH (2008) Ocean warming and late-twentieth-century Sahel drought and recovery. J Clim 21:3797–3814
Hastenrath S, Nicklis A, Greischar L (1993) Atmospheric hydrospheric mechanisms of climate anomalies in the western equatorial Indian Ocean. J Geophys Res 98(C 11):20219–20235
Hendon HH, Thompson DWJ, Wheeler MC (2007) Australian rainfall and temperature variations associated with the southern hemisphere annular mode. J Clim 20:2452–2467
Hendon HH, Lim E, Liu G (2012) The role of air-sea interaction for seasonal prediction of Australian summer monsoon rainfall. J Clim 25:1278–1290
Hewitson BC, Crane R (1996) Climate downscaling: techniques and application. Clim Res 7:85–95
Hewitson BC, Crane RG (2006) Consensus between GCM climate change projections with empirical downscaling: precipitation downscaling over South Africa. Int J Climatol 26:1315–1337
Higgins RW, Ahijevych D, Amador J (2006) The NAME 2004 field campaign and modeling strategy. Bull Am Meteorol Soc 87(1):79–94
Hoerling M, Hurrell J, Eischeid J, Phillips A (2006) Detection and attribution of twentieth-century Northern and Southern African rainfall change. J Clim 19:3989–4008
Hohenegger C, Brockhaus P, Schär C (2008) Towards climate simulations at cloud-resolving scales. Meteorol Z 17(4):383–394(12)
Hu Q, Feng S (2008) Variation of the North American summer monsoon regimes and the Atlantic multidecadal oscillation. J Clim 21:2371–2383
Jakob C (2010) Accelerating progress in global atmospheric model development through improved parameterizations: challenges, opportunities, and strategies. Bull Am Meterol Soc 91:869–875
Janicot S, Caniaux G, Chauvin F, de Coëtlogon G, Fontaine B, Hall N, Kiladis G, Lafore J-P, Lavaysse C, Lavender SL, Leroux S, Marteau R, Mounier F, Philippon N, Roehrig R, Sultan B, Taylor CM (2010) Intraseasonal variability of the West African monsoon. Atmos Sci Lett. doi:10.1002/asl.280
Joly M, Voldoire A, Douville H, Terray P, Royer J-F (2007) African monsoon teleconnections with tropical SSTs: validation and evolution in a set of IPCC4 simulations. Clim Dyn 29:1–20
Jones C, Schemm J (2000) The influence of intraseasonal variations on medium- to extended-range weather forecasts over South America. Mon Weather Rev 128:486–494
Jones C, Giorgi F, Asrar G (2011) The coordinated regional downscaling experiment: an international downscaling link to CMIP5. CLIVAR exchanges no. 56 16(2):34–40
Junquas C, Vera CS, Li L, Le Treut H (2012) Summer precipitation variability over Southeastern South America in a global warming scenario. Clim Dyn 38(9):1867–1883
Kain JS, Fritsch JM (1990) A one-dimensional entraining/detraining plume model and its application in convective parameterization. J Atmos Sci 47(23):2784–2802
Kang I-S, Kim HM (2010) Assessment of MJO predictability for boreal winter with various statistical and dynamical models. J Clim 23:2368–2378

Kang I-S, Shukla J (2006) Dynamical seasonal prediction and predictability of the Asian monsoon (Chap 15). In: Wang B (ed) The Asian monsoon. Praxis/Springer, Berlin/Heidelberg, pp 585–612
Kayano MT, Andreoli RV (2007) Relations of South American summer rainfall interannual variations with the Pacific decadal oscillation. Int J Climatol 27:531–540
Kelly P, Mapes B (2010) Land surface heating and the North American monsoon anticyclone: model evaluation from diurnal to seasonal. J Clim 23:4096–4106
Khairoutdinov M, Randall D (2006) High-resolution simulation of shallow-to-deep convection transition over land. J Atmos Sci 63(12):3421–3436
Kim D, Sperber K, Stern W, Waliser D, Kang I-S, Maloney E, Wang W, Weickmann K, Benedict J, Khairoutdinov M, Lee M-I, Neale R, Suarez M, Thayer-Calder K, Zhang G (2009) Application of MJO simulation diagnostics to climate models. J Clim 22:6413–6436. doi:10.1175/2009JCLI3063.1
Kitoh A (2011) Impact of climate change on Asian monsoon characteristics (Chap 32). In: Change CP et al (eds) The global monsoon system: research and forecast, 2nd edn. World Scientific Publication Company, Singapore, pp 557–568
Kitoh A (2012) Projections of future climate change in the Asian monsoon region. In: OCHAMP-12, 21–25 Feb 2012, Pune. Available in www.tropmet.res.in/ochamp/
Knight JR, Folland CK, Scaife AA et al (2006) Climate impacts of the Atlantic multidecadal oscillation. Geophys Res Lett 33:L17706. doi:10.1029/2006GL026242
Kniveton DR, Layberry R, William CJR, Peck M (2009) Trends in the start of the wet season over Africa. Int J Climatol 29:1216–1225
Konare A, Zakey AS, Solmon F, Giorgi F, Rauscher S, Ibrah S, Bi X (2008) A regional climate modeling study of the effect of desert dust on the West African monsoon. J Geophys Res 113:D12206. doi:10.1029/2007JD009322
Koster RD et al (2004) Regions of strong coupling between soil moisture and precipitation. Science 305:1138–1140
Koster RD et al (2011) The second phase of the global land-atmosphere coupling experiment: soil moisture contributions to subseasonal forecast skill. J Hydrometeorol 15:805–822
Kothawale DR, Rupa KK (2005) On the recent changes in surface temperature trends over India. Geophys Res Lett 32:L18714. doi:10.1029/2005GL023528
Krishna Kumar K, Kamala K, Rajagopalan B, Hoerling MP, Eischeid JK, Patwardhan SK, Srinivasan G, Goswami BN, Nemani R (2011) The once and future pulse of Indian monsoonal climate. Clim Dyn. doi:10.1007/s00382-010-0974-0
Krishnan R, Kumar V, Sugi M, Yoshimura J (2009) Internal feedbacks from monsoon-midlatitude interactions during droughts in the Indian summer monsoon. J Atmos Sci 66(3):553–578
Kucharski F, Bracco A, Yoo JH, Molteni F (2008) Atlantic forced component of the Indian monsoon interannual variability. Geophys Res Lett 35, L04706. doi:10.1029/2007GL033037
Lafore J-P (2007) La mousson africaine. Atmosphériques 33:8–9
Landman WA, Beraki A (2012) Multi-model forecast skill for mid-summer rainfall over Southern Africa. Int J Climatol 32:303–314. doi:10.1002/joc.2273
Landman WA, DeWitt D, Lee D, Beraki A, Lotter D (2012) Seasonal rainfall prediction skill over South Africa: one-versus two-tiered forecasting systems. Weather Forecast 27(2):489–501. doi:10.1175/WAF-D-11-00078.1
Lavaysse C, Flamant C, Janicot S, Parker DJ, Lafore J-P, Sultan B, Pelon J (2009) Seasonal evolution of the West African heat low: a climatological perspective. Clim Dyn 33(2–3):313–330. doi:10.1007/s00382-009-0553-4
Lavender SL, Matthews AJ (2009) Response of the West African monsoon to the Madden–Julian oscillation. J Clim 22:4097–4116
Lee S-J, Berbery EH (2012) Land cover change effects on the climate of the La Plata Basin. J Hydrometeorol 13:84–102
Lee M-I, Schubert SD, Suarez MJ, Bell TL, Kim K-M (2007) Diurnal cycle of precipitation in the NASA seasonal to interannual prediction project atmospheric general circulation model. J Geophys Res 112:D16111. doi:10.1029/2006JD008346
Lemos MCTJ, Finan RW, Fox DRN, Tucker J (2002) The use of seasonal climate forecasting in policymaking: lessons from Northeast Brazil. Clim Chang 55:479

Lenters JD, Cook KH (1999) Summertime precipitation variability over South America: role of the large-scale circulation. Mon Weather Rev 127:409–431

Liebmann B, Kiladis G, Vera CS, Saulo AC (2004) Subseasonal variations of rainfall in South America in the vicinity of the low- level jet east of the Andes and comparison to those in the South Atlantic convergence zone. J Clim 17:3829–3842

Liu C, Haines K, Iwi A, Smith D (2012) Comparing t he UK met office climate prediction system DePreSys with idealized predictability in the HadCM3 model. Q J R Meteorol Soc 138:81–90

Losada T, Rodríguez-Fonseca B, Janicot S, Gervois S, Chauvin F, Ruti P (2009) A multi-model approach to the Atlantic equatorial mode: impact on the West African monsoon. Clim Dyn 35:29–43

Losada T, Rodríguez-Fonseca B, Mohino E, Bader J, Janicot S, Mechoso CR (2012) Tropical SST and Sahel rainfall: a non-stationary relationship. Geophys Res Lett 39:L12705. doi:10.1029/2012GL052423

Lu J, Vecchi GA, Reichler T (2007) Expansion of the Hadley cell under global warming. Geophys Res Lett 34(L06805)

Ma H-Y, Mechoso CR (2007) Submonthly variability in the South American monsoon system. J Meteorol Soc Jpn 85:385–401

Mackellar N, Hewitson B, Tadross MA (2010) Effects of vegetation map change in MM5 simulations of Southern Africa's summer climate. Int J Climatol 29:885–898

Maloney ED, Hartmann DL (2000) Modulation of Eastern North Pacific hurricanes by the Madden-Julian oscillation. J Clim 13:1451–1460

Mani NJ, Suhas E, Goswami BN (2009) Can global warming make Indian monsoon weather less predictable? Geophys Res Lett 36(L08811):1–5. doi:10.1029 /2009GL037989

Marengo JA, Calvalcanti IFA, Satyamurty P, Trosnikov I, Nobre CA, Bonatti JP, Camargo H, Sampaio G, Sanches MB, Manzi AO, Castro CAC, D'Almeida C, Pezzi LP, Candido L (2003) Assessment of regional seasonal rainfall predictability using the CPTEC/COLA atmospheric GCM. Clim Dyn 5–6:459–475

Marengo JA, Ambrizzi T, Rocha RP, Alves LM, Cuadra SV, Valverde MC, Ferraz SET, Torres RR, Santos DC (2009) Future change of climate in South America in the late XXI century: intercomparison of scenarios from three regional climate models. Clim Dyn. doi:10.1007/s00382-009-0721-6

Marengo JA, Chou SC, Kay G, Alves LM, Pesquero JF, Soares WR, Santos DC, Lyra A, Sueiro G, Betts R, Chagas DJ, Gomes JL, Bustamante JF, Tavares P (2011) Development of regional future climate change scenarios in South America using the Eta CPTEC/HadCM3 climate change projections: climatology and regional analyses for the Amazon, São Francisco and the Paraná River Basins. Clim Dyn. doi:10.1007/s00382-011-1155-5

Marengo JA, Liebmann B, Grimm AM, Misra V, Silva Dias PL, Cavalcanti IFA, Carvalho LMV, Berbery EH, Ambrizzi T, Vera CS, Saulo AC, Nogues-Paegle J, Zipser E, Seth A, Alves LM (2012) Recent developments on the South American monsoon system. Int J Climatol 32:1–21

Matthews AJ (2004) Intraseasonal variability over tropical Africa during northern summer. J Clim 17:2427–2440

Mearns LO, Gutowski WJ, Jones R, Leung L-Y, McGinnis S, Nunes AMB, Qian Y (2009) A regional climate change assessment program for North America. EOS 90:311–312

Mearns LO, Arritt R, Biner S, Bukovsky MS, McGinnis S, Sain S, Caya D, Correia J Jr, Flory D, Gutowski W, Takle ES, Jones R, Leung R, Moufouma-Okia W, McDaniel L, Nunes AMB, Qian Y, Roads J, Sloan L, Snyder M (2012) The North American regional climate change assessment program: overview of phase I results. Bull Am Meteorol Soc 93(9):1337–1362

Mechoso CR (2000) Introduction to VAMOS. CLIVAR Exch 5(2):4–6

Meehl GA, Tebaldi C, Teng H, Peterson TC (2007a) Current and future U.S. weather extremes and El Niño. Geophys Res Lett 34, L20704. doi:10.1029/2007GL031027

Meehl GA, Covey C, Taylor KE, Delworth T, Stouffer RJ, Latif M, McAvaney B, Mitchell JFB (2007b) The WCRP CMIP3 multimodel dataset: a new era in climate change research. Bull Am Meteor Soc 88:1383–1394

Meehl GA, Arblaster JM, Collins WD (2008) Effects of black carbon aerosols on the Indian monsoon. J Climate 21(12):2869–2882. doi:10.1175/2007JCLI1777.1

Meehl GA, and Coauthors (2009) Decadal prediction. Bull Am Meteor Soc 90:1467–1485
Menendez CG et al (2010) Downscaling extreme month-long anomalies in Southern South America. Clim Chang 98(3–4):379–403
Meywerk J, Ramanathan V (1999) Observations of the spectral clear-sky aerosol forcing over the tropical Indian Ocean. J Geophys Res 104:24 359–24 370
Mohino E, Janicot S, Bader J (2010) Sahel rainfall and decadal to multi-decadal sea surface temperature variability. Clim Dyn. doi:10.1009/s00382-010-0867-2
Mohino E, Rodriguez-Fonseca B, Mechoso CR, Gervois S, Ruti P, Chauvin F (2011) Impacts of the tropical Pacific/Indian Oceans on the seasonal cycle of the West African monsoon. J Clim 24:3878–3891
Monaghan AJ, Rife DL, Pinto JO, Davis CA, Hannan JR (2010) Global precipitation extremes associated with diurnally varying low-level jets. J Clim 23(19):5065–5084. doi:10.1175/2010JCLI3515.1
Mounier F, Janicot S (2004) Evidence of two independent modes of convection at intraseasonal timescale in the West African monsoon. Geophys Res Lett 31, L16116
Mounier F, Janicot S, Kiladis GN (2008) The West African monsoon dynamics. Part III: the quasi-biweekly zonal dipole. J Clim 21:1911–1928
Ndiaye O, Ward MN, Thiaw WM (2011) Predictability of seasonal Sahel rainfall using GCMs and lead-time improvements through the use of a coupled model. J Clim 24(7):1931–1949. doi:10.1175/2010JCLI3557.1
Neena JM, Suhas E, Goswami BN (2011) Leading role of internal dynamics in the 2009 Indian summer monsoon drought. J Geophys Res. doi:10.1029/2010JD015328
Nikulin G, Jones C, Giorgi F, Asrar G, Büchner M, Cerezo-Mota R, Christensen O, Déqué M, Fernandez J, Haensler A, van Meijgaard E, Samuelsson P, Sylla M, Sushama L (2012) Precipitation climatology in an ensemble of CORDEX-Africa regional climate simulations. J Clim 25:6057–6078. doi:10.1175/JCLI-D-11-00375.1
Nogués-Paegle J, Mo K-C (1997) Alternating wet and dry conditions over South America during summer. Mon Weather Rev 125:279–291
Nogués-Paegle J, Mo K-C (2002) Linkages between summer rainfall variability over South America and sea surface temperature anomalies. J Clim 15:1389–1407
Ogallo LJ (1988) Relationship between seasonal rainfall in East Africa and Southern Oscillation. J Climatol 8:34–43
Pasquero C, Emanuel K (2008) Tropical cyclones and transient upper-ocean warming. J Clim 21(1):149–162
Philippon NE, Mougin LJ, Frison P-L (2005) Analysis of the linkages between rainfall and land surface conditions in the West African monsoon through CMAP, ERS-WSC, and NOAA-AVHRR data. J Geophys Res 110:D24115. doi:10.1029/2005JD006394
Rajeevan M, Sridhar L (2008) Inter-annual relationship between Atlantic sea surface temperature anomalies and Indian summer monsoon. Geophys Res Lett 35:L21704. doi:10.1029/2008GL036025
Rajeevan M, Unnikrishnan C, Preethi B (2011) Evaluation of the ENSEMBLES multi-model seasonal forecasts of Indian summer monsoon variability. Clim Dyn. doi:10.1007/s00382-011-1061-x
Ramanathan V, Crutzen PJ, Kiehl JT, Rosenfield D (2001) Aerosols, climate, and the hydrological cycle. Science 7, 294(5549):2119–2124. doi:10.1126/science.1064034
Rashid HA, Hendon HH, Wheeler MC, Alves O (2010) Prediction of the Madden-Julian Oscillation with the POAMA dynamical prediction system. Clim Dyn. doi:10.1007/s00382-010-0754-x
Re M, Barros VR (2009) Extreme rainfalls in SE South America. Clim Chang 96(1–2):119–136
Reason CJC (2001) Subtropical Indian Ocean SST dipole events and Southern African rainfall. Geophys Res Lett 28:225–228
Reason CJC (2002) Sensitivity of the Southern African circulation to dipole SST patterns in the South Indian Ocean. Int J Climatol 22:377–393
Reason CJC, Mulenga HM (1999) Relationships between South African rainfall and SST anomalies in the South West Indian Ocean. Int J Climatol 19:1651–1673
Reason CJC, Hachigonta S, Phaladi RF (2005) Interannual variability in rainy season characteristics over the Limpopo region of Southern Africa. Int J Climatol 25:1835–1853

Reason CJC, Landman W, Tennant W (2006) Seasonal to decadal prediction of Southern African climate and its links with variability of the Atlantic Ocean. Bull Am Meteorol Soc 87:941–955
Redelsperger J-L, Diedhiou A, Janicot S, Sonneville A, Thorncroft C (2010) 2nd international science plan of AMMA. (http://www.amma-international.org/IMG/pdf/ISP2_v2.pdf)
Rio C, Hourdin F, Grandpeix J-Y, Lafore J-P (2009) Shifting the diurnal cycle of parameterized deep convection over land. Geophys Res Lett 36:L07809. doi:10.1029/2008GL036779
Robertson AW, Mechoso CR (2000) Interannual and interdecadal variability of the South Atlantic convergence zone. Mon Weather Rev 128:2947–2957
Rodríguez-Fonseca B, Polo I, García-Serrano J, Losada T, Mohino E, Mechoso CR, Kucharski F (2009) Are Atlantic Niños enhancing Pacific ENSO events in recent decades? Geophys Res Lett 36:L20705. doi:10.1029/2009GL040048
Rodríguez-Fonseca B, Janicot S, Mohino E, Losada T, Bader J, Caminade C, Chauvin F, Fontaine B, García-Serrano J, Gervois S, Joly M, Polo I, Ruti P, Roucou P, Voldoire A (2011) Interannual and decadal SST-forced responses of the West African monsoon. Atmos Sci Lett 12:67–74
Rogers RF, Fritsch JM (1996) A general framework for convective trigger functions. Mon Weather Rev 124(11):2438–2452
Rouault M, Florenchie P, Fauchereau N, Reason CJC (2003) South East Atlantic warm events and Southern African rainfall. Geophys Res Lett 30(5):8009. doi:10.1029/2002GL014840
Saji NH, Goswami BN, Vinayachandran PN, Yamagata T (1999) A dipole mode in the tropical Indian Ocean. Nature 401:360–363
Saulo AC, Ferreira L, Nogués-Paegle J, Seluchi ME, Ruiz J (2010) Land–atmosphere interactions during a Northwestern Argentina low event. Mon Weather Rev 138:2481–2498. doi:10.1175/2010MWR3227.1
SDWG (2012) Community earth systems model societal dimensions working group. [http://www.cesm.ucar.edu/working_groups/Societal/]
Shen X, Kimoto M, Sumi A (1998) Role of land surface processes associated with interannual variability of broad-scale Asian summer monsoon as simulated by the CCSR/NIES AGCM. J Meteorol Soc Jpn 76(2):217–236
Shukla J, Hagedorn R, Hoskins B, Kinter J, Marotzke J, Miller M, Palmer TN, Slingo J (2009) Strategies: revolution in climate prediction is both necessary and possible: a declaration at the world modelling summit for climate prediction. Bull Am Meteorol Soc 90:175–178
Silvestri GE, Vera CS (2003) Antarctic Oscillation signal on precipitation anomalies over Southeastern South America. Geophys Res Lett 30:2115. doi:10.1029/2003GL018277
Silvestri GE, Vera CS (2009) Nonstationary impacts of the southern annular mode on southern hemisphere climate. J Clim 22:6142–6148
Slingo J, Spencer H, Hoskins B, Berrisford P, Black E (2005) The meteorology of the Western Indian Ocean, and the influence of the East African highlands. Philos Trans R Soc A 363(1826): 25–42. doi:10.1098/rsta.2004.1473
Soares WR, Marengo JA (2009) Assessments of moisture fluxes east of the Andes in South America in a global warming scenario. Int J Climatol 29(10):1395–1414
Sperber KR, Yasunari T (2006) The 1st Pan-WCRP workshop on monsoon climate systems: toward better prediction of the monsoons. University of California, Irvine, 15–17 June 2005. WCRP Informal Report No.2/2006
Stechmann SN, Stevens B (2010) Multiscale models for cumulus cloud dynamics. J Atmos Sci 67:3269–3285. doi:10.1175/2010JAS3380.1
Sultan B, Janicot S (2003) The West African monsoon dynamics, part II: the "pre-onset" and the "onset" of the summer monsoon. J Clim 16:3407–3427
Tadross M, Suarez P, Lotsch A, Hachigonta S, Mdoka M, Unganai L, Lucio F, Kamdonyo D, Muchinda M (2009) Growing-season rainfall and scenarios of future change in Southeast Africa: implications for cultivating maize. Clim Res 40:147–161
Takle ES, Gutowski WJ, Arritt RA, Pan Z, Anderson CJ, da Silva RR, Caya D, Chen S-C, Christensen JH, Hong S-Y, Juang H-MH, Katzfey J, Lapenta WM, Laprise R, Lopez P, McGregor J, Roads JO (1999) Project to intercompare regional climate simulations (PIRCS): description and initial results. J Geophys Res 104:19,443–19,461

Takle ES, Roads J, Rockel B, Gutowski WJ, Arritt RW, Meinke I, Jones CG, Zadra A (2007) Transferability intercomparison: an opportunity for new insight on the global water cycle and energy budget. Bull Am Meteorol Soc 88:375–384. doi:10.1175/BAMS-88-3-375

Tall A, Mason SJ, van Aalst M, Suarez P, Ait-Chellouche Y, Diallo AA, Braman L (2012) Using seasonal climate forecasts to guide disaster management: the red cross experience during the 2008 West Africa floods. Int J Geophys. doi:10.1155/2012/986016

Tao WK, Moncrieff MW (2009) Multiscale cloud system modeling. Rev Geophys 47, Article No. RG4002. doi:10.1029/2008RG000276

Taylor CM, Parker DJ, Lloyd CR, Thorncroft CD (2005) Observations of synoptic-scale land surface variability and its coupling with the atmosphere. Q J R Meteorol Soc 131:913–938

Taylor KE, Stouffer RJ, Meehl GA (2012) An overview of CMIP5 and the experiment design. Bull Atmos Soc 4:485–498

Tebaldi C, Hayhoe K, Arblaster JM, Meehl GA (2006) Going to the extremes. An intercomparison of model simulated historical and future changes in extreme events. Clim Chang 79(3–4): 185–211

Todd MC, Washington R (2004) Climate variability in central equatorial Africa: influence from the Atlantic sector. Geophys Res Lett 31:L23202. doi:10.1029/2004GL020975

Turner AG, Slingo JM (2009) Subseasonal extremes of precipitation and active-break cycles of the Indian summer monsoon in a climate change scenario. Q J R Meteorol Soc 133:548–567

Ventrice MJ, Thorncroft CD, Roundy PE (2011) The Madden Julian Oscillation's influence on African easterly waves and downstream tropical cyclogenesis. Mon Weather Rev 139:2704–2722

Vera C, Baez J, Douglas M, Emmanuel CB, Marengo J, Meitin J, Nicolini M, Nogues-Paegle J, Paegle J, Penalba O, Salio P, Saulo C, Silva Dias MA, Silva Dias P, Zipser E (2006a) The South American Low-Level Jet Experiment (SALLJEX). Bull Am Meteorol Soc 87:63–77

Vera C, Higgins W, Amador J, Ambrizzi T, Garreaud R, Gochis D, Gutzler D, Lettenmaier D, Marengo J, Mechoso CR, Nogues-Paegle J, Silva Dias PL, Zhang C (2006b) A unified view of the American monsoon systems. J Clim 19:4977–5000

Vera C, Silvestri G, Liebmann B, Gonzalez P (2006c) Climate change scenarios for seasonal precipitation in South America from IPCC-AR4 models. Geophys Res Lett 33, L13707. doi:10.1029/2006GL025759

Vera C, Barange M, Dube OP, Goddard L, Griggs D, Kobysheva N, Odada E, Parey S, Polovina J, Poveda G, Seguin B, Trenberth K (2010) Needs assessment for climate information on decadal timescales and longer. Procedia Environ Sci 1(1):275–286

Wang B, Ding QH, Fu XH, Kang IS, Jin K, Shukla J, Doblas-Reyes F (2005) Fundamental challenges in simulation and prediction of summer monsoon rainfall. Geophys Res Lett 32(L15711)

Wang B, Lee J-Y, Shukla J, Kang I-S, Park C-K (2009) Advance and prospectus of seasonal prediction: assessment of APCC/CliPAS 14-model ensemble retrospective seasonal prediction (1980–2004). Clim Dyn 33:93–117

Webster PJ, Moore AM, Loschnigg JP, Leben RR (1999) Coupled ocean-atmosphere dynamics of the Indian Ocean during 1997–98. Nature 401:356–360

Wheeler MC, Hendon HH, Cleland S, Meinke H, Donald A (2009) Impacts of the MJO on Australian rainfall and circulation. J Clim 22:1482–1497

Wood R, Mechoso CR (2008) Southeastern pacific coupled climate field experiment. EOS Trans AGU 89(33):303

Wood R, Mechoso CR, Bretherton CS, Weller RA, Huebert B, Straneo F, Albrecht BA, Coe H, Allen G, Vaughan G, Daum P, Fairall C, Chand D, Gallardo Klenner L, Garreaud R, Grados C, Covert DS, Bates TS, Krejci R, Russell LM, de Szoeke S, Brewer A, Yuter SE, Springston SR, Chaigneau A, Toniazzo T, Minnis P, Palikonda R, Abel SJ, Brown WOJ, Williams S, Fochesatto J, Brioude J, Bower KN (2011) The VAMOS Ocean-Cloud-Atmosphere-Land Study Regional Experiment (VOCALS-REx): goals, platforms, and field operations. Atmos Chem Phys 11:627–654

Wu G, Liu Y, Wang T, Wan R, Liu X, Li W, Wang Z, Zhang Q, Duan A, Liang X (2007) The influence of mechanical and thermal forcing by the Tibetan Plateau on Asian climate. J Hydrometeorol 8:770–789

Wyant MC, Wood R, Bretherton CS, Mechoso CR, Bacmeister J, Balmaseda MA, Barrett B, Codron F, Earnshaw P, Fast J, Hannay C, Kaiser JW, Kitagawa H, Klein SA, Köhler M, Manganello J, Pan HL, Sun F, Wang S, Wang Y (2010) The PreVOCA experiment: modeling the lower troposphere in the Southeast Pacific. Atmos Chem Phys 10:4757–4774

Xie S-P, Miyama T, Wang Y, Xu H, de Szoeke SP, Justin R, Small O, Richards KJ, Mochizuki T, Awaji T (2007) A regional ocean–atmosphere model for Eastern Pacific climate: toward reducing tropical biases. J Clim 20(8):1504–1522

Xing YL, Majda AJ, Grabowski WW (2009) Efficient sparse space-time algorithms for superparameterization on mesoscales. Mon Weather Rev 137:4307–4324. doi:10.1175/2009MWR2858.1

Xue Y, de Sales F, Li W-P, Mechoso CR, Nobre CA, Juang H-M (2006) Role of land surface processes in South American monsoon development. J Clim 19:741–762. doi:http://dx.doi.org/10.1175/JCLI3667.1

Yang G-Y, Slingo J (2001) The diurnal cycle in the tropics. Mon Weather Rev 129(4):784–801

Yu W, Han WQ, Maloney ED, Gochis D, Xie SP (2011) Observations of eastward propagation of atmospheric intraseasonal oscillations from the Pacific to the Atlantic. J Geophys Res Atmos 116

Zamboni L, Kucharski F, Mechoso CR (2012) Seasonal variations of the links between the interannual variability over South America and the South Pacific. Clim Dyn 38:2115–2129

Zhou J, Lau WK-M (2001) Principal modes of interannual and decadal variability of summer rainfall over South America. Int J Climatol 21:1623–1644

Ziervogel G (2004) Targeting seasonal climate forecasts for integration into household level decisions: the case of smallholder farmers in Lesotho. Geogr J 170:6

Zuidema P, Chang P, Mechoso CR, Terray L (2011) Coupled ocean-atmosphere-land processes in the tropical Atlantic. CLIVAR Exchanges 16, 1(55):12–14

Attribution of Weather and Climate-Related Events

Peter A. Stott, Myles Allen, Nikolaos Christidis, Randall M. Dole, Martin Hoerling, Chris Huntingford, Pardeep Pall, Judith Perlwitz, and Dáithí Stone

Abstract Unusual or extreme weather and climate-related events are of great public concern and interest, yet there are often conflicting messages from scientists about whether such events can be linked to climate change. There is clear evidence that climate has changed as a result of human-induced greenhouse gas emissions, and that across the globe some aspects of extremes have changed as a result. But this does not imply that human influence has significantly altered the probability of occurrence or risk of every recently observed weather or climate-related event, or that such events are likely to become significantly more or less frequent in the future. Conversely, it is sometimes stated that it is impossible to attribute any individual

P.A. Stott (✉) • N. Christidis
Met Office/Hadley Centre, FitzRoy Road, Exeter EX1 3PB, UK
e-mail: peter.stott@metoffice.gov.uk; nikos.christidis@metoffice.gov.uk

M. Allen
School of Geography and the Environment, University of Oxford, South Parks Road, OX1 3QY Oxford, UK
e-mail: myles.allen@ouce.ox.ac.uk

R.M. Dole • M. Hoerling
NOAA Earth System Research Laboratory, 325 Broadway, Boulder, CO 80305, USA
e-mail: randall.m.dole@noaa.gov; martin.hoerling@noaa.gov

C. Huntingford
Centre for Ecology and Hydrology, Maclean Lane, Crowmarsh Gifford, Wallingford, Oxon OX10 8BB, UK
e-mail: chg@ceh.ac.uk

P. Pall
ETH Zurich, Institute for Atmospheric and Climate Science, Zurich, Switzerland

Lawrence Livermore National Laboratory, Berkeley Lab, MS 50B4230, 1 Cyclotron Road, Berkeley, CA 94720-8150, USA
e-mail: ppall@llbl.gov

G.R. Asrar and J.W. Hurrell (eds.), *Climate Science for Serving Society: Research, Modeling and Prediction Priorities*, DOI 10.1007/978-94-007-6692-1_12,

weather or climate-related event to a particular cause. Such a statement can be interpreted to mean that human-induced climate change could never be shown to be at least partly responsible for any specific weather event, either the probability of its occurrence or its magnitude. There is clear evidence from recent case studies that individual event attribution is a feasible, if challenging, undertaking.

We propose a way forward, through the development of carefully calibrated physically-based assessments of observed weather and climate-related events, to identify changed risk of such events attributable to particular factors including estimating the contributions of factors to event magnitude. Although such event-specific assessments have so far only been attempted for a relatively small number of specific cases, we describe research under way, coordinated as part of the international Attribution of Climate-related Events (ACE) initiative, to develop the science needed to better respond to the demand for timely, objective, and authoritative explanations of extreme events. The paper considers the necessary components of a prospective event attribution system, reviews some specific case studies made to date (Autumn 2000 UK floods, summer 2003 European heatwave, annual 2008 cool US temperatures, July 2010 Western Russia heatwave) and discusses the challenges involved in developing systems to provide regularly updated and reliable attribution assessments of unusual or extreme weather and climate-related events.

Keywords Attribution • Extreme weather • Climate variability • Climate change

1 Introduction

Episodes of extreme weather or unusual climatic conditions often cause major economic and human losses. In the aftermath of such events, the scientific community is often faced with the challenge of generating and communicating scientifically robust and timely information about their causes, quantifying their links to human-induced climate change, if any, and evaluating the prospects for better early warning of any enhanced risk of such events a month or more in advance.

In this paper we refer to such episodes under the general nomenclature of "weather and climate-related events". Such events are discrete episodes of extreme

J. Perlwitz
Physical Sciences Division, NOAA Earth System Research Laboratory,
325 Broadway, Boulder, CO 80305, USA

University of Colorado, Boulder, CO, USA
e-mail: judith.perlwitz@noaa.gov

D. Stone
Lawrence Livermore National Laboratory, Berkeley Lab, MS 50B4230, 1 Cyclotron Road, Berkeley, CA 94720-8150,, USA

University of Cape Town, Rondebosch, Western Cape, South Africa
e-mail: dstone@lbl.gov

weather or unusual climate conditions, often associated with deleterious impacts on society or natural systems, defined using some metric to characterize either the meteorological characteristics of the event or the consequent impact. Examples in the literature considering attribution of weather and climate-related events include the flooding of Vicarage Road in a suburb of Oxford, England (Allen 2003), the relatively cool annual mean temperatures across North America during 2008 (Perlwitz et al. 2009), and the extreme summer temperatures in Europe in 2003 (Stott et al. 2004). Events are often defined as occurring when some relevant threshold is crossed. For example, in their study of the 2003 European heatwave, Stott et al. (2004) chose a threshold for mean summer temperatures averaged over a large region of Europe that before 2003 had not been exceeded since the start of the instrumental record in 1851. An attribution analysis of the event in question usually requires a consideration of aspects of the atmospheric and ocean conditions in addition to that captured by a simple metric, but the latter serves to identify the occurrence of the event in question as a discrete meteorological episode. An event, therefore, has specificity in place and time, and event attribution is concerned with determining the changed probability of the event's occurrence, due to various factors, or is concerned with determining how various factors contribute to the intensity of the event.

The demand for information is often at its greatest in an event's immediate aftermath, requiring a rapid response from the scientific community. But apparently conflicting views can confuse the public, for example that all weather events are affected by climate change (Trenberth 2011), or that it is not possible to attribute an extreme weather event to climate change (an oft quoted statement as for example in an online answer to how many people have died from heat waves per ton of carbon emissions, http://wiki.answers.com/Q/How_many_people_have_died_from_heat_waves_per_ton_of_carbon_emissions).

The risk is that such potential confusion could undermine the credibility of the science of climate change. As a result there is a need for climate science to better inform decision makers, keenly aware of the need to protect life and property from the impacts of extreme weather and climate, and who wish to know whether any enhanced risk of such events could have been anticipated regardless of whether there has been a human influence (Stott et al. 2011), and whether they are likely to become more or less frequent owing to future climate change.

A main contributor to any adaptation strategy therefore are reliable assessments of the probabilities of such events, the likely magnitudes that climate-related events might be expected to acquire in a stationary climate, and how those might be changing in time due to human-induced climate change.

Climate science has already provided robust evidence that human influence, dominated by emissions of greenhouse gases, has altered the climate system (e.g. Hegerl et al. 2007; Stott et al. 2010), in such a way as to change the occurrence of extreme temperatures (Zwiers et al. 2011; Christidis et al. 2011a; Morak et al. 2011, 2012) and to lead to an intensification of heavy precipitation events over a large fraction of northern hemisphere continents (Min et al. 2011). However, notwithstanding the assertion that all weather events are affected to some extent by climate change (Trenberth 2012), and recognizing that all events are in fact affected by large-scale climate conditions whether natural or anthropogenic, it is clearly not the case that the occurrence of a specific weather or climate-related event should, *a priori,* be assumed attributable unambiguously to human influence. Many types of such events could happen in a stationary climate, and indeed have happened in pre-industrial times (Büntgen et al. 2011) and, some types of events are set to become less not more likely in future (Massey et al. 2012; Christidis and Stott 2012). Therefore, attribution assessments that relate to the specific weather or climate-related event in question are required before a conclusion can be drawn about the links between that event and climate change.

Whereas detection is concerned with determining whether or not climate or a system affected by climate has changed in such a way that the change's likelihood of occurrence by chance due to internal variability alone is small, attribution is the process of evaluating the relative contributions of multiple drivers of climate to a change or event with an assignment of statistical confidence (Hegerl et al. 2010). Consequently, all attribution analyses compare what has actually happened with what would have happened if a particular climate driver had not been present, and therefore requires models as well as observations (Hegerl and Zwiers 2011). An attribution analysis of a weather or climate-related event focuses on a specific region and time period and in the case of an extreme event focuses on the tails of distributions of variables (e.g. Stott et al. 2004). Attribution is inherently probabilistic and an attribution analysis applied to a specific event is no exception. While in most cases it is not possible to determine that the weather or climate-related event in question could only have happened because of a particular climate driver, it is possible to calculate how the climate driver has changed the likelihood of the event (Allen 2003).

While regular and reliable observational assessments of recent weather and climate-related events are regularly produced (e.g. the annual State of the Climate report published in the Bulletin of the American Meteorological Society; Blunden and Arndt 2012), such a regular attribution assessment has only recently been launched for a few selected events (Peterson et al. 2012). However, attribution of extreme weather and climate-related events in this way severely stretches the current state of climate science (Stott et al. 2012a). Furthermore, mistakenly attributing an increased risk of an extreme event to climate change could, if natural variability is playing the major role, lead to poor adaptation decisions; for example, through allocating resources toward preparing for a greater frequency of such events when in fact they have become less likely.

The overarching challenge for the community is to move beyond research-mode case studies and to develop systems that can deliver regular, reliable and timely assessments in the aftermath of notable weather and climate-related events, typically in the weeks or months following (and not many years later as is the case with some research-mode studies; e.g. Pall et al. 2011). In this paper, potential stakeholders are identified who could benefit from such assessments and illustrations provided of specific case studies that have been carried out so far. Progress in developing attribution systems is described. We draw lessons from the research work to date and propose some future research needs.

2 Relevance of Attribution Assessments of Weather and Climate-Related Events

Here we discuss six reasons why the development of reliable attribution assessments of weather and climate-related events could be relevant to different groups of stakeholders. We describe the potential benefits to climate science, to informing the public, to litigation, for adaptation to ongoing climate change, to geoengineering and for insurance.

2.1 Improved Climate Science

The regular assessment of weather and climate-related extremes is central to a rigorous process that seeks to improve understanding and to ensure the provision of better prediction systems. This involves identifying gaps in how such events are described from the existing observing systems, in better identifying the physical processes by which extremes arise, and in evaluating the suitability of existing models that are used for near-term predictions and long-term projections (Trenberth 2008).

Predicting, with known accuracy, the statistics of occurrence of weather or climate-related events that pose imminent and/or long-term threats to lives, property, and overall environmental health and sustainability represents a frontier of climate science (e.g. Smith et al. 2012; Knutti et al. 2010). At present a wide class of extreme events are not fully understood, including the physics of their causes and how those may link to human-induced climate change, nor are some classes of extremes well represented in many climate models (Seneviratne et al. 2012). The development of a carefully calibrated physically based assessment of observed weather and climate-related events must therefore occur in tandem with the appraisal of models, an ongoing evaluation of their suitability, and a quest to improve their representation of physical processes.

A key element in using a model for event attribution is to assess how reliably the model captures the real-world predictability of the events in question (Christidis et al. 2012). Attribution assessments are thus also central to the ongoing evaluation of predictability. A routine question regarding extremes is whether the event could have been anticipated a month or more in advance (Dole et al. 2011). Due to the chaotic nature of the climate system many extreme events are inherently unpredictable but this does not prevent attribution provided the model is capable of reliably capturing the statistics of the event. Central to the challenge of attribution is the identification of the forced climate change signal and therefore the extent to which the signal of the climate forcing can be identified above the noise of natural chaotic weather variability (Hegerl and Zwiers 2011). A grand scientific challenge is to improve capacities to quantify the climate change signal at regional scales, to determine the extent to which emerging trends are a forced signal or internal variability, and to assess how the probability of extreme events is sensitive to mean climate changes.

2.2 General Public

There is growing public awareness of climate change (Leiserowitz et al. 2012a) and that this might result in not just changes in averages, but that the frequency or intensity of extremes might vary (Sampei and Aoyagi-Usoi 2009; Leiserowitz et al. 2012b). Given climate model projections for changes in some extremes, including in some parts of the world more frequent and intense heatwaves and heavy daily rainfall, and in other places less frequent and intense cold spells and snowfall (Seneviratne et al. 2012), there is often considerable public interest in the possible link between a particular extreme weather or climate-related event (such as a very cold or hot season or year) and climate change, interest that is often at its greatest during or in the immediate aftermath of such events (Schiermeier 2011).

Recent examples in the UK include, in 2010, the coldest December in the UK national temperature record from 1910 (BBC 2011; Met Office 2010 http://www.metoffice.gov.uk/climate/uk/ineresting/dec2010/), with considerable adverse consequences including closed airports and schools and large economic losses (http://www.guardian.co.uk/uk/2010/dec/17/snow-closes-roads-airports-travel-misery), and the particularly hot spell the following Spring, that included the warmest April in the Central England temperature record stretching back to 1659 (http://www.metoffice.gov.uk/climate/uk/interesting/2011-spring/). In the aftermath, many people were interested in knowing whether such events are expected to happen more often in the future and whether they should be seen as a sign of a changing climate or an unusual occurrence of natural weather. In such circumstances the public often receives equivocal answers (Nature 2011).

Given that reliable attribution of extreme weather and climate-related events is important for the public's understanding of the effects of climate change, and can affect their willingness to support measures to mitigate greenhouse gas emissions (Schiermeier 2011), what is often lacking in the aftermath of an extreme weather event is a fully informed and timely response based on the best available climate science that enables the public and decision makers to put such an event into the context of both natural variability and climate change. Ideally one would wish such assessments to be issued regularly, applying a pre-defined methodology, and in a timely fashion, as in the case of weather forecasting (although not necessarily to be issued as frequently). This would limit the scope for ad-hoc structural biases, post-hoc reasoning and politicization of scientific information. While such rapid attribution assessments may be superseded by later more detailed analyses, they are nonetheless potentially of great value, and, like weather forecasts which offer great user value despite the remaining inherent forecast errors, timely probabilistic event attribution assessments should not necessarily be embargoed barring definitive conclusions, providing appropriate validation procedures are put in place and there is careful communication of the remaining uncertainties (Stott et al. 2012b).

As with weather forecasting, a regular attribution process would potentially lead to a continued improvement in reliability and could enhance the prospects for early warning of extreme events through enhanced understanding of predictability (Dole et al. 2011).

In addition, such a regular process meets an important need to be proactive in the attribution activity and not solely reactive to specific events, which may give an unwarranted impression of selectivity and bias (Stott et al. 2012a).

2.3 Litigation

The extent to which a specific damaging weather event could be blamed on greenhouse gas emissions is of relevance in legal contexts. There will almost certainly be attempts to seek redress for harm caused by emissions or, as in the case of six states before the US Supreme Court, to force power companies to cut their emissions of greenhouse gases under environmental protection legislation (Adam 2011). This would require robust evidence presented on the extent to which emissions can be linked to harmful effects (David 2003). Allen et al. (2007) argue that an objective operational attribution approach would be of considerable benefit to the courts since it would reduce the extent to which courts rely on expert judgment in legal contexts where the outcome often depends delicately on the exact question being asked. For example, even the same expert might agree that "human influence on climate played a substantial role in causing the European heatwave of 2003" and that "it is impossible to attribute any single weather event to human influence on climate", positions that could initially appear to be contradictory to a court. Therefore Allen et al. (2007) argue the need for agreed objective operational assessments that could, like routine operational weather forecasts, be used by courts. These would contribute as objective testimony requiring a more minor role for expert judgment in interpretation. A number of questions need to be considered by the legal community including what a court might consider as natural climate, over what time scales are damages relevant, and what levels of reliability, neutrality and acceptability are required for attribution assessments to be successfully used in legal contexts (Allen et al. 2007). These are difficult issues and not those traditionally considered by the climate modeling community.

2.4 Adaptation

The character of societal responses to extreme climate events often reveals details of the society's resilience and vulnerability, potentially exposing major "adaptability gaps" (IPCC et al. 2007). Activities designed for adaptation to climate change can be concerned with time frames ranging from the present through to many decades into the future, while those activities designed to better deal with the rare extremes associated with natural variability may have a different character. Attribution studies can thus be usefully tailored to inform adaptation strategies encompassing natural hazard mitigation as well as to help reduce the vulnerability of societies to human induced climate change (Hoegh-Guldberg et al. 2011).

By determining the causes of extreme weather events being observed now, robust information can be provided on the extent to which a specific extreme event is a harbinger of the future (e.g. Beniston 2004). If a recent extreme weather or climate-related event has shown a society to be vulnerable, that society may want to develop further resilience; alternatively, if an attribution assessment concludes that the event is either likely to remain extremely rare or become less likely in the future, the society may adjust policies, for instance, by judging that such events and their impacts do not constitute a long-term adaptation priority (Hoegh-Guldberg et al. 2011). While it has been argued that attribution studies should not play a role in informing adaptation policies (Hulme et al. 2011), incorporating attribution assessments as part of a regular suite of climate services alongside weather and climate prediction systems should help avoid the misuse of attribution results through politicization and bias.

2.5 *Geoengineering*

With greenhouse gas emissions still following a "business-as-usual" trajectory, geoengineering, as an option for reducing the risk of dangerous climate change, is an issue rising up the climate change agenda and attribution will likely become an important component of any research and development of such technologies. A recent report by the Royal Society provided a comprehensive assessment of all geoengineering options that are actively being considered (Royal Society 2009). It raised a number of concerns about the possible adverse consequences on climate in some regions as a result of attempting to reduce the risks of crossing dangerous thresholds of climate change were society unable to constrain global emissions. Given that geoengineering could be beneficial to some stakeholders and damaging to others, were a damaging weather or climate-related event to occur following a geoengineering intervention, attribution assessments would be of interest to stakeholders seeking compensation for any unwelcome effects as a consequence of attempts at delivering collective overall planetary benefits. Event attribution studies could also inform the design of experiments to test geoengineering options at a local scale before planetary scale implementation, and would also be needed post-implementation to determine whether geoengineering is working in the sense of reducing the occurrence of dangerous extreme weather and climate-related events.

2.6 *Insurance*

The insurance industry relies heavily on observed records and extreme value theory to assess the probability of occurrence of rare weather events (e.g. Smith 2003) but both tools may yield incorrect conclusions if the location or shape of the distribution of a particular class of weather event is changing as a result of some external driver on a timescale comparable to the length of record used for model-calibration

(Räisänen and Ruokolainen 2008). Faced with uncertainty about how these risks are changing, but knowing that human influence on climate has altered the occurrence of climate extremes (as discussed above) and therefore that they should distrust historical probabilities, insurers can respond simply by withdrawing cover entirely, as has been observed in certain sectors of the US hurricane insurance market (Haufler 2009). Providing information on how the risks are changing would thus provide insurers with the possibility to continue cover by altering premiums appropriately.

3 Development of Event Attribution

In recent years event attribution has developed considerably, with a number of studies having been published that quantify the role of human and natural influences on specific weather and climate-related events.

The approach of using model experiments to calculate how a particular climate driver has changed the probability of an event occurring (Allen 2003; Stone and Allen 2005) has been applied to a number of different cases (e.g. Stott et al. 2004; Christidis et al. 2010; Pall et al. 2011). The probability of a particular event happening in an ensemble of model simulations representing current conditions is compared with a parallel ensemble of model simulations representing an alternative world that might have occurred had the particular driver been absent. Although many detection and attribution studies that analyze long term changes in climate variables do not require the climate model to simulate the correct amplitude of the responses to forcings, since they include scaling terms that can compensate for under- or over-responsive models (Hegerl and Zwiers 2011), event attribution studies typically make stronger assumption about the correctness of models, the validity of which needs to be carefully tested (Christidis et al. 2012).

The approach for calculating the change in likelihood of an event attributable to a particular climate driver is illustrated schematically in Fig. 1. The distribution shown in red represents the current probability distribution of a particular climate variable, and that in green, the equivalent probability distribution of that variable in the world that might have been in the absence of the climate driver. Then for a particular threshold, the probabilities, P1, of exceeding that threshold currently, and P0, of exceeding the threshold in the absence of the climate driver, can be calculated.

From these two probabilities the Fraction Attributable Risk (FAR) can be calculated, where FAR = 1-P0/P1 (Allen 2003). FAR expresses the fraction of risk of a particular threshold being exceeded (e.g., an extreme temperature threshold associated with a heat wave) that can be attributed to a particular driver. For example if the probability that a particular threshold being exceeded has increased by a factor of 4 as a result of human influence on climate, FAR = 0.75, and hence three quarters of the risk of that event is attributable to human influence. In this case, under the current climate, on average ¾ of such events could be blamed on human influence. Such a result does not indicate that human influences were responsible for 75 % of the observed magnitude of the particular metric being used to define the event.

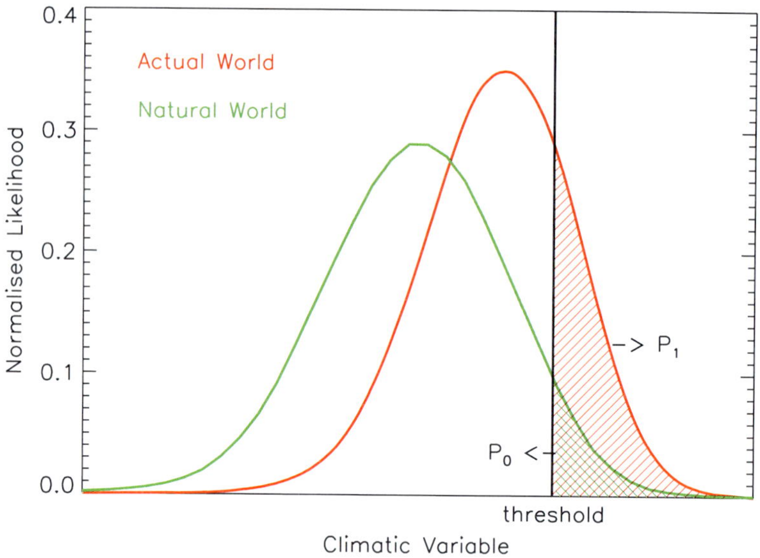

Fig. 1 A schematic illustration of the distributions of a climatic variable with (*red*) and without (*green*) the effect of anthropogenic forcings. *The hatched areas* mark the probability of exceeding a threshold value in the two climates. The FAR is the fractional change in the probability $1-(P_0/P_1)$

By their very definition, the nature of most extreme events means that their probabilities need to be estimated by statistical extrapolation or modeling unless they occur sufficiently commonly that their probabilities can be estimated directly from their observed frequencies. If events are sufficiently frequent it can be possible to carry out a "single-step" attribution analysis (Hegerl et al. 2010), in which observed and modeled changes are compared directly. In this way, Stott et al. (2011) detected a significant increase in the observed frequencies of warm seasonal temperatures in many regions that were attributable to human influence.

It is always necessary to use models to generate simulations of the counterfactual world that did not happen in order to estimate P0, the probability of the event in the absence of a particular climate driver. It is helpful to express findings in statements that are robust to modeling and observational uncertainties. A number of studies (e.g. Stott et al. 2004; Pall et al. 2011) have employed a particular formulation to characterize uncertainty in FAR, which states the minimum value that FAR is expected to exceed at some level of likelihood. Thus in the case of European summer temperatures, Stott et al. (2004) concluded that despite uncertainty in the precise value of FAR for the threshold chosen in that study as being relevant to the 2003 European heatwave, it was "very likely" (using the IPCC definition of a >90 % chance of the statement being correct) that FAR was greater than 0.5, i.e. that the probability of the threshold being exceeded had more than doubled as a result of human influence. Such statements are also more closely aligned with interests of many potential stakeholders listed in Sect. 2 who are concerned whether iconic thresholds, such as a doubling of the probability, as in Stott et al. (2004), have been passed (e.g. Grossman 2003). For the same reason, it may also be helpful to determine whether the probability of the event has likely not changed substantially.

Characterizing changes in the shape of tails of extreme event distributions is challenging, so it is important not to read too much into heuristic examples based on idealized distributions such as Fig. 1. In some cases (e.g. Stott et al. 2004; Pall et al. 2011), the change in the distribution of a particular climate variable can be consistent with a constant ratio of exceedance probabilities, P0/P1, over a broad range of thresholds, a point that is important for the robustness of results from such analyses where there is ambiguity about the actual threshold that was exceeded during the event in question and where models have biases that affect the probability of exceeding absolute thresholds in the model. While some such biases can be addressed by an adjustment of the model's baseline statistics to an observed climatology (e.g. Otto et al. 2012), others can be harder to correct. For example if a climate model's representation of phenomena such as the variability of the El Niño/Southern Oscillation is inaccurate this could lead to systematic biases that may only be corrected through model improvements.

Recent attribution studies have also begun to pose the question how various factors, including human-induced climate change, contributed to the *magnitude* of an event (e.g. Perlwitz et al. 2009; Dole et al. 2011; Hoerling et al. 2012). These investigations are broadly aligned with a class of studies that assess the intensity of the signal of anthropogenic climate change relative to the intensity of the background climate variability (e.g. Hawkins and Sutton 2012). Of specific interest has been to diagnosis the relative magnitude of specific naturally occurring climate conditions within which an event developed, relative to estimates of the regional impact of human-induced climate change.

The predictability of an event and therefore the potential for early warning (including both its amplitude and temporal characteristics) are also investigated in some attribution studies. Diagnosing the ability of a model to capture the physical and statistical properties of a particular event is an important test of its capabilities if it is to be used for attribution. While attribution is possible in the absence of predictability (from initial conditions or from knowledge of the sea-surface temperatures or from other conditions internal to the climate system), a high level of confidence in an attribution assessment can only be justified if the models used are capable of capturing the relevant processes, since anomalous atmospheric flow and unusual oceanic and land surface conditions are often associated with extreme weather (Perlwitz et al. 2009; Dole et al. 2011). An example is that climate models are often criticized for their shortcomings in representing atmospheric blocking although much of this may be related to climatological biases in models (Scaife et al. 2010). But blocking is not the only phenomenon that can challenge models and hence limit confidence in attribution assessments. Therefore attribution assessment should assess the extent to which conclusions drawn about the metrics used to represent a particular event are robust to inadequacies in the representation of such phenomena in the models used; conversely such studies can be valuable in suggesting needs for model improvements.

In many cases, the most damaging weather events will be those that are least predictable, in the sense that, even with a perfect seasonal forecasting system, there would be little basis for anticipating that they would occur when they did a month or more in advance. The 2010 Russian heatwave may be a case in point, with Dole et al. (2011) observing that there was no basis for climate predictability (month or longer lead times) of the extreme blocking associated with the heatwave in two models forced by observed sea surface temperatures and sea ice or in the NOAA coupled forecast model used for seasonal predictions. This apparent lack of predictability need not imply lack of understanding: in a chaotic system, certain low-probability events occur infrequently with little warning. It also does not necessarily imply inability to attribute to human influence: as Rahmstorf and Coumou (2011) observe, the probability of occurrence of the 2010 Russian heatwave, while still low, may have been substantially increased by the large-scale warming that has occurred since the 1960s. This conclusion is not inconsistent with the statement that there was little basis for anticipating the heatwave a month or more in advance.

The crucial test of whether models are capable of simulating specific weather events with a view to using them for attribution is the *reliability* of seasonal forecasts or hindcasts of the event in question based on these models: that is, when the forecasting system predicts a particular class of event will occur 10 % of the time, it is observed to occur 10 % of the time (Christidis et al. 2012). As noted above, attaining reliability may often require correction of biases. Whether or not forecasts of the event have any *resolution* (that is, whether the combination of initial and boundary conditions that obtain at the time make the forecast probability of the event any different from its climatological probability) is a separate question. Some events are intrinsically unpredictable because they are not affected either by initial conditions or by short-term changes in boundary conditions. These events will often be the most damaging precisely because of this lack of predictability. Because they occur infrequently, direct detection of a trend in occurrence-frequency is also very difficult. Given adequate models and computing resources, however, it may still be possible to assess reliably how these low probabilities are changing and hence how much of the present-day risk can be attributed to external climate drivers.

To illustrate further the challenges involved in event attribution and to describe the main progress made so far in this area, published studies carrying out attribution of specific events are outlined in the next section.

4 Examples of Event Attribution for Specific Cases

4.1 2003 Central European Summer Temperatures

In their study following the 2003 European heatwave, Stott et al. (2004) (and later Christidis et al. 2010a, using additional data) analyzed the temperature changes averaged over summer for a large part of continental Europe and the Mediterranean. Using an optimal detection analysis of simulations of the HadCM3 coupled climate model with and without anthropogenic forcings, Stott et al. (2004) showed that there had been a significant anthropogenic contribution to the observed warming of regional summer mean temperatures. They then used the model to infer the probabilities in the current world (P1) and the world without human influence on climate (P0) of exceeding a particular seasonal mean temperature threshold associated with the year 2003 event. The threshold they chose was the summer mean temperature that was exceeded in 2003 but in no other year (before that) since the start of the instrumental temperature record in 1851, a threshold somewhat lower than that reached in 2003 and a formulation designed to minimize the selection effect of choosing a threshold too closely associated with what actually occurred in 2003.

This multi-step attribution approach yielded an estimate for the Fraction Attributable Risk (FAR) of 2003 European mean summer temperatures where in the first step, a change in the decadal background summer temperature was attributed to human influence, and then in the second step the relationship between year-to-year variability and the decadal background variability in summer temperatures was attributed to processes simulated in a climate model, allowing an inference of the probability of exceeding the threshold in that particular summer to be made. Figure 2a shows the calculated distributions of P0 (green) and P1 (red) expressed as number of occurrences per thousand years where the likelihood distribution represents their uncertainty, a combination of uncertainty in the estimate of the anthropogenic warming in the region and uncertainty in the probability of exceeding the chosen temperature threshold given a particular level of anthropogenic warming (Stott et al. 2004). The derived distribution of FAR is shown in Fig. 2b (estimated from the two probability distributions shown in Fig. 2a) with the median value also shown. Based on the result that the 10th percentile of the distribution (as shown by the leftmost grey band in Fig. 2b) is greater than 0.5, Stott et al. (2004) concluded that the probability of seasonal mean temperatures as warm as those observed in Europe in 2003 had very likely at least doubled as a result of human influence. Their conclusion that FAR for their metric is very likely greater than 0.5 serves as the first practical example of how to make a scientifically robust attribution assessment about a specific extreme event. Many of the excess deaths in summer 2003 were associated with the period when the heatwave was at its most intense in early August in central Europe (Schar and Jendritzky 2004). Attribution of the impacts of the heatwave of 2003 would therefore require consideration of a shorter period and a more geographically restricted region than analyzed by Stott et al. (2004).

4.2 2000 UK Floods

Pall et al. (2011) considered the extensive floods that occurred during the record-wet Autumn of year 2000 in England and Wales, and estimated the change in probability of such floods occurring at that time as a result of twentieth-century anthropogenic greenhouse gas emissions. This study again followed a multi-step approach, with the first step attributing the bulk of warming in global sea surface temperatures to

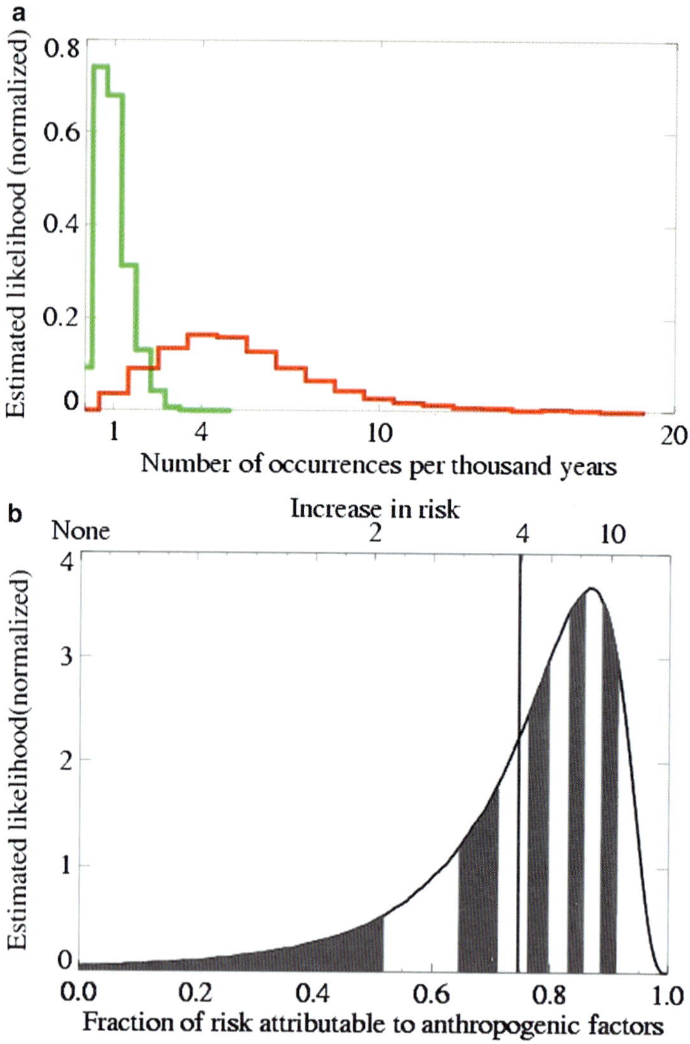

Fig. 2 Change in risk of mean European summer temperatures exceeding the 1.6 K threshold. (**a**) Histograms of instantaneous return periods under late-twentieth-century conditions in the absence of anthropogenic climate change (*green line*) and with anthropogenic climate change (*red line*). (**b**) Fraction attributable risk (FAR). Also shown, as the *vertical line*, is the 'best estimate' FAR, the mean risk attributable to anthropogenic factors averaged over the distribution. From Stott et al. (2004). Bands of *white* and *shade underneath the curve* represent ten percent bands of the distribution (i.e. 0–10 %, 10–20 % etc.)

anthropogenic greenhouse gas emissions through the use of an established "optimal fingerprinting" regression analysis (Stott et al. 2006; Nozawa et al. 2005). Because of the lack of observations of events this rare and recognizing the largely atmospheric seasonal-timescale nature of the event, the second step used a seasonal-forecast-resolution atmospheric climate model to generate simulations of possible

Autumn 2000 weather. These simulations were conducted both under conditions observed at that time and under parallel conditions that might have been obtained at that time in the absence of increased greenhouse gases, as determined from the first attribution step.

The observed event was relatively rare and unpredictable, so ensembles of several thousand weather simulations were generated under these two conditions (via climate*prediction*.net public volunteer distributed computing) to sufficiently capture such unusual flood-producing weather and its change. Results were fed into a precipitation-runoff model for England and Wales to then simulate a measure of flooding, with the probability of floods of a specific magnitude counted directly from the simulations. The atmospheric model simulations acted as pseudo-observations for investigating the role of various mechanisms that could lead to changes in flood frequency between the two climates, noting that the reliability of the model in delineating mechanisms therefore becomes critical to the correctness of the findings of such a study. It was found that almost all differences were due to a simple thermodynamic increase in precipitable water, a mechanism that is well understood, although this conclusion is dependent on the reliability of the model in discounting non-thermodynamic changes, such as circulation changes, as major factors. Quite a large uncertainty was found in the magnitude by which the greenhouse gases increased flood risk at the threshold relevant to autumn 2000. Thus there is quite a large spread in the return times of particular values of daily runoff shown in (Fig. 3a) under conditions that would have been obtained in the absence of anthropogenic greenhouse warming over the twentieth century, the wide spread being largely driven by uncertainty in the change in sea surface temperatures attributable to greenhouse warming. While Pall et al. (2011) found that the precise magnitude of the anthropogenic contribution to flood risk was uncertain, in nine out of ten cases their analysis indicated that twentieth-century anthropogenic greenhouse gas emissions increased the risk of floods occurring in England and Wales in autumn 2000 by more than 20 %, and in two out of three cases by more than 90 %.

In contrast, Kay et al. (2011) showed, using the same climate model experiments, that there was a decrease in the risk of flooding in Spring as a result of a reduced risk of snow-melt-induced run off (Fig. 3b). This provides an example of a hypothetical weather-related event that has been made less likely as a result of human influence on climate (and which did not occur in Spring 2001).

4.3 2008 Cool US

Perlwitz et al. (2009) studied the nature of the very cool year 2008 climate conditions in North America that diverted strongly from the long term warming trend observed over previous years. Based on a suite of model experiments their study showed that an anthropogenic warming of North American temperature was overwhelmed by a particularly strong bout of naturally induced cooling resulting from the continent's sensitivity to widespread coolness of the tropical and northeastern

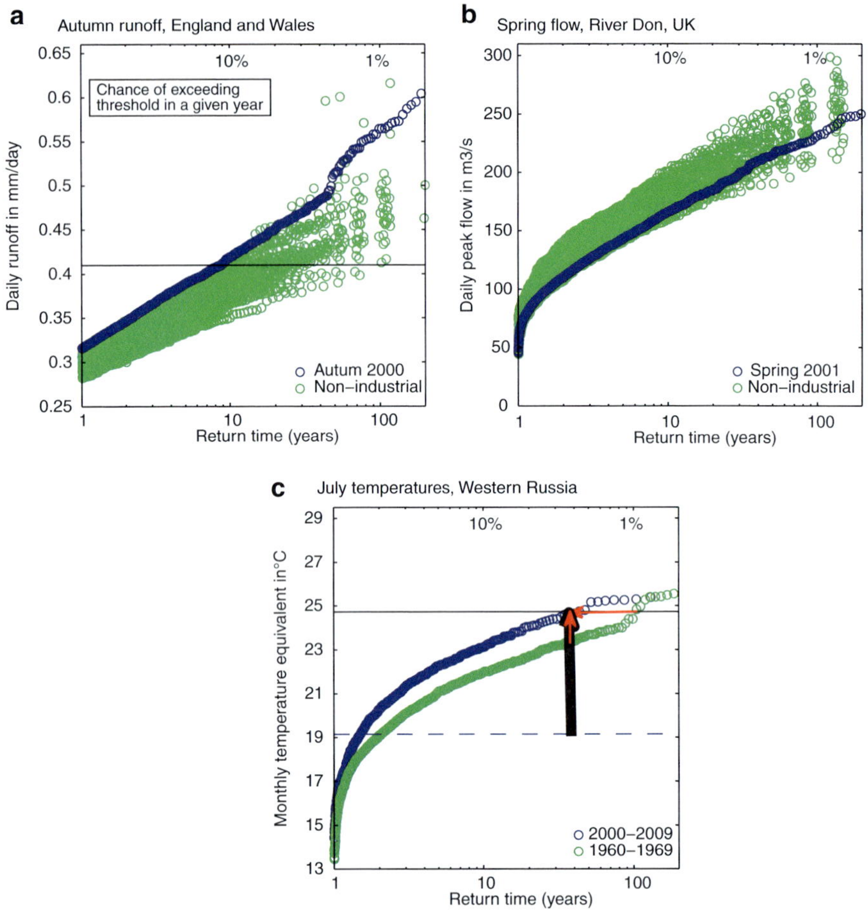

Fig. 3 (**a**) and (**b**) Return times for precipitation-induced floods aggregated over England and Wales for (**a**) conditions corresponding to October to December 2000 and (**b**) conditions corresponding to January to March 2001 with (for both panels) boundary conditions as observed (*blue*) and under a range of simulations of the conditions that would have obtained in the absence of anthropogenic greenhouse warming over the twentieth century (*green*) (Adapted from Pall et al. 2011; Kay et al. 2011). (**c**) Return periods of temperature-geopotential height conditions estimated for the 1960s (*green*) and the 2000s (*blue*). The *vertical black arrow* shows the anomaly of the Russian heatwave 2010 (*black horizontal line*) compared to the July mean temperatures of the 1960s (*dashed line*). The *vertical red arrow* gives the increase in temperature for the event whereas the *horizontal red arrow* shows the change in the return period (From Otto et al. 2012)

Pacific sea surface temperatures. Figure 4 shows North American surface temperature change for 1970–2007 (left) and departures for 2008 (right) as observed and as simulated by coupled models (CMIP) and atmosphere only models (AMIP) forced with the observed 2008 sea surface temperatures. The observed pattern of temperatures in 2008 (Fig. 4b) is much closer to the pattern of temperatures from the ensemble

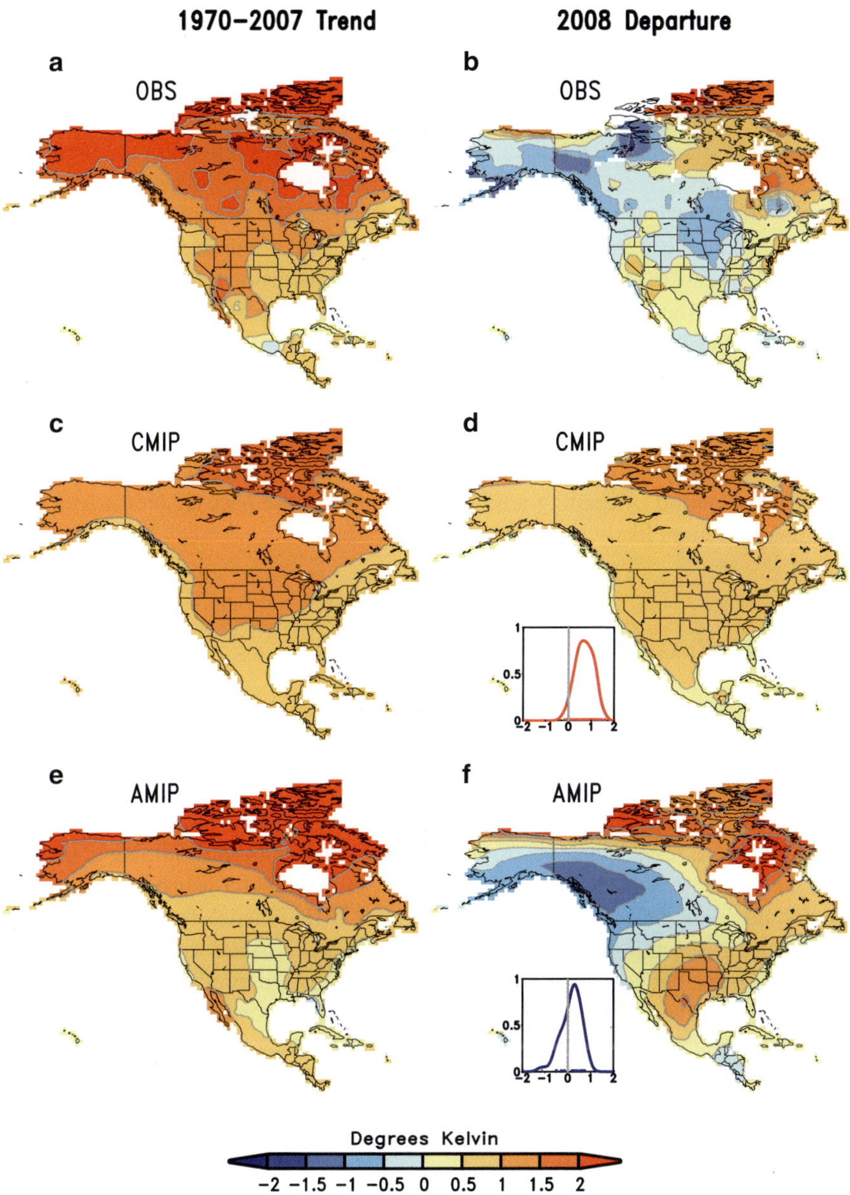

Fig. 4 (*left*) North American surface temperature change for 1970–2007 [K/38 year] and (*right*) departures for 2008 (in [K] relative to 1971–2000 mean) based on (**a** and **b**) observations, (**c** and **d**) ensemble CMIP simulations, and (**e** and **f**) ensemble AMIP simulations. Inset in Fig. 1d, f are probability distribution functions of the individual simulated annual 2008 surface temperature departures area-averaged over North America. The observed 2008 departure was near zero (Fig. 1 from Perlwitz et al. 2009) (Reproduced by permission of American Geophysical Union)

mean of the AMIP models forced with observed SSTs than the observed or modeled trends over recent decades or the 2008 departure as simulated by the mean of the CMIP models (which by averaging will have eliminated most natural internal variability components). There is expected to be a spread of patterns of temperature anomalies in different coupled model ensemble members reflecting the combined effects of external forcings and internal variability. However, the greater agreement between the observed 2008 departure and the spread of simulations in the AMIP models than between the observed 2008 departure and the spread of simulations in the CMIP models led Perlwitz et al. (2009) to the conclusion that a large part of the departure from the long term trend during 2008 could be attributed to a particular anomalous state of SST conditions in 2008. Further, it was found that such a strongly anomalous ocean state and its North American impacts are well simulated by models, indicating their potential for skillful predictions of fluctuations in large scale temperature anomalies over North America. Perlwitz et al. (2009) concluded that the cool year in 2008 did not indicate that the climate was likely to embark upon a prolonged period of cooling and, on the contrary, the pace of North American warming was more likely to resume in coming years.

4.4 2010 Russian Heatwave

Dole et al. (2011) considered the relative importance of various physical factors contributing to the extreme heat wave affecting Moscow and adjacent regions, using both observational analyses and model experiments. They showed that, in contrast to the region affected by the 2003 European heat wave, the primary region affected by the 2010 heat wave had not experienced significant long-term warming in summer over the prior 130-year period, although other studies find a warming trend in the past few decades (Rahmstorf and Coumou 2011). CMIP-3 models forced by increasing greenhouse gases and other external forcings showed a small mean warming over the same period, but no significant change in variability. Ensemble model simulations forced by observed global sea surface temperatures and sea ice conditions also showed no statistically significant response over the heat wave region. Dole et al. concluded that the extreme magnitude of the 2010 Russian heat wave was caused primarily by internal dynamical processes that led to a very strong and persistent blocking pattern over the heat wave region. They also concluded that regional land surface feedbacks were not important in explaining the heat wave's intensity. While these results did not support increasing greenhouse gases having contributed substantially to the magnitude of the 2010 Russian heat wave, model projections suggest that western Russia is on the cusp of a period in which the probability of such extreme heatwaves will increase rapidly. Many other regions around the world are already experiencing a rapid increase in the frequency of more moderate extreme temperatures (Jones et al. 2008; Rahmstorf and Coumou 2011; Fig. 5).

No more basic question exists in climate science than "What caused the event?" Yet, this obvious and simple query can invite a multitude of answers, which though perhaps scientifically consistent with each other, sometimes bear very different

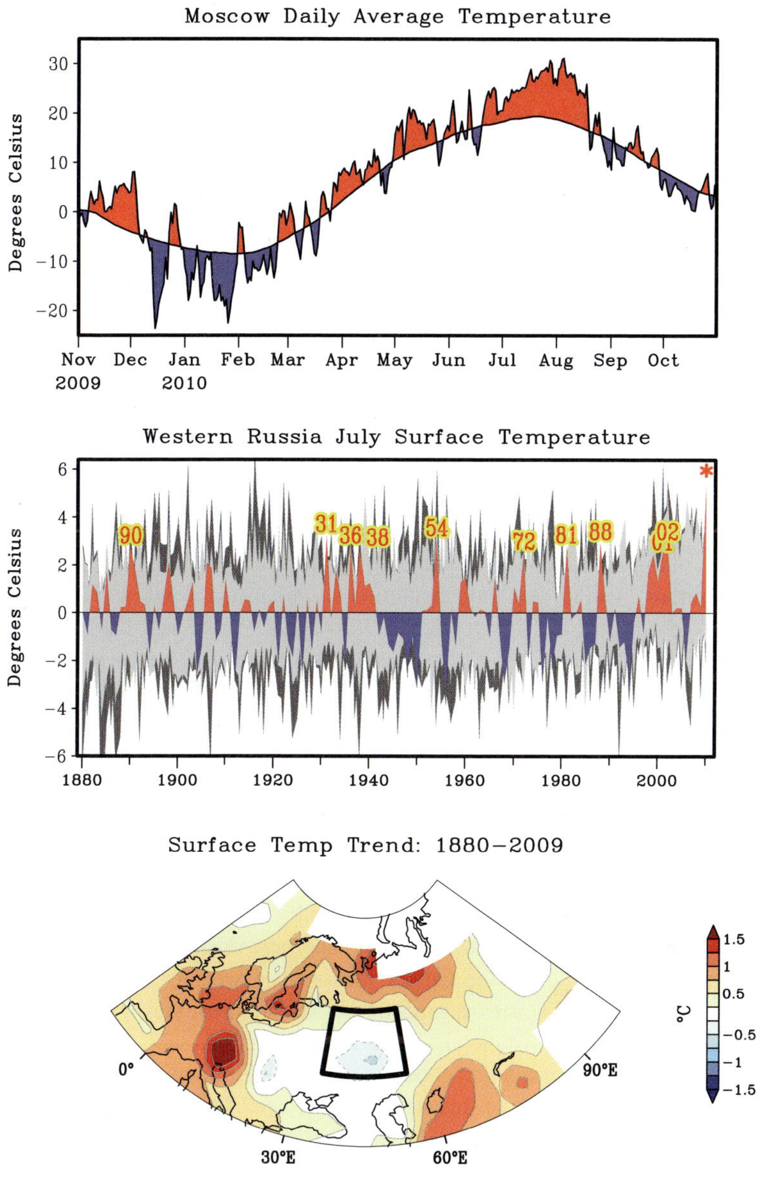

Fig. 5 (*top*) Daily Moscow temperature record from November 1 2009 to October 31 2010, with daily departures computed with respect to the climatological seasonal cycle. Data are from the Global Summary of the Day produced by National Climatic Data Center. (*middle*) Observed time series of western Russia July temperature anomalies for the period 1880–2010 indicated as positive (*red*) and negative (*blue*) temperature anomalies relative to the base period from 1880 to 2009. Numbers indicate the years of the ten most extreme positive anomalies. The *red asterisk* indicates year 2010. The *light and dark shaded areas* represents the envelopes of positive and negative monthly mean temperature extremes based on 22 CMIP3 model simulations for normalized and non-normalized anomaly time series respectively. (*bottom*) Map of observed July temperature trend [o C/130 years] for July 1880–2009. *Box* shows the area used to define "western Russia" surface temperatures (Fig. 1 from Dole et al. 2011) (Reproduced by permission of American Geophysical Union)

meanings. Whereas Dole et al. (2011) concluded that the Russian heatwave event was mainly caused by natural internal variability, Rahmstorf and Coumou (2011) founded a greatly increased probability of such a heatwave. Otto et al. (2012) showed that these outwardly opposite conclusions are in fact reconcilable upon clear distinction of the different questions that each study asked. While Dole et al. (2011) asked what factors caused the full magnitude of that heat wave, estimated by some to have been +10 °C above normal over Moscow during July 2010 (e.g. Barriopedro et al. 2011), Coumou and Rahmstorf (2012) asked what was the probability that a record-breaking heat wave, could have occurred in a stationary climate (see Fig. 3c). The fact that human-induced climate change very likely has increased the odds of breaking a prior record is thus consistent with the fact that most of the magnitude of the event was nonetheless the consequence of natural variability.

5 Attribution of Climate-Related Events Group

The studies outlined above indicate the potential for event attribution, but the remaining uncertainties, even for the very small number of specific events so far considered, demonstrate that there are many scientific challenges to be faced in developing a robust assessment process for extreme events. It should also be noted that there is often an appreciable delay between the occurrence of the event in question and the appearance of the peer-reviewed studies. Meanwhile scientists continue to face demands for more rapid reaction on the links between climate change and unusual weather events, and the number of climate litigation cases is increasing (Adam 2011). As a result of this demand, interested scientists from around the world have recently joined together to coordinate their efforts and consider research needs as part of the Attribution of Climate-related Events activity (ACE). The first full meeting was held in August, 2010 (hosted by NOAA and supported by funding from the UK FCO).

In the run up to the meeting, there was no shortage of illustrations regarding the question of who cares about the causes for extreme weather and climate-related events with many media stories regarding the Russian heat wave, Pakistan floods, and China floods, and of the concerns about the implications such events held for the immediate future, for example, on food supplies and commodity prices. The events impressed upon the attendees of this workshop the need for rapid, yet accurate, attribution information. The attendees also concurred about the links between the development of attribution information and the development of prediction systems.

The meeting attendees agreed that a comprehensive and authoritative attribution activity, one that meets user needs for coverage and for trustworthiness, will demand enhanced collaboration and coordination of numerous partners in order to provide a test bed for evaluating and applying data, theories, and computational methods. In this regard, the underpinning of a strong and sustained research base to provide the best possible operational systems for attribution was emphasized. The foundations of an authoritative explanation of extreme events begin with a real-time monitoring and climate analysis capability, and availability of historical data sets, such that current events can be placed into a reliable and physically consistent historical context. Model simulations and experimentation, including the use of multiple models and perturbed physics ensembles, were likewise seen as core elements that provide an essential tool in "connecting the dots" so as to establish plausible cause-effect relationships. The workshop attendees also emphasized that society and decision makers also need to be provided with a clear statement of the meaning and implications of the scientific findings.

A concrete outcome of the meeting was a proposal for modeling centers to carry out a coordinated set of experiments in order to explore the importance of the experimental design, climate model design, event location and timing, and the inclusion of various anthropogenic factors to increase the robustness of attribution assessments. The experiment has also been endorsed by the CLIVAR C20C group (Kinter and Folland 2011). Under C20C, modeling centers have already tested the ability of atmosphere only models to simulate observed climate events (Scaife et al. 2008). Under this new attribution component of C20C, several modeling centers around the world will perform time-slice experiments with atmospheric modeling, following the method developed by Pall et al. (2011). These experiments will cover the period from 1960 to the near-future, allowing an evaluation of how anthropogenic contributions have been changing and also how estimation of these contributions depends on various aspects of the experimental setup.

6 Development of Near-Real Time Weather and Climate Event Attribution

6.1 Coupled Model Approaches

The analysis of the 2003 European heatwave by Stott et al. (2004) provided a template for how a multi-step attribution analysis using coupled models could be carried out in which changes in the frequency of a specific event under human influences can be inferred based first on an attribution analysis of mean temperatures over a region, and second, on an analysis of the expected change in frequency of exceeding thresholds of extreme temperature due to the effects of internal variability around the baseline of mean temperatures attributed in the first step. This same approach can be applied using regularly updated data to provide assessments of attributable risk immediately following a particular season. Such an approach using coupled models has also been used to estimate the change in the probabilities of exceeding a pre-defined temperature threshold for every season in Europe (Christidis et al. 2010a) and extended to sub-continental scale regions throughout the world (Christidis et al. 2010b, 2011b). Unlike previous studies that carried out an optimal fingerprinting analysis over the region of interest, these studies employed constraints from a global analysis to estimate the regional temperature distributions. Look-up tables, as shown in Fig. 6, of the FAR of a threshold exceedance, can be computed for each region over a range of thresholds thereby providing the potential

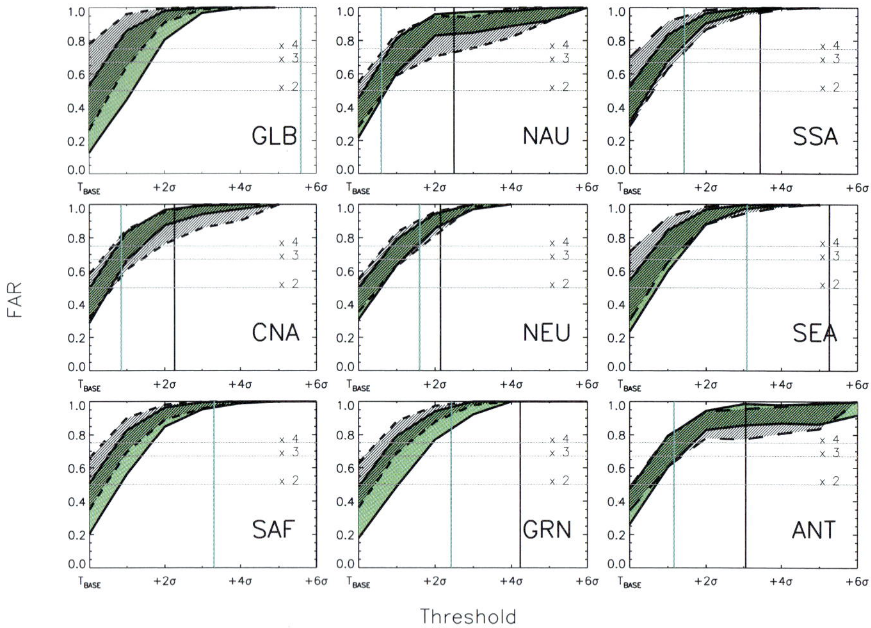

Fig. 6 Estimates of the FAR in nine different regions measuring how much anthropogenic forcings have increased the likelihood of exceeding a pre-specified annual mean temperature anomaly threshold during 2000–2009. Results are shown for a range of thresholds increasing from zero by multiples of the standard deviation which represents the effect of internal climate variability. The *green* and the *black* hatched areas illustrate the 5–95 % range of the FAR computed with HadGEM1 and MIROC fingerprints respectively. The *vertical blue line* marks the annual mean temperature anomaly in 2000–2009. The *vertical black line* corresponds to the maximum annual mean temperature anomaly since 1900. The *horizontal grey lines* mark the FAR values which correspond to an increase in the likelihood of exceeding a threshold by a factor of 2, 3 and 4. Regions are: Global (*GLB*), Northern Australia (*NAU*), Southern South America (*SSA*), Central North America (*CAN*), Northern Europe (*NEU*), Southeast Asia (*SEA*), Southern Africa (*SAF*), Greenland (*GRN*), Antarctica (*ANT*) (From Christidis et al. 2011b)

for regularly updated attribution information to be provided alongside monitoring information about the season just finished and seasonal forecasting information of the season to come.

6.2 Very Large Ensembles Using Distributed Computing Experiments

Attribution approaches that are appropriate for regional precipitation are being developed that use larger ensembles of higher resolution models. In particular the use of atmosphere-only models in which sea surface temperatures are prescribed allows a better discrimination between ensembles of models with each being tied to a particular evolution of SSTs, the approach described in Sect. 4.2 (Pall et al. 2011).

An important consideration here is that such an experimental design investigates the change in risk conditional on certain aspects of variability being tied to those observed (e.g. the state of ENSO) whereas ensembles of coupled models with different climate forcings are able to estimate the overall change in risk associated with the presence of a particular forcing.

In many cases interest may be in weather extremes that are extremely rare and therefore for which either very large ensembles of model simulations need to be made in order to capture the occurrence of such events, or else statistical extrapolation techniques need to be used to make inferences about changes in such extremes based on changes in that part of the distribution that can be modeled. An important aspect of research therefore is to understand the benefits of the very large ensembles made possible by distributed computing, as well as the limitations of smaller ensembles that could be run more regularly as part of a near-real time attribution service. The climateprediction.net project which has pioneered the use of large ensembles for attribution and prediction has recently included a regional modeling component (the weatherathome.org project) to downscale global models for the Western US, Southern African and European regions. Regional models enable a better representation than global models of small scale processes that are relevant to attribution assessments, particularly of extreme rainfall events.

6.3 Analogue Methods for Diagnosing the Influence of Circulation Characteristics

An approach that can be applied routinely to provide information on recent climate-related events in a timely fashion is to quantify the contribution of large-scale circulations to temperature anomalies, thereby discerning whether recent temperatures are warmer or colder than would be expected from flow-analogues from previous years (Cattiaux et al. 2010a, b; Vautard and Yiou 2009). Such an approach has been used to show that the cold northwestern European winter of 2009/2010, associated with a very negative value of the North Atlantic Oscillation (NAO) index, would have been even colder without the effects of long-term warming (Cattiaux et al. 2010a). Such approaches do not fully answer the attribution question because they do not quantify the link to human emissions (in explaining the long term warming trend in a region) but they are helpful in putting extreme events into a climate perspective (Peterson et al. 2012).

6.4 A Near-Real Time Attribution Capability Linked to Seasonal Forecasting

A pilot version of an attribution forecast system has been implemented that runs in parallel with an existing seasonal forecasting service, following a simplified version of the Pall et al. (2011) experimental design (http://www.csag.uct.ac.za/~daithi/forecast). Along with the real seasonal forecast, a parallel forecast is run of a non-greenhouse gas world in which human activities had never released greenhouse gases to the atmosphere and the ocean had not warmed in response to those emissions. Currently this simple implementation mainly serves as an apparatus for learning about how such a linked system can function when using multiple models and as a demonstration to aid in ascertaining the requirements and characteristics that potential users of such a system might demand. It has however revealed seasonal and regional variations in attributable risk, as well as some apparently robust similarities and differences between attribution and seasonal forecasting products, for instance in that the relative predictability of temperature versus precipitation events can be different in an attribution system when sea surface temperatures are known than in a standard seasonal forecast when sea surface temperatures are predicted. This system has also provided insights into the advantages and limitations of a pro-active approach. Pre-defined event definitions have inevitably not conformed to what users might wish to know about after the fact, revealing the need to refine definitions in a balance between relevance and systematic objectivity. The use of confidence statements involving thresholds of attributable risk, such as that the chance of the event has at least doubled or halved, have provided a clear framework in communication sessions with potential users of such information. Just as important for such users, though, is the provision of statements of when we can say that any influence is less than a certain threshold.

The development of a more rigorous near-real time attribution capability is underway at the Met Office Hadley Centre with a focus on system evaluation and validation, and based on ensembles of simulations of the atmosphere version of the seasonal forecasting model with atmosphere-only GCMs and prescribed SSTs (Arribas et al. 2009; Christidis et al. 2012). The ensembles using the HadGEM3-A atmospheric model are generated using (a) random perturbations that represent the uncertainty in a number of parameters (Murphy et al. 2004; Collins et al. 2006) and (b) vorticity perturbations that counteract the damping of small scale features introduced by the semi-Lagrangian advection scheme (Bowler et al. 2009). Simulations of the 'actual' world employ the anthropogenic and natural forcings used in previous experiments with the HadGEM1 model (Stott et al. 2006) and sea surface temperature (SST) and sea-ice data from the HadISST dataset (Rayner et al. 2003). Simulations for the climate without human influences include natural forcings only (from changes in solar output and due to explosive volcanic eruptions) and remove an estimate of the anthropogenic change in the SSTs and sea-ice from the prescribed HadISST data. Experiments to date have been carried out using an estimate of SST change from the HadGEM1, HadGEM2 and HadCM3 models, while the change in the sea-ice is computed based on empirical relationships derived with HadISST data, a similar approach to Pall et al. (2011). Notably, the setup of this real-time system is closely aligned to the design of the ACE C20C experiments.

A useful tool in the validation of ensembles for seasonal forecasting is the reliability diagram which plots the observed frequency of an event against the forecast probabilities (Wilks 1995). Reliability is indicated by the proximity of the plotted curve to the diagonal. Points above/below the diagonal indicate that the forecast model is under/over-forecasting at the respective probability threshold. Reliability diagrams can be constructed for a number of regions to examine the model skill in simulating high and low temperature and precipitation events. An illustrative

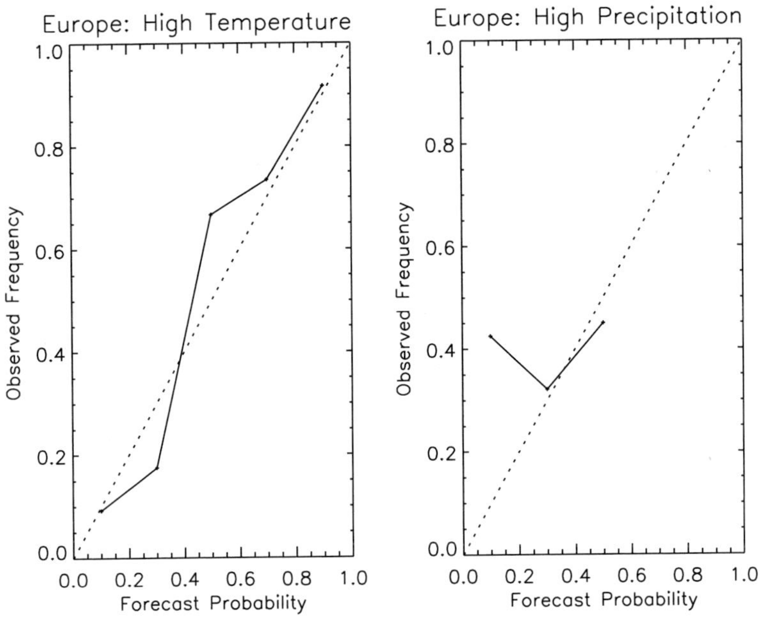

Fig. 7 Reliability diagrams that assess the forecast skill of the model in predicting seasonal mean temperature (*left panel*) and precipitation (*right panel*) values averaged over Europe above the upper tercile of the 1971–2000 climatology. The NCEP/NCAR reanalysis was used to compute the observed frequency (Kalnay et al. 1996)

example is shown in Fig. 7 for forecasts of upper tercile temperature and precipitation events in Europe. The results suggest that the model has much better skill in capturing the predictability of temperature than of precipitation in the chosen region. Such reliability diagrams, supplemented by comparisons of observed and modeled distributions of the variable being attributed, can be used to inform the confidence that can be placed in attribution assessments. In situations where there is a high degree of predictability, for example in the case of the large scale pattern of North American temperatures in 2008 as analyzed by Perlwitz et al. (2009) and discussed above, a necessary component of model fidelity is that it is able to capture the predictable features of the weather or climate-related event in question. Where there is little predictability, the model should be able to capture the main impacts that attributable changes in climate could have on the statistics of the event in question.

7 Discussion: Lessons Learned and Future Research Needs

Based on the requirements for attribution assessments outlined in Sect. 2 it is clear that attribution forms a key part of any climate service, the essential bridge between monitoring and prediction services, that puts recent weather and climate-related events into a long term context. An important benefit of well defined attribution assessments is in avoiding the apparent discrepancies that can arise between the conclusions of different attribution studies, for example those of the Russian heatwave of 2010 discussed above. While a variety of stakeholders require reliable information in the immediate aftermath of extreme events (Sect. 2), for other purposes, for example to inform litigation, more complete information could be required at a later date. It will be important to identify the tolerance of potential decision-making processes to uncertainties and errors in observations and models, and therefore the levels of confidence attached to attribution assessments. Some extreme events are amenable for *a priori* analysis of their probability of occurrence (for instance, droughts and heat waves), given expectations of near-term changes in sea surface conditions or external radiative forcings. Thus attribution assessments for the coming season, year, or decade could be conducted for regional scales with suitable methods.

Observational records of sufficient length and quality are required to define extreme events in relevant contexts and to characterize the range of variability in the particular climate variable and region of interest. In many regions observational data are not available over multi-decadal timescales or the data contains unphysical jumps or trends (inhomogeneities) related to measurement errors, changes in observational systems, or other non-physical factors (e.g. Stott and Thorne 2010). In many regions therefore extending the type of study made by Pall et al. (2011) of extreme river flow in the UK will be very challenging given the large gaps in many observational datasets. For example, the study showing human contribution to more intense precipitation extremes by Min et al. (2011) was restricted to limited regions of the Northern Hemisphere. Improvements in the robustness of the climate observing system, including improvements in in-situ observations, supported by remote observations and weather-forecast-related products (e.g. climate quality reanalyses), will be required to develop more reliable monitoring and attribution systems (Trenberth 2008). Estimates of remaining observational uncertainties including their time dependence will also be a very important ingredient for robust event attribution assessments (Thorne et al. 2011).

Physical understanding complements a statistical modeling approach and is essential for developing confidence in modeling and statistically based approaches. The 2010 Russian heat wave and the 2011 Texas heat wave/drought were examples of climate extremes in which dynamical processes played a dominant role in the event's origin, with its extreme intensity likely aided by land surface feedbacks (Dole et al. 2011; Hoerling et al. 2012). Many climate extremes typically reflect regional climate controls (e.g. Alexander 2011) as well as remote linkages (teleconnections) to global climate. Rigorous attribution assessments will therefore require validation that physical and dynamical processes are sufficiently well represented in the models being used.

Models are required in order to generate the counterfactual worlds in which particular factors are absent. Multiple models help assess structural modeling uncertainty and large ensembles help sample the tails of distributions. Ensemble sizes need to be tailored for the application, recognizing that not all event attribution requires very large ensembles. For example Perlwitz et al. (2009) and Dole et al. (2011) find that 50 member ensembles are adequate for identifying a shift in the mean. However, larger ensembles or statistical models may be required to indentify changes in the tails of distributions. To the extent that causal factors can be identified and model deficiencies addressed, the development of event attribution could lead to improved predictions of such events in the future.

Pre-agreed procedures and regular assessments conducted regardless of whether extreme weather events occur in a particular region or not favor largely objective results not distorted by selection effects (Stott et al. 2012a). Nevertheless, some level of expert judgment will always be necessary for defining those procedures and in interpreting the robustness of and confidence in results. Any such judgment can only be based on a careful consideration of the reliability of an attribution assessment and therefore on the extent to which there is a good physical understanding of the causal links behind the event in question.

Human influences have increased the risk of some extreme weather- and climate-related events, reduced the risk of others, and for some may not have affected the risk substantially. A few published studies have made assessments of particular events, reporting an attributable human influence on the probability of some (including the Autumn 2000 flooding in the UK, the 2003 European heatwave), the magnitude of others as being attributable to natural variability (the Russian heatwave of 2010 and the Texas heat wave of 2011) and showing that some cold events are consistent with the interplay of on-going global warming and internal variability (e.g. the cold North American temperatures in 2008, the cold European winter of 2009–2010). While such initial studies demonstrate the potential for event attribution they also highlight many of the challenges still to be faced, as discussed in this article. An important consideration is that regional attribution resulting from one region is not necessarily portable to another region even when the two regions are relatively close geographically. Therefore future research will need to consider a wider range of regions and event types as well as investigate the robustness of attribution results for events already considered.

The potential of weather and climate-related event attribution to societies can only be realized with the further underpinning of research needed to develop physical understanding, and improve the observational and modeling basis. As the scientific underpinning develops, it will be important to have realistic expectations of what can be achieved. While it is possible for an event attribution service to provide quantitative results, it is much harder to provide carefully assessed results that include sufficiently well calibrated information that would enable a user to fully understand the qualities and limitations of the information provided. Therefore future progress in serving the needs of the public, policy makers and other stakeholders depends on further development of the underpinning climate science and effective communication of attribution results, including their remaining uncertainties.

Acknowledgements The authors acknowledge the support of the UK FCO in funding meetings of the Attribution of Climate-related Events initiative. The authors acknowledge the support of the International Detection and Attribution Group by the US Department of Energy's Office of Science, Office of Biological and Environmental Research and the US National Oceanic and Atmospheric Administration's Climate Program Office. DAS received funding from Microsoft Research. PAS and NC were supported by the Joint DECC/Defra Met Office Hadley Centre Climate Programme (GA01101). CH acknowledges the Centre for Ecology and Hydrology Science Budget.

References

Adam D (2011) Climate change in court. Nat Clim Chang 1:127–130

Alexander LV (2011) Climate science: extreme heat rooted in dry soils. Nat Geosci 4:12–13. doi:10.1038/ngeo1045

Allen MR (2003) Liability for climate change. Nature 421:891–892

Allen M, Pall P, Stone D, Stott P, Frame D, Min S-K, Nozawa T, Yukimoto S (2007) Scientific challenges in the attribution of harm to human influence on climate. Univ Pa Law Rev 155:1353–1400

Arribas A, Glover M, Maidens A, Peterson K, Gordon M, MacLachlan C (2009) Towards a new generation of seasonal forecasting systems. Fisica de la Tierra 21:219–224

Barriopedro D, Fischer EM, Luterbacher J, Trigo R, Ricardo M (2011) The hot summer of 2010: redrawing the temperature record map of Europe. Science 332:220–224

BBC (2011) Last December UK's coldest for 100 years. Online piece available at http://www.bbc.co.uk/news/science-environment-12119329

Beniston M (2004) The 2003 heat wave in Europe: a shape of things to come? An analysis based on Swiss climatological data and model simulations. Geophys Res Lett 31:L02202. doi:10.1029/2003GL018857

Blunden J, Arndt DS (2012) State of the climate in 2011. Bull Am Meteor Soc 93(7):S1–S264

Bowler NE, Arribas A, Beare SE, Mylne KR, Shutts GJ (2009) The local ETKF and SKEB: upgrades to the MOGREPS short-range ensemble prediction system. Q J R Meteorol Soc 135:767–776

Büntgen U, Brázdil R, Heussner K-U, Hofmann J, Kontic R, Kyncl T, Pfister C, Chromá K, Tegel W (2011) Combined dendro-documentary evidence of central European hydroclimatic springtime extremes over the last millennium. Q Sci Rev 30:3947–3959

Cattiaux J, Vautard R, Cassou C, Yiou P, Masson-Delmotte V, Codron F (2010a) Winter 2010 in Europe: a cold extreme in a warming climate. Geophys Res Lett 37:L20704. doi:10.10129/2010GL044613

Cattiaux J, Vautard R, Yiou P (2010b) North-Atlantic SST amplified recent wintertime European land temperature extremes and trends. Clim Dyn 26:2113–2128. doi:10.1007/s00382-010-0869-0

Christidis N, Stott PA (2012) Lengthened odds of the cold UK winter of 2010/11 attributable to human influence. In: Peterson TC, Stott PA, Herring S (eds) Explaining extreme events of 2011 from a climate perspective. Bull Am Meteor Soc 93:1041–1067. doi:10.1175/BAMS-D-12-00021.1

Christidis N, Stott PA, Brown S, Hegerl GC, Caesar J (2010) Detection of changes in temperature extremes during the second half of the 20th century. Geophys Res Lett 32:L20716. doi:10.1029/2005GL023885

Christidis N, Stott PA, Jones GS, Shiogama H, Nozawa T, Luterbacher J (2010a) Human activity and anomalously warm seasons in Europe. Int J Climatol. doi:10.1002/joc.2262

Christidis N, Stott PA, Zwiers FW, Shiogama H, Nozawa T (2010b) Probabilistic estimates of recent changes in temperature: a multi-scale attribution analysis. Clim Dyn 34:1139–1156. doi:10.1007/s00382-009-0615-7

Christidis N, Stott PA, Brown SJ (2011a) The role of human activity in the recent warming of extremely warm days. J Clim 24:1922–1930. doi:10.1175/2011JCLI4150.1

Christidis N, Stott PA, Zwiers FW, Shiogama H, Nozawa T (2011b) The contribution of anthropogenic forcings to regional changes in temperature during the last decade. Clim Dyn. doi:10.1007/s00382-011-1184-0

Christidis N, Stott PA, Scaife AA, Arribas A, Jones GS, Copsey D, Knight JR, Tennant WJ (2012) A new HadGEM3-A based system for attribution of weather and climate-related extreme events. J Climate. doi:10.1175/JCLI-D-12-00169.1

Collins M, Booth BBB, Harris GR, Murphy JM, Sexton DMH, Webb MJ (2006) Towards quantifying uncertainty in transient climate change. Clim Dyn 27:127–147

Coumou D, Rahmstorf S (2012) A decade of weather extremes. Nat Clim Chang. doi:10.1038/NCLIMATE1452

David A Grossman (2003) Warming up to a not-so-radical idea: tort-based climate change litigation, 28 COLUM. J Envtl L 1:23
Dole R et al (2011) Was there a basis for anticipating the 2010 Russian heat wave? Geophys Res Lett 38:L06702. doi:10.1029/2010GL046582
Haufler V (Lead Author), Maggie L. Walser (Topic Editor) (2009) Insurance and reinsurance in a changing climate. In: Cleveland CJ (ed) Encyclopedia of earth. Environmental Information Coalition, National Council for Science and the Environment. Washington, DC. First published in the encyclopedia of earth July 24, 2009; Last revised Date August 30, 2011. http://www.eoearth.org/article/Insurance_and_reinsurance_in_a_changing_climate. Retrieved 18 May 2012
Hawkins E, Sutton R (2012) Time of emergence of climate signals. Geophys Res Lett 39:L01702. doi:10.1029/2011GL050087
Hegerl GC, Zwiers FW (2011) Use of models in detection and attribution of climate change. WIREs Clim Chang 2:570–591
Hegerl GC, Zwiers FW, Barconnot P, Gillett NP, Luo Y, Merengo Orsini JA, Nicholls N, Penner JE, Stott PA (2007) Understanding and attributing climate change. In: Climate change 2007: the physical science basis. Contribution of Working Group I to the Fourth Assessment Report of the Intergovernmental Panel on Climate Change, Cambridge
Hegerl GC, Hoegh-Guldberg O, Casassa G, Hoerling MP, Kovats RS, Parmesan C, Pierce DW, Stott PA (2010) Good practice guidance paper on detection and attribution related to anthropogenic climate change. In: Stocker TF, Field CB, Qin D, Barros V, Plattner G-K, Tignor M, Midgely PM, Ebi KL (eds). Meeting report of the intergovernmental panel on climate change expert meeting on detection and attribution of anthropogenic climate change. IPCC working group I Technical Support Unit, University of Bern, Bern
Hoegh-Guldberg O, Hegerl G, Root T, Zwiers F, Stott P, Pierce D, Allen M (2011) Difficult but not impossible. Nat Clim Change 1:72
Hoerling MP et al (2012) Anatomy of an extreme event. J Climate (submitted)
Hulme M, O'Neill SJ, Dessai S (2011) Is weather event attribution necessary for adaptation funding? Science 334:764
IPCC (2007) Climate Change 2007: impacts, adaptation and vulnerability. Contribution of Working Group II to the Fourth Assessment Report of the Intergovernmental Panel on Climate Change, Parry ML, Canziani OF, Palutikof JP, van der Linden PJ, Hanson CE (eds). Cambridge University Press, Cambridge, 976 pp
Jones GS, Stott PA, Christidis N (2008) Human contribution to rapidly increasing frequency of very warm northern hemisphere summers. J Geophys Res 113:D02109. doi:10.1029/2007JD008914
Kalnay E, Kanamitsu M, Kistler R, Collins W, Deaven D, Gandin L, Iredell M, Saha S, White G, Woollen J, Zhu Y, Leetmaa A, Reynolds R, Chelliah M, Ebisuzaki W, Higgins W, Janowiak J, Mo KC, Ropelewski C, Wang J, Jenne R, Joseph D (1996) The NCEP/NCAR 40-year reanalysis project. Bull Am Meteorol Soc 77:437–470
Kay AL, Crooks SM, Pall P, Stone DA (2011) Attribution of autumn/winter 2000 flood risk in England to anthropogenic climate change: a catchment-based study. J Hydrol 406:97–112
Kinter J, Folland C (2011) The international CLIVAR Climate of the 20th Century project: report of the Fifth Workshop. CLIVAR Exchanges 57:39–42
Knutti R, Furrrer R, Tebaldi C, Cermak J, Meehl GA (2010) Challenges in combining projections from multiple climate models. J Clim 23:2739–2758
Leiserowitz A, Maibach E, Roser-Renouf C, Hmielowski JD (2012a) Climate change in the American mind: Americans' global warming beliefs and attitudes in March 2012. Yale University and George Mason University. Yale Project on Climate Change Communication, New Haven, CT. Available at http://environment.yale.edu/climate/files/Climate-Beliefs-March-2012.pdf
Leiserowitz A, Maibach E, Roser-Renouf C, Hmielowski JD (2012b) Extreme weather, climate & preparedness in the American mind. Yale University and George Mason University. Yale Project on Climate Change Communication, New Haven. Available at http://environment.yale.edu/climate/files/Extreme-Weather-Climate-Preparedness.pdf

Massey N, Aina T, Rye, C, Otto FEL, Wilson S, Jones RG, Allen MR (2012) Have the odds of warm November temperatures and of cold December temperatures in Central England changed? In: Peterson TC, Stott PA, Herring S (eds) 2012. Explaining extreme events of 2011 from a climate perspective. Bull Am Meteor Soc 93:1041–1067. doi:10.1175/BAMS-D-12-00021.1
Min S-K, Zhang X, Zwiers FW, Hegerl GC (2011) Human contribution to more intense precipitation extremes. Nature 470:378–381
Morak S, Hegerl G, Kenyon J (2011) Detectable regional changes in the number of warm nights. Geophys Res Lett 38:L17703
Morak S, Hegerl GC, Christidis N (2012) Detectable changes in the frequency of temperature extremes. J Climate 26:1561–1574
Murphy JM, Sexton DMH, Barnett DN, Jones GS, Webb MJ, Collins M, Stainforth DA (2004) Quantification of modelling uncertainties in a large ensemble of climate change simulations. Nature 430:768–772
Nature editorial (2011) Heavy weather. Nature 477:131–132
Nozawa T, Nagashima T, Shiogama H, Crooks S (2005) Detecting natural influence on surface air temperature in the early twentieth century. Geophys Res Lett 32:L20719. doi:10.1029/2005GL023540
Otto FEL, Massey N, van Oldenborgh GJ, Jones RG, Allen MR (2012) Reconciling two approaches to attribution of the 2010 Russian heatwave. Geophys Res Lett 39:L04702. doi:10.1029/2011GL050422
Pall P, Aina T, Stone DA, Stott PA, Nozawa T, Hilberts AGJ, Lohmann D, Allen MR (2011) Anthropogenic greenhouse gas contribution to flood risk in England and Wales in autumn 2000. Nature 470:382–385
Perlwitz J, Hoerling M, Eischeid J, Xu T, Kumar A (2009) A strong bout of natural cooling in 2008. Geophys Res Lett 36
Peterson TC, Stott PA, Herring S (eds) (2012) Explaining extreme events of 2011 from a climate perspective. Bull Am Meteor Soc 93:1041–1067. doi:10.1073/pnas.1101766108
Rahmstorf S, Coumou D (2011) Increase of extreme events in a warming world. Proc Natl Acad Sci. doi:10.1073/pnas.1101766108
Räisänen J, Ruokolainen L (2008) Estimating present climate in a warming world: a model-based approach. Clim Dyn 31:573–585
Rayner NA, Parker DE, Horton EB, Folland CK, Alexander LV, Rowell DP, Kent EC, Kaplan A (2003) Global analyses of sea surface temperature, sea ice, and night marine air temperature since the late nineteenth century. J Geophys Res 108(D14):4407. doi:10.1029/2002JD002670
Royal Society (2009) Geoengineering the climate: science, governance and uncertainty (Report), London, UK. p 1. ISBN 978-0-85403-773-5
Sampei Y, Aoyagi-Usoi M (2009) Mass-media coverage, its influence on public awareness of climate-change issue, and implications for Japan's national campaign to reduce greenhouse gas emissions. Glob Environ Chang 19:203–212
Scaife AA et al (2008) The CLIVAR C20C project: selected twentieth century climate events. Clim Dyn. doi:10.1007/s00382-008-0451-1
Scaife AA, Woolings T, Knight J, Martin G, Hinton T (2010) Atmospheric blocking and mean biases in models. J Clim 23:6143–6152
Schar C, Jendritzky G (2004) Hot news from summer 2003. Nature 432:559–560
Schiermeier Q (2011) Extreme measures. Nature 477:148–149
Seneviratne SI, Nicholls N, Easterling D, Goodess CM, Kanae S, Kossin J, Luo Y, Marengo J, McInnes K, Rahimi M, Reichstein M, Sorteberg A, Vera C, Zhang X (2012) Changes in climate extremes and their impacts on the natural physical environment. In: Field CB, Barros V, Stocker TF, Qin D, Dokken DJ, Ebi KL, Mastrandrea MD, MachKJ, Plattner G-K, Allen SK, Tignor M, Midgley PM (eds) Managing the Risks of Extreme Events and Disasters to Advance Climate Change Adaptation. A special report of working groups I and II of the Intergovernmental Panel on Climate Change (IPCC). Cambridge University Press, Cambridge/New York, pp 109–230

Smith RL (2003) Statistics of extremes, with applications in environment, insurance and finance. In: Finkenstadt B, Rootzen H (eds) Extreme values in finance, telecommunications and the environment. Chapman and Hall/CRC Press, London

Smith DM, Scaife AA, Kirtman BP (2012) What is the current state of scientific knowledge with regard to seasonal and decadal forecasting ? Environ Res Lett 7:015602. doi:10.1088/1748-9326/7/1/015602

Stone DA, Allen MR (2005) Tne end-to-end attribution problem: from emissions to impacts. Clim Change 71:303–318

Stott PA, Thorne PW (2010) How best to log local temperatures? Nature 465:158–159

Stott PA, Stone DA, Allen MR (2004) Human contribution to the European heatwave of 2003. Nature 432:610–614

Stott PA, Jones GS, Lowe JA, Thorne P, Durman CF, Johns TC, Thelen J-C (2006) Transient climate simulations with the HadGEM1 climate model: causes of past warming and future climate change. J Clim 19:2763–2782

Stott PA, Gillett NP, Hegerl GC, Karoly D, Stone D, Zhang X, Zwiers F (2010) Detection and attribution of climate change: a regional perspective. WIREs Clim Chang 1:192–211

Stott PA, Christidis N, Betts RA (2011) Changing return periods of weather-related impacts: the attribution challenge. Clim Chang 109:263–268

Stott PA, Peterson TC, Herring S (2012a) Introduction. In: Peterson T, Stott PA, Herring S (eds) Explaining extreme events of 2011 from a climate perspective. BAMS, 93, 1041-1047, 10.1175/BAMS-D-12-00021.1

Stott PA, Peterson TC, Herring S (2012b) Conclusions. In: Peterson T, Stott PA, Herring S (eds) Explaining extreme events of 2011 from a climate perspective. BAMS, 93, 1041-1047, 10.1175/BAMS-D-12-00021.1

Thorne PW et al (2011) Guiding the creation of a comprehensive temperature resource for twenty-first century climate science. Bull Am Meteorol Soc 92(11):ES40–ES47

Trenberth KE (2012) Framing the way to relate climate extremes to climate change. Clim Change 115(2):283–290. doi:10.1007/s10584-012-0441-5

Trenberth KE (2008) Observational needs for climate prediction and adaptation. WMO Bull 57:17–21

Trenberth KE (2011) Attribution of climate variations and trends to human influences and natural variability. WIREs Clim Change 2(6):925–930. doi:10.1002/wcc.142, Wiley-Blackwell

Vautard R, Yiou P (2009) Control of recent European surface climate change by atmospheric flow. Geophys Res Lett 36, L22702. doi:10.1029/2009GL040480

Wilks DS (1995) Statistical methods in the atmospheric sciences: an introduction. Academic, San Diego, 467pp

Zwiers FW, Zhang X, Feng Y (2011) Anthropogenic influence on long return period daily temperature extremes at regional scales. J Clim 24:881–892

Climate Extremes: Challenges in Estimating and Understanding Recent Changes in the Frequency and Intensity of Extreme Climate and Weather Events

Francis W. Zwiers, Lisa V. Alexander, Gabriele C. Hegerl, Thomas R. Knutson, James P. Kossin, Phillippe Naveau, Neville Nicholls, Christoph Schär, Sonia I. Seneviratne, and Xuebin Zhang

Abstract This paper focuses primarily on extremes in the historical instrumental period. We consider a range of phenomena, including temperature and precipitation extremes, tropical and extra-tropical storms, hydrological extremes, and transient extreme sea-level events. We also discuss the extent to which detection and attribution research has been able to link observed changes to external forcing of the climate system. Robust results are available that detect and often attribute changes in frequency and intensity of temperature extremes to external forcing. There is also some evidence that on a global scale, precipitation extremes have intensified due

---

Chapter Contributors M Donat, O Krueger, S Morak, TQ Murdock, M Schnorbus, V Ryabin, C Tebaldi, XL Wang

F.W. Zwiers (✉)
Pacific Climate Impacts Consortium, University of Victoria,
P.O. Box 1700 STN CSC, Victoria, BC V8W 2Y2, Canada
e-mail: fwzwiers@uvic.ca

L.V. Alexander
Climate Change Research Centre and ARC Centre of Excellence for Climate System Science,
The University of New South Wales UNSW, Sydney, NSW 2052, Australia
e-mail: l.alexander@unsw.edu.au

G.C. Hegerl
School of Geosciences, Grant Institute, University of Edinburgh,
The King's Buildings, West Mains Road, Edinburgh, EH9 3JW, UK
e-mail: Gabi.Hegerl@ed.ac.uk

T.R. Knutson
Geophysical Fluid Dynamics Laboratory, NOAA, Princeton University,
Forrestal Campus 201 Forrestal Road, Princeton, NJ 08540-6649, USA
e-mail: Tom.Knutson@noaa.gov

J.P. Kossin
NOAA's National Climatic Data Center, 151 Patton Avenue,
Asheville, NC 28801-5001, USA
e-mail: kossin@ssec.wisc.edu

G.R. Asrar and J.W. Hurrell (eds.), *Climate Science for Serving Society: Research, Modeling and Prediction Priorities*, DOI 10.1007/978-94-007-6692-1_13,

to forcing. However, robustly detecting and attributing forced changes in other important extremes, such as tropical and extratropical storms or drought remains challenging.

In our review we find that there are multiple challenges that constrain advances in research on extremes. These include the state of the historical observational record, limitations in the statistical and other tools that are used for analyzing observed changes in extremes, limitations in the understanding of the processes that are involved in the production of extreme events, and in the ability to describe the natural variability of extremes with models and other tools.

Despite these challenges, it is clear that enormous progress is being made in the quest to improve the understanding of extreme events, and ultimately, to produce predictive products that will help society to manage the associated risks.

Keywords Extremes • Extremes indices • Detection and attribution • Temperature and precipitation extremes • Extratropical storms • Tropical cyclones • Flood • Drought • Sea level

1 Introduction

This paper reviews some aspects of the current status of research on changes in climate extremes, identifying gaps and issues that warrant additional work. It focuses primarily on the historical instrumental period, giving a sense of the nature of the results that have been obtained and the challenges that arise from observational, methodological and climate modeling uncertainties. It also discusses the extent to which detection and attribution research has been able to link observed changes to external forcing of the climate system. In addition, the paper also very briefly

P. Naveau
Laboratoire des Sciences du Climat et l'Environnement (LSCE), CNRS,
Orme des Merisiers/Bat. 701 C.E. Saclay, Gif-sur-Yvette, 91191 Paris, France
e-mail: Philippe.Naveau@lsce.ipsl.fr

N. Nicholls
School of Geography and Environmental Science, Monash University,
Building 11, Clayton Campus, Wellington Road, Clayton, VIC 3800, Australia
e-mail: Neville.Nicholls@monash.edu

C. Schär • S.I. Seneviratne
Institute for Atmospheric and Climate Science, ETH Zurich,
Universitätstrasse 16, 8092, Zurich, Switzerland
e-mail: christoph.schaer@env.ethz.ch; sonia.seneviratne@env.ethz.ch

X. Zhang
Climate Research Division, Environment Canada, 4905 Dufferin Street,
Toronto, ON M3H 5T4, Canada
e-mail: Xuebin.Zhang@ec.gc.ca

discusses some aspects of projections for the twenty-first century, although this is not its primary focus. Extremes are not discussed on paleo time scales, in the context of the present (i.e., short term forecasting), or in the context of climate surprises (tipping points). These choices reflect our desire not to attempt too broad a review of the topic due to space constraints, as well as a view that high priority should be given to reducing uncertainty in the understanding of historical changes in extremes over the instrumental period as a prerequisite to confidently predicting changes over the next century. This includes the development of improved and comprehensive observational records, the development of better physical models, forcing data sets and more powerful statistical techniques, the development and refinement of the understanding of the physical processes that produce extremes, and continued improvement in the ability to attribute causes to those changes. Overall progress on understanding implications of ongoing and future changes in extremes will be strongly dependent upon the ability to document and understand changes in extremes during the period of history that has been (and continues to be) the most comprehensively and directly observed, which is why this is the topic of the present paper. While it is not the focus of this paper, it is clearly also very important to understand changes in extremes over longer periods of history, particularly where proxy data indicate larger extremes than observed during the modern instrumental period, such as for regional drought (e.g., Woodhouse and Overpeck 1998; Woodhouse et al. 2010).

Before beginning our review, it is worth taking a few minutes to think about the terminology that is used to describe extremes in climate science (see also Seneviratne et al. 2012, Box 3-1). Considerable confusion results from the various definitions of extremes that are in use. Part of this confusion occurs because the word *extreme* can be used to describe either a characteristic of a climate variable or that of an impact. In the case of a climate variable, such as surface air temperature or precipitation, the notion of an extreme is reasonably well defined and refers to values in the tails of the variable's distribution that would be expected to occur relatively infrequently. However, even in this case, there can be ambiguity concerning the definition of extremes. For example, a great deal of climate research on "extremes" deals with indicators of the frequency or intensity of events that, in fact, describe parts of the distribution that are not very extreme, such as warm events that occur beyond the 90th percentile of daily maximum temperature. Such events lie well within the observations that are collected each season, and they are typically studied by determining whether there are trends in their rates of occurrence. They are often referred to as "moderate extremes" in the literature (and we will also use that term occasionally below), but this term is not one that is used in statistical science to describe the upper part of a distribution, since the 90th percentile of daily values, for example, while in the upper tail would not necessarily be considered extreme in a statistical sense. The mechanisms involved in these 'moderate extremes' nevertheless should be similar to those involved in truly extreme events, and they are affected by different model biases from those for mean values (Hanlon et al. 2012a). There are also instances when the distribution of exceedances above the 90th percentile can be well approximated by an extreme value distribution. Nor does the term "moderate extremes" comprehensively describe the collection of ETCCDI[1] indices (Klein Tank et al. 2009) since they characterize various points in the distributions of daily temperature and precipitation observations, including diagnostics of daily variability that is not extreme, at least not everywhere, such as the annual number of frost days.

In addition to the literature on indices, or "moderate extremes" of climate variables, there is also a body of work that deals with rare values of climate variables that are generally not expected to recur each year. In this case the concept corresponds well to that used in the statistical sciences, and thus powerful statistical tools based on extreme value theory are available to aid in the analysis of historical and future extremes (e.g., Coles 2001; Katz et al. 2002). Such tools were originally developed to make statements about what might happen outside the range of the observed sample, such as the problem of estimating the 100-year return value on the basis of a 30- or 40-year sample. Hence, the notion of "extremes" in that context is defined as very high quantiles, such as the 95th, 99th or 99.9th percentiles of annual maximum values. An important aspect of this theory is to quantify the uncertainty of such extrapolations through the computation of suitably constructed confidence intervals. Increasingly, these tools are being used in the evaluation extreme events simulated in climate models (e.g., Kharin et al. 2007; Wehner et al. 2010; Wehner 2013). These tools are being further developed in the statistical sciences, and there is currently a very high level of interaction between that community and the climate sciences community on the development and application of methods that can be used in the climate sciences, such as the ExtREmes toolkit (see http://cran.r-project.org/web/packages/extRemes/).

In the case of extremes defined by their impacts, the concept of what constitutes an extreme may be less well defined, and this may affect the approaches that are available for analysis. For example, all tropical cyclones that are classified as Category 1–5 storms on the Saffir-Simpson scale are considered to be extreme because of their high potential to cause damage from high winds, rainfall, and/or storm surge flooding. These storms are an important component of the energetics climate system and occur in more or less constant numbers (globally) each year. They are more difficult to characterize statistically than, for example, extreme temperature events that are identified relative to variability recorded at fixed locations. The numbers of tropical cyclones within a region are not constant, the regions affected vary with time, and historical data that might be used to locate tropical cyclones in the tails of an appropriate probability distribution, while being constantly improved, often remain subject to substantial inhomogeneities due to the evolution of our observing systems (Knutson et al. 2010; Seneviratne et al. 2012).

---

[1] The joint World Meteorological Organization Commission on Climatology (CCl), World Climate Research Program Climate Variability and Predictability project (CLIVAR), and Joint Commission on Marine Meteorology (JCOMM) Expert Team on Climate Change Detection and Indices. See http://www.clivar.org/category/panels/etccdi

For the purpose of this article we consider "extreme events" to be <u>well-defined</u> weather or climate events (including tropical cyclones) that are <u>rare</u> within the current climate. With the term "well-defined" it is understood that these events may be defined in terms of measurable physical quantities such as temperature, precipitation, wind speed, runoff levels or similar; and the term "rare" is used to refer to values in the tails of the variable's distribution as discussed above, starting from the 90th percentile of the distribution to capture research on 'moderate' extremes.

It is important to note that the linkage between extreme events and extreme impacts (i.e. natural disasters) is not straightforward. Events that are rare from a statistical perspective may not necessarily lead to impacts if there is either no exposure or no vulnerability to the particular event. Also, the impact of an extreme event may depend on its season, its duration, and co-occurrence of further extremes, such as drought conditions with heat waves (Seneviratne et al. 2012). The occurrence of an extreme event does not necessarily imply monetary damages. Rather the occurrence of damages also depends upon whether there is any infrastructure at risk and its characteristics, population density, factors affecting the vulnerability of the population including whether emergency response measures are in place, etc (IPCC 2012). Conversely, not all damages from weather or climate events are related to extreme events as defined above. For instance, poor building practices may allow a "normal" or moderate event to generate extreme damages. For example, while the 2011 Thailand flood caused more than eight billion US dollars in insured damages, the amount of rain that fell in the region was not very unusual (van Oldenborgh et al. 2012). This issue is very familiar to the re-insurance industry, which uses damage models to link extreme events to impacts (e.g. Klawa and Ulbrich 2003; Watson and Johnson 2004). Extreme impacts in ecosystems may also occur following moderate events, e.g. when these are compounded with other climate events (see discussion in Hegerl et al. 2011 and Seneviratne et al. 2012).

The structure of the remainder of this paper is as follows. The paper begins in Sect. 2 with a discussion of the status of research on simple indices that are derived from daily (or occasionally more frequent) observations that are collected primarily at operational meteorological stations. The main focus here is on temperature and precipitation extremes, but wind extremes derived from station data are also discussed. Section 3 discusses storms (extra-tropical cyclones, tropical cyclones and tornadoes). This is followed by a discussion of hydrological extremes (droughts and floods) in Sect. 4, and extreme sea-levels (e.g., storm surge events) in Sect. 5. A summary and recommendations are presented in Sect. 6. Amongst other sources, the paper draws upon the IPCC 4th Assessment Report (IPCC 2007a, b), the US Global Change Program Special Assessment Product on extremes (i.e., CCSP 3.3, Karl et al. 2008), the recent WMO assessment on tropical cyclones (Knutson et al. 2010), a recently completed review of research on indices by Zhang et al. (2011), and on the IPCC Special Report on Extremes (Seneviratne et al. 2012).

2 Simple Indices Derived from Daily Data

2.1 Introduction

The indices that are discussed in this section are generally derived from daily observations of individual meteorological variables, such as temperature or precipitation. Indices calculated from daily data have appeal for a number of reasons, including the fact that they are relatively easy to calculate and that they summarize information on changes in variability compactly, and in a way that is accessible to a broad range of users.

Indices have been designed to characterize different parts of the distribution of a given variable. The indices that are of interest here are those that characterize aspects of the tails of the distribution (the "extremes") since these tend to be more relevant to society and natural systems than indices that characterize aspects of the distribution that occur more frequently, since extreme events are more likely to cause societal or environmental damage. However, a benefit of 'moderate' extremes is that they are better sampled and hence estimates of change in these kinds of extremes are less uncertain than estimates of changes in extremes that are further out in the tail of the distribution (Frei and Schär 2001).

Most indices of extremes tend to represent only "moderate extremes," i.e. those that typically occur at least once a year. In many cases, changes in the tails of the distribution, as indicated by changes in the indices, are essentially similar to those in other parts of the distribution (Fig. 1). However, even for temperature, changes may be seen that are not consistent between means and extremes, minimum and maximum, and upper and lower tail (e.g., Hegerl et al. 2004; Kharin et al. 2007) due to soil freezing, alterations in feedback processes, or energy balance constraints that may affect different parts of the distribution differently (e.g., Fischer and Schär 2009; Zazulie et al. 2010; Hirschi et al. 2011; Mueller and Seneviratne 2012). This can lead, for example, to strong changes where ice and snow-cover changes (Kharin and Zwiers 2005). Some indices for climate extremes can also be used for secondary inference; for example, statistical extreme value theory can be used to estimate long return period precipitation amounts from long time series of annual maximum

Fig. 1 (continued) Extremes are denoted by the *shaded areas*. In the case of temperature, changes in the frequencies of extremes are strongly affected by changes in the mean; a relatively small shift of the distribution to the right would substantially increase warm extremes and decrease cold extremes. In addition, the frequency of extremes can also be affected by changes in the shape of the tails of the temperature distribution, which could become wider or narrower, or could become somewhat skewed rather than being symmetric as depicted. In a skewed distribution such as that of precipitation, a change in the mean of the distribution generally affects its variability or spread, and thus an increase in mean precipitation would also likely imply an increase in heavy precipitation extremes, and vice-versa. In addition, the shape of the right hand tail could also change, affecting extremes. Furthermore, climate change may alter the frequency of precipitation and the duration of dry spells between precipitation events (From Zhang and Zwiers (2013), after Folland et al. (1995) and Peterson et al. (2008))

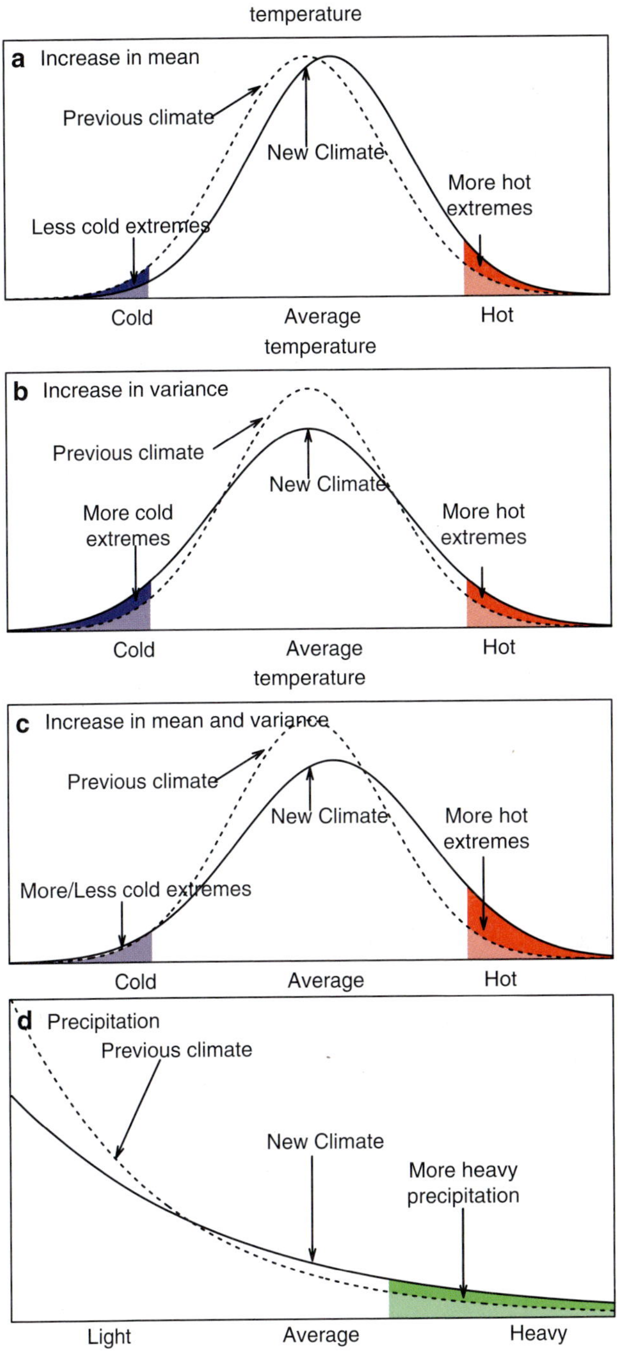

Fig. 1 Schematic representations of the probability distributions of daily temperature, which tends to be approximately Gaussian (exceptions can be caused by soil freezing, feedbacks, or energy balance constraints, see text), and daily precipitation, which has a skewed distribution.

daily precipitation amounts (Klein Tank et al. 2009). It should be noted that the estimation of return levels is often based on the assumption of spatial and/or temporal independence among sites or grid points (either on the raw data or conditionally on their distributional parameters). Consequently, uncertainties can be underestimated or these assumptions can be challenged. On the other hand, many studies also employ schemes that borrow information from adjacent locations to improve local parameter and return value estimates. Approaches range from simple averaging of key parameters across nearby grid points (e.g., Kharin and Zwiers 2000) to regional analysis approaches that derive spatial trends in distributional parameters estimated at different locations (e.g., Hanel et al. 2009).

In addition to indices that summarize various aspects of the tails of the daily variability of individual meteorological parameters, there have also been a variety of attempts to build indices that incorporate information from multiple parameters to summarize information related to impacts, such as fire weather indices that were first developed for operational use in wild fire risk management (e.g., Van Wagner 1987) and subsequently used to study the potential impacts of climate change on wild fire frequency and extent (e.g., Flannigan et al. 2005). Similar types of development are seen in a variety of indices (another example being health-related heat indices such as that described by Steadman 1979; Karl and Knight 1997; Fischer and Schär 2010; Sherwood and Huber 2010). Since these types of indices are impact specific, their construction must ultimately be informed by the characteristics and functioning of the system (ecological, social, or economic) or biological organism that is impacted (health, agriculture). This requires inter- and trans-disciplinary collaboration, and involves a range of potential compound indices far greater than would be required to monitor and understand change in the physical climate system.

2.2 *Status*

2.2.1 Temperature and Precipitation Indices

Many indices have been defined (e.g., Frich et al. 2002; Klein Tank et al. 2009) for the purpose of monitoring changes in the moderately far tails of surface variables such as temperature and precipitation that are routinely observed on a daily, or more frequent, basis. These indices include: (i) absolute quantities such as the annual maximum and minimum temperature and the annual maximum precipitation; (ii) the frequency of exceedance beyond a fixed absolute threshold, such as the annual count of the number of days with precipitation amounts greater than 20 mm; (iii) the frequency of exceedance above or below fixed relative thresholds such as the 90th percentile of daily maximum temperature or the 10th percentile of daily minimum temperature where the threshold is determined from a climatological base period such as 1961–1990; and (iv) dimensionless indices, such as the proportion of annual precipitation that is produced by events larger than the 95th percentile of daily precipitation amounts, where the threshold is again determined from a fixed base period.

These indices are studied because they describe aspects of temperature and precipitation variability that have been linked, to greater or lesser degrees, to societal or ecological impacts. Relative indices also have the advantage that they can be applied across different climatic zones. Their calculation is actively coordinated by the CLIVAR/CCl/JCOMM Joint Expert Team on Climate Change Detection and Indices (ETCCDI). The state of development of these indices has recently been reviewed comprehensively by Zhang et al. (2011). Further, Sillmann et al. (2013a, b) have recently described the performance of climate models participating in the Coupled Model Intercomparison Project Phase 3 (Meehl et al. 2007b) and Phase 5 (Taylor et al. 2012) in simulating observed and projected changes in the suite of ETCCDI indices.

The calculation of indices requires high quality, high frequency (daily or better), homogeneous meteorological data. High quality data are available from hydro-meteorological services in many parts of the world, and are often freely available for scientific research at least nationally, if not on a fully open basis internationally, though various limitations to (mostly raw) data access remain an issue (see also point i below). Data availability is generally greater in developed countries than in developing countries, where resources and/or mandate sometimes limit the collection and dissemination of daily meteorological observations, although restricted data access also remains a problem in some developed countries. The ETCCDI has an ongoing program of open source software development and international workshops that are intended to train developing world scientists in the homogenization of data that are collected by their hydro-meteorological services, and in the subsequent calculation of indices (Peterson and Manton 2008). The calculated indices are published in the peer-reviewed literature (e.g., Aguilar et al. 2009) and are subsequently contributed to global scale index datasets such as HadEX (Alexander et al. 2006) and its updates (e.g. Donat and Alexander 2011; Alexander and Donat 2011), thereby helping to improve the global coverage of these datasets and consequently enabling more confident global scale monitoring and detection and attribution.

While the ETCCDI type of approach is helpful, there are nevertheless ongoing challenges. These include:

(i) Concerns about the reproducibility of the entire chain of index production. Currently the reproducibility of the full processing sequence cannot be guaranteed because, while methods and codes are freely available, the underlying daily station data are not always openly accessible to the international scientific community since regional data gathering organizations may not have the capacity or mandate to support open data dissemination.

(ii) Lack of access to daily station data also implies a lack of access to metadata describing the history of observing stations. This is an important concern because small changes in observing station location or exposure can affect both the mean and variability of the recorded data, leading to large artificial changes in extremes (Katz and Brown 1992). In the absence of station metadata, it is often difficult to determine if such issues have affected indices derived from the underlying data.

(iii) Lack of real-time updating, particularly for regions that are unable to contribute to the Global Historical Climate Network (GHCN, see http://www.ncdc.noaa.gov/oa/climate/ghcn-daily/). This is a concern because maintaining and monitoring indices is not always part of the primary mandate of the developing world scientists who participate in the ETCCDI workshops and are involved in index development for their countries or regions. It should be noted however, that the Asia Pacific Network (APN; Manton et al. 2001), which has focused on a specific region, has been successful in running repeat workshops that have allowed for the updating of indices in that region.

(iv) While the indices provide much needed information on daily variability, some scientific information is lost when providing only a limited number of pieces of information about the distribution of daily temperature and precipitation. This is ameliorated somewhat by approaches to the analyses of indices (such as the annual extremes of daily minimum and maximum temperature) that are based on extreme value theory. Such methods can be used to make inferences about changes in extremes over time and are able to provide results for thresholds more extreme than that used to define the underlying index.

(v) Potential difficulties in characterizing the statistical distributions of some indices, particularly where extreme value theory cannot be directly applied, which makes it more difficult to make reliable statistical inferences about things such as the presence or absence of trend in a time series of annual indices.

(vi) Consideration of specific impacts often requires information that relies upon simultaneous values of several climate variables. For instance, health impacts from heat waves depend upon temperature and humidity (and additional factors), information that cannot be recovered from standard indices.

An additional challenge is that the spatial coverage of index datasets remains far from being truly global, with significant fractions of the globe still under-sampled, for example, in Africa and South America (see Fig. 2a–c). Further challenges in the production of global datasets are also related to the choice of gridding framework in addition to parameter choices that are made within a chosen gridding method (e.g. Donat and Alexander 2011). This adds additional uncertainty to long term variability measures and trend estimates. Nevertheless, even when different choices are made, trends are broadly similar, at least on a global scale and particularly for temperature extremes. Large differences in observed trends can be associated with data processing choices, such as whether the daily data are gridded first before the indices are calculated, as occurs when indices are derived from HadGHCND (red curve in Fig. 2d), or vice-versa as in HadEX2 (blue curve in Fig. 2d) or GHCNDEX (green curve in Fig. 2d). These sensitivities are addressed in some studies by using data that are processed in more than one way (Morak et al. 2011).

The index approach also has several scientific limitations. One such limitation, for which a solution has been found, is the possibility that inhomogeneities can be introduced into index time series unintentionally, such as can occur in the case of threshold crossing frequency indices when thresholds representative of the far tails are estimated from a fixed observational base period (e.g., Zhang et al. 2005).

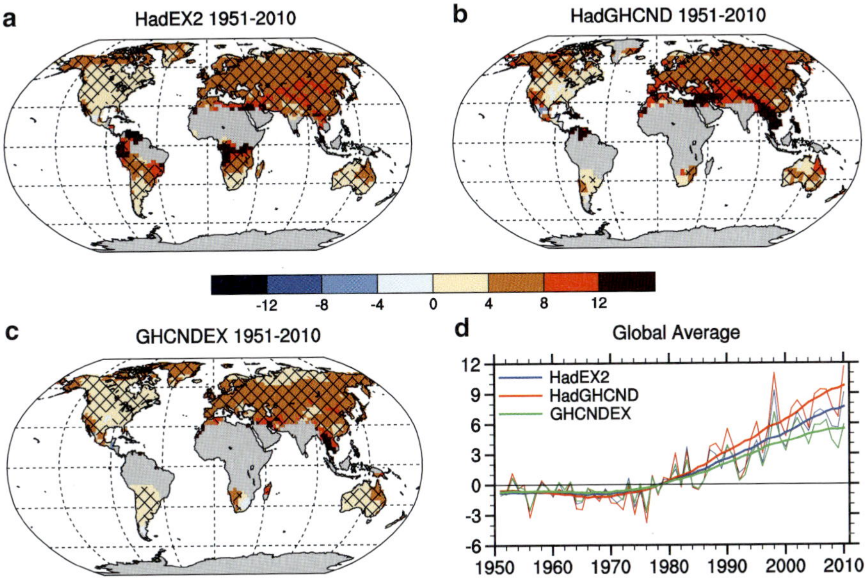

Fig. 2 Annual trends in warm nights (TN90p) using different datasets for the period 1951–2010 where at least 40 years are available. The datasets are (**a**) HadEX2 (Alexander and Donat 2011), (**b**) HadGHCNDEX (ETCCDI indices calculated from an updated version of HadGHCND (Caesar et al. 2006)) and (**c**) GHCNDEX (Donat and Alexander 2011). *Panel* (**d**) represents the global average time series plots for each of the three datasets presented as anomalies relative to the 1961–1990 with associated 21-year Gaussian filters

Another limitation, which can also be circumnavigated, is that differences in the recording resolution of observational data can cause non-climatic spatial variations in threshold crossing frequency and trends (e.g., Zhang et al. 2009). A third limitation is that in a changing climate, the number of exceedances of thresholds based on a climatological base climate may saturate, e.g. exceedances may never or almost always occur under strong climate change. Thus, percentage exceedance indices are only useful for characterizing change in a distribution that is not too far from the base period (see e.g. Portmann et al. 2009). A further limitation is that the nature of index data, which typically provides only one value per month (Alexander and Donat 2011), and in the earlier data, only one per year (Alexander et al. 2006), may limit the range of possible approaches that can be used to analyze change in certain types of extremes. For example, long return period extremes (e.g., the intensity of the 20-year extreme daily precipitation event) can be estimated from the annual extremes that are recorded in HadEX, but the analyst can only do so using the so-called block-maximum approach to extreme value analysis, which only considers the most extreme of a series of values observed within a block of a defined length (e.g. the annual maximum). In contrast, it is often argued by statisticians that the so-called peaks-over-threshold approach, by which all values exceeding a given threshold are used in the analysis, may result in more confident estimates of long period return values since it has the potential to utilize the information about extremes that is available in a long time series of daily values more effectively than the block-maximum approach. Dupuis (2012) gives a recent example of a peaks-over-threshold analysis for temperature extremes in several US cities. It should be noted however, that the peaks-over-threshold approach remains difficult to apply to large gridded datasets, such as the output from global climate models, because of the challenges associated with finding an automated procedure for reliably determining the appropriate threshold at each location in the grid. A further consideration is that most available index datasets do not currently provide the date (or dates) on which the extreme values were recorded. This creates a limitation when attempting to study the association between the occurrences of extremes in different variables or between climate extremes on the one hand and impacts on the other, and limits process based analyses of the conditions leading to recorded extremes. In contrast, the availability of monthly indices now makes it possible to study changes in the seasonality of extremes (see, for example, Morak et al. 2013).

As noted, methods have been developed to prevent inhomogeneities in indices that count exceedances beyond quantile based thresholds and to account for the effects of different data reporting resolutions (Zhang et al. 2005, 2009). Other limitations could be overcome by adding a modest number of additional indices to the "standard" ETCCDI list. For example, one could include within the suite of indices the r most extreme values observed annually for some small number $r>1$ and not just the most extreme value annually, thereby enabling the application of the more efficient "r-largest" extreme value analysis techniques (e.g., Smith 1986; Zhang et al. 2004). Another example would be to store the dates of the annual occurrence of indices. In addition, it would be appropriate to redefine the ETCCDI indices such that they describe annual extremes and counts that pertain to a year that is defined in a climatologically appropriate manner, where the definition of the year would depend upon location and parameter, taking into account the form of the annual cycle for the specific aspect of the parameter that is relevant for each index. This may be challenging in regions with complex annual cycles, such as those with multiple wet and dry seasons. It should also be noted that the definition of the year has implications for many types of indices and not just annual extremes as discussed above. A specific example is CDD (consecutive dry days, see Klein Tank et al. 2009), an index that can show very large changes in climate models under future emissions scenarios (e.g. Tebaldi et al. 2006; Orlowsky and Seneviratne 2012). CDD calculated on the basis of the calendar year has a different interpretation in places where the climatological dry period spans the year boundary as opposed to places where the climatological dry period occurs in the middle of the year; while dry periods may be of comparable length in both types of places, CDD will tend to report them as being substantially shorter in the former. In contrast, a CDD index that was calculated from years that are defined locally in such a way that the climatological dry period occurs everywhere in the middle of the year would have a more uniform interpretation across different locations.

There are a number of factors that limit our ability to evaluate how well models simulate indices in comparison to observed indices. These include observational limitations, such as limited spatial and temporal coverage of observing stations, and the likelihood that there are few regions in the world where precipitation station density is sufficient to reliably estimate daily grid box mean precipitation on GCM and RCM scales (see discussion in Zhang et al. 2007). As a consequence, model evaluation often relies on proxies for direct observations, such as reanalysis products. This is a reasonable approach for variables such as surface temperature that are reasonably well constrained by observations in reanalyses, but it is more problematic in the case of variables such as precipitation (e.g. Lorenz and Kunstmann 2012) that are generally not observationally constrained in reanalyses (the North American Regional Reanalysis, Mesinger et al. 2006, is an exception; it uses precipitation observations to adjust latent heating profiles). Furthermore, the observational data streams assimilated in reanalysis data products are not consistent over time, e.g. because of the relatively short length of satellite data, which may affect their use for the assessment of climatic trends (e.g. Bengtsson et al. 2004; Grant et al. 2008; Lorenz and Kunstmann 2012; Sillmann et al. 2013a). Taking these various limitations into account, models are found to simulate the climatology of surface temperature extremes with reasonable fidelity (Kharin et al. 2007; Sillmann et al. 2013a) on global and regional scales when compared against reanalyses, although there are uncertainties associated with, for example, the representation of land-atmosphere feedback processes in models (Seneviratne et al. 2006). In contrast, intercomparisons between models, reanalyses, and large scale observational precipitation products such as CMAP (Xie et al. 2003) suggest large uncertainties in all three types of precipitation products; particularly in the tropics (e.g., see Figure 6 in Kharin et al. 2007).

Scaling issues (e.g., differences between the statistical characteristics and spatial representativeness of point observations from rain gauges or gridded observed precipitation versus that of grid box mean quantities simulated by climate models; Klein Tank et al. 2009; Chen and Knutson 2008), uncertainties in observational gridded products (Donat and Alexander 2011), and incomplete process understanding continue to limit the extent to which direct quantitative comparison can be made between station observations and models (Mannshardt-Shamseldin et al. 2010). It should be noted, however, that models of sufficiently high resolution may be capable of simulating precipitation extremes of comparable intensity to observed extremes. For example, Wehner et al. (2010) show the global model that they study produces precipitation extremes comparable to observed extremes at a horizontal resolution of approximately 60 km. In contrast, most global models continue to operate at lower resolutions, leading to ambiguities in the interpretation of projected changes in extremes. Nevertheless, precipitation change at large scales is determined primarily by changes in the global hydrological cycle that are reflected in changes in evaporation, atmospheric moisture content, circulation (which affects moisture transport and convergence), and energy and moisture budgets, providing a fundamental basis for the qualitative (in terms of the direction of change and its large scale features), if not quantitative (in terms of the absolute values of the changes and their detailed geographic patterns), interpretation of modeled precipitation changes. The scaling issue can sometimes be circumnavigated by transforming observed and simulated precipitation into dimensionless quantities that can more readily be intercompared, such as has been done by Min et al. (2011). A disadvantage of such transformations, however, is that the translation of extremes onto a probability or other type of relative scale may impede the physical interpretation of trends and variability. Also, the application of such transforms requires strong assumptions concerning the physical processes that generate extremes at different scales that are difficult to evaluate.

2.2.2 Wind Indices

To date, temperature and precipitation indices have been studied most intensively. Indices of wind extremes, while of enormous importance in engineering applications, have received less attention, in part because of the greater difficulty in obtaining homogeneous high-frequency wind data. Wind records are often affected by non-climatic influences, such as development in the vicinity of an observing station that alters surface roughness over time. It has also been postulated by Vautard et al. (2010) that large scale changes in vegetative cover over many land areas has altered surface roughness and that this may be an important contributor to the apparent stilling (reduction) of surface wind speeds in many mid-latitude regions (e.g., see also Zwiers 1987; Roderick et al. 2007).

An alternative to using direct anemometer observations of wind speeds is to consider a proxy that is based on pressure readings that are usually more homogeneous than wind speed observations. Several storm proxies currently being used are derived from pressure readings at single stations, such as the statistics of 24-hourly local pressure changes or of the frequency of low pressure readings. These single station proxies relate to synoptic experience and reflect storminess indirectly as they seek to detect atmospheric disturbances (e.g. Schmith et al. 1998; Hanna et al. 2008; Allan et al. 2009; Bärring and von Storch 2004; Bärring and Fortuniak 2009). Another approach to explore past storminess is to make use of the statistics of geostrophic wind speeds. Geostrophic wind speeds can be derived by considering mean sea-level pressure gradients in networks of reliable surface pressure records over homogenous mid-latitude domains, such as the north-east Atlantic and western Europe (e.g., Schmidt and von Storch 1993; Alexandersson et al. 1998). These records, which continue to be developed in the North Atlantic and European region (e.g., Wang et al. 2011) and are also being developed for south-eastern Australia (e.g., Alexander et al. 2011), are available for much longer periods of record than the more limited anemometer network. For the North Atlantic region for which they have been most extensively developed, they show predominately the effects of natural low frequency variability in atmospheric circulation on variations in storminess and extreme geostrophic wind speeds.

Recently Krueger and von Storch (2011) used a regional climate model to evaluate the underlying assumption that the extremes of geostrophic wind speed are indeed representative of surface wind speed extremes, and found good correspondence between the two. They also considered the sensitivity of the proxy to the density of stations in the network, concluding that higher density networks should give more reliable estimates of wind speed extremes. Work is currently underway to evaluate the robustness of such proxies to instrumental error in pressure readings and to inhomogeneity in one or more of the surface pressure records that are used to derive the geostrophic winds. Further, a study that evaluates how well a number of single-station pressure proxies represent storminess has recently been completed (Krueger and von Storch 2012) and concludes that all single-station pressure proxies considered were linearly related to storm activity, with absolute pressure tendency being most strongly correlated.

Another possibility for the construction of wind speed and storminess indices is provided by reanalyses, such as the NCEP (Kistler et al. 2001), ERA-40 (Uppala et al. 2005), or the twentieth century (20CR) reanalysis of Compo et al. (2011), which is based only on surface observations and covers the period 1871–2010. In contrast with wind speed observations and recent extreme wind speed reconstructions from surface pressure readings (e.g., Wang et al. 2011), all reanalyses appear to show an increase in European storm indicators during the last few decades of the twentieth century (Smits et al. 2005; Donat et al. 2011). For tropical cyclones, the intensities of the storms (i.e., maximum near-surface sustained 1-min wind speeds) can also be estimated globally using satellite data, at least since the early 1980s (Kossin et al. 2007; Elsner et al. 2008).

2.3 *Role of External Influences*

2.3.1 Temperature Extremes

Considerable progress has been made in the detection and attribution of externally forced change in surface temperature extremes since the feasibility of such studies was first demonstrated by Hegerl et al. (2004). Studies that detect human influence on surface temperature extremes are available on the global and regional scale and use a range of indices that probe different aspects of the tails of the surface temperature distribution. This includes studies of changes in the frequency moderately extreme temperature events (e.g., Morak et al. 2011; Fig. 3, which also shows that human influence can be detected in the frequency of warm nights in most regions; Morak et al. 2013) and the magnitude (e.g., Christidis et al. 2005, 2011; Zwiers et al. 2011) of extreme surface temperature events. Results are robust across a range of methods and across both types of indices. Some studies use methods that rely on extreme value theory (e.g., Christidis et al. 2011; Zwiers et al. 2011), and are therefore best suited for studying change in the far tails of the temperature distribution, whereas other studies that consider less extreme parts of the distribution (Christidis et al. 2005; Morak et al. 2011, 2013) appropriately use standard fingerprinting approaches (e.g., Hegerl et al. 2007). Some studies (e.g., Christidis et al. 2011) are also able to separate and quantify the responses to anthropogenic and natural external forcing from observed changes in surface temperature extremes, thereby increasing

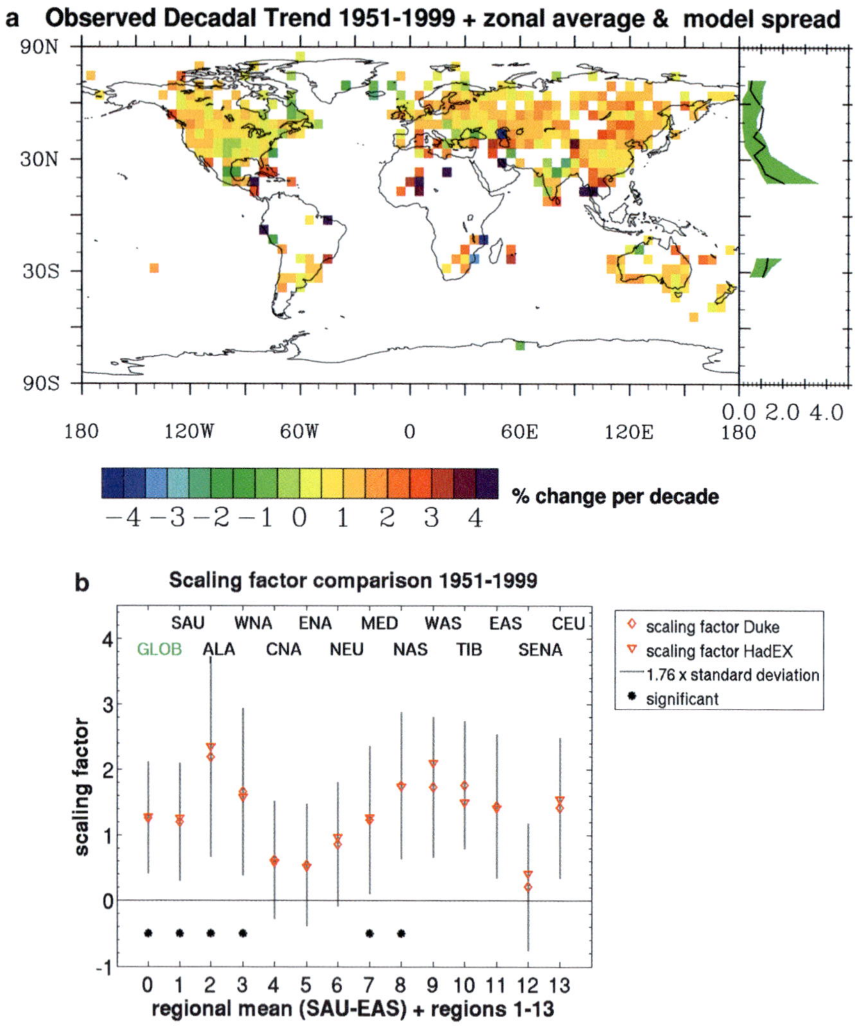

Fig. 3 (**a**) 1951–1999 observed decadal trend of TN90 (in % change per decade) based on a combination of HadEX (Alexander et al. 2006) and additional index data from Kenyon and Hegerl (2008). The zonal average of the observations (*black line*) and the spread of trends in an ensemble of CMIP3 (Coupled Model Intercomparison Project Phase 3, see http://www-pcmdi.llnl.gov/ipcc/about_ipcc.php) "ALL" forcings model simulations for the same period (*green shaded area*) is shown on the side of the plot. (**b**) The scaling factors (*red markers*) of observed changes projected onto the multi-model mean fingerprint for the period 1951–1999. The "*diamonds*" indicate scaling-factors based the Kenyon and Hegerl (2008) dataset (labeled Duke in the legend), and the "*triangles*" indicate scaling-factors based on HadEX. *Grey bars* indicate 5–95 % uncertainty ranges. Regions in which results are detectable at the 5 % significance level and where model simulated internal variability is consistent with regression residuals are indicated with an asterisk. Results indicate broad increases in the frequency of warm nights, as well as the detection of anthropogenic influence in the pattern of observed increases globally and in several regions (From Morak et al. 2011)

confidence in the attribution of a substantial part of the observed changes to external forcing on global scales. Other studies use indirect evidence for attributing significant changes to forcing, such as the tight link between changes in mean and extreme temperatures in a multi-step attribution method (Morak et al. 2011; see Hegerl et al. 2010).

There is the potential to further develop techniques in order to be able to conduct the analysis more fully within the framework of extreme value theory and more confidently separate signals by utilizing recent developments in the statistical modeling of extremes that account for their spatial dependence properties. One approach would be to model extremes spatially via so-called max-stable processes (e.g., Smith 1990; Schlather 2002; Vannitsem and Naveau 2007; Blanchet and Davison 2011).[2] Other approaches are also actively being considered. By working within the framework of extreme value theory, as has already been done in the recent studies of Christidis et al. (2011) and Zwiers et al. (2011), it should become possible to attribute changes in the *likelihood* of extreme events to external causes, thereby contributing to the scientific underpinnings that will be required for event attribution (see Stott et al. 2012). For example, Zwiers et al. (2011) provide rough estimates of circa 1990s expected waiting times for events that nominally had a 20-year expected waiting time in the 1960s, showing that cool temperature extremes have become substantially less frequent globally, whereas warm temperature extremes have become modestly more frequent. Approaches such as that of Zwiers et al. (2011), which considers grid points or stations independently of each other, could be made more efficient if the spatial dependence between extremes could be taken into account. Statistical space-time modeling can account for spatial dependence between parameters of extreme value distributions, for example, by setting prior expectations of spatial dependence that are updated with data. These methods can account for complex space-time structure of extremes and make use of information in data more completely (e.g., Sang and Gelfand 2009, 2010; Heaton et al. 2010). Climatologists will need the assistance of statisticians to fully realize the benefits from these types of approaches. It should be noted that several of the detection and attribution techniques currently applied to extremes are able to take spatial dependence into account (e.g., Hegerl et al. 2004; Christidis et al. 2005, 2011; Min et al. 2011; Morak et al. 2011) by casting the problem in such a way that the Gaussian assumption should hold approximately.

A limitation of many studies that have been conducted to date is that they have been confined to the twentieth century, in part due to the design of the CMIP3 experiment which ended the historical simulations and the single forcing runs at 1999 or 2000, but more importantly, because suitable observational datasets providing broad

[2] A probability distribution is said to be *stable* when the average of a sample of independently drawn values from that distribution has a distribution belonging to the same family of distributions (Feller 1971). The Gaussian distribution is an example of a stable distribution. Stability can also be defined in terms of some other types of operations that may be applied to a sample. In particular, *max-stable* distributions have the property that the maximum value of such a sample again has a distribution within the same family of distributions. The generalized extreme value distribution is max-stable.

coverage of annual temperature extremes have not yet been updated to the more recent decade (e.g., Alexander et al. 2006), although recent studies extend into the twenty-first century (e.g. Morak et al. 2013). Initiatives to expand these datasets, including updating them in near-real time are currently underway or finished (Donat and Alexander 2011; Alexander and Donat 2011). Also, modeling groups participating in CMIP3 generally were not able to make available large volumes of high frequency (daily or higher) output or ensembles of historical single forcing runs (e.g., runs with historical greenhouse gases or aerosol forcing only). Consequently, currently available studies that separate signals have only been performed with single climate models rather than with multi-model ensembles. All of these problems are presently being alleviated at least to some extent with the advent of updated research quality datasets, such as HadEX2 (Alexander and Donat 2011), and the growing availability of CMIP5 simulations (Taylor et al. 2012) that are currently being analyzed by the climate modeling community and are making available high frequency output more broadly than their predecessors in CMIP3, enabling a more thorough exploration of model uncertainties (for example, Hanlon et al. 2012b show results for a multi-model detection analysis for temperature extremes over Europe).

The studies available to date use only a limited number of models. Across many of these studies results suggest that the climate model simulated pattern of the warming response to historical anthropogenic forcing in cold extremes fits observations best when its amplitude is scaled by a factor greater than one (i.e., when the simulated warming signal is scaled up). Conversely, the expected warming signal in warm daily maximum temperature extremes generally needs to be scaled down, and in fact, has only recently been detected in observations through the use of more sophisticated statistical techniques (Christidis et al. 2011; Zwiers et al. 2011). These results point to the possibility that the forcing and/or response mechanisms, including the possibility of feedbacks that operate differently during the warm and cold seasons and during different parts of the diurnal cycle (day versus night), may not be fully understood (e.g. Portmann et al. 2009) or accurately modeled. Recent examples include work by Seneviratne et al. (2006, 2010) and Nicholls and Larsen (2011) concerning the role of land-atmosphere feedbacks in the development of temperature extremes, by Sillmann et al. (2011) on the role of blocking in the development of cold temperature extremes in winter over Europe, and by Hohenegger et al. (2009) on the role of the soil-moisture precipitation feedback.

2.3.2 Precipitation Extremes

As is also the case with change in the mean state, in comparison with surface air temperature only limited progress has been made in determining the extent to which external influences on the climate system have influenced changes in the intensity or frequency of heavy or extreme precipitation. Various observational studies have found that extreme precipitation can have heavy tailed behavior (with a shape parameter in the range of approximately 0–0.2 when annual maxima of daily precipitation are fitted with a generalized extreme value distribution, e.g., Fowler et al. 2010).

While climate models simulate substantial precipitation extremes, it is not clear that they simulate daily intensities that are as heavy-tailed as observed, nor is it clear that they do so given the different scales represented by observed point values and simulated grid-box values. For example, Kharin and Zwiers (2005) do not find strong evidence for heavy tailed behavior in the model that they studied, estimating shape parameters that are positive, but near zero. Fowler et al. (2010) similarly find a discrepancy in tail behavior between observed and climate model simulated extreme precipitation in the model they study. Averaging in space and time smoothes the tail behavior recorded at weather stations but this reduces the applicability for impact studies. In addition, it is a real challenge to detect and attribute changes whenever the variable of interest has a positive shape parameter, indicating unbounded growth in return values as return periods become very long. In such cases, uncertainties grow rapidly with a slight change in the shape parameter and consequently very long time series are necessary. Thus there are substantial statistical challenges associated with the detection and attribution of the precipitation response to external forcing.

Nevertheless, there is a modest body of literature that has investigated whether there is evidence that natural or anthropogenic forcing has affected global land mean precipitation (e.g., Gillett et al. 2004; Lambert et al. 2005), the zonal distribution of precipitation over land (e.g., Zhang et al. 2007; Noake et al. 2011; Polson et al. 2013) and the quantity of precipitation received at high northern latitudes (Min et al. 2008). Since the variability of precipitation is related to the mean (there is greater short term precipitation variability in regions that receive more precipitation), the detection of human influence on the mean climatological distribution of precipitation should imply that there has also been an influence on precipitation variability, and thus extremes. Hegerl et al. (2004) found in a model-study that changes in moderately extreme precipitation may be more robustly detectable than changes in mean precipitation since models robustly expect extreme precipitation to increase across a large part of the globe while the pattern of increase and decrease in annual total precipitation is more sensitive to model uncertainty.

Min et al. (2011) recently investigated this possibility, finding evidence for a detectable human influence in observed changes in precipitation extremes during the latter half of the twentieth century. This was accomplished by transforming the tails of observed and simulated distributions of annual maximum daily precipitation amounts into a probability based index (PI) before applying an optimal detection formalism, thereby partly circumnavigating the scaling issues that are associated with precipitation. It should be noted however, that some strong assumptions are implicit in such transformations that are not necessarily verifiable. For example, it is implicitly assumed that forced changes in precipitation extremes result in comparable changes in PI at different scales, even though the mechanisms that generate extreme precipitation locally may be quite different from those that determine extreme events on climate model grid box scales and larger. Even with the transformation, it was found that a best fit with observations required that the magnitude of the large-scale climate model simulated responses to external forcing be increased by a considerable factor, with a greater increase in magnitude being required in the

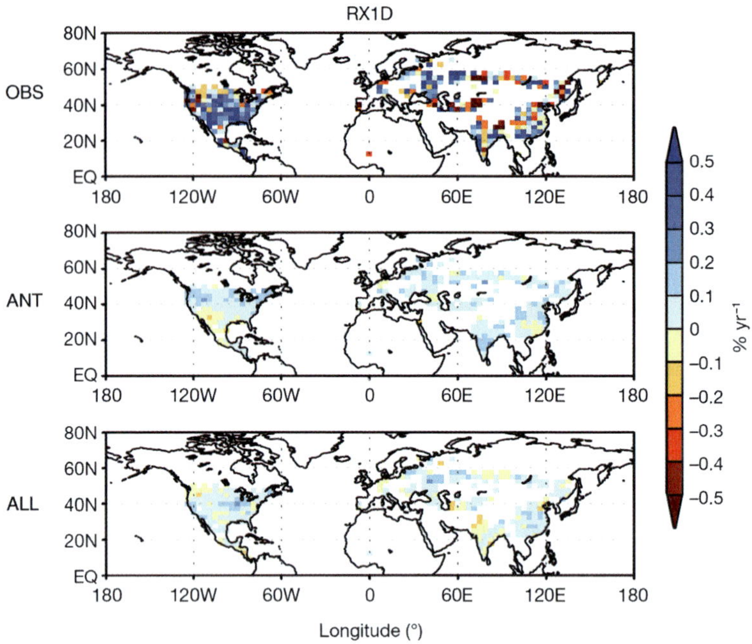

Fig. 4 Geographical distribution of trends of extreme precipitation indices (*PI*) for annual maximum daily precipitation amounts (*RX1D*) during 1951–1999. Observations (*OBS*); model simulations with anthropogenic (*ANT*) forcing; model simulations with anthropogenic plus natural (*ALL*) forcing. For models, ensemble means of trends from individual simulations are displayed. Units: per cent probability per year (From Min et al. (2011; see paper for details))

case of historical simulations that take into account a combination of anthropogenic and natural forcing (ALL forcing), than for simulations accounting only for the former (ANT forcing; see Fig. 4). The discrepancy between scaling factors for ALL and ANT forcing is understandable given that the anthropogenically forced signal is still small, and that natural forcing (from changes in solar and volcanic activity) would have offset some of the response to ANT forcing, thereby weakening the ALL signal during the latter part of the twentieth century. This leads to smaller expected changes in the ALL fingerprint, which are more strongly affected by noise and thus more difficult to detect, than the 'cleaner' signal from ANT forcing. The on-line supplementary information accompanying Min et al. (2011) includes an extensive set of sensitivity analyses that consider a broad range of uncertainties affecting their results.

The cause of the discrepancies between observed and simulated changes in both mean and extreme precipitation remains to be fully understood. Explanations could include uncertainties in observations, forcing, or the representation of moist processes in models. The observations used in detection studies to date have been limited to the twentieth century, extending to the early twenty-first century in some recent cases, and have been based exclusively on station data. Noake et al. (2011)

suggest that the scale problem (see below) may be part of the model-data mismatch, as it reduces when precipitation changes are expressed in percent. Polson et al. (2013) find that while detection in some seasons is robust to data uncertainty, CMIP5 models and data agree within data and sampling uncertainty for most seasons. Nevertheless, coverage is limited to land areas only and in many regions, is inadequate due to limitations in observing network density, access to existing observations for the purposes of scientific research, or lack of capacity or mandate to facilitate the dissemination of observations. Remote sensing products may eventually solve these problems, but they have not yet been used in detection and attribution studies due to homogeneity concerns and lack of sufficiently long records, although they have been used in some cases for model evaluation (e.g., Kharin et al. 2007). Without broader coverage it is difficult to assess, for example, whether discrepancies in changes between models and observations are a global phenomenon or whether they are regional in nature, reflecting, for example, differences in moisture transport between models and the observed world. Topography, land-atmosphere coupling, and the representation of teleconnected patterns of variability all affect precipitation and are subject to uncertainty due to limited resolution in climate models or lack of complete process knowledge. In addition, wide uncertainty also remains in aerosol forcing (e.g., Forster et al. 2007), aerosol transport, the effect of aerosols upon the production of precipitation, and so on, which may affect both temperature extremes and precipitation extremes. Further, there are differences in the mechanisms of response to long- and short-wave forcing (e.g., Mitchell et al. 1987; Allen and Ingram 2002) and thus the possibility that models may over- or under-simulate the response to one or the other type of forcing.

3 Storms

High energy cyclonic phenomena driven by latent heat release occur in the atmosphere on a number of scales, ranging from individual tornadoes to mesoscale convective complexes to extra-tropical and tropical cyclones. They often cause extensive damage directly by high wind speeds and/or heavy precipitation, and this may be compounded by the effects of flying debris, drifting snow, storm surges and high waves, and wind driven ice movements and other associated events.

3.1 Extra-Tropical Cyclones

Extratropical cyclones (synoptic-scale low pressure systems) exist throughout the mid-latitudes and are associated with extreme winds, sea levels, waves and precipitation. Climate models project changes in the large scale flow and reduced meridional temperature gradients as a consequence of greenhouse gas forcing, both of which affect extra-tropical cyclone development, and consequently produce changes in their number distribution (Lambert and Fyfe 2006) and in the positioning of extra-tropical storm tracks (Bengtsson et al. 2006).

Climate models represent the general structure of the storm track pattern reasonably well (Bengtsson et al. 2006; Greeves et al. 2007; Ulbrich et al. 2008; Catto et al. 2010) although models tend to have excessively zonal storm tracks (Randall et al. 2007). Detecting changes in extra-tropical cyclone numbers, intensity, and activity based on reanalysis remains challenging due to concerns about inhomogeneity that is introduced through changes over time in the observing system, particularly in the southern hemisphere (Hodges et al. 2003; Wang et al. 2006, 2012). Even though different reanalyses correspond well in the Northern Hemisphere (Hodges et al. 2003; Hanson et al. 2004; Wang et al. 2012), changes in the observing system over time may also have affected the fidelity with which cyclone characteristics are represented in reanalyses there as well (Bengtsson et al. 2004).

Numerous studies using reanalyses suggest that the main northern and southern hemisphere storm tracks have shifted polewards during the last 50 years (e.g., Trenberth et al. 2007). Idealized modeling studies (e.g., Brayshaw et al. 2008; Butler et al. 2010) suggest that radiative forcing from increases in well mixed greenhouse gases and decreases in stratospheric ozone may have played a role in these shifts. However, Sigmond et al. (2007) note that the response of the extratropical circulation to global warming is not necessarily robust across different models even for a common SST change pattern, and for a given model and SST change the extratropical response can depend on the horizontal resolution and on certain poorly constrained tuning parameters. For the moment, observational studies of pressure-based indices (discussed above; e.g., Wang et al. 2011 for the European/North Atlantic region, see Fig. 5; Alexander et al. 2011 for south-eastern Australia) are not able to provide corroborating evidence of a poleward shift in the principal storm track locations, since in both hemispheres, the domain over which pressure triangles needed to produce these indices is rather limited. Ongoing work with single station pressure proxies may help to alleviate this situation in the future. For example, a regional study over Canada that considered changes in observed cyclone deepening rates based on pressure tendencies at stations (Wang et al. 2006) found qualitative agreement between reanalyses and station data suggesting a northward shift of the winter storm track over Canada.

Detection and attribution studies examining whether human influence has played a role in changes in cyclone number, intensity or distribution have not yet been conducted. However, human influence has been detected in the global sea level pressure (Gillett et al. 2005; Gillett and Stott 2009) and in one study, in geostrophic wind energy derived from sea level pressure records (Wang et al. 2009b). Gillett and Stott (2009) show that observed patterns of trends, which indicate decreases in high latitude sea level pressure and increases elsewhere, are robust when calculated from data for 1949–2009. Observed changes were consistent with expectations based on the model (HadGEM1) used in that study, suggesting that anthropogenic influence has contributed to both pressure decreases at high latitudes and increases at low latitudes. The mechanism for the latter is not well understood. Using an approach

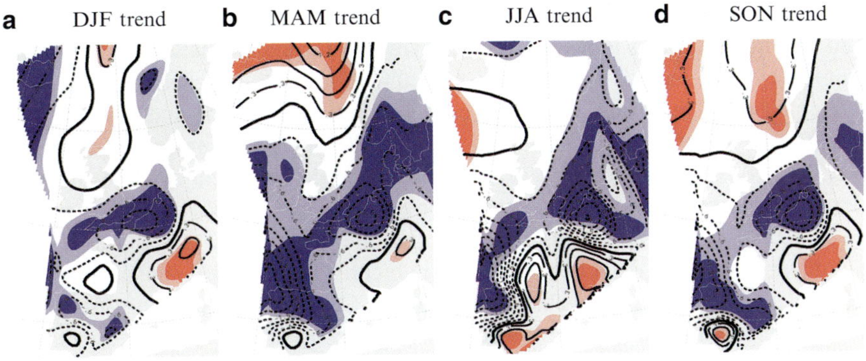

Fig. 5 Example of an analysis of trends in seasonal storm indices derived from long surface pressure records. This figure shows contour maps of Theil-Sen (also sometimes know as Kendall's) linear trend estimates (in unit per century) in seasonal storm indices defined as the 99th percentile of sub-daily geostrophic wind speed estimated from pressure triangles for the period 1902–2007 in a domain that covers western Europe and the eastern North Atlantic. The contour interval is 0.3. The zero contours are shown in *bold*. Positive trends are shown in *thin solid contours*, and *reddish* shadings indicate at least 20 % significance; and negative trends in *dashed contours* and *bluish shadings*. The *darker shadings* indicate areas with trends that are significant at the 5 % level or lower. Significance is determined using the Mann-Kendall trend test (From Wang et al. (2011). The statistical methods are described in Wang and Swail (2001))

that would not formally be considered as a detection and attribution method, Fogt et al. (2009) find that both coupled climate model simulated trends and observed trends in the Southern Annular Mode (SAM) lie outside the range of internal climate variability during the austral summer, suggesting that human influence has contributed to the observed SAM trends.

3.2 Tropical Cyclones

About 90 tropical cyclones have been observed annually since the introduction of geostationary satellites. The global frequency has remained more or less constant over this period, albeit with substantial variability in the frequency of tropical cyclones and locations of their tracks within individual ocean basins (e.g., Webster et al. 2005; Kossin et al. 2010).

Tropical cyclones are typically classified in terms of their intensity according to the Saffir-Simpson scale as indicated by near-surface wind speed or central pressure. Long-term records of the strongest storms are potentially less reliable than those of tropical cyclones in general (Landsea et al. 2006). In addition to intensity, other impact-relevant characteristics of tropical cyclones include frequency, duration, track, precipitation, and the structure and areal extent of the wind field in tropical cyclones, the latter of which can be very important for damage through storm surge as well as the direct wind-related damage.

Forming robust physical links between changes in tropical cyclone characteristics and natural or human-induced climate changes is a major challenge. Historical tropical cyclone records are known to be heterogeneous due to changing observing technology and reporting protocols (e.g., Landsea et al. 2004) and because data quality and reporting protocols vary substantially between regions (Knapp and Kruk 2010). The homogeneity of the global record of tropical cyclone intensity derived from satellite data has been improved (Knapp and Kossin 2007; Kossin et al. 2007), but these records represent only the past 30–40 years. Statistically significant trends have not been observed in records of the global annual frequency of tropical cyclones (e.g., Webster et al. 2005). Century-scale trends in frequency have been identified in the North Atlantic, but are contested (see below). Increasing century-scale frequency trends have not been identified in other basins although a declining trend in the frequency of land-falling tropical cyclones has recently been identified in a new long-term dataset for eastern Australia (Callaghan and Power 2011). Power dissipation has increased sharply in the North Atlantic and more weakly in the western North Pacific over the past 25 years (Emanuel 2007), but the interpretation of longer-term trends is constrained by data quality concerns as well as uncertainties on the potential role of natural climate variability in the observed increases. Satellite-based records of extreme precipitation associated with tropical cyclones also appear to have substantial homogeneity issues due to satellite changes (Lau et al. 2008). It remains difficult to robustly place tropical cyclone metrics for recent decades into a longer historical context (Knutson et al. 2010) because pre-satellite records are incomplete and therefore require the use of methods to estimate storm undercounts and other biases; these methods have provided mixed conclusions to date (e.g., for the North Atlantic basin, see Holland and Webster 2007; Landsea 2007; Mann et al. 2007; Vecchi and Knutson 2008; Landsea et al. 2009; Knutson et al. 2010; see also Fig. 6).

Our understanding of the factors that affect tropical cyclone metrics and their variation is improving but remains incomplete. Anthropogenic forcing has been identified as a cause of SST warming in tropical cyclogenesis regions (e.g., Santer et al. 2006; Gillett et al. 2008). Potential intensity theory (Bister and Emanuel 1998) links changes in the mean thermodynamic state of the tropics to cyclone potential intensity and implies that a greenhouse warming could induce a shift towards greater intensities. This has received some support from dynamical hurricane model simulations (summarized in Knutson et al. 2010, Table S2). Results suggest that human influence could have altered tropical cyclone intensities over the twentieth century. However, as noted above, the available evidence concerning historical trends and detectable anthropogenic influence on tropical cyclone characteristics is mixed. A global analysis of trends in satellite-based tropical cyclone intensities has identified an increasing trend that is largest in the upper quantiles of the distribution (Elsner et al. 2008), and most pronounced in the Atlantic basin. However, this record extends back only to 1981 which is regarded as too short to distinguish a long-term trend from the pronounced multi-decadal variability in the Atlantic basin. Historical data show that tropical cyclone power dissipation is related to sea surface temperatures (SSTs), near-tropopause temperatures and vertical wind shear (Emanuel 2007),

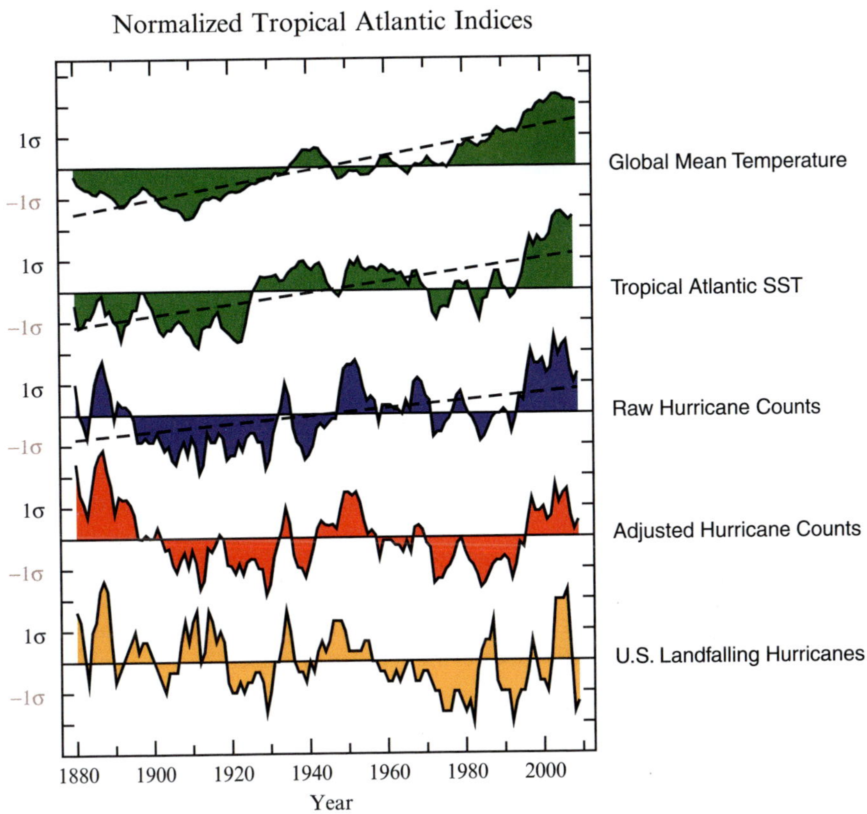

Fig. 6 Five-year running means of tropical Atlantic indices. *Green curves* depict global annual-mean temperature anomalies (*top*) and August–October Main Development Region (MDR, defined as 20–80 W, 10–20 N) SST anomalies (second from *top*). *Blue curve* shows unadjusted Atlantic hurricane counts. *Red curve* shows adjusted Atlantic hurricane counts that include an estimate of "missed" hurricanes in the pre-satellite era. *Orange curve* depicts annual U.S. landfalling hurricane counts. *Vertical axis tic marks* denote one standard deviation intervals (shown by the σ symbol). *Dashed lines* show linear trends. Only the top three curves have statistically significant trends (Source: Adapted from Vecchi and Knutson 2011)

but it has been suggested that the spatial pattern of SST variation in the tropics may exert an even stronger influence on Atlantic hurricane activity than absolute local SSTs (Swanson 2008; Vecchi and Soden 2007; Ramsay and Sobel 2011). This would have important implications for the interpretation of climate model projections (Vecchi et al. 2008). Related to this, a growing body of evidence suggests that the SST threshold for tropical cyclogenesis (currently about 26 °C) would increase at about the same rate as tropical SSTs due to greenhouse gas forcing (e.g., Ryan et al. 1992; Knutson et al. 2008; Johnson and Xie 2010). This means, for example, that the areas of simulated tropical cyclogenesis would not expand along with the 26 °C isotherm in climate model projections. The most recent assessment by the World

Meteorological Organization (WMO) Expert Team on Climate Change Impacts on Tropical Cyclones (Knutson et al. 2010) concluded that it remains uncertain whether past changes in any measure of tropical cyclone activity (frequency, intensity, rainfall) exceeds the variability expected through natural causes, after accounting for changes in observing capabilities over time. Seneviratne et al. (2012) drew essentially the same conclusion, stating that "The uncertainties in the historical tropical cyclone records, the incomplete understanding of the physical mechanisms linking tropical cyclone metrics to climate change, and the degree of tropical cyclone variability provide only *low confidence* for the attribution of any detectable changes in tropical cyclone activity to anthropogenic influences". However, recent advances in understanding and phenomenological evidence for shorter-term effects on tropical cyclones from aerosol forcing are providing increasing confidence that anthropogenic forcing has had a measurable effect on tropical cyclone activity in certain regions (Mann and Emanuel 2006; Evan et al. 2009, 2011; Booth et al. 2012; Villarini and Vecchi 2013) although the relative influence of aerosols vs. natural variability on recent multidecadal variability in the Atlantic basin remains uncertain (e.g., Ting et al. 2009; Zhang and Delworth 2009; Camargo et al. 2013; Villarini and Vecchi 2013). Thus, when assessing changes in tropical cyclone activity, it is clear that detection and attribution aimed simply at long-term linear trends forced by increasing well-mixed greenhouse gasses is not adequate to provide a complete picture of the potential anthropogenic contributions to the changes in tropical cyclone activity that have been observed.

Based on a variety of model projections of late twenty-first century climate, it is expected that global tropical cyclone frequency will either decrease or display little change as a consequence of greenhouse warming, but that there will be an increase in mean wind speed intensity and in tropical cyclone rainfall rates over the twenty-first century (Meehl et al. 2007a; Knutson et al. 2010). Projected changes for individual basins are more uncertain than global mean projections, as they show large variations between different modeling studies. Studies that have compared tropical cyclone projections downscaled from different climate models using a single downscaling framework (e.g., Zhao et al. 2009; Sugi et al. 2009) suggest that at the regional scale, the uncertainties in tropical cyclone projections due to differences in projected SST patterns are substantial. Concerning detection and attribution of tropical cyclone changes, in addition to the substantial uncertainty in historical records, a further challenge for identifying such an anthropogenic change signal in observations is that the projected changes are typically small compared to estimated observed natural variability. Modeling studies (e.g. Knutson and Tuleya 2004; Bender et al. 2010) suggest, on the basis of idealized simulations, that unambiguous detection of the effect of greenhouse gas forcing on Atlantic tropical cyclone characteristics may still be decades off. Other studies that have considered projected changes in tropical cyclone-related damage and loss under the A1B emissions scenario (Crompton et al. 2011; Emanuel 2011; Mendelsohn et al. 2012) predict a broad range of emergence time-scales from decades to centuries. However, it should again be emphasized that regional forcing by agents other than greenhouse gases, such as anthropogenic aerosols, is known to affect the regional climatic conditions differently (e.g. Villarini and Vecchi 2013), and that there is evidence that anthropogenic aerosol pollution has affected tropical cyclone activity in some regions. Thus it seems likely that the emergence time-scales projected under A1B warming are sensitive to the A1B aerosol forcing projections, which are known to be highly uncertain (Forster et al. 2007; Haerter et al. 2009).

3.3 *Tornadoes and Other Types of Small Scale Severe Weather*

Tornadoes typically occur during severe thunderstorms in which rapid vertical motion and the resulting convergence of angular momentum produces the potential for very high local vorticity. While our understanding of tornadoes has increased in recent years (e.g., Trapp et al. 2005), the body of research that is available globally on changes in tornado frequency and intensity remains limited. This is in part because the available data are inhomogeneous in time (e.g., Brooks 2004) due to changes in reporting practices as well as changes in population and public awareness, and the introduction of technology such as Doppler radar, all of which undoubtedly affect detection rates. The assessments of Trenberth et al. (2007) and Karl et al. (2008) contain brief sections summarizing available research on tornadoes and other types of small scale severe weather. The scale of these phenomena implies that there are only limited opportunities for interpretation of the observed record using models. At present, any change in their likelihood of occurrence can only be inferred indirectly from models by considering changes in atmospheric conditions such as stability and vertical shear that affect their occurrence. For this reason, as well as the inadequacy of the observational record, detection and attribution studies have not been attempted. Projections of future changes in the incidence and intensity of tornadoes due to greenhouse warming and other climate forcings also remain uncertain, partly because competing influences on tornado occurrence and intensity might change in different ways. Thus, on the one hand, greenhouse gas induced warming may lead to greater atmospheric instability due to increases in temperature and moisture content, suggesting a possible increase in severe weather, but on the other hand, vertical shear may decrease due to reduced pole-to-equator temperature gradients (Diffenbaugh et al. 2008).

4 Hydrological Extremes

We discuss here floods and droughts, which are complex phenomena with large impacts that affect large numbers of people each year. Space and time scales can be large, particularly in the case of droughts which can occur on sub-continental to continental scales and have extended durations of years or longer. In contrast, some types of flooding can be localized and of short duration, although flooding may also occur in large basins over an extended period of time (months). While floods and droughts generally represent opposite ends of the spectrum of variability in a region's hydrological balance, it should be noted that the two phenomena are not completely mutually exclusive. For example, extreme precipitation events, with the possibility of local flash flooding, can occur during drought (e.g., Hanesiak et al. 2011).

4.1 Floods

Floods are affected by various characteristics of precipitation. For example, freshet flooding is driven by meteorological and synoptic characteristics that control the timing and magnitude of energy fluxes into the snowpack, possibly confounded by the occurrence of rainfall. The frequency and intensity of floods can be altered by natural and human engineered and non-engineered land use effects on drainage basins, which makes the detection of climatic influences difficult. Human engineering-induced effects include the possibility that the impoundment of water may alter the local precipitation climatology (Hossain et al. 2009). Storm surge events can cause coastal flooding, which may be exacerbated in estuaries if a storm surge event coincides with heavy discharge. Sea level rise (Sect. 5) can also interact with storm surge events to increase the risk of coastal flooding (Abeysirigunawardena et al. 2009).

The IPCC AR4 (Rosenzweig et al. 2007) and the IPCC Technical Paper VI based on the AR4 (Bates et al. 2008) concluded that documented trends in floods show no evidence for a globally widespread change in flooding (see also, for example, Kundzewicz et al. 2005), although there was abundant evidence for earlier spring peak flows and increases in winter base flows in basins characterized by snow storage. They also noted that there was some evidence of a reduction in ice-jam floods in Europe (Svensson et al. 2006). As highlighted in the SREX (Seneviratne et al. 2012), subsequent research, which continues to be hampered by the limited availability and coverage of river gauge data, provides mixed results. Some studies suggest that there has been an increase in flooding over time in some basins (e.g., some basins in south-east Asia, Delgado et al. 2009; Jiang et al. 2008; and South America, Barros et al. 2004). Another study tentatively concluded that a significant increase was detectable in "great floods"—referring to floods with discharges exceeding 100-year levels in basins larger than 200,000 km^2 (Milly et al. 2002). However, many other studies suggest no climate-driven change (e.g., in northern Asia, Shiklomanov et al. 2007; North America, Cunderlik and Ouarda 2009; Villarini et al. 2009) or provide regionally inconsistent findings (e.g., in Europe, Allamano et al. 2009; Hannaford and Marsh 2008; Mudelsee et al. 2003; and Africa, Di Baldassarre et al. 2010), or a change in the characteristics of flooding such as might be expected when a snowmelt driven flood regime switches, with warming, to a mixed snowmelt-rainfall regime (e.g., Cunderlik and Ouarda 2009).

River discharge simulation under a changing climate scenario is generally undertaken by driving a hydrological model with downscaled, bias-corrected climate model outputs. However, bias-correction and statistical downscaling tend to ignore the energy closure of the climate system, which could be a non-negligible source of uncertainty in hydrological projections (Milly and Dunne 2011). Most hydrological models must first be tuned on a basin-by-basin basis to account for sub-grid-scale characteristics such as basin hypsometry, the degree of watercourse meander and other channel characteristics. Hydrologic modeling is therefore subject to a cascade of uncertainties from climate forcing, climate models, downscaling approach, tuning, and hydrological model uncertainty that remain difficult to quantify comprehensively.

Recently, several studies have detected the influence of anthropogenically-induced climate change in variables that may affect floods. These include Zhang et al. (2007), Noake et al. (2011) and Polson et al. (2013), who detected human influence in observed changes in zonally averaged land precipitation, Min et al. (2008), who detected human influence in northern high-latitude precipitation and Min et al. (2011), who detected human influence in observed global scale change in precipitation extremes. Nevertheless, the extent to which such changes in precipitation may lead to changes in flooding depend on the regional climate characteristics of the respective river catchments, as well as on changes in other climate variables such as soil moisture content. While human influence has not yet been detected in the magnitude/frequency of floods, at least two studies using detection and attribution methodologies that incorporated output from hydrologic models driven with downscaled climate model output have suggested that human influences have had a discernable effect on the hydrology of the regions that they studied. Barnett et al. (2008) detected anthropogenic influence in western US snowpack and the timing of peak-flow (see also Hidalgo et al. 2009), and Pall et al. (2011) estimated that human influence on the climate system increased the likelihood of a fall 2000 flooding event that occurred in the southern part of the UK.

Uncertainty is still large in the projected changes in the magnitude and frequency of floods. The largest source of uncertainties in hydrological projections is from differences between the driving climate models, but the choice of future emission scenarios, downscaling method, and hydrologic model also contribute uncertainty (e.g., Kay et al. 2009; Prudhomme and Davies 2009; Shrestha et al. 2011; Taye et al. 2011). The relative importance of downscaling, bias-correction and the choice of hydrological models as sources of uncertainty may depend on the selected region/catchment, the selected downscaling and bias-correction methods, and the selected hydrological models (Wilby et al. 2008). Chen et al. (2011) demonstrated considerable uncertainty was caused by the choice of downscaling method used to make hydrological projections for a snowmelt-dominated Canadian catchment. Downscaling and bias-correction are also a major source of uncertainty in rain-dominated catchments (van Pelt et al. 2009).

4.2 *Droughts*

Drought is affected by multiple climate variables on multiple times scales, including atmospheric circulation, precipitation, temperature, wind speed, solar radiation, and antecedent soil moisture and land surface conditions. It can feed back upon the atmosphere via land-atmosphere interactions, potentially affecting the extremes of temperature, precipitation and other variables (e.g., Seneviratne et al. 2010; Nicholls and Larsen 2011). It can take multiple forms including meteorological drought (lack of precipitation), agricultural (or soil moisture) drought and hydrological drought (runoff or streamflow). There are few direct observations of drought-related variables (e.g., Trenberth et al. 2007), including soil moisture, and hence drought proxies such as the Palmer Drought Severity Index (PDSI – Palmer 1965; Dai et al. 2004; Heim 2002), the Standardized Precipitation Index (SPI – McKee et al. 1993; Heim 2002) and the Standardized Precipitation Evapotranspiration Index (SPEI – Vicente-Serrano et al. 2010) are often used to monitor and study changes in drought conditions. However, the use of these indirect indices results in substantial uncertainties in the resulting analyses; in particular the PDSI has been criticized as having several limitations (see discussion in Seneviratne et al. 2012). In contrast, hydrologic drought can be observed/analyzed via statistical analysis of discharge records (see e.g., Fleig et al. 2006).

Global assessments of changes in drought remain uncertain. Trenberth et al. (2007), using the Dai et al. (2004) dataset, found large increases in dry areas as indicated by the PDSI. However, it has been noted that the PDSI may not be comparable between diverse climatological regions (e.g., Karl 1983; Alley 1984). The self-calibrating (sc-) PDSI introduced by Wells et al. (2004) attempts to alleviate this problem by replacing fixed empirical constants with values based on the local climate. Using the sc-PDSI, van der Schrier et al. (2006) show that twentieth century soil moisture trends in Europe are not statistically significant. Using a more comprehensive land surface model than that implicit in either the PDSI or sc-PDSI, together with observation-based forcing, Sheffield and Wood (2008) inferred that decreasing trends in drought duration, intensity and severity were prevalent globally during 1950–2000 (Fig. 7). However, they also noted strong regional variation and increases in drought indicators in some regions, consistent with some regional studies. For example, Andreadis and Lettenmaier (2006), using a similar approach, found increasing trends in soil moisture and runoff in much of US in the latter half of twentieth century. On the other hand, Dai (2011) found a global tendency for increases in drought based on various versions of the PDSI including the sc-PDSI and soil moisture from a land surface model driven with observation-based forcing. Patterns of change obtained with those different techniques were largely consistent, with substantial spatial variability being a dominant characteristic. Nevertheless, inconsistencies between studies and indicators demonstrate that there remain large uncertainties with respect to global assessments of past changes in droughts, making it difficult to confidently attribute observed changes to external forcing on the climate system (Seneviratne et al. 2012).

Characterizing hydrologic (i.e. runoff and streamflow) drought globally and regionally is also challenging due to difficulties in establishing robust and/or standardized quantitative drought descriptions over varied hydrologic regimes (e.g., Fleig et al. 2006). Some recent examples regarding analysis of streamflow records for detection of possible trends in low flow include work in Europe (Stahl et al. 2010), Canada (Ehsanzadeh and Adamowksi 2007) and the UK (Hannaford and Marsh 2006).

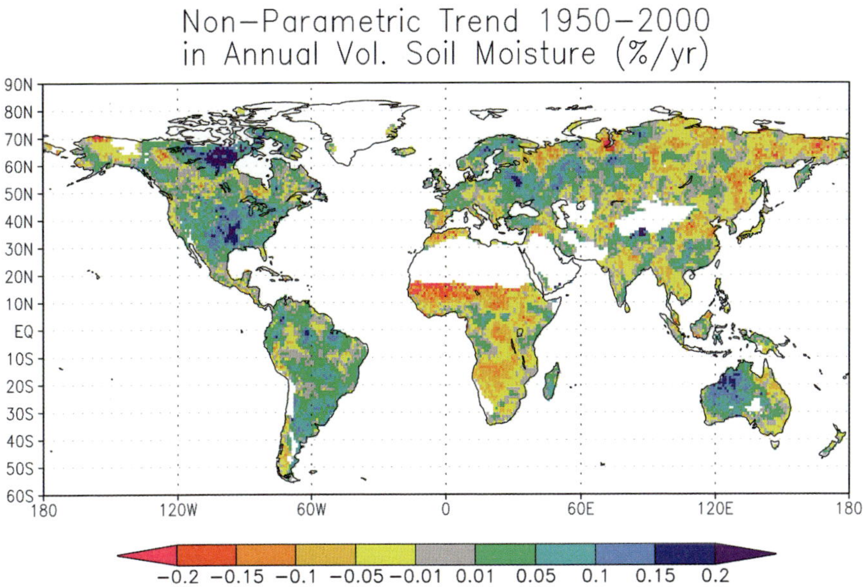

Fig. 7 Global distribution of linear trends in annual mean volumetric soil moisture for 1950–2000 obtained from the Variable Infiltration Capacity (*VIC*) hydrologic model when driven with observationally based forcing. The trends are calculated using the Theil-Sen estimator and evaluated with the Mann–Kendall nonparametric trend test. Regions with mean annual precipitation less than 0.5 mm day^{-1} have been masked out because the VIC model simulates small drying trends in desert regions that, despite being essentially zero, are identified by the nonparametric test (From Sheffield and Wood (2008; Fig. 1))

Despite these uncertainties in global scale studies, there is often more agreement amongst regional studies of historical and current drought, consistent with the notion that circulation changes should induce regionally coherent shifts in drought regimes. For example, precipitation is strongly affected by the El Niño/Southern Oscillation in many parts of the world (Ropelewski and Halpert 1987), including extremes (Alexander et al. 2009; Kenyon and Hegerl 2010; Zhang et al. 2010), and the resulting teleconnected circulation responses are often linked to the occurrence of precipitation deficits and drought in different regions (e.g., Folland et al. 1986; Hoerling and Kumar 2003; Held et al. 2005; Hoerling et al. 2006; Giannini et al. 2008; Schubert et al. 2009) although internal atmospheric variability that is not forced by slowly changing boundary conditions can also create drought (e.g., Hoerling et al. 2009). Also, progress is being made in understanding the role of land-atmosphere feedbacks that affect surface conditions (e.g., Koster et al. 2004; Seneviratne et al. 2006, 2010; Fischer et al. 2007), although the rate of advance is limited by the availability of observational data.

Christensen et al. (2007) provide an assessment of regional drought projections based on simulations that were performed for CMIP3, noting consistency across models in projected increases in droughts particularly in subtropical and mid-latitude areas. Uncertainty in drought projections stems from multiple sources. Perhaps the most fundamental of these is the uncertainty in the pattern of sea-surface temperature response to forcing, which is "El Niño like" in many models (Meehl et al. 2007a), and which therefore cascades to other aspects of model behavior through the teleconnected responses to SST change. A second source of uncertainty is associated with the possible alteration of land-atmosphere feedback processes, both as a consequence of change in the physical climate system and change in the terrestrial biosphere. A third source of uncertainty arises because the complexities of drought are at best incompletely represented in commonly used drought indices, leading to potential discrepancies of interpretation. For example, Orlowsky and Seneviratne (2012) show, using a more complete ensemble of CMIP3 simulations than was available at the time of Christensen et al. (2007), that ensemble projections based on meteorological and agricultural drought indices can be quite different, particularly at higher latitudes. Also, Burke and Brown (2008), considering several drought indices and two different ensembles of climate model simulations, show little change in the proportion of the land surface that is projected to be in drought based on the SPI, whereas indices that account for change in the atmospheric demand for moisture showed significant increases in the global land area affected by drought. It has been suggested that inferences based on climate model simulated soil moisture may be more robust than those based on other types of drought indicators. This is because model results are often found to be consistent after simple scaling (e.g., Koster et al. 2009; Wang et al. 2009a).

5 Sea Level

Transient sea level extremes caused by severe weather events such as tropical or extratropical cyclones can produce storm surges and extreme wave heights at the coast. Extreme sea levels may change in the future as a result of both changes in atmospheric storminess and mean sea level rise, neither of which will be spatially uniform across the globe. Sea level change along coast lines may also be affected by some additional factors including glacial isostatic adjustment, coastal engineering, and changes in the Earth's gravitational field (e.g., Mitrovica et al. 2010) arising from glacial and ice-sheet melting. Global mean sea level rose at an average rate of 1.7 [1.2–2.2] mm year^{-1} over the twentieth century, 1.8 [1.3–2.3] mm year^{-1} over 1961–2003, and at a rate of 3.1 [2.4–3.8] mm year^{-1} over 1993–2003 (Bindoff et al. 2007). Externally induced sea level rise occurs against a backdrop of natural variability in sea level that must be taken into account when attributing causes to observed changes. For example, natural modes of variability such as the El Niño/Southern Oscillation (Menéndez and Woodworth 2010), the Pacific Decadal Oscillation (Abeysirigunawardena and Walker 2008), the North Atlantic Oscillation (Marcos et al. 2009) and the position of the South Atlantic high (Fiore et al. 2009) all have transient effects on extreme sea levels. It is *very likely* that humans contributed to sea level rise during the latter half of the twentieth century (Hegerl et al. 2007),

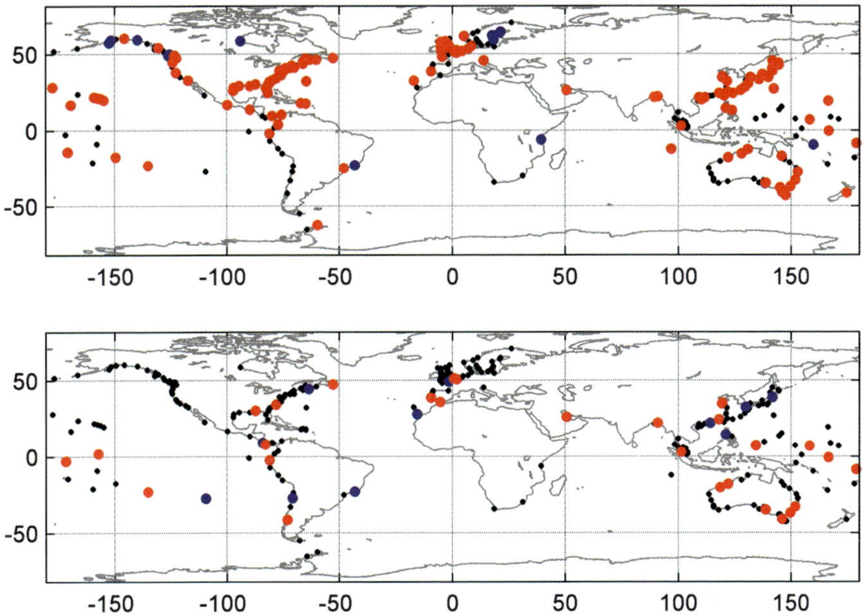

Fig. 8 Estimated trends in (*upper*) annual 99th percentile of sea level based on monthly maxima of hourly tide gauge readings from 1970 onwards, and (*lower*) 99th percentile after removal of the annual medians of hourly readings. Only trends significant at the 5 % level are shown in color: *red* for positive trends and *blue* for negative trends. Linear trends were estimated via least-squares regression taking the interannual perigean tidal influence into account (From Menéndez and Woodworth 2010). The figure shows that extreme sea levels have risen broadly, and that the dominate influence on that rise is from the increase in mean sea level

and therefore *more likely than not* that humans contributed to the trend in extreme high sea levels (Solomon et al. 2007). Both mean and extreme sea level has continued to rise since the AR4 (Church et al. 2011; Menéndez and Woodworth 2010; Woodworth et al. 2011; see Fig. 8).

Meehl et al. (2007a) projected model based 90 % ranges for sea level rise for 2090–2099 relative to 1980–1999 that varied from 18 to 38 cm in the case of the SRES B1 scenario to 26–59 cm in the case of the A1FI scenario. These estimates accounted for ocean thermal expansion, glaciers and ice caps, and modeled aspects of ice sheets. It was also estimated that an acceleration of the flow of ice from Greenland and Antarctic could increase the upper ends of these ranges by 10–20 cm, and it was noted that insufficient understanding of ice sheet dynamics meant that a larger contribution could not be ruled out. Subsequent studies that use statistical models to extrapolate sea level changes based on historical relationships between temperature and sea level have suggested somewhat higher ranges, for example, 0.75–1.90 m (Vermeer and Rahmstorf 2009, based on SRES B1 to A1FI scenarios), and 0.90–1.30 m (Grinsted et al. 2010, based on the SRES AIB scenario only).

Projections of extreme sea level can be produced regionally in several ways. Often, such studies involve a combination of downscaling and hydrodynamic modeling (e.g., Debernard and Roed 2008, who consider the European region and projected both decreases and increases depending upon location). Lin et al. (2012) used a statistical-dynamical hurricane simulation model together with a dynamical model of storm surge to project large reductions in the return periods of tropical cyclone-related surge events in New York City over the twenty-first century. Such approaches may not be feasible in all locations if the driving climate model does not simulate the phenomena that are likely to cause storm surge in a given region (e.g., tropical cyclones). In such cases it may be possible to construct statistical or idealized models of tropical cyclone characteristics from observations that can then be perturbed to represent future conditions and to drive hydrodynamic models (e.g., McInnes et al. 2003; Harper et al. 2009; Mousavi et al. 2011). A further approach is to conduct sensitivity analyses to assess the relative impacts on mean sea level rise and wind speed increase (e.g., McInnes et al. 2009).

6 Summary and Recommendations

In this paper we have reviewed some, but not all, aspects of the current status of research on changes in climate extremes. We have focused primarily on the historical instrumental record, noting results and challenges that arise from observational, methodological and climate modeling uncertainties. The choice to focus on the historical instrumental record reflects our view that high priority should be given to reducing uncertainty in our understanding of historical changes in extremes over the instrumental period as a prerequisite to confidently predicting changes over the next century. This includes the development of improved and comprehensive observational records, improvement in our ability to confidently detect changes in observations through the development of better physical models, forcing data sets and more powerful statistical techniques, the development and refinement of our understanding of the physical processes that produce extremes, and continued improvement in our ability to attribute causes to those changes. This does not imply that research on other aspects of extremes is of lesser importance, but rather that overall progress on understanding the implications of ongoing and future changes in extremes will be strongly dependent upon our ability to document and understand changes in extremes during the period of history that has been (and continues to be) the most comprehensively and directly observed.

Despite the limited scope of this review, it is apparent that a number of substantive challenges remain that impede the advancement of our understanding of extreme phenomena. We will discuss several in the following paragraphs.

The most fundamental of all of these challenges is simply *the state of the historical observational record* itself. Irrespective of the state of our process knowledge and our ability to integrate that knowledge into climate and weather prediction models, it is difficult to have confidence in predictions or projections if we do not have adequate historical data to reliably document how the extremes behavior of the climate system has changed over the past century and to evaluate both model variability and model behavior under historical forcing. While progress has been made in improving datasets, much remains to be done to improve access to even basic daily meteorological observations. The current situation, improved somewhat through the efforts of the ETCCDI and APN[3] (but at the loss of complete reproducibility of all calculations involved in the derivation of extremes indices, and at the cost of large delays in the construction of research-quality datasets), is far from satisfactory as is clearly evident by the far less than global coverage of available datasets of temperature and precipitation extremes. We cannot state strongly enough the importance of continuing and enhancing such efforts to develop datasets of high-frequency in situ observations that are as spatially and temporally complete as possible, as homogenous as possible, and that are accompanied by as much metadata as possible concerning the history of each observing system or station. The lack of metadata describing changes in the exposure and location of observations and in observing procedures is arguably the greatest uncertainty in any work regarding instrumentally observed changes in extremes. With such metadata we know we can remove many of the non-climate influences of changes in instrumentation or location – but these metadata are simply not available for most of the world. This applies to floods, droughts, extreme temperature and precipitation, and tropical cyclones. An additional concern is that there remains a great deal of historical high-frequency data in hard-copy that has yet to be digitized. Much of this data is under threat, thus additional programs (such as the US NOAA Forts Program[4]) are needed to ensure the archival and digitization of such data (see also Page et al. 2004). The limitations of current datasets, whether they are derived directly from the available observational record or interpret observations using models of various complexities (e.g., drought indicators), severely limit our ability to answer key policy-relevant questions about the historical record, such as whether humans have influenced the intensity of extreme precipitation, or whether they have contributed to any perceived change in tropical cyclone behavior.

An important effort with regard to surface temperature is the International Surface Temperature Initiative[5] which seeks to assemble a comprehensive, open, transparent and traceable international data base of surface temperature observations with temporal resolution ranging from hourly upwards, and including associated metadata. A similar effort for precipitation observations, and other key variables such as surface pressure and wind observations, would also be exceedingly valuable. An innovative and promising development with regard to the improvement of climate datasets is the use of "crowd-sourcing"[6] for the digitization and analysis

[3] Asia-Pacific Network for Global Change Research.
[4] See http://www.ncdc.noaa.gov/oa/climate/cdmp/forts.html
[5] http://www.surfacetemperatures.org/
[6] The use of unpaid volunteers, often solicited via the internet.

of climate data, as is being done at US National Climatic Data Centre for both surface temperature data rescue and ongoing tropical cyclone reanalysis.[7]

A second set of challenges concerns *the state of our tools for analyzing observed changes in extremes*. It should be acknowledged that a great deal of progress has been achieved using available tools. For example, there is now a large body of research on more "moderate" extremes because more data tend to be available, signal-to-noise ratios tend to be higher, and because changes in their characteristics can often be successfully studied with more or less standard statistical techniques. However, further progress could be made by improving our tools.

One basic tool is the language that is used to describe extremes, and in this case it is clear that there is a lack of precision in the language that is used in climatology. This lack of precise language hinders advances in research on extremes because it makes the job of clearly articulating hypotheses and objects for analysis all the more difficult. In climatology, the term "extreme" can refer to occurrences of high impact phenomena (e.g., droughts, floods, tropical cyclones) that may or may not be characterized by rare values of the underlying meteorological variables, events that are in fact not very rare (e.g., exceedance of the 90th percentile of temperature or precipitation), or rare events that occur in the far tails of the distributions of clearly defined hydro-meteorological variables such as temperature, precipitation, wind speed, stream flow, and so on. While statistical reasoning and methods are useful in all three cases, the powerful extreme value theory of statistical science can only be brought to bear on the latter, and even in this case, there are clear limitations in practice and in the available theory that impede progress in the analysis of climatological extreme values. Some of these challenges include,

- The need for improvements in the reliability of estimators of the attributes of heavy-tailed variables, and in methods to determine whether these attributes are changing over time.
- A need for the further development of methods or concepts to realistically represent the spatial dependence of extreme values. Currently available approaches based on max-stable processes (e.g., Smith 1990; Schlather 2002) remain difficult to use, do not appear to provide a sufficient broad set of models to represent the heterogeneity and anisotropy of the spatial dependence of extremes that is seen in the real world, and do not provide an obvious approach to dimension reduction, which is a more or less essential component of standard detection and attribution methods.
- The development of methods that would allow for the automated application of so-called peaks-over-threshold approaches to extreme value analysis. If this could be achieved with suitable statistical rigor, it would represent a highly desirable development for the analysis of large collections of station data and gridded datasets since peaks-over-threshold approaches arguably use the available data

[7] http://www.cicsnc.org/corp/presentations/Scott%20Hausman.pdf (presentation made to the 30th Conference on Hurricanes and Tropical Meteorology, Ponte Verde, Florida, USA, 15–20 April 2012).

more efficiently than the more frequently used block-maximum approach. It should be noted however, that such a development would only be beneficial if the underlying high-frequency weather data were available for analysis; indices defined on fixed thresholds or annual blocks, such as those that result from the work of the ETCCDI, would not be suitable.

- Development of methods that are able to combine information on extremes from observations and models, suitably representing uncertainty in the analysis that arises from multiple sources, including uncertainties in the responses to external forcing that are present in extremes and uncertainty associated with the forcing, the climate models themselves, and the internal variability that they simulate.

A third set of challenges concerns *continuing deficiencies in the state of our understanding of the processes that are involved in the production of extreme events*, which limits our confidence in the interpretation of observed extreme events and in observed changes in the frequency and intensity of extreme events. This type of challenge is evident in a number of different ways. A very fundamental aspect is apparent when comparing observed and model-simulated precipitation extremes; due to limited resolution, current global climate models do not simulate precipitation extremes that are of the same intensity as those that are observed in station data (Chen and Knutson 2008). Climatologists refer to this in the literature as a "scaling" issue, and statisticians refer to it as a "change of support" problem. One approach that has been used in detection and attribution research (e.g., Min et al. 2011) is to use probability integral transforms to convert model-simulated and observed precipitation extremes to a common dimensionless scale. While this formally allows comparison between the two, it does not at all resolve the question of whether the physical processes that lead to extreme precipitation on a climate model grid-point scale are the same as those that lead to extreme precipitation at the local scale. While this problem will become less severe as climate model resolution improves, it will still challenge, particularly the interpretation of warm season convective heavy precipitation.

Another area in which the importance of process knowledge is increasingly apparent is in the understanding and interpretation of temperature extremes, where there is a growing understanding of the role of feedback processes in determining the amplitude, duration and extent of extreme events (e.g., Seneviratne et al. 2006; Fischer et al. 2007; Sillmann et al. 2011; Mueller and Seneviratne 2012). It is also increasingly apparent that large scale low-frequency variability plays an important role in altering the likelihood of extreme events, including the effects of ENSO on the intensity and frequency of extreme precipitation (e.g., Alexander et al. 2009; Kenyon and Hegerl 2010; Zhang et al. 2010) and the effects of tropical SST anomalies on drought in regions such as the Sahel (e.g., Held et al. 2005; Hoerling et al. 2006) and southwestern North America (Cook et al. 2007). As is evident from the example of North American drought, it is often only through the study of paleoclimate data that we become aware of the role of low-frequency climate variability in the occurrence of extremes. In the case of tropical cyclones, there are some very specific improvements in process knowledge that would increase our confidence in both historical changes and future projections. These include improvements in the understanding of historical and future changes in tropical tropospheric lapse rates, up to and including the tropopause transition layer, which is important for determining tropical cyclone potential intensity (Emanuel 2010). An important question that remains unresolved is whether projections of relative SST (i.e., regional SST relative to the tropical mean) can be used as proxy for future potential intensity (Emanuel et al. 2013), since relative SST is generally not shown to increase substantially in the next century (Vecchi and Soden 2007). Another presently unresolved question is what portion of the observed multi-decadal climate variability in the tropical Atlantic (which tropical cyclones are observed to substantially respond to) is due to natural variability versus external forcing by greenhouse gasses and anthropogenic aerosols. Understanding changes in the frequency and intensity of extremes both due to external forcing and internal climate variability is further only possible if seasonally resolved information on changes in extremes is available and analyzed. For example, circulation (some of aspects of which are predicted to change in a changing climate) impacts both temperature and precipitation extremes differently in different seasons (Kenyon and Hegerl 2008, 2010). This can only be captured if indices of extremes are resolved at seasonal or shorter time scales.

A topic that has not been explicitly discussed in this paper, which poses a challenge that cuts across definitional issues, our state of process understanding in the physical climate system, and our state of understanding of the impacts of extremes, is the analysis of compound or multi-variable climate extremes; that is, events where the combined effect of, for example, temperature, wind speed and precipitation produces extreme impacts where perhaps the individual temperature, wind or precipitation readings would not be considered to be particularly extreme. While much discussed, there has as yet been relatively little research to investigate such events. That said, research on recognized phenomena such as heat waves, drought, or tropical and extra-tropical cyclones does fit into this category, as does recent event attribution research (e.g., see Stott et al. 2004; Fischer et al. 2007; Pall et al. 2011; Stott et al. 2012; see also Peterson et al. 2012 and Otto et al. 2012). Also, there have been a few attempts to develop multi-indicator extremes indices for monitoring the extent to which a large region is being affected by extremes (e.g., such as introduced by Karl et al. 1996 and revised by Gleason et al. 2008). This situation comes about in part because of the state of available data resources, which remains limited, but also because there is insufficient process and impacts knowledge to rigorously describe multi-variable events in a manner that avoids selection bias.

Finally, the reliable detection and attribution of changes in extremes, regardless of the specific type of phenomenon of interest, depends heavily upon *the ability of models to simulate the natural background variability of the climate system*. In the case of tropical cyclones, this means simulating tropical SSTs patterns and their variability correctly, as well as simulating the variability of the vertical structure of the tropical atmosphere correctly. More generally, it means ensuring that the large scale modes of variability, such as the El-Niño/Southern Oscillation, the Pacific Decadal Oscillation and the Atlantic Multi-decadal Oscillation, are well understood from an observational perspective and well simulated from a modeling perspective.

While extremes represent the tail behavior of climate and weather variables, a growing body of research indicates that their likelihood and intensity is very much influenced by behavior that is more central to the distribution of climate and weather states.

While we have focused on the challenges that are faced by those who attempt to undertake research on extremes, it is also evident that this is an area in which enormous progress has been made, as is discussed by Nicholls and Alexander (2007) and as is clearly evident from recent assessments, including IPCC (2007a), Karl et al. (2008) and particularly Seneviratne et al. (2012). This is an area with very significant momentum and in which the potential exists for the development of applied climate science in terms of predicting or identifying the predictability of extremes. There is considerable potential for developing useful products, for example, which may be able to provide predictions or projections of changes in the likelihood of extremes, either through modeling the influence of seasonal to multi-decadal climate variability on the frequency and/or intensity of extremes, or modeling the direct or indirect impact of external forcing on the properties of extremes. Their interpretation and possible predictive utility may be instrumental for the development of useful climate services and the user interface for those services, for example, as envisioned through the WMO Global Framework for Climate Services.

References

Abeysirigunawardena DS, Walker IJ (2008) Sea level responses to climatic variability and change in northern British Columbia. Atmosphere-Ocean 46:277–296

Abeysirigunawardena DS, Gilleland E, Bronaugh D, Wong W (2009) Extreme wind regime responses to climate variability and change in the inner south coast of British Columbia, Canada. Atmosphere-Ocean 47:41–62

Aguilar E et al (2009) Changes in temperature and precipitation extremes in western Central Africa, Guinea Conakry and Zimbabwe, 1955–2006. J Geophys Res 114, D02115. doi:10.1029/2008JD011010

Alexander L, Donat M (2011) The CLIMDEX project: creation of long-term global gridded products for the analysis of temperature and precipitation extremes. In: WCRP Open Science conference, Denver, Oct 2011

Alexander LV et al (2006) Global observed changes in daily climate extremes of temperature and precipitation. J Geophys Res 111, D05109. doi:10.1029/2005JD006290

Alexander LV, Uotila P, Nicholls N (2009) The influence of sea surface temperature variability on global temperature and precipitation extremes. J Geophys Res-Atmos 114, D18116. doi:10.1029/2009JD012301

Alexander LV, Wang XL, Wan H, Trewin B (2011) Significant decline in storminess over southeast Australia since the late 19th century. Aust Meteorol Oceanogr J 61:23–30

Alexandersson H, Schmith T, Iden K, Tuomenvirta H (1998) Long term variations of the storm climate over NW Europe. Glob Atmos Ocean Syst 6:97–120

Allamano P, Claps P, Laio F (2009) Global warming increases flood risk in mountainous areas. Geophys Res Lett 36, L24404

Allan R, Tett S, Alexander LV (2009) Fluctuations of autumn-winter severe storms over the British Isles: 1920 to present. Int J Climatol 29:357–371. doi:10.1002/joc.1765

Allen MR, Ingram WJ (2002) Constraints on future changes in climate and the hydrologic cycle. Nature 419:224–232
Alley WM (1984) The palmer drought severity index: limitations and assumptions. J Clim Appl Meteorol 23:1100–1109
Andreadis KM, Lettenmaier DP (2006) Trends in 20th century drought over the continental United States. Geophys Res Lett 33, L10403. doi:10.1029/2006GL025711
Barnett TP et al (2008) Human-induced changes in the hydrology of the western United States. Science 319:1080–1083
Bärring L, Fortuniak K (2009) Multi-indices analysis of southern Scandinavian storminess 1780–2005 and links to interdecadal variations in the NW Europe-North Sea region. Int J Climatol 29:373–384
Bärring L, von Storch H (2004) Scandinavian storminess since about 1800. Geophys Res Lett 31:1790–1820
Barros V, Chamorro L, Coronel G, Baez J (2004) The major discharge events in the Paraguay river: magnitudes, source regions, and climate forcings. J Hydrometeorol 5:1161–1170
Bates BC, Kundzewics ZW, Wu S, Palutikof JP (2008) Climate change and water. Technical paper of the Intergovernmental Panel on Climate Change. IPCC Secretariat, Geneva, 210 pp
Bender MA, Knutson TR, Tuleya RE, Sirutis JJ, Vecchi GA, Garner ST, Held IM (2010) Modeled impact of anthropogenic warming on the frequency of intense Atlantic hurricanes. Science 327:454–458
Bengtsson L, Hagemann S, Hodges KI (2004) Can climate trends be calculated from reanalysis data? J Geophys Res 109, D11111. doi:10.1029/2004JD004536
Bengtsson L, Hodges KI, Roeckner E (2006) Storm tracks and climate change. J Clim 19:3518–3543
Bindoff NL et al (2007) Observations: oceanic climate change and sea level. In: Solomon S et al (eds) The physical science basis. Contribution of working group I to the fourth assessment report of the Intergovernmental Panel on Climate Change. Cambridge University Press, Cambridge/New York
Bister M, Emanuel KA (1998) Dissipative heating and hurricane intensity. Meteorol Atmos Phys 65:233–240
Blanchet J, Davison AC (2011) Spatial modeling of extreme snow depth. Ann Appl Stat 5:1699–1725. doi:10.1214/11-AOAS464
Booth BBB, Dunstone NJ, Halloran PR, Andrews T, Bellouin N (2012) Aerosols implicated as a prime driver of twentieth-century North Atlantic climate variability. Nature 484:228–232
Brayshaw DB, Hoskins B, Blackburn M (2008) The storm-track response to idealized SST perturbations in an aquaplanet GCM. J Atmos Sci 65:2842–2860
Brooks HE (2004) On the relationship of tornado path length and width to intensity. Weather Forecast 19:310–319
Burke EJ, Brown SJ (2008) Evaluating uncertainties in the projection of future drought. J Hydrometeorol 9:292–299
Butler AH, Thompson DW, Heikes R (2010) The steady-state atmospheric circulation response to climate change-like thermal forcings in a simple general circulation model. J Clim 23:3474–3496. doi:10.1175/2010JCLI3228.1
Caesar J, Alexander L, Vose R (2006) Large-scale changes in observed daily maximum and minimum temperatures: creation and analysis of a new gridded data set. J Geophys Res 111, D05101. doi:10.1029/2005JD006280
Callaghan J, Power SB (2011) Variability and decline in the number of severe tropical cyclones making land-fall over eastern Australia since the late nineteenth century. Clim Dyn 37:647–662. doi:10.1007/s00382-010-0883-2
Camargo S, Ting M, Kushnir Y (2013) Influence of local and remote SST on North Atlantic tropical cyclone potential intensity. Clim Dyn 40:1515–1529
Catto JL, Shaffrey LC, Hodges KJ (2010) Can climate models capture the structure of extratropical cyclones? J Clim 23:1621–1635
Chen C-T, Knutson TR (2008) On the verification and comparison of extreme rainfall indices from climate models. J Clim 21:1605–1621. doi:10.1175/2007JCLI1494.1
Chen J, Brissette FP, Leconte R (2011) Uncertainty of downscaling method in quantifying the impact of climate change on hydrology. J Hydrol 401:190–202

Christensen JH et al (2007) Regional climate projections. In: Solomon S, Qin D, Manning M, Chen Z, Marquis M, Averyt KB, Tignor M, Miller HL (eds) Climate change 2007: the physical science basis. Contribution of working group I to the fourth assessment report of the Intergovernmental Panel on Climate Change. Cambridge University Press, Cambridge, pp 847–940

Christidis N, Stott PA, Brown SJ, Hegerl GC, Caesar J (2005) Detection of changes in temperature extremes during the second half of the 20th century. Geophys Res Lett 32, L20716

Christidis N, Stott PA, Brown SJ (2011) The role of human activity in the recent warming of extremely warm daytime temperatures. J Clim 24:1922–1930

Church JA, Gregory JM, White NJ, Platten SM, Mitrovica XJ (2011) Understanding and projecting sea level change. Oceanography 24:130–143

Coles S (2001) An introduction to the statistical modeling of extreme values. Springer, London, 208 pp. ISBN ISBN 1-85233-459-2

Compo GP, Whitaker JS, Sardeshmukh PD, Matsui N, Allan RJ et al (2011) The twentieth century reanalysis project. Q J R Meteorol Soc 137:1–28. doi:10.1002/qj.776

Cook ER, Seager R, Cane MA, Stahle DW (2007) North American drought: reconstructions, causes and consequences. Earth Sci Rev 81:93–134. doi:10.1016/j.earscirev.2006.12.002

Crompton RP, Pielke RA Jr, McAneney KJ (2011) Emergence timescales for detection of anthropogenic climate change in US tropical cyclone loss data. Environ Res Lett 6:014003. doi:10.1088/1748-9326/6/1/014003

Cunderlik JM, Ouarda TBMJ (2009) Trends in the timing and magnitude of floods in Canada. J Hydrol 375:471–480

Dai A (2011) Drought under global warming: a review. Wiley Interdiscip Rev Clim Change 2:45–65

Dai A, Trenberth KE, Qian T (2004) A global data set of palmer drought severity index for 1870–2002: relationship with soil moisture and effects of surface warming. J Hydrometeorol 5:1117–1130

Debernard JB, Roed LP (2008) Future wind, wave and storm surge climate in the Northern Seas: a revisit. Tellus A 60:427–438

Delgado JM, Apel H, Merz B (2009) Flood trends and variability in the Mekong river. Hydrol Earth Syst Sci 6:6691–6719

Di Baldassarre G, Montanari A, Lins H, Koutsoyiannis D, Brandimarte L, Blöschl G (2010) Flood fatalities in Africa: from diagnosis to mitigation. Geophys Res Lett 37, L22402

Diffenbaugh NS, Trapp RJ, Brooks H (2008) Does global warming influence tornado activity? Eos Trans Am Geophys Union 89:553

Donat M, Alexander L (2011) Uncertainties related to the production of gridded global data sets of observed climate extreme indices. In: Abstract for WCRP Open Science conference, Denver, Oct 2011

Donat MG, Renggli D, Wild S, Alexander LV, Leckebusch GC, Ulbrich U (2011) Reanalysis suggests long term upward trends in European storminess since 1871. Geophys Res Lett 38, L14703. doi:10.1029/2011GL047995

Dupuis DJ (2012) Modeling waves of extreme temperature: the changing tails of four cities. J Am Stat Assoc 107(497):24–39. doi:10.1080/01621459.2011.643732

Ehsanzadeh E, Adamowksi K (2007) Detection of trends in low flows across Canada. Can Water Resour J 32:251–264

Elsner JB, Kossin JP, Jagger TH (2008) The increasing intensity of the strongest tropical cyclones. Nature 455:92–95

Emanuel KA (2007) Environmental factors affecting tropical cyclone power dissipation. J Clim 20:5497–5509

Emanuel K (2010) Tropical cyclone activity downscaled from NOAA-CIRES reanalysis, 1908–1958. J Adv Model Earth Syst 2:1. doi:10.3894/JAMES.2010.2.1

Emanuel K (2011) Global warming effects on U.S. hurricane damage. Weather Clim Soc 3:261–268

Emanuel K, Solomon S, Folini D, Davis S, Cagnazzo C (2013) Influence of tropical tropopause layer cooling on Atlantic hurricane activity. J Clim 26:2288–2301. doi:10.1175/JCLI-D-12-00242.1

Evan AT, Vimont DJ, Heidinger AK, Kossin JP, Bennartz R (2009) The role of aerosols in the evolution of tropical North Atlantic ocean temperature anomalies. Science 324:778–781
Evan A, Kossin J, Chung C, Ramanathan V (2011) Arabian Sea tropical cyclones intensified by emissions of black carbon and other aerosols. Nature 479:94–97
Feller W (1971) An introduction to probability theory and its applications, vol 2. Wiley, New York, 704 pp
Fiore MME, D'Onofrio EE, Pousa JL, Schnack EJ, Bertola GR (2009) Storm surges and coastal impacts at Mar del Plata, Argentina. Cont Shelf Res 29:1643–1649
Fischer EM, Schär C (2009) Future changes in daily summer temperature variability: driving processes and role for temperature extremes. Clim Dyn 33:917–935
Fischer EM, Schär C (2010) Consistent geographical patterns of changes in high-impact European heatwaves. Nat Geosci 3:398–403
Fischer EM, Seneviratne SI, Vidale PL, Lüthi D, Schär C (2007) Soil moisture – atmosphere interactions during the 2003 European summer heatwave. J Clim 20:5081–5099
Flannigan M, Logan K, Amiro B, Skinner W, Stocks B (2005) Future area burned in Canada. Clim Chang 72:1–16
Fleig AK, Tallasken LM, Hisdal H, Demuth S (2006) A global evaluation of streamflow drought characteristics. Hydrol Earth Syst Sci 10:535–552
Fogt RL, Perlwitz J, Monaghan AJ, Bromwich DH, Jones JM, Marshall GJ (2009) Historical SAM variability. Part II: twentieth-century variability and trends from reconstructions, observations, and the IPCC AR4 models. J Clim 22:5346–5365
Folland CK, Palmer TN, Parker DE (1986) Sahelian rainfall and worldwide sea temperatures 1901–1985. Nature 320:602–607
Folland CK et al (1995) Observed climate variability and change. In: Houghton JT et al (eds) Climate change 1995: the science of climate change. Cambridge University Press, Cambridge, pp 99–181
Forster P et al (2007) Changes in atmospheric constituents and in radiative forcing. In: Solomon S et al (eds) Climate change 2007: the physical science basis. Contribution of working group I to the fourth assessment report of the Intergovernmental Panel on Climate Change. Cambridge University Press, Cambridge/New York
Fowler HJ, Cooley D, Sain SR, Thurston M (2010) Detecting change in UK extreme precipitation using results from the climateprediction.net BBC climate change experiment. Extremes 13:241–267
Frei C, Schär C (2001) Detection probability of trends in rare events: theory and application to heavy precipitation in the Alpine region. J Clim 14:1568–1584
Frich P, Alexander LV, Della-Marta P, Gleason B, Haylock M, Klein Tank AMG, Peterson T (2002) Observed coherent changes in climatic extremes during the second half of the twentieth century. Clim Res 19:193–212
Giannini A, Biasutti M, Verstraete MM (2008) A climate model-based review of drought in the Sahel: desertification, the re-greening and climate change. Glob Planet Chang 64:119–128
Gillett NP, Stott PA (2009) Attribution of anthropogenic influence on seasonal sea level pressure. Geophys Res Lett 36, L23709
Gillett NP, Weaver AJ, Zwiers FW, Wehner MF (2004) Detection of volcanic influence on global precipitation. Geophys Res Lett 31, L12217. doi:10.1029/2004GL020044
Gillett NP, Allan RJ, Ansell TJ (2005) Detection of external influence on sea level pressure with a multi-model ensemble. Geophys Res Lett 32, L19714
Gillett NP, Stott PA, Santer BD (2008) Attribution of cyclogenesis region sea surface temperature change to anthropogenic influence. Geophys Res Lett 35, L09707
Gleason KL, Lawrimore JH, Levinson DH, Karl TR, Karoly DJ (2008) A revised U.S. climate extremes index. J Clim 21:2124–2137. doi:10.1175/2007JCLI1883.1
Grant AN, Bronnimann S, Haimberger L (2008) Arctic warming vertical structure contested. Nature 455:E2–E3. doi:10.1038/nature07257
Greeves CZ, Pope VD, Stratton RA, Martin GM (2007) Representation of Northern Hemisphere winter storm tracks in climate models. Clim Dyn 28:683–702

Grinsted A, Moore AJ, Jevrejeva S (2010) Reconstructing sea level from paleo and projected temperatures 200 to 2100 AD. Clim Dyn 34:461–472
Haerter JO, Roeckner E, Tomassini L, von Storch JS (2009) Parametric uncertainty effects on aerosol radiative forcing. Geophys Res Lett 36, L15707
Hanel M, Buishand TA, Ferro CAT (2009) A nonstationary index flood model for precipitation extremes in transient regional climate model simulations. J Geophys Res 114, D15107. doi:10.1029/2009JD011712
Hanesiak JM et al (2011) Characterization and summary of the 1999–2005 Canadian Prairie drought. Atmosphere-Ocean 49:421–452. doi:10.1080/07055900.2011.626757
Hanlon H, Hegerl GC, Tett SFB, Smith DM (2012a) Can a decadal forecasting system predict temperature extreme indices? J Clim, in press. doi10.1175/JCLI-D-12-00512.1
Hanlon H, Morak S, Hegerl GC (2012b) Detection and prediction of mean and extreme European summer temperatures with a CMIP5 multi-model ensemble. J Geophys Res, submitted
Hanna E, Cappelen J, Allan R, Jonsson T, Le Blancq F, Lillington T, Hickey K (2008) New insights into North European and North Atlantic surface pressure variability, storminess, and related climatic change since 1830. J Clim 21:6739–6766
Hannaford J, Marsh T (2006) An assessment of trends in UK runoff and low flows using a network of undisturbed catchments. Int J Climatol 26:1237–1253
Hannaford J, Marsh TJ (2008) High-flow and flood trends in a network of undisturbed catchments in the UK. Int J Climatol 28:1325–1338
Hanson CE, Palutikof JP, Davies TD (2004) Objective cyclone climatologies of the North Atlantic – a comparison between the ECMWF and NCEP reanalyses. Clim Dyn 22:757–769
Harper B, Hardy T, Mason L, Fryar R (2009) Developments in storm tide modelling and risk assessment in the Australian region. Nat Hazard 51:225–238
Heaton MJ, Katzfuss M, Ramachandar S, Pedings K, Gilleland E, Mannshardt-Shamseldin E, Smith RL (2010) Spatio-temporal models for large-scale indicators of extreme weather. Environmetrics 22:294–303
Hegerl GC, Zwiers FW, Stott PA, Kharin VV (2004) Detectability of anthropogenic changes in annual temperature and precipitation extremes. J Clim 17:3683–3700
Hegerl GC, Zwiers FW, Braconnot P, Gillett NP, Luo Y, Marengo Orsini JA, Nicholls N, Penner JE, Stott PA (2007) Understanding and attributing climate change. In: Solomon S et al (eds) The physical science basis. Contribution of working group I to the fourth assessment report of the Intergovernmental Panel on Climate Change. Cambridge University Press, Cambridge/New York
Hegerl GC, Hanlon H, Beierkuhnlein C (2011) Elusive extremes. Nat Geosci 4:142–143
Hegerl GC, Hoegh-Guldberg O, Casassa G, Hoerling MP, Kovats RS, Parmesan C, Pierce DW, Stott PA (2010) Good practice guidance paper on detection and attribution related to anthropogenic climate change. In: Stocker TF, Field CB, Qin D, Barros V, Plattner G-K, Tignor M, Midgley PM, Ebi KL (eds) Meeting report of the Intergovernmental Panel on Climate Change expert meeting on detection and attribution of anthropogenic climate change. IPCC working group I technical support unit, University of Bern, Bern
Heim RR (2002) A review of twentieth-century drought indices used in the United States. Bull Am Meteorol Soc 83:1149–1165
Held IM, Delworth TL, Lu J, Findell KL, Knutson TR (2005) Simulation of Sahel drought in the 20th and 21st centuries. Proc Natl Acad Sci USA 102:17891–17896. doi:10.1073_pnas.0509057102
Hidalgo HG et al (2009) Detection and attribution of streamflow timing changes to climate change in the Western United States. J Clim 22:3838–3855
Hirschi M, Seneviratne SI, Alexandrov V, Boberg F, Boroneant C, Christensen OB, Formayer H, Orlowsky B, Stepanek P (2011) Observational evidence for soil-moisture impact on hot extremes in southeastern Europe. Nat Geosci 4:17–21. doi:10.1038/ngeo1032
Hodges KI, Hoskins BJ, Boyle J, Thorncroft C (2003) A comparison of recent reanalysis datasets using objective feature tracking: storm tracks and tropical easterly waves. Mon Weather Rev 131:2012–2037
Hoerling MP, Kumar A (2003) The perfect ocean for drought. Science 299:691–694

Hoerling MP, Hurrell J, Eischeid J, Phillips A (2006) Detection and attribution of twentieth-century Northern and Southern African rainfall change. J Clim 19:3989–4008
Hoerling M, Quan X, Eischeid J (2009) Distinct causes for two principal US droughts of the 20th century. Geophys Res Lett 36, L19708. doi:10.1029/2009GL039860
Hohenegger C, Brockhaus P, Bretherton CS, Schär C (2009) The soil moisture-precipitation feedback in simulations with explicit and parameterized convection. J Clim 22:5003–5020
Holland GJ, Webster PJ (2007) Heightened tropical cyclone activity in the North Atlantic: natural variability or climate trend? Philos Trans R Soc A 365:2695–2716
Hossain F, Jeyachandran I, Pielke R Jr (2009) Have large dams altered extreme precipitation patterns. Eos Trans Am Geophys Union 90:453
IPCC (2007a) Climate change 2007: the physical science basis. In: Solomon S et al (eds) Contribution of working group I to the fourth assessment report of the Intergovernmental Panel on Climate Change. Cambridge University Press, Cambridge/New York, 996 pp
IPCC (2007b) Climate change 2007: impacts, adaptation and vulnerability. In: Parry ML et al (eds) Contribution of working group II to the fourth assessment report of the Intergovernmental Panel on Climate Change. Cambridge University Press, Cambridge, 976 pp
IPCC (2012) Summary for policymakers. In: Field CB, Barros V, Stocker TF, Qin D, Dokken DJ, Ebi KL, Mastrandrea MD, Mach KJ, Plattner G-K, Allen SK, Tignor M, Midgley PM (eds) Managing the risks of extreme events and disasters to advance climate change adaptation. A special report of working groups I and II of the Intergovernmental Panel on Climate Change. Cambridge University Press, Cambridge/New York, pp 1–19
Jiang T, Kundzewicz ZW, Su B (2008) Changes in monthly precipitation and flood hazard in the Yangtze River Basin, China. Int J Climatol 28:1471–1481
Johnson NC, Xie S-P (2010) Changes in the sea surface temperature threshold for tropical convection. Nat Geosci 3:842–845
Karl TR (1983) Some spatial characteristics of drought duration in the United States. J Clim Appl Meteorol 22:1356–1366
Karl TR, Knight RW (1997) The 1995 Chicago heat wave: how likely is a recurrence? Bull Am Meteorol Soc 78:1107–1119
Karl TR, Knight RW, Easterling DR, Quayle RG (1996) Indices of climate change for the United States. Bull Am Meteorol Soc 77:279–292
Karl TR, Meehl GA, Christopher DM, Hassol SJ, Waple AM, Murray WL (2008) Weather and climate extremes in a changing climate. Regions of focus: North America, Hawaii, Caribbean, and U.S. Pacific Islands. A report by the U.S. Climate Change Science Program and the Subcommittee on Global Change Research, Washington, DC, 164 pp
Katz RW, Brown BG (1992) Extreme events in a changing climate: variability is more important than averages. Clim Chang 21:289–302
Katz R, Parlange M, Naveau P (2002) Extremes in hydrology. Adv Water Resour 25:1287–1304
Kay AL, Davies HN, Bell VA, Jones RG (2009) Comparison of uncertainty sources for climate change impacts: flood frequency in England. Clim Chang 92:41–63
Kenyon J, Hegerl GC (2008) Influence of modes of climate variability on global temperature extremes. J Clim 21:3872–3889
Kenyon J, Hegerl GC (2010) Influence of modes of climate variability on global precipitation extremes. J Clim 23:6248–6262
Kharin VV, Zwiers FW (2000) Changes in the extremes in an ensemble of transient climate simulation with a coupled atmosphere–ocean GCM. J Clim 13:3760–3788
Kharin VV, Zwiers FW (2005) Estimating extremes in transient climate change simulations. J Clim 18:1156–1173
Kharin VV, Zwiers FW, Zhang X, Hegerl GC (2007) Changes in temperature and precipitation extremes in the IPCC ensemble of global coupled model simulations. J Clim 20:1419–1444
Kistler R et al (2001) The NCEP-NCAR 50-year reanalysis: monthly means CD-ROM and documentation. Bull Am Meteorol Soc 82:247–267
Klawa M, Ulbrich U (2003) A model for the estimation of storm losses and the identification of severe winter storms in Germany. Nat Hazard Earth Syst Sci 3:725–732

Klein Tank AMG, Zwiers FW, Zhang X (2009) Guidelines on analysis of extremes in a changing climate in support of informed decisions for adaptation (WCDMP-72, WMO-TD/No.1500), 56 pp
Knapp KR, Kossin JP (2007) A new global tropical cyclone data set from ISCCP B1 geostationary satellite observations. J Appl Remote Sens 1:013505
Knapp KR, Kruk MC (2010) Quantifying inter-agency differences in tropical cyclone best track wind speed estimates. Mon Weather Rev 138:1459–1473
Knutson TR, Tuleya RE (2004) Impact of CO2-induced warming on simulated hurricane intensity and precipitation: sensitivity to the choice of climate model and convective parameterization. J Clim 17:3477–3495
Knutson TR, Sirutis JJ, Garner ST, Vecchi GA, Held IM (2008) Simulated reduction in Atlantic hurricane frequency under twenty-first- century warming conditions. Nat Geosci 1:359–364
Knutson TR, McBride JL, Chan J, Emanuel K, Holland G, Landsea C, Held I, Kossin JP, Srivastava AK, Sugi M (2010) Tropical cyclones and climate change. Nat Geosci 3:157–163
Kossin JP, Knapp KR, Vimont DJ, Murnane RJ, Harper BA (2007) A globally consistent reanalysis of hurricane variability and trends. Geophys Res Lett 34, L04815. doi:10.1029/2006GL028836
Kossin JP, Camargo SJ, Sitkowski M (2010) Climate modulation of North Atlantic hurricane tracks. J Clim 23:3057–3076
Koster RD et al (2004) Regions of strong coupling between soil moisture and precipitation. Science 305:1138–1140
Koster RD, Guo ZC, Yang RQ, Dirmeyer PA, Mitchell K, Puma MJ (2009) On the nature of soil moisture in land surface models. J Clim 22:4322–4335
Krueger O, von Storch H (2011) Evaluation of an air pressure based proxy for storm activity. J Clim 24:2612–2619
Krueger O, von Storch H (2012) The informational value of pressure-based single-station proxies for storm activity. J Atmos Ocean Technol 29(4):569–580
Kundzewicz ZW et al (2005) Summer floods in central Europe: climate change track? Nat Hazard 36:165–189
Lambert SJ, Fyfe JC (2006) Changes in winter cyclone frequencies and strengths simulated in enhanced greenhouse warming experiments: results from the models participating in the IPCC diagnostic exercise. Clim Dyn 26:713–728
Lambert FH, Gillett NP, Stone DA, Huntingford C (2005) Attribution studies of observed land precipitation changes with nine coupled models. Geophys Res Lett 32, L18704. doi:10.1029/2005GL023654
Landsea CW (2007) Counting Atlantic tropical cyclones back to 1900. Eos Trans Am Geophys Union 88:197–202
Landsea CW, Anderson C, Charles N, Clark G, Dunion J, Fernandez-Partagas J, Hungerford P, Neumann C, Zimmer M (2004) The Atlantic hurricane database re-analysis project: documentation for the 1851–1910 alterations and additions to the HURDAT database. In: Murnane RJ, Liu KB (eds) Hurricanes and Typhoons: past, present and future. Columbia University Press, New York, pp 177–221
Landsea CW, Harper BA, Hoarau K, Knaff JA (2006) Can we detect trends in extreme tropical cyclones? Science 313:452–454
Landsea CW, Vecchi GA, Bengtsson L, Knutson TR (2009) Impact of duration thresholds on Atlantic tropical cyclone counts. J Clim 23:2508–2519. doi:10.1175/2009JCLI3034.1
Lau KM, Zhou YP, Wu HT (2008) Have tropical cyclones been feeding more extreme rainfall? J Geophys Res 113, D23113. doi:10.1029/2008JD009963
Lin N, Emanuel K, Oppenheimer M, Vannarcke E (2012) Physically based assessment of hurricane surge threat under climate change. Nat Clim Change 2:462–467. doi:10.1038/NCLIMATE1389
Lorenz C, Kunstmann H (2012) The hydrological cycle in three state-of-the-art reanalyses: intercomparison and performance analysis. J Hydrometeorol 13(5):1397–1420. doi:10.1175/JHM-D-11-088.1
Mann ME, Emanuel KA (2006) Atlantic hurricane trends linked to climate change. Eos Trans Am Geophys Union 87:233–241

Mann ME, Emanual KA, Holland GJ, Webster PJ (2007) Atlantic tropical cyclones revisited. Eos Trans Am Geophys Union 88:349–350

Mannshardt-Shamseldin EC, Smith RL, Sain SR, Mearns LD, Cooley D (2010) Downscaling extremes: a comparison of extreme value distributions in point-source and gridded precipitation data. Ann Appl Stat 4:484–502

Manton MJ et al (2001) Trends in extreme daily rainfall and temperature in southeast Asia and the South Pacific: 1916–1998. Int J Climatol 21:269–284

Marcos M, Tsimplis MN, Shaw AGP (2009) Sea level extremes in southern Europe. J Geophys Res 114, C01007

McInnes KL, Walsh KJE, Hubbert GD, Beer T (2003) Impact of sea-level rise and storm surges on a coastal community. Nat Hazard 30:187–207

McInnes KL, Macadam I, Hubbert GD, O'Grady JG (2009) A modelling approach for estimating the frequency of sea level extremes and the impact of climate change in southeast Australia. Nat Hazard 51:115–137

McKee TB, Doesken NJ, Kleist J (1993) The relationship of drought frequency and duration to time scales. Preprints, Eighth conference on applied climatology, American Meteorological Society, Anaheim, pp 179–184

Meehl GA et al (2007a) Global climate projections. In: Solomon S et al (eds) The physical science basis. Contribution of working group I to the fourth assessment report of the Intergovernmental Panel on Climate Change. Cambridge University Press, Cambridge/New York

Meehl GA et al (2007b) The WCRP CMIP3 multi-model dataset: a new era in climate change research. Bull Am Meteorol Soc 88:1383–1394

Mendelsohn R, Emanuel K, Chonabayashi S, Bakkensen L (2012) The impact of climate change on global tropical cyclone damage. Nat Clim Change 2:205–209. doi:10.1038/nclimate1357

Menéndez M, Woodworth PL (2010) Changes in extreme high water levels based on a quasi-global tide-gauge dataset. J Geophys Res 115, C10011

Mesinger F et al (2006) North American regional reanalysis. Bull Am Meteorol Soc 87:343–360. doi:10.1175/BAMS-87-3-343

Milly PCD, Dunne KA (2011) On the hydrologic adjustment of climate-model projections: the potential pitfall of potential evapotranspiration. Earth Interact 15:1–14

Milly PCD, Wetherald RW, Dunne KA, Delworth TL (2002) Increasing risk of great floods in a changing climate. Nature 415:514–517

Min SK, Zhang X, Zwiers FW (2008) Human induced Arctic moistening. Science 320:518–520

Min SK, Zhang X, Zwiers FW, Hegerl GC (2011) Human contribution to more-intense precipitation extremes. Nature 470:378–381. doi:10.1038/nature09763

Mitchell JFB, Wilson CA, Cunnington WM (1987) On CO_2 sensitivity and model dependence of results. Q J R Meteorol Soc 113:293–322

Mitrovica JX, Tamisiea ME, Ivins ER, Vermeersen LLA, Milne GA, Lambeck K (2010) Surface mass loading on a dynamic earth: complexity and contamination in the geodetic analysis of global sea-level trends. In: Church JA et al (eds) Understanding sea-level rise and variability. Wiley-Blackwell, Chichester, pp 285–325

Morak S, Hegerl GC, Kenyon J (2011) Detectable regional changes in the number of warm nights. Geophys Res Lett 38, L17703. doi:10.1029/2011GL048531

Morak S, Hegerl GC, Christidis N (2013) Detectable changes in temperature extremes. J Clim 26:1561–1574

Mousavi ME, Irish JL, Frey AE, Olivera F, Edge BL (2011) Global warming and hurricanes: the potential impact of hurricane intensification and sea level rise on coastal flooding. Clim Chang 104:575–597. doi:10.1007/s10584-009-9790-0

Mudelsee M, Borngen M, Tetzlaff G, Grunewald U (2003) No upward trends in the occurrence of extreme floods in central Europe. Nature 425:166–169

Mueller B, Seneviratne SI (2012) Hot days induced by precipitation deficits at the global scale. Proc Natl Acad Sci USA 109:12398–12403. doi:10.1073/pnas.1204330109

Nicholls N, Alexander L (2007) Has the climate become more variable or extreme? Progress 1992–2006. Prog Phys Geogr 31:1–11

Nicholls N, Larsen S (2011) Impact of drought on temperature extremes in Melbourne, Australia. Aust Meteorol Oceanogr J 61:113–116

Noake K, Polson D, Hegerl GC, Zhang X (2011) Changes in seasonal land precipitation during the latter twentieth century. Geophys Res Lett 39, L03706

Orlowsky B, Seneviratne SI (2012) Global changes in extremes events: regional and seasonal dimension. Clim Chang 110:669–696. doi:10.1007/s10584-011-0122-9

Otto FE, Massey N, van Oldenburg GJ, Jones RG, Allen MR (2012) Reconciling two approaches to attribution of the 2010 Russian heat wave. Geophys Res Lett 39, L04702. doi:10.1029/2011GL050422

Page CM, Nicholls N, Plummer N, Trewin BC, Manton MJ, Alexander L, Chambers LE, Choi Y, Collins DA, Gosai A, Della-Marta P, Haylock MR, Inape K, Laurent V, Maitrepierre L, Makmur EEP, Nakamigawa H, Ouprasitwong N, McGree S, Pahalad J, Salinger MJ, Tibig L, Tran TD, Vediapan K, Zhai P (2004) Data rescue in the South-east Asia and South Pacific region: challenges and opportunities. Bull Amer Meteorol Soc 85:1483–1489

Pall P, Aina T, Stone DA, Stott PA, Nozawa T, Hilbberts AGJ, Lohmann D, Allen MR (2011) Anthropogenic greenhouse gas contribution to flood risk in England and Wales in autumn 2000. Nature 470:382–385

Palmer WC (1965) Meteorological drought. Research paper 45, US Department of Commerce, Weather Bureau, Washington, DC, 58 pp. Available from NOAA Library and Information Services Division, Washington, DC, 20852

Peterson TC, Manton MJ (2008) Monitoring changes in climate extremes: a tale of international collaboration. Bull Am Meteorol Soc 89:1266–1271

Peterson TC, Anderson D, Cohen SJ, Cortez M, Murname R, Parmesan C, Phillips D, Pulwarty R, Stone J (2008) Why weather and climate extremes matter. In: Karl TR et al (eds) Weather and climate extremes in a changing climate. Regions of focus: North America, Hawaii, Caribbean, and U.S. Pacific Islands. U.S. Climate Change Science Program and the Subcommittee on Global Change Research, Washington, DC, pp 11–33

Peterson TC, Stott PA, Herring S (eds) (2012) Explaining extreme events of 2011 from a climate perspective. Bull Am Meteorol Soc 93:1041–1066. doi:10.1175/BAMS-D-12-00021.1

Polson D, Hegerl GC, Zhang X, Osborn T (2013) Causes of robust seasonal land precipitation changes. J Clim, in press, doi:10.1175/JCLI-D-12-00474.1

Portmann RW, Solomon S, Hegerl GC (2009) Spatial and seasonal patterns in climate change, temperatures, and precipitation across the United States. Proc Natl Acad Sci USA 106:7324–7329. doi:10.1073_pnas.0808533106

Prudhomme C, Davies H (2009) Assessing uncertainties in climate change impact analyses on the river flow regimes in the UK. Part 2: future climate. Clim Chang 93:197–222

Ramsay HA, Sobel AH (2011) The effects of relative and absolute sea surface temperature on tropical cyclone potential intensity using a single column model. J Clim 24:183–193

Randall DA et al (2007) Climate models and their evaluation. In: Solomon S et al (eds) The physical science basis. Contribution of working group I to the fourth assessment report of the Intergovernmental Panel on Climate Change. Cambridge University Press, Cambridge/New York

Roderick ML, Rotstayn LD, Farquhar GD, Hobbins MT (2007) On the attribution of changing pan evaporation. Geophys Res Lett 34, L17403. doi:10.1029/2007GL031166

Ropelewski CF, Halpert MS (1987) Global and regional scale precipitation patterns associated with the El Niño/Southern Oscillation. Mon Weather Rev 115:1606–1626

Rosenzweig C et al (2007) Assessment of observed changes and responses in natural and managed systems. In: Parry ML et al (eds) Impacts, adaptation and vulnerability. Contribution of working group II to the fourth assessment report of the Intergovernmental Panel on Climate Change. Cambridge University Press, Cambridge/New York

Ryan BF, Watterson IG, Evans JL (1992) Tropical cyclone frequencies inferred from Gray's yearly genesis parameter – validation of GCM tropical climates. Geophys Res Lett 19:1831–1834

Sang H, Gelfand AE (2009) Hierarchical modeling for extreme values observed over space and time. Environ Ecol Stat 16:407–426

Sang H, Gelfand AE (2010) Continuous spatial process models for spatial extreme values. J Agric Biol Environ Stat 15:49–65

Santer BD et al (2006) Forced and unforced ocean temperature changes in Atlantic and Pacific tropical cyclogenesis regions. Proc Natl Acad Sci USA 103:13905–13910

Schlather M (2002) Models for stationary max-stable random fields. Extremes 5:33–44

Schmidt H, von Storch H (1993) German Bight storms analysed. Nature 365:791

Schmith T, Kaas E, Li T-S (1998) Northeast Atlantic winter storminess 1875–1995 re-analysed. Clim Dyn 14:529–536

Schubert S et al (2009) A US CLIVAR project to assess and compare the responses of global climate models to drought-related SST forcing patterns: overview and results. J Clim 22:5251–5272

Seneviratne SI, Lüthi D, Litschi M, Schär C (2006) Land-atmosphere coupling and climate change in Europe. Nature 443:205–209

Seneviratne SI, Corti T, Davin EL, Hirschi M, Jaeger E, Lehner I, Orlowsky B, Teuling AJ (2010) Investigating soil moisture-climate interactions in a changing climate: a review. Earth Sci Rev 99:125–161

Seneviratne SI, Nicholls N, Easterling D, Goodess CM, Kanae S, Kossin J, Luo Y, Marengo J, McInnes K, Rahimi M, Reichstein M, Sorteberg A, Vera C, Zhang X (2012) Changes in climate extremes and their impacts on the natural physical environment. In: Field CB, Barros V, Stocker TF, Qin D, Dokken D, Ebi KL, Mastrandrea MD, Mach KJ, Plattner G-K, Allen SK, Tignor M, Midgley PM (eds) Intergovernmental Panel on Climate Change special report on managing the risks of extreme events and disasters to advance climate change adaptation. Cambridge University Press, Cambridge/New York

Sheffield J, Wood EF (2008) Global trends and variability in soil moisture and drought characteristics, 1950–2000, from observation-driven simulations of the terrestrial hydrologic cycle. J Clim 21:432–458

Sherwood SC, Huber M (2010) An adaptability limit to climate change due to heat stress. Proc Natl Acad Sci USA 107(21):9552–9555

Shiklomanov AI, Lammers RB, Rawlins MA, Smith LC, Pavelsky TM (2007) Temporal and spatial variations in maximum river discharge from a new Russian data set. J Geophys Res 112:G04S53

Shrestha RR, Berland AJ, Schorbus MA, Werner AT (2011) Climate change impacts on hydroclimatic regimes in the Peace and Columbia watersheds, British Columbia, Canada. In: Pacific climate impacts consortium, University of Victoria, Victoria, 37 pp. Available from http://pacificclimate.org/sites/default/files/publications/Shrestha.Synthesis.FinalReport.Apr2011.pdf

Sigmond M, Kushner PJ, Scinocca JF (2007) Discriminating robust and non-robust atmospheric circulation responses to global warming. J Geophys Res 112, D20121. doi:10.1029/2006JD008270

Sillmann J, Croci-Maspoli M, Kallache M, Katz RW (2011) Extreme cold winter temperature in Europe under the influence of North Atlantic atmospheric blocking. J Clim 24:5899–5913. doi:10.1175/2011JCLI4075.1

Sillmann J, Kharin VV, Zhang X, Zwiers FW, Bronaugh D (2013a) Climate extreme indices in the CMIP5 multi-model ensemble. Part 1: model evaluation in the present climate. J Geophys Res 118, doi:10.1002/jgrd.50203

Sillmann J, Kharin VV, Zwiers FW, Zhang X, Bronaugh D (2013b) Climate extreme indices in the CMIP5 multi-model ensemble. Part 2: future climate projections. J Geophys Res 118, doi:10.1002/jgrd.50188

Smith RL (1986) Extreme value theory based on the r-largest annual events. J Hydrol 86:27–43

Smith RL (1990) Max-stable processes and spatial extremes. Unpublished manuscript. Available at http://www.stat.unc.edu/postscript/rs/spatex.pdf

Smits A, Klein Tank AMG, Können GP (2005) Trends in storminess over the Netherlands, 1962–2002. Int J Climatol 25:1331–1344. doi:10.1002/joc.1195

Solomon S et al (2007) Technical summary. In: Solomon S et al (eds) Climate change 2007: the physical science basis. Contribution of working group I to the fourth assessment report of the Intergovernmental Panel on Climate Change. Cambridge University Press, Cambridge/New York

Stahl K, Hisdal H, Hannaford J, Tallaksen LM, van Lanen HAJ, Sauquet E, Demuth S, Fendekova M, Jodar J (2010) Streamflow trends in Europe: evidence from dataset of near-natural catchments. Hydrol Earth Syst Sci 14:2367–2382

Steadman RG (1979) The assessment of sultriness. Part I: a temperature-humidity index based on human physiology and clothing science. J Appl Meteorol 18:861–873

Stott PA, Stone DA, Allen MR (2004) Human contribution to the European heatwave of 2003. Nature 432:610–614. doi:10.1038/nature0308

Stott PA et al (2012) Attribution of weather and climate-related extreme events. In: WCRP Open Science conference: climate research in service to society, World Climate Research Programme, Denver (this volume)

Sugi M, Murakami H, Yoshimura J (2009) A reduction in global tropical cyclone frequency due to global warming. SOLA 5:164–167. doi:10.2151/sola.2009-042

Svensson C, Hannaford J, Kundzewicz ZW, Marsh TJ (2006) Trends in river floods: why is there no clear signal in observations? IAHS/UNESCO Kov Colloq: Front Flood Res 305:1–18

Swanson KL (2008) Non-locality of Atlantic tropical cyclone intensitites. Geochem Geophys Geosyst 9:Q04V01

Taye MT, Ntegeka V, Ogiramoi NP, Willems P (2011) Assessment of climate change impact on hydrological extremes in two source regions of the Nile River Basin. Hydrol Earth Syst Sci 15:209–222

Taylor KE, Stouffer RJ, Meehl GA (2012) An overview of CMIP5 and the experiment design. Bull Am Meteorol Soc 93(4):485–498

Tebaldi C, Hayhoe K, Arblaster JM, Meehl GA (2006) Going to the extremes—an intercomparison of model-simulated historical and future changes in extreme events. Clim Change 79:185–211

Ting M, Kushnir Y, Seager R, Li C (2009) Forced and internal 20th century SST trends in the North Atlantic. J Clim 22:1469–1481. doi:10.1175/2008JCLI2561.1

Trapp RJ et al (2005) Tornadoes from squall lines and bow echoes. Pt I: climatological distribution. Weather Forecast 20:23–34

Trenberth KE et al (2007) Observations: surface and atmospheric climate change. In: Solomon S et al (eds) The Physical science basis. Contribution of working group I to the fourth assessment report of the Intergovernmental Panel on Climate Change. Cambridge University Press, Cambridge/New York

Ulbrich U, Pinto JG, Kupfer H, Leckebusch GC, Spangehl T, Reyers M (2008) Changing northern hemisphere storm tracks in an ensemble of IPCC climate change simulations. J Clim 21:1669–1679

Uppala SM et al (2005) The ERA-40 re-analysis. Q J R Meteorol Soc 131:2961–3012. doi:10.1256/qj.04.176

van der Schrier G, Briffa KR, Jones PD, Osborn TJ (2006) Summer moisture variability across Europe. J Clim 19:2818–2834

van Oldenborgh GJ, van Urk A, Allen M (2012) The absence of a role of climate change in the 2011 Thailand floods. In: Peterson TC et al (eds) Explaining extreme events of 2011 from a climate perspective. Bull Am Meteor Soc 93:1041–1067. doi:10.1175/BAMS-D-12-00021.1

van Pelt SC, Kabat P, ter Maat TW, van den Hurk BJJM, Weerts AH (2009) Discharge simulations performed with a hydrological model using bias corrected regional climate model input. Hydrol Earth Syst Sci 13:2387–2397

Van Wagner C (1987) Development and structure of the Canadian forest fire weather index system. Technical report #35, Canadian Forest Service

Vannitsem S, Naveau P (2007) Spatial dependences among precipitation maxima over Belgium. Nonlinear Process Geophys 14:621–630

Vautard R, Cattiaux J, Yiou P, Thepaut J-N, Ciais P (2010) Northern Hemisphere atmospheric stilling partly attributed to an increase in surface roughness. Nat Geogr 3:756–761. doi:10.1038/NGEO979

Vecchi GA, Knutson TR (2008) On estimates of historical North Atlantic tropical cyclone activity. J Clim 21:3580–3600

Vecchi GA, Knutson TR (2011) Estimating annual numbers of Atlantic hurricanes missing from the HURDAT database (1878–1965) using ship track density. J Clim 24:1736–1746. doi:10.1175/2010JCLI3810.1
Vecchi GA, Soden BJ (2007) Effect of remote sea surface temperature change on tropical cyclone potential intensity. Nature 450:1066–1070
Vecchi GA, Swanson KL, Soden BJ (2008) Whither hurricane activity. Science 322:687–689
Vermeer M, Rahmstorf S (2009) Global sea level linked to global temperature. Proc Natl Acad Sci USA 106:21527–21532
Vicente-Serrano SM, Begueria S, Lopez-Moreno JI (2010) A multiscalar drought index sensitive to global warming: the standardized precipitation evapotranspiration index. J Clim 23(7): 1696–1718
Villarini G, Vecchi GA (2013) Projected increases in North Atlantic tropical cyclone intensity from CMIP5 models. J Clim 26:3231–3240. doi:10.1175/JCLI-D-12-00441.1
Villarini G, Serinaldi F, Smith JA, Krajewski WF (2009) On the stationarity of annual flood peaks in the continental United States during the 20th century. Water Resour Res 45, W08417
Wang XL, Swail VR (2001) Changes of extreme wave heights in Northern Hemisphere oceans and related atmospheric circulation regimes. J Clim 14:2204–2221
Wang XL, Wan H, Swail VR (2006) Observed changes in cyclone activity in Canada and their relationships to major circulation regimes. J Clim 19:896–915
Wang AH, Bohn TJ, Mahanama SP, Koster RD, Lettenmaier DP (2009a) Multimodel ensemble reconstruction of drought over the continental United States. J Clim 22:2694–2712
Wang XL, Swail VR, Zwiers FW, Zhang X, Feng Y (2009b) Detection of external influence on trends of atmospheric storminess and northern oceans wave heights. Clim Dyn 32:189–203
Wang XL, Wan H, Zwiers FW, Swail VR, Compo GP, Allan RJ, Vose RS, Jourdain S, Yin X (2011) Trends and low-frequency variability of storminess over western Europe, 1887–2007. Clim Dyn 37:2355–2371. doi:10.1007/s00382-011-1107-0
Wang XL, Feng Y, Compo GP, Swail VR, Zwiers FW, Allan RJ, Sardeshmukh PD (2012) Trends and low frequency variability of extra-tropical cyclone activity in the ensemble of twentieth century reanalysis. Clim Dyn. doi:10.1007/s00382-012-1450-9
Watson CC, Johnson ME (2004) Hurricane loss estimation models: opportunities for improving the State of the Art. Bull Am Meteorol Soc 85:1713
Webster PJ, Holland GJ, Curry JA, Chang HR (2005) Changes in tropical cyclone number, duration, and intensity in a warming environment. Science 309:1844–1846
Wehner MF (2013) Very extreme seasonal precipitation in the NARCCAP ensemble: model performance and projections. Clim Dyn 40(1–2):59–80. doi:10.1007/s00382-012-1393-1
Wehner MF, Smith RL, Bala G, Duffy P (2010) The effect of horizontal resolution on simulation of very extreme precipitation events in a global atmospheric model. Clim Dyn 34:241–247. doi:10.1007/s00382-009-0656-y
Wells N, Goddard S, Hayes MH (2004) A self-calibrating Palmer drought severity index. J Clim 17:2335–2351
Wilby RL, Beven KJ, Reynard NS (2008) Climate change and fluvial flood risk in the UK: more of the same? Hydrol Process 22:2511–2523
Woodhouse CA, Overpeck JT (1998) 2000 years of drought variability in the central United States. Bull Am Meteorol Soc 79:2693–2714
Woodhouse CA, Meko DM, MacDonald GM, Stahle DW, Cook ER (2010) A 1200-year perspective on 21st century drought in southwestern North America. Proc Natl Acad Sci USA 107:21283–21288. doi:10.1175/2011JCLI4075.1
Woodworth PL, Gehrels WR, Nerem RS (2011) Nineteenth and twentieth century changes in sea level. Oceanography 24:80–93. doi:10.5670/oceanog.2011.29
Xie P, Janowiak JE, Arkin PA, Adler R, Gruber A, Ferraro R, Huffman GJ, Curtis S (2003) GPR pentad precipitation analyses: an experimental dataset based on gauge observations and satellite estimates. J Clim 16:2197–2214
Zazulie N, Rusticucci M, Solomon S (2010) Changes in climate at high southern latitudes: a unique daily record at orcadas spanning 1903–2008. J Clim 23:189–196. doi:10.1175/2009JCLI3074.1

Zhang R, Delworth T (2009) A new method for attributing climate variations over the Atlantic Hurricane Basin's main development region. Geophys Res Lett 36, L06701. doi:10.1029/2009GL037260

Zhang X, Zwiers FZ (2013) Statistical indices for diagnosing and detecting changes in extremes. In: Agha Kouchak et al (eds) Extremes in a changing climate. Water Science and Technology Library 65. Springer, Dordrecht. doi:10.1007/978-94-007-4479-0-1

Zhang X, Zwiers FW, Li G (2004) Monte Carlo experiments on the detection of trends in extreme values. J Clim 17:1945–1952

Zhang X, Hegerl GC, Zwiers FW (2005) Avoiding inhomogeneity in percentile-based indices of temperature extremes. J Clim 18:1641–1651

Zhang X, Zwiers FW, Hegerl GC, Lambert FH, Gillett NP, Solomon S, Stott PA, Nozawa T (2007) Detection of human influence on twentieth-century precipitation trends. Nature 448:461–465

Zhang X, Zwiers FW, Hegerl G (2009) The influence of data precision on the calculation of temperature percentile indices. Int J Climatol. doi:10.1002/joc.1738

Zhang X, Wang J, Zwiers FW, Groisman PY (2010) The influence of large scale climate variability on winter maximum daily precipitation over North America. J Clim 23:2902–2915

Zhang X, Alexander LV, Hegerl GC, Klein Tank A, Peterson TC, Trewin B, Zwiers FW (2011) Indices for monitoring changes in extremes based on daily temperature and precipitation data. Wiley Interdiscip Rev Clim Change 2:851–870. doi:10.1002/wcc.147

Zhao M, Held I, Lin S-J, Vecchi GA (2009) Simulations of global hurricane climatology, interannual variability, and response to global warming using a 50 km resolution GCM. J Clim 22:6653–6678. doi:10.1175/2009JCLI3049.1

Zwiers FW (1987) An extreme value analysis of wind speed at 5 Canadian locations. Can J Stat 15:317–327

Zwiers FW, Zhang X, Feng Y (2011) Anthropogenic influence on long return period daily temperature extremes at regional scales. J Clim 24:881–892

Carbon Dioxide and Climate: Perspectives on a Scientific Assessment

Sandrine Bony, Bjorn Stevens, Isaac H. Held, John F. Mitchell, Jean-Louis Dufresne, Kerry A. Emanuel, Pierre Friedlingstein, Stephen Griffies, and Catherine Senior

Abstract Many of the findings of the Charney Report on CO_2-induced climate change published in 1979 are still valid, even after 30 additional years of climate research and observations. This paper considers the reasons why the report was so prescient, and assesses the progress achieved since its publication. We suggest that emphasis on the importance of physical understanding gained through the use of theory and simple models, both in isolation and as an aid in the interpretation of the results of General Circulation Models, provided much of the authors' insight at the time. Increased emphasis on these aspects of research is likely to continue to be productive in the future, and even to constitute one of the most efficient routes towards improved climate change assessments.

S. Bony (✉) • J.-L. Dufresne
Laboratoire de Météorologie Dynamique/IPSL, CNRS, Pierre & Marie Curie University, case 99, 4 place Jussieu, Paris 75252, France
e-mail: bony@lmd.jussieu.fr; jldufres@lmd.jussieu.fr

B. Stevens
Max-Planck Institute for Meteorology, Hamburg 20146, Germany
e-mail: bjorn.stevens@mpimet.mpg.de

I.H. Held • S. Griffies
Geophysical Fluid Dynamics Laboratory, Princeton University, 201 Forrestal Road, Princeton, NJ 08540-6649, USA
e-mail: Isaac.Held@noaa.gov; Stephen.Griffies@noaa.gov

J.F. Mitchell • C. Senior
MetOffice/Hadley Center, FitzRoy Road, Exeter, Devon EX1 3PB, United Kingdom
e-mail: john.f.mitchell@metoffice.gov.uk; cath.senior@metoffice.gov.uk

K.A. Emanuel
Massachusetts Institute for Technology, 77 Massachusetts Avenue, Cambridge 02139, MA, USA
e-mail: emanuel@mit.edu

P. Friedlingstein
College of Engineering, Mathematics and Physical Sciences, University of Exeter, Harrison Building, Streatham Campus, North Park Road, Exeter EX4 4QF, UK
e-mail: P.Friedlingstein@exeter.ac.uk

G.R. Asrar and J.W. Hurrell (eds.), *Climate Science for Serving Society: Research, Modeling and Prediction Priorities*, DOI 10.1007/978-94-007-6692-1_14,

Keywords Climate change assessment • Charney report • Hierarchical climate modeling • Physical understanding • Climate projections • Climate processes, forcings and feedbacks • General circulation models

1 Introduction

In 1896 Svante Arrhenius first suggested that increased CO_2 in the atmosphere might affect climate. Observational evidence that CO_2 concentrations were actually increasing in the atmosphere became available in the 1960s, thanks to the continuous measurement begun by Charles D. Keeling in 1958. In 1979 the US National Academy of Sciences asked a small work group of scientists led by Jule Charney to undertake a scientific assessment of the possible effects of CO_2 on climate (Charney et al. 1979: "*Carbon Dioxide and Climate: A Scientific Assessment*"). Owing to the striking consistency of most of its conclusions with those of current assessments on climate change, the report (which became known as the "Charney Report") arouses admiration but also inevitably makes us wonder: since then, what progress have we made in assessing the effects of CO_2 on climate in the last 30 years? Where are the gaps? What are the implications for community efforts to improve assessments of future long-term climate change? This paper addresses these issues based on the personal reflections of a small group of scientists from a range of backgrounds and specialities.

After a brief presentation of the Charney report (Sect. 2), we discuss the scientific progress (or lack of progress) addressed in the key disciplines identified by this report (Sect. 3). In Sect. 4, we highlight lessons drawn from climate research over the last decades, and make some suggestions for further progress.

2 The Charney Report

In the foreword to the Charney report, Vern Suomi noted that scientists had known for more than a century that changing atmospheric composition could affect climate, that they now had "incontrovertible evidence" that atmospheric composition was indeed changing and that this had prompted a number of recent investigations of the implications of increasing CO_2. Thus the Charney Report was written at an auspicious moment: 20 years of measurements at Mauna Loa had established beyond doubt that CO_2 concentrations were rising, and general circulation models were just beginning to be applied to understanding the consequences.

The relatively high impact of the Charney Report might be partially attributable to its succinctness. The whole report is 16½ small pages long, and the main conclusions are summarized in an introductory section only 2¼ pages in length. The authors begin by estimating that CO_2 concentrations would double by some time in the first half of the twenty-first century, and proceed to estimate the resultant change in equilibrium global mean surface temperature *to be near 3°C* with larger increases at higher latitudes. After discussing the uncertainties inherent in such an estimate, they state that *it is significant, however, that none of the model calculations predicts negligible warming*. While the report focuses on changes in global mean temperature, the authors note that

> The evidence is that the variations in these anomalies with latitude, longitude, and season will be at least as great as the globally averaged changes themselves, and it would be misleading to predict regional climatic changes on the basis of global or zonal averages alone.

While the authors make it clear that their conclusions are based primarily on the results of three-dimensional general circulation models, they state that

> Our confidence in our conclusion that a doubling of CO_2 will eventually result in significant temperature increases and other climate changes is based on the fact that the results of the radiative-convective and heat-balance model studies can be understood in purely physical terms and are verified by the more complex GCM's. [General Circulation Models]

The authors' philosophy in using GCMs is emphasized again, later in the report:

> In order to assess the climatic effects of increased atmospheric concentrations of CO_2, we consider first the primary physical processes that influence the climatic system as a whole. These processes are best studied in simple models whose physical characteristics may readily be comprehended. The understanding derived from these studies enables one better to assess the performance of the three-dimensional circulation models on which accurate estimates must be based.

The authors discussed what they considered to be the primary obstacles to better projections of climate change, including the rates at which heat and CO_2 are mixed into the deep ocean and the feedback effect of changing clouds. They also discussed their inability to say much about the regional patterns of climate change, given the large uncertainties associated with regional climate projections from GCMs. Such issues remain very much alive today.

What made the *Charney Report* so prescient? The emphasis on the importance of physical understanding gained through theory and simple models, both for its own sake, to facilitate the distillation of scientific knowledge, and to help interpret and check the results of GCMs, proved highly productive and led to a projection of the global mean temperature increase that is virtually identical to current projections, even though the authors did not have the benefit of a clear signal of warming in the observations at their disposal. For instance, the authors used a variety of approaches to estimate climate feedbacks, starting with simple physical principles and assumptions, working through one-dimensional models to make an initial quantification of feedbacks, and using full general circulation models to refine or extend that assessment. This meant that they had a good understanding of the main processes governing climate sensitivity, and could defend their range of answers without having to rely on complex models. This may be why their findings were accepted and have stood the test of time.

3 Key Areas of Progress (or Lack of Progress) Since the Charney Report

The importance of non-CO_2 forcings such as methane or other long-lived greenhouse gases, ozone and aerosols, has been emphasized since the publication of the Charney Report, especially for interpreting the evolution of the twentieth century climate. However, we expect the increase in CO_2 concentration to dominate the acceleration of the anthropogenic forcing over the next decades. Therefore, anticipating the effects of CO_2 on climate remains a key issue. The progress achieved on that issue over the last three decades is discussed here by considering the different components of the CO_2-induced climate change problem considered by the Charney report: the evolution of carbon in the atmosphere (Sect. 3.1), the CO_2 radiative forcing (Sect. 3.2), climate sensitivity (Sect. 3.3), the physical processes important for climate feedbacks (Sect. 3.4), the role of the ocean (Sect. 3.5), and the credibility of GCM projections (Sect. 3.6).

3.1 Carbon in the Atmosphere

The Charney Report presented little new information on the global carbon cycle, only briefly summarizing its key features, based on a SCOPE review book published on the same year (Bolin et al. 1979). This includes comments that the "proper role of the deep sea as a potential sink for fossil-fuel CO_2 has not been accurately assessed" and "whether some increase of carbon in the remaining world forests has occurred is not known". Nevertheless, the report concluded that "Considering the uncertainties, it would appear that a doubling of atmospheric carbon dioxide will occur by about 2030 if the use of fossil fuels continues to grow at a rate of about 4 percent per year, as was the case until a few years ago. If the growth rate were 2 percent, the time for doubling would be delayed by 15 to 20 years, while a constant use of fossil fuels at today's levels shifts the time for doubling well into the twenty-second century." Although they do not say so explicitly, their main assumption appears to be that the ocean acts as the sole sink of anthropogenic carbon, and that the terrestrial biosphere remains neutral. Also, they report that "it has been customary to assume to begin with that about 50 percent of the emissions will stay in the atmosphere".

We now have a clearer and much more quantitative picture of the global carbon cycle. Although deforestation is still recognized as a source of CO_2 (LeQuéré et al. 2009; Friedlingstein et al. 2010), terrestrial ecosystems overall are now understood to be net sinks of anthropogenic CO_2, absorbing about the same amount of CO_2 as the global oceans. This is now well known from observations of combined changes in atmospheric CO_2 and O_2, top-down inversions of atmospheric CO_2, and bottom-up modeling of ocean and terrestrial biogeochemistry (see Denman et al. 2007 for a review of these different methods).

Over the last decade, work has also shown how climate change might affect the ability of both oceans and land ecosystems to absorb atmospheric CO_2. Modeling studies performed this last decade have suggested a positive feedback between climate change and the global carbon cycle (Cox et al. 2000; Dufresne et al. 2002). Increased stratification of the upper-ocean due to warming at the surface reduces the export of carbon from the surface to the deep ocean, and hence limits the air-sea exchange of CO_2. Declining productivity in tropical forests and a general increase in the rate of soil carbon decomposition (heterotrophic respiration) partially offset the land carbon uptake due to the CO_2 fertilization effect. Despite the large uncertainty in the magnitude of the climate carbon cycle feedback (Friedlingstein et al. 2006), analysis of proxy-based temperature and CO_2 from ice cores indicates that it is likely to be positive (Frank et al. 2010). The airborne fraction is expected to increase in the future as a result of sinks saturating with increasing CO_2 and declining in a warmer world. Coupled climate carbon cycle models suggest the airborne fraction could rise from the current value of 45–62 % (median estimate). Analysis of the past 50 years seems to indicate that the airborne fraction has already increased (LeQuéré et al. 2009). In the context of the Charney report, this finding would not alter the estimate of the climate sensitivity, as it is based on the climate response to a prescribed doubling of the CO_2 concentration, however, it would accelerate the timing of the CO_2 doubling (Fig. 1).

Perhaps the most important development has simply been the ice core CO_2 records, which began to appear shortly after the Charney report (Delmas et al. 1980). The remarkable glacial-interglacial fluctuations of the CO_2 provide constraints on climate sensitivity and pose a challenge to our understanding of the controls on the background carbon cycle that is being perturbed by anthropogenic emissions.

3.2 *Radiative Forcing*

The concepts of radiative forcing and equilibrium climate sensitivity were well established at the time of the Charney report. The major issues in estimating the radiative forcing for an atmosphere with fixed clouds and water vapor had already been addressed in the literature on which the Charney report is based (e.g. Ramanathan et al. 1979; Manabe and Wetherald 1967, 1975). The importance of using radiative fluxes at the tropopause rather than the surface, the stratospheric adjustment, the dependence of CO_2 absorption on CO_2 concentration, and the overlap between the H_2O and CO_2 absorbing bands were all discussed. The radiative forcing for a doubling of CO_2 concentration was estimated in the report to be about 4 W m^{-2} within an uncertainty of ±25 %. The authors anticipated some of the difficulty of computing this forcing, and rejected much larger values in the available literature (*e.g.*, MacDonald et al. (1979) estimated a radiative forcing of 6–8 W m^{-2}) on methodological grounds.

Since the report, the radiative calculations underlying this computation have been regularly improved, with the number of absorption lines used in radiative

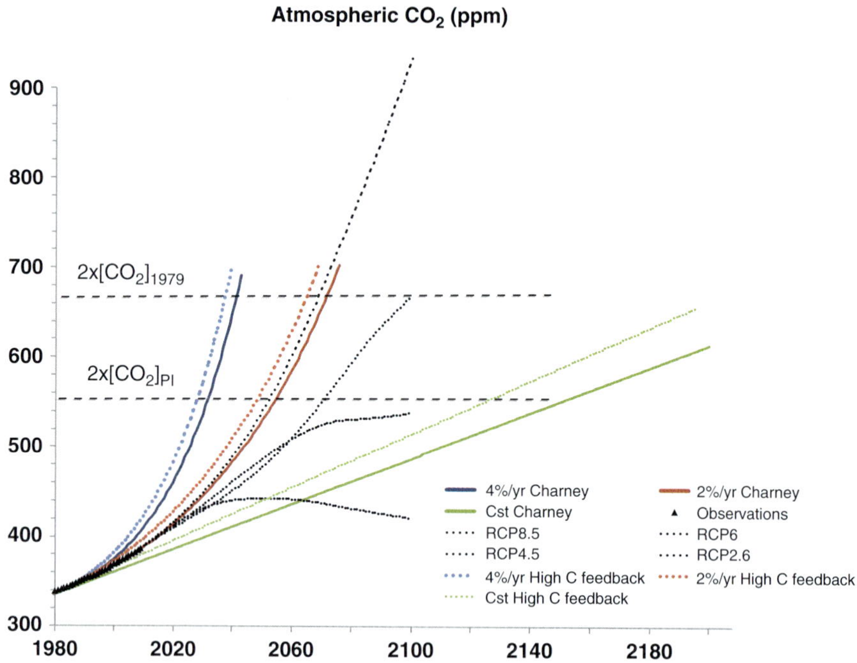

Fig. 1 Atmospheric CO_2 concentration future projections assuming, as in the Charney report, future anthropogenic emissions to increase at a rate of 4 % per year (*blue*), 2 % per year (*red*) or to remain constant (*green*). Also shown (*dotted lines*) are the projected concentrations for these three cases accounting for a positive climate-carbon feedback, absent from the Charney's calculations. The observed CO_2 concentrations, and the twenty-first century CO_2 concentrations projected for the four Representative Concentration Pathways (*RCPs*) used in CMIP5 are shown in *black symbols*

transfer calculations increasing by a factor of several tens and a larger number of gas species taken into account, while the water vapor absorption continuum is better if still incompletely understood. For standard atmospheric profiles, the value of the CO_2 radiative forcing estimated with different line-by-line radiation codes vary with only about a 2 % standard deviation, while estimates from GCM codes exhibit a larger standard deviation of about 10 % (Collins et al. 2006). These differences increase if one takes into account uncertainties in the specified cloud distribution and the fuzziness in the definition of the tropopause. Yet the current best estimate for this "classic" radiative forcing, 3.7 ± 0.3 W m^{-2} (Myhre et al. 2001; Gregory and Webb 2008), is fully consistent with the estimate in the Charney report, while the uncertainty has been considerably reduced.

However, the concept of radiative forcing continues to evolve, particularly owing to the recognition that the fast responses to a change in CO_2 (responses that occur before the oceans and troposphere warm significantly) include not only the stratospheric adjustment but also tropospheric changes, particularly in cloud. This alters the definitions of both forcing and feedback (e.g., Hansen et al. 2002; Shine et al. 2003; Gregory et al 2004; Andrews and Forster 2008). These new concepts are proving valuable in sharpening our understanding of the spread of model responses (Gregory and Webb 2008; Williams et al. 2008), but in the process one loses the clean distinction between a "forcing" that can be computed from radiative processes alone and "feedbacks" that are model dependent.

3.3 Climate Sensitivity

The Charney report produced a range in equilibrium climate sensitivity of 1.5–4.5 °C, with a best guess of 3 °C. As is well known, the large range has proven difficult to reduce. IPCC AR4 (Meehl et al. 2007) states that the equilibrium climate sensitivity is "likely to be in the range 2–4.5 °C, with a best estimate of 3 °C".

Since the Charney report, it has been emphasized how the definition of "equilibrium" depends on which relatively slow processes are considered, including the evolution of the Greenland and Antarctic ice sheets as well as the carbon and other biogeochemical cycles. It has been argued, in particular, that albedo feedback from the ice sheets can increase climate sensitivity substantially above that estimated from the relatively fast feedbacks considered in the Charney report (e.g., Hansen and Sato 2011).

A number of issues that dominate many current discussions of climate sensitivity do not appear in the Charney report. There is no discussion of transient climate sensitivity or appreciation of the multi-century time scales required to approach these equilibrium responses (see Sect. 3.4). There is also little discussion of observational constraints on climate sensitivity – such as the response to volcanic aerosol in the stratosphere, the response to the 11 year solar cycle, and the glacial-interglacial responses to orbital parameter variations (and many other paleoclimate observations), and most, obviously, the warming trends over the past century itself – and the role of models in interpreting these observations, for example, by determining how a response to the Pinatubo volcano relates to responses to more slowly evolving greenhouse gas forcings. And the report reads very differently from recent assessments in that there is no discussion of detection and attribution, and consistently, no discussion of non-CO_2 anthropogenic forcings (greenhouse gases other than CO_2, aerosols, land-use changes). Nevertheless, the power of the climate sensitivity concept highlighted by the report is likely to have influenced the current thinking about the effect of non-CO_2 forcing agents on climate.

Finally, there is little or no attempt to discuss the hydrological cycle or regional climate changes or climate extremes. Was this a flaw in the report? Why should we care about global mean climate sensitivity? We return to this question in Sect. 4 below.

3.4 Principal Feedbacks

The Charney Report clearly outlined the main feedback mechanisms within the physical climate system and endeavored to estimate the climate sensitivity through their quantification. The report's focus was on the water vapor and surface albedo changes, as these were the best known feedback mechanisms (e.g. Manabe and Wetherald 1967), and the nature or sign of each could be inferred based on simple physical arguments; one expects the absolute humidity to increase as the atmosphere warms while maintaining an approximately constant relative humidity, and the surface albedo to decrease as snow and ice retreat with surface warming. Based on model studies that incorporated this reasoning, the Charney Report estimated the magnitude of the water-vapor feedback to be 2.0 W m^{-2} K^{-1} and gave 0.3 W m^{-2} K^{-1} as the most likely value for the surface albedo feedback. For reference the water vapor and lapse rate feedbacks as most recently assessed by the IPCC are 1.8 ± 0.18 and 0.26 ± 0.08 W m^{-2} K^{-1} respectively (Randall et al. 2007). Thus while our best estimate of the magnitude of these important feedbacks has changed little since the Charney Report, considerable effort and progress has been made in establishing the robustness of the physical reasoning that underpinned their assessment, and in assessing it using observations (e.g. Soden et al. 2005).

The Charney Report also recognized possible changes in cloudiness, relative humidity, and temperature lapse rates as the leading sources of uncertainty in their estimate of climate sensitivity, associating a feedback strength of 0 ± 0.5 W m^{-2} K^{-1}, with the combined effects of such processes. The report is not at all clear as to how its authors arrived at this number, although it seems likely that the magnitude of the water vapor feedback which was and is generally believed to be "the most important and obvious of the feedback effects", and a desire to maintain consistency with the general circulation model studies, may have played a role in their thinking. For reference, the IPCC most recently assessed the combined effect of the lapse rate and cloud feedbacks, each of which is estimated as somewhat stronger than 0.5 W m^{-2} K^{-1} but of opposing sign, as 0.15 ± 0.46 W m^{-2} K^{-1}.

Admittedly little progress has been made in narrowing the uncertainty the Charney Report ascribed to the net effects of these climate feedbacks. Discussions about the potential role of cloud-aerosols interactions in these feedbacks have even complicated the issue. But this does not imply that progress in our understanding and estimation of climate feedbacks is out of reach (Bony et al. 2006; see also Hannart et al 2009 for a response to the argument of Roe and Baker (2007) that reducing this uncertainty will be very difficult for fundamental statistical reasons). Actually, important strides have been made towards developing better physical understanding of physical mechanisms associated with climate feedbacks. At the time of the Charney Report there seems to have been little more than a vague idea as to why cloudiness should change with either increasing concentrations of greenhouse gases or surface temperatures. The intervening decades have seen an articulation of a wide variety of mechanisms, ranging from the tendency for clouds to shift upward as the climate warms (e.g. Hansen et al. 1984; Wetherald and Manabe 1988; Mitchell and Ingram 1992), hypotheses that link cloud liquid water to the lapse rate of liquid water (Somerville and Remer 1984), cloud amounts in the subtropics to the tropical temperature lapse rates (Klein and Hartmann 1993), these lapse-rates themselves having been linked to the behavior of deep convection (Zhang and Bretherton 2008). Ideas have also emerged as to why the storm tracks can be expected to migrate poleward in a warmer climate, and how this effect may redistribute clouds relative to the distribution of solar radiation, or how the increased surface fluxes and changing profiles of moist static energy demanded by an atmosphere that maintains a constant relative humidity might be expected to produce more precipitation, but fewer clouds (Held and Soden 2006; Brient and Bony 2013; Rieck et al. 2012).

3.5 Role of the Ocean

The Charney Report considered the primary role of the ocean in climate change as setting the timescale over which heat and carbon are sequestered into the ocean interior, and there was little appreciation for the role of the ocean in climate dynamics at decadal to centennial time scales. From a modern perspective, its treatment of the oceans is likely its weakest aspect.

While the report correctly anticipated the role of ocean intermediate and mode waters in controlling the rate at which the ocean takes up heat, there was little understanding of the physical mechanisms involved in this control (Fig. 2). Ocean heat content may change through passive ventilation, whereby a water parcel interacting with the atmosphere carries heat into the interior largely through isopycnal transport (e.g., Church et al. 1991). Additionally, ocean heat may be modified as stratification increases and overturning circulation decreases, so that interior ocean properties accumulate (Banks and Gregory 2006).

Ocean observations and modeling capabilities were very rudimentary 30 years ago. The observational network, which formerly consisted of measurements by ship-based platforms, has been revolutionized by satellite measurements and profiling floats (Freeland et al. 2010). The density of the measurements in the upper 700 m of the ocean, while not covering the mode waters that ventilate at high latitudes, have nonetheless begun to make it possible to track changes in ocean heat content on decadal scales (Lyman et al. 2010). However, large uncertainties remain in current observational estimates of the ocean heat content. It is likely that difficulties in closing the Earth's global heat budget (Trenberth and Fasullo 2010) partly result from these uncertainties, although Meehl et al. (2011) suggest that deep-ocean heat uptake may explain the apparent 'missing heat'.

The oceanic component of climate models, though still possessing errors and limitations, has advanced greatly over the last decades. A new generation of models is now able to represent important processes such as mesoscale eddies (e.g., Farneti et al. 2010) and high latitude shelf and overflow processes (e.g., Legg et al. 2009) that regulate how the ocean transports heat and mass from the surface to its interior.

The incorporation of the new generation of measurements into both process and realistic ocean climate models now facilitates mechanistic interpretations of observations and physically based evaluation of more complex models (see, *e.g.,* Griffies et al. 2010), thereby developing the type of robust understanding that must underlie our confidence in estimates of the ocean's role in climate change.

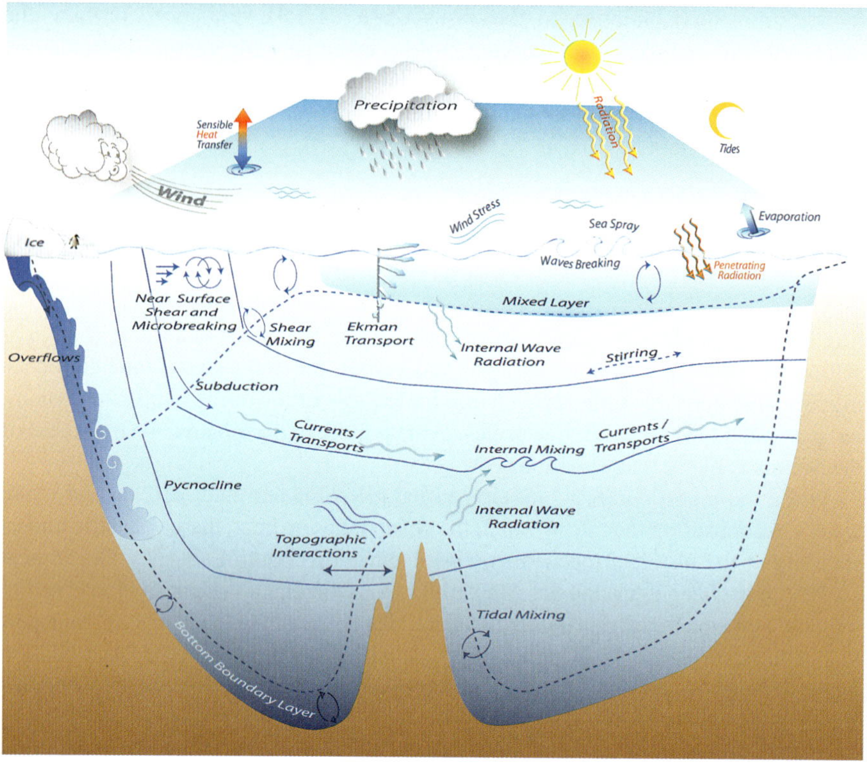

Fig. 2 Understanding and quantifying the ocean's role in climate change involves a variety of questions related to how physical processes impact the movement of tracers (e.g., heat, salt, carbon, nutrients) across the upper ocean interface and within the ocean interior. In general, processes move tracers across density surface (dianeutrally) or along neutral surfaces (epineutrally), with epineutral processes dominant in the interior, yet dianeutral processes directly impacting vertical stratification. This figure provides a schematic of such processes, including turbulent air-sea exchanges and upper ocean wave breaking and Langmuir circulations; gyre-scale, mesoscale, and submesoscale transport; high latitude convective and downslope shelf ventilation; and mixing induced by breaking internal gravity waves energized by winds and tides. Nearly all such processes are subgrid scale for present day global ocean climate simulations. The formulation of sensible parameterizations, including schemes that remain relevant under a changing climate (e.g., modifications to stratification), remains a key focus of oceanographic research efforts

Through this process, the role of stratification has emerged as a particularly important one. In addition to its role in the carbon cycle and net ocean heat uptake (mentioned in the Charney report), the stratification of the ocean may also modify much shorter time-scale processes ranging from decadal climate fluctuations, to ENSO, to the life-cycle of tropical cyclones which depend crucially on their ability to extract heat from the upper ocean. This contributes to our increasing appreciation of the importance of characterizing climate variability on the decadal to century time scales, and the potential for internal variability to complicate the attribution of observed climate changes to specific anthropogenic forcing agents.

3.6 *Credibility of GCM Projections*

The Charney Report considered only five models, and examined the key physical features of each to assess the most realistic and robust outcome. For example, in a model simulation with excessive sea-ice extents, it was assumed that the ice-albedo effect would be exaggerated, and this bias was accounted for in the final assessment. Over time, models have increased in number (model inter-comparisons can now involve more than 20 modeling groups and 40 models) and complexity, advancing opportunities to identify the robust features of complex model simulations, but linking individual model biases to a particular model process or feature has become more difficult. In view of this, intercomparisons increasingly make use of metrics to assess models rather than direct physical interpretation. Since there are so many potential metrics, and since different metrics often tell different stories as to which models are better or worse, a key problem for the field is to tailor metrics to particular predictions. An instructive example is Hall and Qu (2006), who show a clear relationship between simulated snow surface albedo/temperature feedback estimated from the current seasonal cycle and from climate change simulations. The climate feedback can then be calibrated using the observed seasonal cycle feedback. Research on the climatic response to the ozone hole has likewise isolated the persistence time for the Southern Annular mode as a key metric for predictions of the poleward movement of the westerlies and midlatitude storm track (Son et al. 2010).

The report did not consider changes in regional climate. It noted that due to lack of resolution and differences in parameterizations, two models could give very different changes in regional circulations such as the monsoon and related rainfall patterns, and therefore were unreliable. The use of regional models may improve regional detail, but is dependent on the driving model providing the correct change in large-scale circulation and with a few notable exceptions little progress has been made in identifying robust changes in regional circulations.

Higher resolution is invaluable in distinguishing between errors that are dependent on resolution and those that are not, sharpening focus on key physically based errors. The use of ensemble simulations, sampling the structural uncertainties among the world's climate models and also the physical uncertainties obtained by systematically perturbing individual models, has helped identify some robust features of climate change (for example, in changes in precipitation), and prompted further research to explain the robustness in physical terms. These multi-model studies are indispensable for improving the quantification of some sources of uncertainty. However they do not necessarily produce insights into how to reduce uncertainty, unless they help in interpreting and understanding model errors or inter-model differences.

4 Lessons from Past Experience and Recommendations to WCRP

Looking back at the Charney report and at the progress (or lack of progress) in climate research and modeling achieved over the last few decades, several key lessons for the future can be drawn. A selection of them are highlighted below.

4.1 Several Key Fundamental Questions Raised by the Charney Report Remain Burning Issues

If the scope of current climate change assessments has broadened since the Charney report, some of the key questions recognized in 1979 as critical for assessing the effect of CO_2 on climate remain with us. At least two striking examples are worth emphasizing:

(1) Climate sensitivity:
Should global climate sensitivity continue to be a focal point for climate research since impacts of climate change are dependent on regional scale transient responses in hydrology and extreme weather, rather than the globally averaged equilibrium response? We argue that it should and that this emphasis continues to be justified. The estimate of climate sensitivity matters for the evaluation of the economic cost of climate change and the design of climate stabilization scenarios (Caldeira et al. 2003; Yohe et al. 2004). It also conditions many other aspects of climate change.

Imagine that we aggregate our estimates of the impacts of climate change on societies and ecosystems into a globally aggregated cost function, *C(R)*. Given an ensemble of model outputs *R*, it is reasonable to assume that *C(R)* will increase with increasing climate sensitivity, as climates are pushed farther into regimes to which societies and ecosystems would adjust with greater and greater difficulty. C(R) will of course also depend on regional changes of the climate system and their specific impacts on societies and ecosystems, but these will certainly scale with climate sensitivity. We do not have to trust detailed regional projections to make this argument, but only to assume that response magnitudes typically increase alongside the global mean temperature response, and that limits in our understanding of processes that control the equilibrium response of the system also influence its transient response (as justified by the analysis of Dufresne and Bony 2008).

There is, in fact, considerable coherence across models in the spatial and seasonal patterns of the temperature response, understandable in part due to the land/ocean configuration, sea ice and snow cover retreat, and (in transient responses) spatial structure in the strength of coupling of shallow to deeper ocean layers. Regional hydrological changes in models are less coherent, but common features still emerge that are understandable in part as responses to the pattern of warming and the accompanying increases in total atmospheric water content, and in part as responses to the CO_2 radiative forcing itself (Bony et al. 2013). Although much research is needed, we can hope to understand changes in weather extremes, in turn, as reactions to these changes in the larger scale temperature and water vapor environment and to changes in surface energy balances. *We conclude that climate sensitivity continues to be a centrally important measure of the size, and significance, of climate response to CO_2. The aggregated impacts of climate change can be expected to scale superlinearly with climate sensitivity.*

(2) *"Inaccuracies of general circulation models are revealed much more in their regional climates owing to shortcomings in the representation of physical processes and the lack of resolution. The modelling of clouds remains one of the weakest links in the general circulation modelling efforts".*

As reaffirmed by a recent survey on "climate and weather models development and evaluation" organized across the World Climate and Weather Research Programmes (Pirani, Bony, Jakob, and van den Hurk, personal communication, 2011), model errors and biases remain a key limitation of the skill of model predictions over a wide range of time (weather to decadal) and space (regional to planetary) scales. It is not a new story, and the increase of model complexity has not solved the problem; on the contrary, shortcomings in the representation of basic fundamental processes such as convection, clouds and precipitation or ocean mixing often amplify the uncertainty associated with more complex processes added to make models more comprehensive. For example, inaccurate representations of clouds and moist processes lead to precipitation errors which may result in inaccurate atmospheric loadings of aerosols or chemical species, inaccurate climate-carbon feedbacks over land, the wrong regional impacts of climate change, and so on.

There is ample evidence that the increase in resolution (horizontal and vertical) is beneficial for some aspects of climate modeling (e.g., the latitudinal position of jets and storm tracks or the magnitude of extreme events) that matter for regional climate projections. However, many model biases turn out to be fairly insensitive to resolution and seem rather rooted in the physical content of models, although separating the role of dynamical errors from physical errors through use of high resolution models or short initialized forecasts (e.g., Boyle and Klein 2010) has helped to elucidate this. Promoting improvements in the representation of basic physical processes in GCMs thus remains a crucial necessity.

Relatively little was known at the time of the Charney report about how clouds and convection couple to the climate system let alone why or how this picture might change. However, coming as it did at the dawn of the satellite era, and in the early days of cloud-resolving modeling studies, it is interesting that the report did not emphasize the importance of these emerging technologies for our understanding of the susceptibility of the climate system to cloud changes (e.g., Hartmann and Short 1980; Held et al. 1993). Indeed the reports oversight in this respect is matched only by its prescience in recognizing the extent to which the modeling of clouds would remain one of the "*weakest links in the general circulation modelling efforts*". To narrow the uncertainty in estimates of the response of the climate system to increasing concentration of greenhouse gases will require a determined effort to address this "*weak link*." Our best hope of doing so is to connect the revolution the Charney report missed with the crisis it anticipated.

4.2 *Improvements of Long-Term Climate Change Assessments Disproportionately Depend on the Development of Physical Understanding*

The pressure put on the scientific community to provide improved assessments of how climate will change in the future, including at the regional scale, has never been as high as it is today. Climate models play a key role in these assessments, and conventional wisdom often suggests that models of highest realism (higher resolution, more complexity) are likely to have wider and better predictive capabilities. Consequently, Earth System Models increasingly contribute to climate change assessments, especially in the 5th round of the Coupled Model Intercomparison Project (CMIP5). However, past experience shows that the spread of GCM projections did not decrease as they became more complex; instead this complexity (e.g. climate-carbon cycle feedbacks) introduces new uncertainties often by amplifying existing uncertainty.

About the large uncertainties associated with regional climate projections from GCMs, the Charney report stated its authors' optimistic belief that "*this situation may be expected to improve gradually as greater scientific understanding is acquired and faster computers are built*". Previous discussion (Sect. 3.6) suggests that increased computing resources (necessary to increase resolution, complexity and the number of ensemble simulations) have helped to confirm inferences from simple models or back-of-the-envelope estimates (e.g. the "dry get drier, wet get wetter" behavior of large-scale precipitation changes or the poleward shift of the storm tracks in a warmer climate), and thus have increased our confidence in the credibility of some robust aspects of the climate change signal. However, the current difficulty of identifying robust changes in regional circulations (e.g. monsoons) or phenomena (e.g. El-Nino) suggests that improved assessments of many aspects of regional climate change will depend more on our ability to develop *greater scientific understanding* than to acquire *faster computers*.

Looking into the future, many hold out hope for global non-hydrostatic atmospheric modeling in which the energy-containing eddies or dominating deep moist convection begin to be resolved explicitly, and for global ocean models with more explicit representations of mesoscale eddy spectrum. These efforts do need to be pushed vigorously, but what we already know of the importance of turbulence within clouds, cloud microphysical assumptions, small-scale ocean mixing, and the biological complexity of land carbon cycling indicate that increasing resolution alone will not be a panacea.

Progress should be measured not by the complexity of our models, but rather the clarity of the concepts they are used to help develop. This inevitably requires the development and sophisticated use of a spectrum of models and experimental frameworks, designed to adumbrate the basic processes governing the dynamics of the climate system (Fig. 4). This point of view gains weight when it is realized that unlike in numerical weather prediction (for which fairly direct evaluations of the predictive abilities of models are possible), observational tests applied to climate

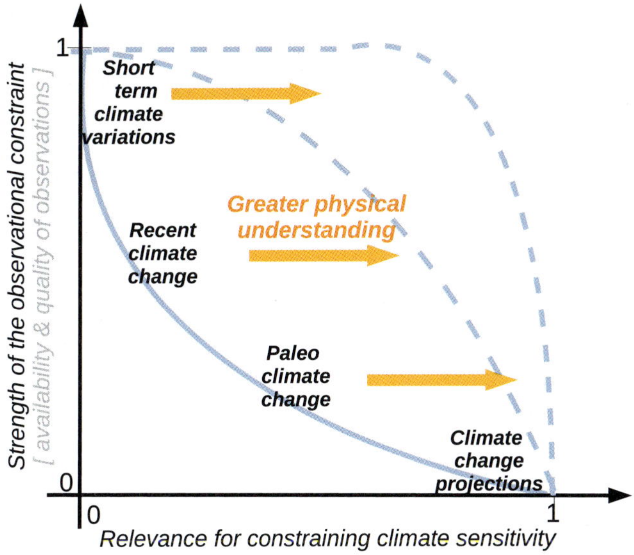

Fig. 3 Unlike weather prediction, there are limited opportunities to evaluate long-term projections (or climate sensitivity as an example) using observations. Multi model analysis show that many of the observational tests applied to climate models are not discriminating of long-term projections and may not be adequate for constraining them. Short-term climate variations may not be considered as an analog of the long-term response to anthropogenic forcings as the processes that primarily control the short-term climate variations may differ from those that dominate the long-term response. By improving our physical understanding of how the climate system works using observations, theory and modeling, we will better identify the processes which are likely to be key players in the long-term climate response. It will help to determine how to use observations for constraining the long-term response

models are not adequate for constraining the long-term climate response to anthropogenic forcings. Indeed observations are generally not fully discriminating of long-term climate projections (Fig. 3). How well a model encapsulates the present state of the climate system, a question to which more 'realistic' models lend themselves, provides an insufficient measure of how well such models can represent hypothesized changes in the climate system. Paleoclimatic studies, while invaluable in providing additional constraints, also do not provide close enough analogues to fully discriminate between alternative futures. The outcome of humanities ongoing and inadvertent experiment on the Earth's climate may come too late help us usefully discriminate among models. *Hence the reliability of our models will remain difficult to establish and the confidence in our predictions will remain disproportionately dependent on the development of understanding.*

The formulation of clear hypotheses about mechanisms or processes thought to be critical for climate feedbacks or climate dynamics helps make complex problems more tractable and encourages the development of targeted observational tests. Moreover, it helps define how the wealth of available observations may be used to address key climate questions and evaluate models through relevant observational tests (Fig. 3). For instance, Hartmann and Larson (2002) formulated the Fixed Anvil Temperature hypothesis to explain and predict the response of upper-level clouds and associated radiative feedbacks in climate change. The support of this hypothesis by several observational (e.g. Eitzen et al. 2009) and numerical investigations with idealized high-resolution process models (Kuang and Hartmann 2007) together with its connection to basic physical principles gives us confidence in at least one component of the positive cloud feedback in models under global warming (Zelinka and Hartmann 2010). Similarly, the recent recognition of the fast response of clouds to CO_2 radiative forcing (Gregory and Webb 2008; Colman and McAvaney 2010) promises progress in our understanding of the cloud response to climate change and our interpretation of inter-model differences in climate sensitivity. Thus we see many reasons for confidence that progress will be made on pieces of the "cloud problem" – as for numerous other problems – seasoned by a realization of many remaining difficulties.

The long-term robustness of the Charney report's conclusions actually demonstrates the power of physical understanding combined with judicious use of simple and complex models in making high-quality assessments of future climate change several decades in advance.

4.3 The Balance Between Prediction and Understanding Should Be Improved in Climate Modeling

With the growing use of numerical modeling in meteorology, a vigorous debate emerged in the 1950s and 1960s (between J. Charney, A. Eliassen and E. Lorentz among others) around the question of whether atmospheric models were to be used mainly for prediction or for understanding (see Dahan-Dalmedico 2001 for an analysis of this debate). A similar debate remains very much alive today with regard to climate change research. As discussed by Held (2005), one witnesses a growing gap between simulation and understanding.

Communication with scientists, stakeholders and society about the reasons for our confidence (or lack of confidence) in different aspects of climate change modeling remains a very difficult task. This level of confidence is based on an elaborate assessment combining physical arguments and a complex appreciation of the various strengths and limits of model capabilities. Improving our physical interpretation of climate change and of the different model results would greatly facilitate this communication. In particular it would help in conveying the idea that the evolution of climate change assessments resembles more the construction of a puzzle in which a number of key pieces are already in place than a house of cards in which a new piece of data can easily destroy the entire edifice.

Consistent with previous discussions recognizing the crucial importance of physical understanding in the elaboration of climate change assessments, our research community should strive to fill this gap. For instance, graduate education in climate science should promote the use of a spectrum of models and theories to address scientific issues and interpret the results from complex models. Besides the basic need to promote fundamental research, filling the gap between simulation and understanding also implies a number of adjustments or practical recommendations to the climate modeling community.

4.4 Recommendations

The lessons discussed above lead us to the following recommendations:

1. **Recognize the necessity of better understanding how the Earth system works in terms of basic physical principles as elucidated through the use of a spectrum of models, theories and concepts of different complexities**. So doing requires the community to avoid the illusion that progress in climate change assessments necessitates the growth in complexity of the models upon which they are based. Thirty years of experience in climate change research suggests that a lack of understanding continues to be the greatest obstacle to our progress, and that often what is left out of a model is a better indication of our understanding than what is put in to it. *In striving to connect our climate projections to our understanding* (what we call the Platonosphere in Fig. 4), *the promotion and inclusion of highly idealized or simplified experiments in model intercomparison projects must play a vital role.* Very comprehensive and complex modeling plays a vital role in this spectrum of modeling activity, but it should not be thought of as an end in itself, subsuming all other climate modeling studies.
2. **Promote research devoted to better understanding interactions between cloud and moist processes, the general circulation and radiative forcings**. Research since the Charney report has shown us that such an understanding is key (i) to better assess how anthropogenic forcings will affect the hydrological cycle, large-scale patterns and regional changes in precipitation, and natural modes of climate variability; (ii) to interpret systematic biases of model simulations at regional and planetary scales; (iii) to understand teleconnection mechanisms and potential sources of climate predictability over a large range of time scales (intraseasonal to decadal); and (iv) to understand and predict biogeochemical feedbacks in the climate system.
3. **Promote research that improves the physical content of comprehensive GCMs, especially in the representation of fundamental processes such as convection, clouds, ocean mixing and land hydrology**. So doing is necessary to address the gaps in our understanding, as in many respects our models remain inadequate to address important questions raised in our first two recommendations. More generally, model failures to simulate observed climate features should be viewed as opportunities to improve our understanding of climate, and to improve our assessment of the reliability of model projections. *WCRP should*

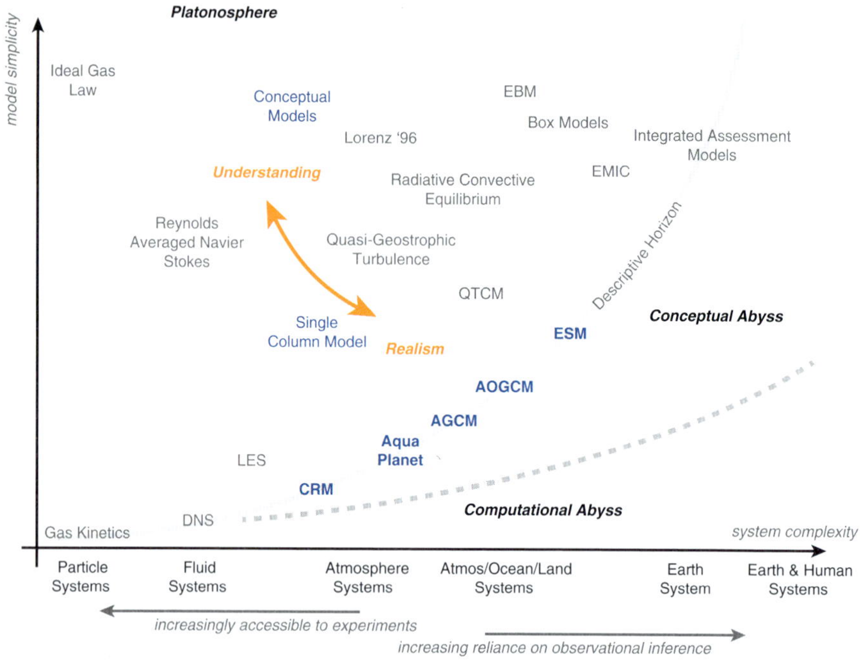

Fig. 4 Distribution of models in a space defined by increasing model simplicity (relative to the system it aims to represent) on the vertical axis and system complexity on the horizontal axis. Our attempt to realistically represent the earth system is both computationally and conceptually limited, and conceptual problems that arise in less realistic models are compounded as we move to complex models, with the result being that adding more complexity to models does not make necessarily make them more realistic, or bring them closer to the earth system. Understanding is developed by working outward from a particular starting point, through a spectrum of models, toward the Platonosphere, which is the realm of the Laws. Reliability is measured by empirical adequacy of our models, which is manifest in the fidelity of their predictions to the world as we know it. To accelerate progress we should work to close conceptual gaps at their source, and try to advance understanding by developing a conceptual framework that allows us to connect behavior among models with differing amounts of realism/simplification. As time and technical capacity evolve models may move around in this abstraction-complexity space

be pro-active in encouraging the community to tackle long-standing, difficult problems in addition to new uncharted problems. A strategy for doing so may include Climate Process Teams now in use in the USA.

4. **Prioritize community efforts and experimental methodologies that help identify which processes are robust vs which lead to the greatest uncertainty in projections and use this information to communicate with society, to guide future research and to identify needs for specific observations**. When analyzing climate projections from multi-model ensembles, a greater emphasis should be placed on identifying robust behaviors and interpreting them based on physical principles. The analysis of inter-model differences should also be encouraged,

particularly to the extent that such analyses advance a physical interpretation of the differences among models. For this purpose, fostering creativity and developing new approaches or analysis methods that connect the behavior of complex models to concepts, theories or the behavior of simpler model results should be strongly encouraged. *This process of distillation is central to the scientific process, and thus vital for our discipline.*

5 Conclusion

Societal demands for useful regional predictions are commensurate to the great scientific challenge that the climate research community has to address. Climate prediction is still very much a research topic. Unlike weather prediction, there are limited opportunities to evaluate predictions against observed changes, and there is little evidence so far that increased resolution and complexity of climate models helps to narrow uncertainties in climate projections. Hence, and as demonstrated by the impressive robustness of the Charney report's conclusions, in the foreseeable future the credibility of model projections and our ability to anticipate future climate changes will depend primarily on our ability to improve basic physical understanding about how the climate system works.

Climate modeling, together with observations and theory, plays an essential role in this endeavor. In particular, our ability to better understand climate dynamics and physics will depend on efforts to improve the physical basis of general circulation models, to develop and use a spectrum of models of different complexities and resolutions, and to design simplified numerical experiments focused on specific scientific questions. Accelerating progress in climate science and in the quality of climate change assessments, should not only benefit scientific knowledge but also climate services and all sectors of our society that need guidance about future climate changes. One aspect of basic research that is often overlooked, is its role in providing a framework for answering questions that policy makers have yet to think of – in this respect the search for understanding is crucial to the general social development.

Finally, and more practically, to ensure that the frequency of assessments is consistent with the rate of scientific progress, which may vary from one topic to another, we suggest that in the future, the World Climate Research Programme play a larger role in organizing focused scientific assessments associated with specific aspects of climate change.

Acknowledgments We thank Hervé Le Treut and Amy Dahan-Dalmedico for valuable comments and discussions with who helped us put the discussions of this opinion paper into an historical context. We are grateful to V. Ramaswamy, two anonymous reviewers and several participants of the WCRP Open Science Conference (Denver, CO, October 2011) for thorough comments and suggestions that helped improve the manuscript. This paper was supported by the French ANR project ClimaConf.

References

Andrews T, Forster PM (2008) CO_2 forcing induces semi-direct effects with consequences for climate feedback interpretations. Geophys Res Lett 35, L04802. doi:10.1029/2007GL032273
Banks HT, Gregory JM (2006) Mechanisms of ocean heat uptake in a coupled climate model and the implications for tracer based predictions of ocean heat uptake. Geophys Res Lett 33(7), L07608. doi:10.1029/2005gl025352
Bolin B, Degens ET, Kempe S, Ketner P (eds) (1979) The global carbon cycle. SCOPE 13, Scientific Committee on Problems of the Environment, International Council of Scientific Unions, Wiley, New York, 491 pp
Bony S, Colman R, Kattsov VM, Allan RP, Bretherton CS, Dufresne JL, Hall A, Hallegatte S, Holland MM, Ingram W, Randall DA, Soden BJ, Tselioudis G, Webb MJ (2006) How well do we understand and evaluate climate change feedback processes ? J Clim 19(15):3445–3482
Bony S, Bellon G, Klocke D, Sherwood S, Fermepin S, Denvil S (2013) Robust direct effect of carbon dioxide on tropical circulation and regional precipitation. Nat Geosci. doi: 10.1038/ngeo1799
Boyle J, Klein SA (2010) Impact of horizontal resolution on climate model forecasts of tropical precipitation and diabatic heating for the TWP-ICE period. J Geophys Res 115, D23113
Brient F, Bony S (2013) Interpretation of the positive low-cloud feedback predicted by a climate model under global warming. Clim Dyn 40(9–10):2415–2431. doi:10.1007/s00382-011-1279-7. ISSN: 0930-7575
Caldeira K, Jain AK, Hoffert MI (2003) Climate sensitivity uncertainty and the need for energy without CO_2 emission. Science 299:2052–2054
Charney JG, Coauthors (1979) Carbon dioxide and climate: a scientific assessment. National Academy of Science, Washington, DC, 22 pp
Church JA, Godfrey JS, Jackett DR, McDougall TJ (1991) A model of sea-level rise caused by ocean thermal expansion. J Cli 4(4):438–456
Collins WD et al (2006) Radiative forcing by well-mixed greenhouse gases: estimates from climate models in the Intergovernmental Panel on Climate Change (IPCC) fourth assessment report (AR4). J Geophys Res 111, D14317. doi:10.1029/2005JD006713
Colman RA, McAvaney BJ (2010) On tropospheric adjustment to forcing and climate feedbacks. Clim Dyn 36(9–10):1649–1658
Cox PM, Betts RA, Jones CD, Spall SA, Totterdell IJ (2000) Acceleration of global warming due to carbon-cycle feedbacks in a coupled climate model. Nature 408:184–187
Dahan-Dalmedico A (2001) History and epistemology of models: meteorology (1946–1963) as a case study. Arch Hist Exact Sci 55:395–422, Springer, pp 395–422
Delmas RJ, Ascencio J-M, Legrand M (1980) Polar ice evidence that atmospheric CO_2 20,000 yr BP was 50% of present. Nature 284:155–157
Denman KL, Brasseur G, Chidthaisong A, Ciais P, Cox PM, Dickinson RE, Hauglustaine D, Heinze C, Holland E, Jacob D et al (2007) Couplings between changes in the climate system and biogeochemistry. In: Solomon S, Qin Q, Manning M, Chen Z, Marquis M, Averyt KB, Tignor M, Miller HL (eds) Climate change, 2007: the physical science basis. Contribution of working group I to the fourth assessment report of the Intergovernmental Panel on Climate Change. Cambridge University Press, Cambridge/New York
Dufresne J-L, Bony S (2008) An assessment of the primary sources of spread of global warming estimates from coupled ocean-atmosphere models. J Clim 21(19):5135–5144
Dufresne J-L, Friedlingstein P, Berthelot M, Bopp L, Ciais P, Fairhead L, Le Treut H, Monfray P (2002) On the magnitude of positive feedback between future climate change and the carbon cycle. Geophys Res Lett 29(10). doi:10.1029/2001GL013777A
Eitzen ZA, Xu KM, Wong T (2009) Cloud and radiative characteristics of tropical deep convective systems in extended cloud objects from CERES observations. J Clim 22:5983–6000

Farneti R, Delworth TL, Rosati AJ, Griffies SM, Zeng F (2010) The role of mesoscale eddies in the rectification of the Southern ocean response to climate change. J Phys Oceanogr 40:1539–1557. doi:10.1175/2010JPO4353.1

Frank DC, Esper J, Raible CC, Buntgen U, Trouet V, Stocker B, Joos F (2010) Ensemble reconstruction constraints on the global carbon cycle sensitivity to climate. Nature 463:527–530

Freeland H et al (2010) Argo – a decade of progress. In: Hall J, Harrison DE, Stammer D (eds) Proceedings of OceanObs'09: sustained ocean observations and information for society, vol 2, Venice, Italy, 21–25 Sept 2009. ESA Publication WPP-306, Roma. doi:10.5270/OceanObs09.cwp.32

Friedlingstein P et al (2006) Climate – carbon cycle feedback analysis, results from the C4MIP model intercomparison. J Clim 19:3337–3353

Friedlingstein P, Houghton RA, Marland G, Hackler J, Boden TA, Conway TJ, Canadell JG, Raupach MR, Ciais P, Le Quéré C (2010) Update on CO_2 emissions. Nat Geosci 3:811–812

Gregory JM, Webb M (2008) Tropospheric adjustment induces a cloud component in CO_2 forcing. J Clim 21:58–71

Gregory JM et al (2004) A new method for diagnosing radiative forcing and climate sensitivity. Geophys Res Lett 31, L03205. doi:10.1029/2003GL018747

Griffies S et al (2010) Problems and prospects in large-scale ocean circulation models. In: Hall J, Harrison D E, Stammer D (eds) Proceedings of OceanObs'09: sustained ocean observations and information for society, vol 2, Venice, Italy, 21–25 Sept 2009. ESA Publication WPP-306, Roma. doi:10.5270/OceanObs09.cwp.38

Hall A, Qu X (2006) Using the current seasonal cycle to constrain snow albedo feedback in future climate change. Geophys Res Lett 33, L03502. doi:10.1029/2005GL025127

Hannart A, Dufresne J-L, Naveau P (2009) Why climate sensitivity may not be so unpredictable. Geophys Res Lett 36, L16707. doi:10.1029/2009GL039640

Hansen JE, Sato M (2011) Paleoclimate implications for human-made climate change. In: Berger A, Mesinger F, Šijački D (eds) Climate change at the eve of the second decade of the century: inferences from paleoclimate and regional aspects: Proceedings of Milutin Milankovitch 130th anniversary symposium. Springer.

Hansen J, Lacis A, Rind D, Russell G, Stone P, Fung I, Ruedy R, Lerner J (1984) Climate sensitivity: analysis of feedback mechanisms. In: Climate processes and climate sensitivity, Geophysical Monograph. American Geophysical Union, Washington, DC, pp 130–163

Hansen J, Sato M, Nazarenko L, Ruedy R, Lacis A, Koch D, Tegen I, Hall T, Shindell D, Santer B, Stone P, Novakov T, Thomason T, Wang R, Wang Y, Jacob D, Hollandsworth S, Bishop L, Logan J, Thompson A, Stolarski R, Lean J, Willson R, Levitus S, Antonov J, Rayner N, Parker D, Christy J (2002) Climate forcings in Goddard Institute for space studies SI2000 simulations. J Geophys Res 107(D18):4347. doi:10.1029/2001JD001143

Hartmann DL, Larson K (2002) An important constraint on tropical cloud-climate feedback. Geophys Res Lett 29:1951–1954

Hartmann DL, Short DA (1980) On the use of earth radiation budget statistics for studies of clouds and climate. J Atmos Sci 37:1233–1250

Held IM (2005) The gap between simulation and understanding in climate modeling. Bull Am Meteorol Soc 86:1609–1614

Held IM, Soden BJ (2006) Robust responses of the hydrological cycle to global warming. J Climate 19:5686–5699. doi:10.1175/JCLI3990.1

Held IM, Hemler RS, Ramaswamy V (1993) Radiative-convective equilibrium with explicit two-dimensional moist convection. J Atmos Sci 50:3909–3927

Klein SA, Hartmann DL (1993) The seasonal cycle of low stratiform clouds. J Clim 6(8):1587–1606

Kuang Z, Hartmann DL (2007) Testing the fixed anvil temperature hypothesis in a cloud-resolving model. J Clim 20:2051–2057

Le Quéré C, Raupach MR, Canadell JG, Marland G, Bopp L, Ciais P, Conway TJ, Doney SC, Feely RA, Foster P, Friedlingstein P, Gurney K, Houghton RA, House J, Huntingford C, Levy PE, Lomas MR, Majkut J, Metzl N, Ometto JP, Peters GPII, Prentice C, Randerson JT, Running SW, Sarmiento JL, Schuster U, Sitch S, Takahashi T, Viovy N, Van Der Werf GR, Woodward FI (2009) Trends in the sources and sinks of carbon dioxide. Nat Geosci 2:831–836

Legg S et al (2009) Improving oceanic overflow representation in climate models: the gravity current entrainment climate process team. Bull Am Meteorol Soc 90:657–670

Lyman JM, Good SA, Gouretski VV, Ishii M, Johnson GC, Palmer MD, Smith DM, Willis JK (2010) Robust warming of the global upper ocean. Nature 465:334–337. doi:10.1038/nature09043

MacDonald GF, Abarbanel H, Carruthers P, Chamberlain J, Foley H, Munk W, Nierenberg W, Rothaus O, Ruderman M, Vesecky J, Zachariasen F (1979) The long term impact of atmospheric carbon dioxide on climate, JASON technical report JSR-78-07, SRI International, Arlington

Manabe S, Wetherald RT (1967) Thermal equilibrium of the atmosphere with a given distribution of relative humidity. J Atmos Sci 24:241–259

Manabe S, Wetherald RT (1975) The effects of doubling the CO_2 concentration on the climate of a general circulation model. J Atmos Sci 32:3–15

Meehl GA et al (2007) Global climate projections. In: Solomon S, Qin D, Manning M, Chen Z, Marquis M, Averyt KB, Tignor M, Miller HL (eds) Climate change 2007: the physical science basis. Contribution of working group 1 to the fourth assessment report of the Intergovernmental Panel on Climate Change. Cambridge University Press, Cambridge, UK/New York

Meehl GA, Arblaster JM, Fasullo JT, Hu A, Trenberth KE (2011) Model-based evidence of deep-ocean heat uptake during surface-temperature hiatus periods. Nat Clim Change 1:360–364

Mitchell JFB, Ingram WJ (1992) Carbon dioxide and climate: mechanisms of changes in cloud. J Clim 5:5–21

Myhre G, Myhre A, Stordal F (2001) Historical time evolution of total radiative forcing. Atmos Environ 35:2361–2373

Ramanathan V, Lian MS, Cess RD (1979) Increased atmospheric CO_2: zonal and seasonal estimates of the effect on radiative energy balance and surface temperature. J Geophys Res 84(C8):4949–4958

Randall DA et al (2007) Climate models and their evaluation. In: Solomon S, Qin D, Manning M, Chen Z, Marquis M, Averyt KB, Tignor M, Miller HL (eds) Climate change 2007: the physical science basis. Contribution of working group 1 to the fourth assessment report of the Intergovernmental Panel on Climate Change. Cambridge University Press, Cambridge, UK/New York

Rieck M, Nuijens L, Stevens B (2012) Marine boundary layer cloud feedbacks in a constant relative humidity atmosphere. J Atmos Sci 69:2538–2550. doi:10.1175/JAS-D-11-0203.1

Roe GH, Baker MB (2007) Why is climate sensitivity so unpredictable? Science 318(5850):629–632. doi:10.1029/2009GL039640

Shine KP, Cook J, Highwood EJ, Joshi MM (2003) An alternative to radiative forcing for estimating the relative importance of climate change mechanisms. Geophys Res Lett 30(20):2047. doi:10.1029/2003GL018141

Soden BJ, Jackson DL, Ramaswamy V, Schwarzkopf MD, Huang X (2005) The radiative signature of upper tropospheric moistening. Science 310:841–844

Somerville RCJ, Remer LA (1984) Cloud optical thickness feedbacks in the CO_2 climate problem. J Geophys Res 89(D6):9668–9672. doi:10.1029/JD089iD06p09668

Son S-W, Gerber EP, Perlwitz J, Polvani LM, Gillett N, Seo K-H, CCMVal Co-authors (2010) Impact of stratospheric ozone on the southern hemisphere circulation change: a multimodel assessment. J Geophys Res 115:D00M07. doi:10.1029/2010JD014271

Trenberth KE, Fasullo JT (2010) Tracking earth's energy. Science 328:316–317

Wetherald R, Manabe S (1988) Cloud feedback processes in a general circulation model. J Atmos Sci 45:1397–1415

Williams KD, Ingram WJ, Gregory JM (2008) Time variation of effective climate sensitivity in GCMs. J Clim 21:5076–5090. doi:10.1175/2008JCLI2371.1

Yohe G, Andronova N, Schlesinger M (2004) To hedge or not against an uncertain climate future? Science 306:416–417

Zelinka MD, Hartmann DL (2010) Why is longwave cloud feedback positive ? J Geophys Res 115, D16117. doi:10.1029/2010JD013817

Zhang M, Bretherton C (2008) Mechanisms of low cloud–climate feedback in idealized single-column simulations with the community atmospheric model, version 3 (CAM3). J Clim 21(18):4859–4878. doi:10.1175/2008JCLI2237.1

Atmospheric Composition, Irreversible Climate Change, and Mitigation Policy

Susan Solomon, Raymond T. Pierrehumbert, Damon Matthews, John S. Daniel, and Pierre Friedlingstein

Abstract The Earth's atmosphere is changing due to anthropogenic increases of gases and aerosols that influence the planetary energy budget. Policy has long been challenged to ensure that instruments such as the Kyoto Protocol or carbon trading deal with the wide range of lifetimes of these radiative forcing agents. Recent research has sharpened scientific understanding of how climate system time scales interact with the time scales of the forcing agents themselves. This has led to an improved understanding of metrics used to compare different forcing agents, and

---

S. Solomon (✉)
Department of Earth, Atmospheric, and Planetary Sciences, Massachusetts Institute of Technology, 77 Massachusetts Av, Cambridge, MA 02139 USA

Department of Atmospheric and Oceanic Sciences,
University of Colorado, Boulder, CO 80309 USA
e-mail: solos@mit.edu

R.T. Pierrehumbert
Department of the Geophysical Sciences, The University of Chicago,
5734 S. Ellis Ave, Chicago, IL 60637 USA
e-mail: rtp1@geosci.uchicago.edu

D. Matthews
Department of Geography, Planning and Environment, Concordia University,
1455 de Maisonneuve Blvd. W, Montreal, QC H3G 1M8, Canada
e-mail: damon.matthews@concordia.ca

J.S. Daniel
Chemical Sciences Division, NOAA Earth System Research Laboratory,
325 Broadway, R/CSD2, Boulder, CO 80305, USA
e-mail: John.S.Daniel@noaa.gov

P. Friedlingstein
College of Engineering, Mathematics and Physical Sciences, University of Exeter,
Harrison Building, Streatham Campus, North Park Road, Exeter EX4 4QF, UK
e-mail: P.Friedlingstein@exeter.ac.uk

G.R. Asrar and J.W. Hurrell (eds.), *Climate Science for Serving Society: Research, Modeling and Prediction Priorities*, DOI 10.1007/978-94-007-6692-1_15,

has prompted consideration of new metrics such as cumulative carbon. Research has also clarified the understanding that short-lived forcing agents can "trim the peak" of coming climate change, while long-lived agents, especially carbon dioxide, will be responsible for at least a millennium of elevated temperatures and altered climate, even if emissions were to cease. We suggest that these vastly differing characteristics imply that a single basket for trading among forcing agents is incompatible with current scientific understanding.

Keywords Climate change • Methane • Carbon dioxide • Global warming potential • Climate policy

1 Introduction

Anthropogenic increases in the concentrations of greenhouse gases and aerosols perturb the Earth's energy budget, and cause a radiative forcing[1] of the climate system. Collectively, greenhouse gases and aerosols can be considered radiative forcing agents, which lead to either increased (positive forcing) or decreased (negative forcing) global mean temperature, with associated changes in other aspects of climate such as precipitation and sea level rise. Here we briefly survey the range of anthropogenic greenhouse gases and aerosols that contribute to present and future climate change, focusing on time scales of the global anthropogenic climate changes and their implications for mitigation options.

Differences in atmospheric residence times across the suite of anthropogenic forcing agents have long been recognized. As decision makers weigh near-term and long-term mitigation actions and tradeoffs, residence times of forcing agents are important along with social, economic, and political issues, such as climate change impacts, costs, and risks sustained by later versus earlier generations (and how these are valued). Recent research has rekindled and deepened the understanding (advanced by Hansen et al. 1997; Shine et al. 2007) that climate changes caused by anthropogenic increases in gases and aerosols can last considerably longer than the gases or aerosols themselves, due to the key role played by the time scales and processes that govern climate system responses. The climate changes due to the dominant anthropogenic forcing agent, carbon dioxide, should be thought of as essentially irreversible on time scales of at least a 1,000 years (Matthews and Caldeira 2008; Plattner et al. 2008; Solomon et al. 2009, 2010).

The largely irreversible nature of the climate changes due to anthropogenic carbon dioxide has stimulated a great deal of recent research, which is beginning to be considered within the policy community. Some research studies have focused on

[1] Radiative forcing is defined (e.g., IPCC 2007) as the change in the net irradiance (downward minus upward, generally expressed in W m^{-2}) at the tropopause due to a change in an external driver of the Earth's energy budget, such as, for example, a change in the concentration of carbon dioxide.

how cumulative carbon dioxide may represent a new metric of utility for policy, as a result of the identification of a near-linear relationship between its cumulative emissions and resulting global mean warming. In this paper, we discuss the use of cumulative carbon to help frame present and future climate changes and carbon policy formulation. We also briefly summarize several other metrics such as e.g., carbon dioxide equivalent concentration, the global warming potential (GWP) and global temperature change potential (GTP). Finally, we examine how current scientific understanding of the importance of time scales not just of different forcing agents, but also of their interactions with the climate system, sharpens the identification of approaches to formulate effective mitigation policies across a range of radiative forcing agents.

2 The Mix of Gases and Aerosols Contributing to Climate Change

A great deal of recent research has focused on understanding changes in atmospheric composition, chemistry, and the individual roles of the range of forcing agents and precursor emissions (leading to the formation of indirect forcing agents after emission) as contributors to observed and future climate change (Forster et al. 2007; Montzka et al. 2011). It is not our goal to review that literature here, but rather to briefly summarize the state of knowledge of contributions of different species to global radiative forcing and time scales of related climate change, and to identify some implications for mitigation policy.

The concentrations of the major greenhouse gases carbon dioxide, methane, and nitrous oxide have increased due to human activities, and ice core data show that these gases have now reached concentrations not experienced on Earth in at least several thousand years (Luthi et al. 2008; Joos and Spahni 2008; MacFarling-Meure et al. 2006). Figure 1 depicts the dramatic increase in carbon dioxide that has taken place over about the past century. The recent rates of increase in CO_2, CH_4, and N_2O are unprecedented in at least 20,000 years (Joos and Spahni 2008). The abundances of CO_2, N_2O and CH_4 are well-mixed over the globe, and hence their concentration changes (and radiative forcings) are well characterized from data such as that shown in Fig. 1; see Table 1.

If anthropogenic emissions of the various gases were to cease, their concentrations would decline at a rate governed by physical and chemical processes that remove them from the global atmosphere. Most greenhouse gases are destroyed by photochemistry in the Earth's atmosphere, including direct photolysis and attack by highly reactive chemical species such as the OH free radical. Many aerosols are removed largely by washout. Carbon dioxide is a unique greenhouse gas that is subject to a series of removal processes and biogeochemical cycling with the ocean and land biosphere, and even the lithosphere, leading to a very long "tail" characterizing a portion of its removal (Archer et al. 1997). While the carbon dioxide concentration changes and anthropogenic radiative forcing since 1750 are very well established,

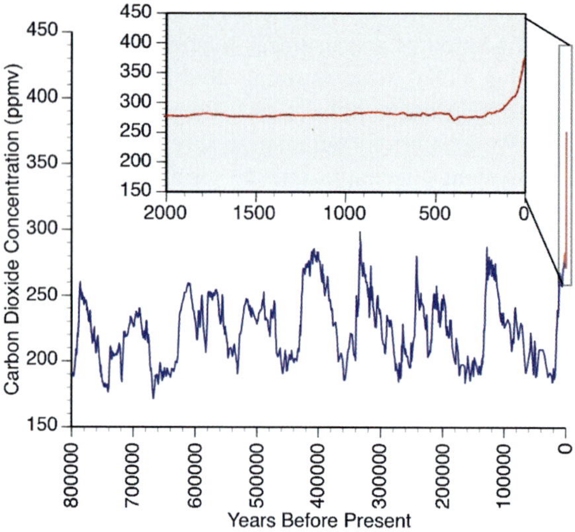

Fig. 1 Carbon dioxide concentrations measured in Antarctic ice cores. The *blue curve* shows the long record from several cores (Available at ftp://ftp.ncdc.noaa.gov/pub/data/paleo/icecore/antarctica/epica_domec/edc-co2-2008.txt), while the *red curve* and inset shows data for 2,000 years prior to 2005 (Available at ftp://ftp.ncdc.noaa.gov/pub/data/paleo/icecore/antarctica/law/law2006.txt)

the relationship of its concentration changes to changes in emission (including those from land use) is much less well characterized, due to the flow of those emissions through the carbon cycle. A few industrial greenhouse gases have lifetimes of many hundreds or even thousands of years, due to their extreme chemical and photochemical stability and represent nearly "immortal" chemicals; in particular, the fully fluorinated compounds such as CF_4, NF_3, and C_2F_6 fall in this category. These gases also are strong absorbers of infrared radiation on a per molecule basis. While these gases are currently present in very small concentrations, like carbon dioxide their contributions to climate change are essentially irreversible on 1,000 year time scales even if policies were to lead to reduced or zero emissions.

Table 1 summarizes the lifetimes (or, in the case of CO_2, multiple removal time scales) that influence the contributions of the range of gases and aerosols to radiative forcing and climate change. Some related uncertainties in lifetimes and distributions are also highlighted.

Direct emissions and other human actions (such as land disturbances, and emissions of precursor gases) have increased the atmospheric burdens of particles, including mineral dust, black carbon, sulfate, and organics. Tropospheric ozone has also increased largely as a result of emissions of precursor gases such as nitric oxide and organic molecules including volatiles as well as methane. Indirect forcings linked to atmospheric aerosols involving changes in clouds may also be very important, and are subject to very large uncertainties (Forster et al. 2007). The short atmospheric lifetimes of aerosols and tropospheric ozone lead to very large variations in their

Table 1 Atmospheric removals and data required to quantify global radiative forcing for a variety of forcing agents

Substance	CO_2	Perfluoro-chemicals (CF_4, NF_3, C_2F_6, etc.)	N_2O	Chlorofluoro-carbons ($CFCl_3$, CF_2Cl_2, etc.)	CH_4	Hydrofluoro-carbons (HFC-134a, HCFC-123, etc.)	Tropospheric O_3	Black carbon	Total all aerosols
Atmospheric removal or lifetime	Multiple processes; most removed in 150 years but ≈ 15–20 % remaining for thousands of years	500–50,000 years, depending on specific gas	≈120 years	≈50–1,000 years, depending on specific gas	≈10 years	One to two decades to years, depending on specific gas (HFC-23 is exceptional with a lifetime of 270 years)	Weeks	Days	Days
Information on past global changes to quantify radiative forcing	Ice core data for thousands of years; in-situ data for half century quantify global changes well	Some ice core for CF_4. In-situ data quantify current amounts and rates of change well	Ice core data for thousands of years; in-situ data for half century quantify global changes well	Snow (firn) data for hundreds of years; in-situ data for more than three decades quantifies the global changes well	Ice core data for thousands of years; in-situ data for half century quantify global changes well	In-situ data quantifies recent global changes well; clear absence of any significant natural sources avoids need for pre-industrial data	Variable distribution poorly sampled at limited sites; uncertain inferences from satellite data since 1979; very few pre-industrial data	Extremely variable distribution poorly sampled at limited sites. Some satellite data in last few decades; a few firn data for pre-industrial amounts	Extremely variable distribution poorly sampled at limited sites; some satellite data in last 1–2 decades; no pre-industrial data

abundances depending upon proximity to local sources and transport, increasing the uncertainty in estimates of their global mean forcing as well as its spatial distribution (see Table 1).

Observations (e.g. of total optical depth by satellites or ground-based methods) constrain the net total optical depth, or the transparency of the atmosphere, and provide information on the total direct radiative forcing due to the sum of all aerosols better than they do the forcing due to individual types of aerosol. Many aerosols are observed to be internal mixtures, i.e., of mixed composition such as sulfate and organics, which substantially affects optical properties and hence radiative forcing (see the review by Kanakidou et al. 2005, and references therein). Aerosols lead to perturbations of the top-of-atmosphere and surface radiation budgets that are highly variable in space, and depend on the place as well as amount of emissions. Limited historical data for emissions or concentrations of aerosols imply far larger uncertainties in their radiative forcings since pre-industrial times than for the well-mixed gases (see Table 1). Current research focuses on understanding the extent to which some regional climate changes may reflect local climate feedbacks to global forcing (e.g., Boer and Yu 2003a, b), versus local responses to spatially variable forcings. For example, increases in black carbon and tropospheric ozone (e.g., Shindell and Faluvegi 2009) may have contributed to the high rates of warming observed in the Arctic compared to other parts of the globe. Sulfate aerosols (which are present in higher concentrations in the northern hemisphere due to industrial emissions) have been suggested as a driver of changes in the north-south temperature gradients and rainfall patterns (e.g., Rotstayn and Lohmann 2002; Chang et al. 2011). Shortwave-absorbing aerosols change the vertical distribution of solar absorption, causing energy that would have been absorbed at the surface and communicated upward by convection to be directly absorbed in the atmosphere instead; this can potentially lead to changes in precipitation and atmospheric circulation even in the absence of warming (e.g. Menon et al. 2002). The large uncertainties in the short-lived forcing terms as well as the regional climate signals they may be inducing have heightened interest in their relevance for mitigation policy, and this is discussed further below (see e.g., Ramanathan and Feng 2008; Jackson 2009; Hansen et al. 1997; Jacobson 2002; UNEP 2011; Shindell et al. 2012).

3 Metrics

Given the very broad diversity of anthropogenic substances with the potential to alter Earth's climate (e.g., CO_2, CH_4, N_2O, SF_6, CFCs, HFC's, absorbing and reflecting aerosols, chemical precursors, etc.), it is a challenging task to compare the climate effect of a unit emission of (for example) carbon dioxide, with one of methane or sulfur dioxide. Nevertheless, there has been a demand for such comparisons, and various metrics have been proposed. The purpose of such metrics is to boil a complex set of influences down to a few numbers that can be used to aid the process of thinking about how different emissions choices would affect future climate. Among other uses, metrics have been used to simplify the formulation of climate-related policy actions, climate-protection treaties and emissions trading schemes. We suggest that to the extent possible, a metric (or set of metrics) should not impose value judgments, least of all hidden value judgments (see Fuglestvedt et al. 2003). Metrics should provide a simplified yet clear set of tools that the policy makers can use to formulate policy implementations to achieve an agreed set of climate protection ends.

3.1 Radiative Forcing and CO_2-Equivalent Concentration

Radiative forcing is one measure of the influence of the burden of a range of forcing agents on the Earth's radiative budget at a given point in time. A closely related metric sometimes used to compare the relative effects of the range of forcing agents is to express them as CO_2-equivalent concentrations, which is the concentration of CO_2 that would cause the same radiative forcing at the chosen time as a given mix of CO_2 and other chemicals (including greenhouse gases and aerosols).

Figure 2 shows the CO_2-equivalent concentration estimates for a range of major forcing agents based on radiative forcing for 2005 from Forster et al. (2007), as given in NRC (2011). The figure shows that among the major forcing agents, by far the largest uncertainties stem from aerosols. Because aerosols represent a substantial negative forcing (cooling effect), this leads to large uncertainty in the net total CO_2-equivalent concentration that is driving current observed global climate change. Current warming represents a transient response that is about half as large as it would become in the long term quasi-equilibrium state if radiative forcing were to be stabilized (NRC 2011). Therefore, uncertainties in today's total CO_2-equivalent concentration imply large uncertainties in how close current loadings of forcing agents may be to eventually warming the climate by more than the 2 °C target noted in the Copenhagen Accord. As Fig. 2 shows, uncertainties in aerosols dominate the uncertainties in total net radiative forcing or total CO_2-equivalent concentration. If aerosol forcing is large, then much of the radiative effect of increases in greenhouse gases is currently being masked by cooling, implying a larger climate sensitivity and far greater risk of large future climate change than if aerosol forcing is small.

A key limitation of radiative forcing or CO_2-equivalent concentrations as metrics is that they do not include any information about the time scale of the impact of the forcing agent. For example, the radiative forcing for a very short-lived forcing agent may be very high at a given time but would drop rapidly if emissions were to decrease, while a longer-lived constituent implies a commitment to further climate change even if emissions were to stop altogether.

Insofar as short-lived aerosols produce a cooling, their masking of a part of the impact of the large load of long-lived warming agents implies that an unseen long-term commitment has already been made to more future warming (e.g. Armour and Roe 2011; Ramanathan and Feng 2008); Hansen describes this as a "Faustian bargain", since short-lived aerosol masking can be accompanied by accumulation of more long-lasting and hence ultimately more dangerous levels of carbon dioxide and other long-lived greenhouse gases in the atmosphere (e.g., Hansen and Lacis 1990).

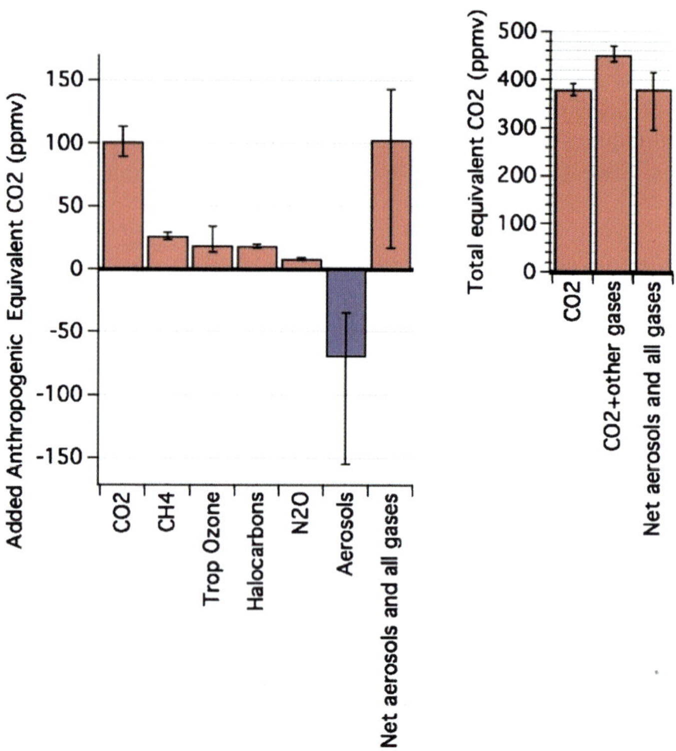

Fig. 2 (*Left*) Best estimates and very likely uncertainty (90 % confidence, as in Forster et al. 2007) ranges for aerosols and gas contributions to CO_2-equivalent concentrations for 2005, based on the concentrations of CO_2 that would cause the same radiative forcing as each of these as given in Forster et al. (2007). All major gases contributing more than 0.15 W m^{-2} are shown. Halocarbons including chlorofluorocarbons, hydrochlorofluorocarbons, hydrofluorocarbons, and perfluorocarbons have been grouped. Direct effects of all aerosols have been grouped together with their indirect effects on clouds. (*Right*) Total CO_2-equivalent concentrations in 2005 for CO_2 only, for CO_2 plus all gases, and for CO_2 plus gases plus aerosols (From Stabilization Targets, NRC 2011)

It is evident that other metrics beyond radiative forcing are needed to capture temporal aspects of the climate change problem. One needs to compare not only the effect of various substances on today's climate change but also how current and past emissions affect future climate change. As will be shown, available metrics all simplify or neglect aspects of temporal information related to individual gases (albeit in different ways), and hence incorporate choices and judgments rather than representing "pure" physical science metrics (Fuglestvedt et al. 2003; Manne and Richels 2001; O'Neill 2000; Manning and Reisinger 2011; Smith and Wigley 2000; Shine 2009).

The problem of formulating a metric for comparing climate impacts of emissions of various greenhouse gases is challenging because it requires consideration of the widely differing atmospheric lifetimes of the gases. Emissions metrics are of most interest, since it is emissions (rather than concentrations) that are subject to direct control. The lifetime affects the way concentrations are related to emissions. For a short-lived gas like CH4, the concentrations are a function of emissions averaged over a relatively short period of time (on the order of a few decades in the case of CH4). For example, while anthropogenic emissions increase, the CH_4 concentration increases but if anthropogenic emissions of CH_4 were to be kept constant, the concentration of the gas would reach a plateau within a few decades. In contrast, for a very persistent gas like CO_2, the concentration is linked to the cumulative anthropogenic emission since the time when emissions first began; concentrations continue to increase without bound so long as emissions are significantly different from zero. In essence, a fixed reduction of emission rate of a short-lived gas yields a step-reduction in radiative forcing, whereas the same reduction of emission rate of a very long-lived gas only yields a reduction in the rate of growth of radiative forcing.

3.2 *GWP$_h$ and GTP$_h$*

The most familiar and widely applied metric for comparing greenhouse gases with disparate atmospheric lifetimes is the Global Warming Potential (GWP). The GWP is defined as the ratio of the time-integrated (over some time horizon) radiative forcing due to a pulse emission of a unit of a given gas, to an emission of the same amount of a reference gas (Forster et al. 2007). This can be expressed as:

$$\mathrm{GWP_h} = \int_0^h \Delta A\ \Delta C(t)dt \Big/ \int_0^h \Delta Ar\ \ \Delta C_r(t)\ dt \tag{1}$$

where h is a specified time horizon, $\Delta C(t)$ is the time series of the change in concentration of the greenhouse gas under consideration (relative to some baseline value), and $\Delta C_r(t)$ that of the reference gas (usually CO_2. as we shall assume throughout the following). ΔA (and ΔAr) represent the radiative efficiencies due to changes in concentration of the greenhouse gas (and reference gas) following a pulse emission at $t=0$. In the remainder of this paper, we refer specifically to GWP_h and GTP_h to emphasize the key role of the time horizon. If the pulse is small enough, the radiative forcing is linear relative to the size of the emission pulse; the conventional assumption is therefore that GWP_h is independent of the size of the pulse. This assumption of linearity can lead to substantial errors when the GWP_h is extrapolated from an infinitesimal pulse to very large emissions. Such errors can arise from nonlinearities in the radiative forcing due to changes in concentration of the emitted gas or that of the reference gas CO_2.

For gases with short atmospheric lifetimes (e.g. methane), the peak of concentration that immediately follows a pulse in emission decays rapidly to zero, leading to a strong dependence of GWP_h on the timescale over which it is calculated (h in Eq. 1). Table 2.14 in Forster et al. (2007) gives GWP_h for a variety of gases, with $h=20$, 100 and 500 years. Methane for example, has a 100-year GWP_h (GWP_{100}) of 25, but a GWP_{500} of only 7.6. The choice of time horizon is crudely equivalent to the imposition of a discount rate, albeit a discount rate that varies with lifetime of the gas (Manne and Richels 2001), and thus represents a value judgment. A choice of small *h* implies that one should not care that CO_2 saddles the future with an essentially permanent alteration of climate, whereas the choice of a very large *h* says that one should not care about the transient warming due to short-lived greenhouse gases. Either assumption embeds a judgment regarding whether the near term future is to be valued above the long term future, or vice versa.

An additional concern with the GWP_h is that it represents only the change in integrated forcing due to the emission of different gases, rather than the change in (for example) global-mean temperature. This has led to the proposal of modified metrics, such as the Global Temperature Potential (GTP_h) put forward by Shine et al. (2005). The GTP_h represents the temperature change at some point *h* in time (rather than time-integrated radiative forcing) resulting from the unit emission of a greenhouse gas, relative to the same emission of carbon dioxide.

In order to illustrate some of the consequences of using GTP_h or GWP_h as climate change metrics for gases of different atmospheric lifetimes, we use a simple two-layer ocean model to translate radiative forcing and surface temperature change over time. This model is a simpler version of the upwelling-diffusion model used in Shine et al. (2005) to critique GWP_h, and has also been proved useful in analyzing the transient climate response in full general circulation models (Winton et al. 2010; Held et al. 2010). The model consists of a shallow mixed layer with temperature anomaly dT'_{mix} and heat capacity μ_{mix} coupled to a deep ocean with temperature anomaly dT'_{deep} and heat capacity $\mu_{deep} >> \mu_{mix}$. The mixed layer loses heat to space (in part via coupling to the atmosphere) at a rate proportional to its temperature. The equations are

$$\mu_{mix}\{dT'_{mix}/dt\} = -\lambda T'_{mix} - \gamma(T'_{mix} - T'_{deep}) + \Delta F(t) \tag{2}$$

$$\mu_{deep}\{dT'_{deep}/dt\} = -\gamma(T'_{deep} - T'_{mix}) \tag{3}$$

For constant radiative forcing ΔF, this model[2] has the steady solution $T'_{mix} = T'_{deep} = \Delta F/\lambda$. Hence $1/\lambda$ gives the quasi-equilibrium climate sensitivity. The model relaxes to this equilibrium state on two time scales. On the short time scale (generally a matter of a few years), the mixed layer relaxes to a near-equilibrium with the atmosphere but the deep ocean has not yet had time to warm up, so $T'_{deep} \approx 0$. The transient climate response during this stage is then $T'_{mix} = \Delta F/(\lambda + \gamma)$. If ΔF is reduced to zero some time after the deep ocean has warmed up to some nonzero value T'_{deep}, then on the short mixed layer time scale T'_{mix} only falls to $T'_{deep}\, \gamma/(\lambda + \gamma)$, and subsequently relaxes to zero on the slow deep ocean time scale. This term is the "recalcitrant warming" due to heat burial in the deep ocean (Held et al. 2010).

[2]The parameters we use in the following are: $\mu_{deep} = 20$, $\mu_{mix} = 200$ J/m^2 K and $\gamma = \lambda = 2$ W/m^2 K.

Figure 3a shows the calculated temperature response of the mixed layer in this model due to pulse emissions of greenhouse gases with various lifetimes and forcing efficiencies. In this calculation, the radiative forcing is assumed to be linear in the concentration, and the concentration is assumed to decay exponentially with the stated lifetime. The magnitude of the emission of each gas is chosen so that all correspond to the same value when weighted by GWP_{100}; i.e., for a pulse emission, the radiative forcing integrated over 100 years is identical in all cases. Figure 3 shows that the GWP_{100} weighted emission for a gas with a 10-year (methane-like) lifetime and radiative efficiency can be the same as for the longer lived gases, since a weaker long-term warming can be compensated by a larger short term warming. If the integrated warming over the 100 year period is all we care about, and the damages are linear in warming, then these cases may indeed all be considered to have identical impact in that the methane-like case produces larger damages for a short time, as opposed to a longer period with smaller damages for the longer-lived gases. However, if the objective is to limit the magnitude of warming when the 100 year time span is reached, the use of GWP_{100} greatly exaggerates the importance of the short-lived gas, since virtually all of the warming has disappeared after 100 years. This is a starting point for considering the value of the alternative concept of Global Temperature-Change Potential (GTP_h) as in Shine et al. (2005). Measured in terms of 100-year GTP_h, the 10-year lifetime gas has only 1/4.5 times the impact of e.g., a 1,000 year gas with identical GWP_{100}. The warming after 100 years even in the 10-year lifetime case has not decayed to zero as quickly as the radiative forcing itself (which has decayed by a factor of 4.5×10^{-5} over this time). The persistent, or recalcitrant warming arises largely from ocean heat uptake (Solomon et al. 2010). But it should also be emphasized that the 100-year GTP_h does not capture the impact of the large short-term warming from the methane-like case. Such short-term warming could be significant if, for example, the near-term rate of temperature change were leading to adaptation stresses.

Although GTP_h may be a superior metric to GWP_h for implementing climate protection goals based on a threshold temperature at a given time, it does not resolve the problem of sensitivity to the time frame chosen when computing the metric. Based on 100-year GTP_h, emitting an amount of a 1,000-year lifetime gas might be considered to be about twice as bad as an emission of a 50-year lifetime gas; however the long lived gas leads to a warming that is nearly constant over the next 200 years whereas the warming due to the 50-year gas has largely disappeared by the end of that time. These two cases result in radically different temperature changes over time and clearly do not represent identical climate outcomes.

An additional problem with both GWP_h and GTP_h is their dependence on the emission scenario. Figure 3a represents the case of a pulse emission while Fig. 3b shows a second case with constant emissions of a methane-like gas with a 10-year lifetime, compared to constant emissions of a gas with an infinite lifetime (see e.g., Shine et al. 2007). In both Fig. 3a and b, the emissions scenarios were selected such that the GWP_{100} values are equivalent. Emissions are sustained for 200 years, and then set to zero at the year 200. In both cases, the warming continues beyond the point at which the concentration of the gas stabilizes; in the case of the methane-like gas, the concentration (not shown) stabilizes after about 10 years but warming continues to

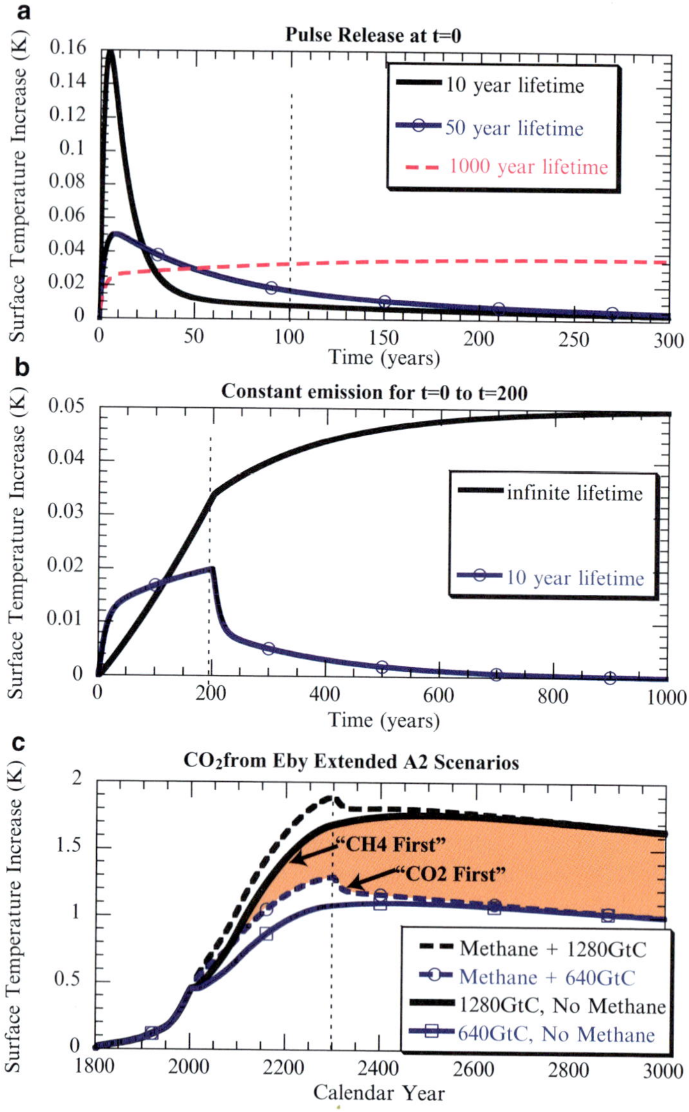

Fig. 3 Surface temperature response of the two-layer ocean model subjected to various time-series of radiative forcing as follows (**a**) Pulse emission of gases with various lifetimes but identical GWP_{100}. The emission corresponds to an initial radiative forcing of 1 W m^{-2} for the shortest-lived gas. (**b**) Constant emission rate up to year 200 for an infinite lifetime CO_2-like gas vs. a short-lived methane-like gas having the same GWP_{100}. The total mass of short-lived gas emitted is the same as in the pulse emission calculation shown in (**a**). (**c**) Temperature increases from the CO_2 time series in test cases in Eby et al.(2009), corresponding to cumulative carbon emissions of 640 or 1,280 GtC between 2,000 and 2,300, alone or with superposed effect of constant-rate methane emissions with total GWP_{100}-weighted emissions equal to the difference in CO_2 emissions between the two cases; all emissions cease by 2,300

increase, illustrating the continuing warming that occurs despite constant atmospheric concentrations, as the deep ocean takes up heat. For the infinitely long-lived gas, concentrations remain elevated even after emission stops, and warming continues to increase (see next section). Indeed, although both cases are equivalent in terms of GWP_{100}-weighted emissions, the infinite-lifetime case leads to a warming that is not only larger at the end of 200 years, but persists for centuries afterwards. The constant-emissions case illustrates the dependence of GTP_h on the emissions scenario. Neither GWP_h nor GTP_h capture what occurs after emissions cease.

As a final example, we have carried out a series of calculations driven by the CO_2 time series computed in Eby et al. (2009). The concentration time series were computed by driving an intermediate-complexity climate-carbon cycle model with historical emissions up to the calendar year 2000, followed by two test scenarios in which the emissions rate rises to a peak after 150 years, and then declines to zero in the subsequent 150 years. The two scenarios shown in Fig. 3c show results corresponding to 640 and 1,280 GtC of post-2000 cumulative carbon emissions (see next section). Note that the warming is fairly constant in the 700 years following cessation of emission, given the realistic atmosphere CO_2 used in this case as compared to the infinite-lifetime case shown in Fig. 3b. Abating cumulative carbon by 640 GtC (the difference between the two emission scenarios shown here) reduces warming by about 0.6 K in the two-box model.

The dashed curves in Fig. 3c show what happens if the radiative forcing from CO_2 is augmented by that from methane released at a constant rate between 2,000 and 2,300, with the total emissions again equivalent to the CO_2 from 640 GtC based on weighting with a GWP_{100} of 25 (Forster et al. 2007). The corresponding methane emission rate is 0.31 Gt per year, which is similar to the current anthropogenic emission rate of about 0.35 Gt per year (see http://cdiac.ornl.gov/trends/meth/ch4.htm). Emissions are stopped entirely in 2300 in this example. One can think of the curve for 640 GtC plus methane (dashed blue line) as the result of deciding to abate CO_2 emissions first and methane later, while the curve with 1,280 GtC and no methane (solid black line) corresponds to abating methane first and carbon later. It is useful to compare the "Methane First" to that for the "CO_2 First" case, recalling that both have the same GWP_{100} weighted emissions. The blue dashed curve ramps up quickly and faster just after 2000 as expected from having more short-lived CH_4. Overall, the two track quite well for the first 100 years (compare the solid black line with the dashed blue line), but thereafter the temperature for "CO_2 First" falls well below that for "Methane First." Moreover, after methane emissions are eliminated, the dashed blue line ("CO_2 First") case converges with the curve for 640GtC alone (solid blue line) within a century, as if methane had never been emitted at all.

Figure 3c highlights the comparison between the two curves representing the "Methane First" vs. "CO_2 First" strategies. The shaded region mirrors the analysis of (Daniel et al. 2011), who used emissions and climate response models that were less idealized. The general lesson to be learned is that over the universe of strategies considered equivalent with regard to GWP_{100}, an emphasis on short-lived forcing agents yields more near-term moderation of warming but comes at the expense of considerably greater long term warming.

A comparison of the bottom two curves in Fig. 3c, in contrast, illustrates the "peak trimming" benefits of reductions in short-lived forcing agents. However, a comparison of the lower two curves alone gives an incomplete picture of the decision framework. One will always get more warming reduction from doing two beneficial things rather than one beneficial thing, but the real question is whether one would get a still better consequence by putting added resources into further reductions of CO_2 versus applying them to short-lived agents.

From the examples in Figure 3, it is clear that emissions of methane (and similarly other short-lived radiative forcing agents) have a strong bearing on the amount of warming during the time over which they are emitted, but have little lasting consequence for the climate system. By contrast, CO_2 and (and to a lesser extent other long-lived forcing agents) are relevant to both short- and long-term climate warming, and in particular generate warming which persists at significant levels long after emissions are eliminated. These fundamental differences between short- and long-lived radiative forcing agents cannot be captured by either GWP_h or GTP_h metrics, which by design can only provide comparisons for the chosen time horizon. Here we have illustrated key limitations of such an approach over time.

3.3 Irreversibility of CO_2-Induced Warming, Climate Commitment, and the Cumulative CO_2 Emissions Metric

As illustrated above, whereas shorter-lived gases and aerosols have a strong bearing on near-future climate changes, warming that persists beyond the twenty-first century, and particularly warming that persists beyond the period of time that humans emit greenhouses gases, will be primarily determined by how much carbon dioxide is emitted over this period of time. Because of the long lifetime of carbon dioxide in the atmosphere compared to other major greenhouse gases, the long-term warming legacy of anthropogenic greenhouse gases will be primarily determined by CO_2-induced warming.

In recent literature, the concept of the irreversibility of climate change due to CO_2 emissions was first highlighted by Matthews and Caldeira (2008) based upon results from an Earth Model of Intermediate Complexity (EMIC). This has led to the recognition that cumulative carbon (the total tonnes of carbon emitted) has particular utility for policy. Matthews and Caldeira (2008) showed that if CO_2 emissions were eliminated, globally-averaged temperature stabilized and remained approximately constant for several hundred years; notably, though CO_2 concentrations decreased in the atmosphere, temperatures remained at a nearly constant level, mainly as a result of a declining rate of heat uptake by the ocean that approximately balances the decline in carbon dioxide levels; for a detailed discussion see Solomon et al. (2010). Several other EMIC studies have also demonstrated the irreversibility of CO_2-induced warming. Solomon et al. (2009) showed that even after 1,000 years of model simulation following the elimination of CO_2 emissions, global temperatures were essentially irreversible, remaining within about half a degree of their peak values for a broad range of emission rates and maximum concentrations. In an intercomparison of eight EMICs, Plattner et al. (2008) showed persistence of high global temperatures for at least several centuries following zero emissions across all the models. More comprehensive global climate models require much more computer time and hence have thus far been run for zero emission tests over multiple centuries rather than millennia, and show similar results (Lowe et al. 2009; Gillett et al. 2011). These studies have confirmed that irreversibility of CO_2-induced warming is a property of the climate system that is driven by basic properties of the system, notably the carbon and ocean heat timescales, and is not limited to intermediate-complexity models.

This body of literature has all contributed to estimating what has been called the "zero-emissions commitment"; that is the anticipated future warming that occurs in the absence of additional future CO_2 emissions. This quantity is distinct from another widely-used definition of committed warming: the "constant-composition commitment," which is defined as the future global temperature change which would be expected under constant concentrations of atmospheric CO_2 (Meehl et al. 2007).

The difference between these two measures of committed future warming was highlighted by Matthews and Weaver (2010), and summarized in Fig. 4a. Under constant atmospheric CO_2 concentrations, temperatures continue to increase as the climate system slowly adjusts to the current atmospheric forcing from CO_2 in the atmosphere. By contrast, if CO_2 emissions were set to zero, atmospheric CO_2 would decrease over time due to removal by carbon sinks, but global temperature would remain approximately constant for several centuries. This difference can also be seen in the example of the simple model shown above: constant composition of an infinite-lifetime gas after year 200 in Fig. 3b leads to increasing global temperatures, whereas zero emissions of CO_2 at the year 2300 in Fig. 3c leads to approximately stable global temperatures. Persistent warming over many centuries is especially relevant for understanding impacts including the large sea level rise that occurs in a warmer world due to slow thermal expansion of the deeper parts of the ocean and the potentially very gradual loss of the great ice sheets of Greenland and Antarctica (Meehl et al. 2007 and references therein).

The difference between the constant-composition and zero-emission commitment can also be understood in terms of the CO_2 emissions associated with each scenario. Figure 4b shows the historical emissions in blue associated with both scenarios, and the future emissions in red required to maintain constant CO_2 concentrations at year-2010 levels. Given the required balance between emissions and removal by carbon sinks to maintain constant atmospheric levels, the future emissions associated with a constant-composition scenario are substantially larger than zero; in this example, the total emissions over 300 years required to maintain constant atmospheric CO_2 amount to about 250 GtC, or close to half of the total historical CO_2 emissions (about 500 GtC). These future emissions are consistent with the continued future warming associated with constant atmospheric CO_2 concentrations. By contrast, zero future emissions is consistent with near-zero additional future warming.

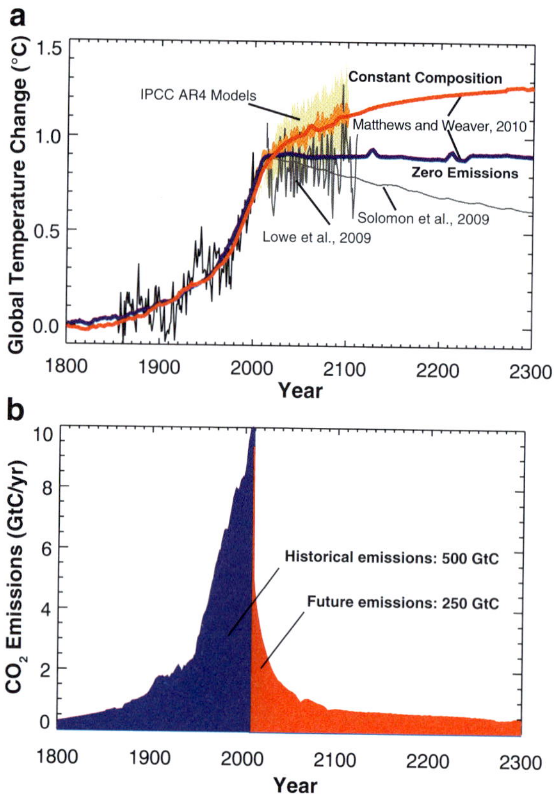

Fig. 4 Climate response to zero CO_2 emissions, compared to the climate response to constant atmospheric CO_2 concentration. *Upper panel* (**a**) shows the global temperature response to zero-emissions from three models (Lowe et al. 2009; Solomon et al. 2009; Matthews and Weaver 2010) and constant-composition scenarios, as in Matthews and Weaver (2010) and references therein. *Lower panel* (**b**) shows the CO_2 emissions scenarios associated with the *red* and *blue lines* in *panel* (**a**), with cumulative emission given for the historical period (*blue shaded area*, corresponding to the historical portion of both scenarios) and the future emissions associated with the constant-composition scenario (*red shaded area*)

As already noted, the removal of anthropogenic CO_2 from the atmosphere involves a multitude of time scales, ranging from a few decades for uptake by the upper ocean and land biosphere, a millennium for uptake by the deep ocean, tens of millennia for carbonate dissolution and weathering to restore ocean alkalinity and allow further uptake, and hundreds of thousands of years for silicate weathering (Archer et al. 1997). The nonlinearity of the carbonate chemistry is important in determining the way climate change relates to larger and larger increases in CO_2. Though the radiative forcing is logarithmic as a function of CO_2 concentration, the carbonate chemistry implies that the fraction of CO_2 that remains in the atmosphere after emission increases with the magnitude of the emission (Eby et al. 2009). Further, the slow decay in radiative forcing due to ocean uptake of carbon following

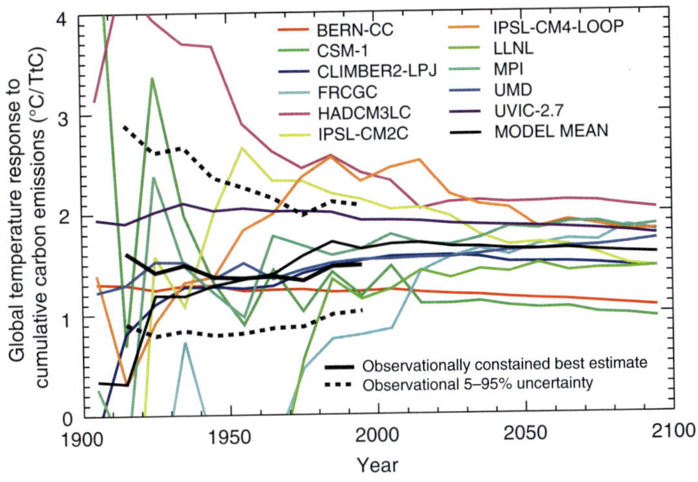

Fig. 5 Climate response to cumulative carbon emissions ("carbon-climate response"), estimated from historical observations of CO_2 emissions and CO_2-attributable temperature changes (*thick black line* with *dashed* uncertainty range), as well as from coupled climate-carbon cycle models (*colored lines*). Both historical observations and model simulations of the twenty-first century show that the carbon-climate response is approximately constant in time, indicating a linear relationship between cumulative carbon emissions and globally-averaged temperature change (See Matthews et al. (2009) for details)

cessation of emissions occurs at roughly the same time scale as the relaxation of the deep ocean temperature towards equilibrium; because these two terms work in opposing directions, the surface temperature attained at the time emissions cease is not only proportional to the cumulative carbon, but is also the temperature which prevails with little change for roughly the next millennium (Matthews and Caldeira 2008; Solomon et al. 2009, 2010; Eby et al. 2009).

The coherence between cumulative emissions of carbon dioxide and global temperature changes has been the subject of several recent studies, and represents a new metric with which to assess the climate response to human CO_2 emissions. Matthews et al. (2009) and Allen et al. (2009) both identified a strong linear relationship between global temperature change and cumulative carbon emissions. Matthews et al. (2009) named this the "carbon-climate response". In this study, they showed the carbon-climate response is well constrained by both coupled climate-carbon models and historical observations to lie between 1 and 2.1 °C per 1,000 GtC emitted (see Fig. 5 above, taken from NRC 2011). Allen et al. (2009) used a simpler climate model, but considered a larger range of possible climate sensitivities; as a result, they estimated that the instantaneous temperature change associated with cumulative carbon emissions fell between 1.4 and 2.5 °C per 1,000 GtC emitted.

Cumulative carbon emissions provides a clear means of estimating the extent of climate warming that will occur from wide range of future CO_2 emissions scenarios. Consequently, the anthropogenic warming that will occur, and which will persist for many subsequent centuries, will be determined to a large extent by the total cumulative

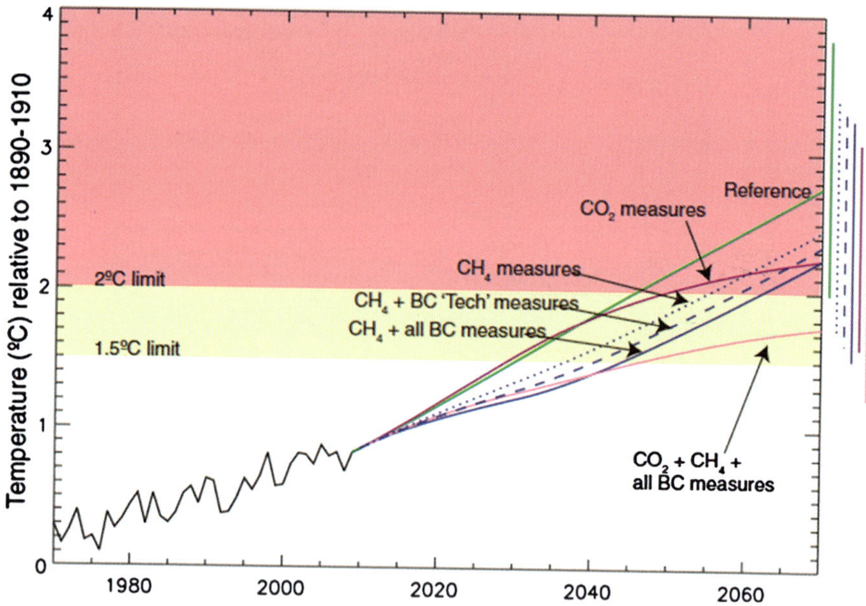

Fig. 6 Observed deviation of temperature to 2009 and projections under various scenarios considered in UNEP (2011) and Shindell et al. (2012); see Shindell et al. (2012) for details. The bulk of the benefits of the assumed CH_4 and black carbon reduction measures are realized by 2040, with the longer term warming being increasingly dependent on carbon dioxide emissions

emissions which occur between now and the time by which humans stop emitting significant amounts of carbon dioxide. If a tipping point (Lenton et al. 2008) in the earth system were to be experienced at any time in the future, even the immediate cessation of CO_2 emissions will be unable to substantially lower the global temperature even on timescales of tens of generations.

4 Policy Outlook

Reducing emissions of shorter-lived gases and aerosols (e.g., black carbon) is indeed a highly effective way to reduce climate forcing or the rate of warming on shorter timescales as shown by many authors (see e.g. UNEP 2011; Jacobson 2002; Shindell et al. 2012 and references therein), and illustrated here in Fig. 6. But Fig. 3 above provides key context to better understand choices among policy options. In particular, Fig. 3c goes beyond the timescale shown in Fig. 6 to illustrate that reductions of short-lived gases or aerosols should be most appropriately thought of as an approach to "trimming the peak" warming (and perhaps the rate of warming) in the near term (but recall the discussion in connection with Fig. 3, bearing on the question of choices between efforts put into peak trimming versus additional CO_2 reductions). Furthermore, delays in the abatement of short lived forcing agents imply greater heat storage in the deep ocean and greater sea level rise; thus, the utility of the peak trimming is affected by when it is implemented as well as by how much. Peak trimming can also reduce the rate of warming, with attendant benefits for the ability of human and natural systems to adapt. Greater benefits in peak trimming are obtained the sooner the emissions are abated (see Held et al. 2010). However, Fig. 3c also shows that the long term climate – i.e. the character of the "Anthropocene" – is determined largely by the cumulative carbon emitted. It is noteworthy that the use of GWP100 in a policy vehicle would consider the "Methane First" scenario to be equivalent to the "CO_2 First" scenario, but the figure makes clear that the latter yields a far better outcome if one is concerned about the climate changes that last beyond 100 years. Thus Fig. 3c demonstrates why trimming the peak cannot substitute for reductions in carbon dioxide emissions that will dominate Earth's climate for many centuries if unabated.

A key policy issue involves the relative reductions to make in the emissions of the range of greenhouse gases. The Kyoto Protocol addressed this issue by placing the regulated greenhouse gases into a single basket and relating their emissions in a common CO_2-equivalent emission determined by multiplying actual emissions with the 100-year GWP_h. Numerous studies have demonstrated that using a single metric in this way has drawbacks arising from the disparity in global lifetimes of the various gases. As we have illustrated here, the choice of a particular time horizon includes value judgments regarding the importance of climate changes at varying times. For example, if a GWP_h with a short time horizon is used in order to better equate short-term climate impacts among gases, the larger relative impact of gases with long lifetimes over long timescales will not be considered. Perhaps more importantly, the use of the GWP_h as the trading metric leads to greenhouse gas trading based on relative integrated radiative forcing, which has a limited connection to temperature change (as shown by the comparison of GTP_h to GWP_h) but probably better represents sea level rise (Smith and Wigley 2000). Many studies have examined ways to more effectively address near-term and long-term warming (e.g., Manne and Richels 2001 and others), but the majority of policy discussions have revolved around greenhouse gas metrics for a given time that cannot account for time-varying policy goals.

The Montreal Protocol regulated ozone-depleting substances (ODSs) that were also characterized by very different lifetimes. This Protocol was highly successful in reducing ozone depletion and took a different approach from that of the Kyoto Protocol. Rather than group all ODSs into a single basket in which production and consumption reductions could be traded using some metric like the ozone depletion potential (ODP), the Montreal Protocol effectively regulated groups of gases (e.g., CFCs, HCFC, halons) and some individual gases (e.g., CH_3CCl_3, CCl_4, CH_3Br) separately. Members of these groups were largely characterized by similar lifetimes. It has been shown that if the Montreal Protocol took an alternative single basket approach, and if trading among ODSs were possible and were performed, the success of the Protocol in limiting short term risks could have been compromised (Daniel et al. 2011).

The principal conclusion of the discussion presented in this paper is that the scientific basis for trading among all greenhouse gases in one single basket is poor, and a more science-based approach for the Kyoto Protocol (and similar regulatory frameworks) would be to abandon the idea of a single-basket approach altogether. As we have shown, short-lived greenhouse gases or aerosols, and CO_2 are knobs that control quite different aspects of the future climate. It does not appear likely that any single metric will be able to fairly represent both. Yet both time scales are clearly important from the policy viewpoint of risks of different types of future climate changes, such as a possibly slow loss of ice from Greenland and Antarctica over millennia and associated massive sea level rise, versus the potential for rapid increases in the area burned by wildfire in the next decade or two. Thus, the research of the past few years shows even more clearly than previous studies that the existing single-basket GWP_h framework is difficult to justify.

Many of the problems with GWP_h and GTP_h are not intrinsic to the metrics themselves, but to the imposition of a single time scale when computing the metric. As a minimum, a two-basket approach seems to be needed. One basket could be CO_2, and the metric used to quantify the climate impact of that basket would be cumulative carbon emission (Matthews et al. 2009). Further work is needed to determine whether perfluorocarbons might also be included in this basket through a suitable adjustment of cumulative carbon. The long-term basket should be recognized as the only path to managing long-term risks to the climate. The second basket would include much shorter-lived forcing agents such as CH_4, tropospheric ozone, and black carbon, which could be grouped together and measured by a metric such as the GTP_h. Carbon dioxide can be considered here as well, since its growth is expected to be important for the rate of climate change in the near term (as well as being not only dominant but controlling the changes in the long-term). Reducing short-lived gases or aerosols does nothing to reduce the long-term risk posed by substances such as carbon dioxide. This second basket would explicitly recognize and manage what can be done to reduce warming in the short-term time scale of decades or so, with the choice of time horizon h being essential. Such an approach would make explicit that reducing short-lived forcing agents can "trim the peak" of global warming but does not, as is sometimes erroneously stated, "buy time" to deal with carbon and other gases (Biello 2012), unless one neglects entirely the longer term impacts of current actions. A two-basket framework would require careful and interactive analysis of the science, risks, and value judgments associated with choosing how much and when to reduce the short-lived and long-lived baskets, and we believe that it would result in a clearer path forward for mitigation policy.

References

Allen MR, Frame DJ, Huntingford C, Jones CD, Lowe JA, Meinshausen M, Meinshausen N (2009) Warming caused by cumulative carbon emissions towards the trillionth tonne. Nature 458(7242):1163–1166. doi:10.1038/nature08019

Archer D, Kheshgi H, Maier-Reimer E (1997) Multiple timescales for neutralization of fossil fuel CO2. Geophys Res Lett 24(4):405–408

Armour KC, Roe GH (2011) Climate commitment in an uncertain world. Geophys Res Lett 38, L01707. doi:10.1029/2010GL045850

Biello D (2012) http://www.scientificamerican.com/article.cfm?id=how-to-buy-time-to-combat-climate-change-cut-soot-methane

Boer GJ, Yu B (2003a) Climate sensitivity and climate state. Clim Dyn 21:167–176

Boer GJ, Yu B (2003b) Climate sensitivity and response. Clim Dyn 20:415–429

Chang C-Y, Chiang JCH, Wehner MF, Friedman AR, Ruedy R (2011) Sulfate aerosol control of tropical Atlantic climate over the twentieth century. J Clim 24:2540–2555. doi:10.1175/2010JCLI4065.1

Daniel JS, Solomon S, Sanford TJ, McFarland M, Fuglestvedt JS, Friedlingstein P (2011) Limitations of single-basket trading: lessons from the Montreal Protocol for climate policy. Clim Chang 111:241–248. doi:10.1007/s10584-011-0136-3

Eby M, Zickfeld K, Montenegro A, Archer D, Meissner KJ, Weaver AJ (2009) Lifetime of anthropogenic climate change: millennial time scales of potential CO2 and surface temperature perturbations. J Clim 22(10):2501–2511. doi:10.1175/2008JCLI2554.1

Forster P et al (2007) Changes in atmospheric constituents and in radiative forcing. In: Solomon S et al (eds) Climate change 2007: the physical science basis. Cambridge University Press, Cambridge, pp 129–234

Fuglestvedt JS, Berntsen TK, Godal O, Sausen R, Shine KP, Skodvin T (2003) Metrics of climate change: assessing radiative forcing and emission indices. Clim Chang 58(3):267–331

Gillett NP, Arora VJ, Zickfeld K, Marshall SJ, Merryfield WJ (2011) Ongoing climate change following a complete cessation of carbon dioxide emissions. Nat Geosci 4:83–87

Hansen JE, Lacis AA (1990) Sun and dust versus greenhouse gases: an assessment of their relative roles in global climate change. Nature 346:713–719. doi:10.1038/346713a0

Hansen J, Sato M, Ruedy R (1997) Radiative forcing and climate response. J Geophys Res-Atmos 102:6831–6864. doi:10.1029/96JD03436

Held IM, Winton M, Takahashi K, Delworth T, Zeng F, Vallis GK (2010) Probing the fast and slow components of global warming by returning abruptly to preindustrial forcing. J Clim 23:24182427. doi:10.1175/2009JCLI3466.1

IPCC (2007) Climate change 2007: the physical science basis. Contribution of Working Group I to the fourth assessment report of the Intergovernmental Panel on Climate Change [Solomon S, Qin D, Manning M, Chen Z, Marquis M, Averyt KB, Tignor M, Miller HL (eds)]. Cambridge University Press, Cambridge/New York

Jackson SC (2009) Parallel pursuit of near-term and long-term climate mitigation. Science 326:526–527

Jacobson MZ (2002) Control of fossil-fuel particulate black carbon and organic matter; possibly the most effective method of slowing global warming. J Geophys Res 107:4410–4431. doi:10.1029/2001JD001376

Joos F, Spahni R (2008) Rates of change in natural and anthropogenic radiative forcing over the past 20000 years. Proc Natl Acad Sci 105:1425–1430. doi:10.1073/pnas.0707386105

Kanakidou M, Seinfeld JH, Pandis SN, Barnes I, Dentener FJ, Facchini MC, Van Dingenen R, Ervens B, Nenes A, Nielsen CJ, Swietlicki E, Putaud JP, Balkanski Y, Fuzzi S, Horth J, Moortgat GK, Winterhalter R, Myhre CEL, Tsigaridis K, Vignati E, Stephanou EG, Wilson J (2005) Organic aerosol and global climate modelling: a review. Atmos Chem Phys 5:1053–1123. doi:10.5194/acp-5-1053-2005

Lenton TM, Held H, Kriegler E, Hall JW, Lucht W, Rahmstorf S, Schellnhuber HJ (2008) Tipping elements in the Earth's climate system. Proc Natl Acad Sci 105(6):1787–1793

Lowe JA, Huntingford C, Raper SCB, Jones CD, Liddicoat SK, Gohar LK (2009) How difficult is it to recover from dangerous levels of global warming? Environ Res Lett 4:014,012

Luthi D et al (2008) High-resolution carbon dioxide concentration record 650,000–800,000 years before present. Nature 453:379–382. doi:10.1038/nature06949

MacFarling-Meure C, Etheridge D, Trudinger C, Steele P, Langenfelds R, van Ommen T, Smith A, Elkins J (2006) Law Dome CO2, CH4 and N2O ice core records extended to 2000 years BP. Geophys Res Lett 33, L14810

Manne AS, Richels RG (2001) An alternative approach to establishing trade-offs among greenhouse gases. Nature 410:675–677. doi:10.1038/35070541

Manning M, Reisinger A (2011) Broader perspectives for comparing different greenhouse gases. Philos Trans R Soc A 369:1891–1905. doi:10.1098/rsta.2010.0349
Matthews HD, Caldeira K (2008) Stabilizing climate requires near-zero emissions. Geophys Res Lett 35:L04,705
Matthews HD, Weaver AJ (2010) Committed climate warming. Nat Geosci 3:142–143. doi:10.1038/ngeo813
Matthews HD, Gillett N, Stott PA, Zickfeld K (2009) The proportionality of global warming to cumulative carbon emissions. Nature 459:829–832
Meehl GA et al (2007) Global climate projections. In: Solomon S et al (eds) Climate change 2007: the physical science basis, contribution of working group I to the fourth assessment report of the Intergovernmental Panel on Climate Change. Cambridge University Press, Cambridge
Menon S, Hansen J, Nazarenko L, Luo Y (2002) Climate effects of black carbon aerosols in China and India. Science 297:2250–2253. doi:10.1126/science.1075159
Montzka SA, Dlugencky EJ, Butler JH (2011) Non-CO2 greenhouse gases and climate change. Nature 476:43–50
National Research Council (2011) Climate stabilization targets: emissions, concentrations and impacts over decades to millennia. The National Academies Press, Washington, DC
O'Neill BC (2000) The jury is still out on global warming potentials. Clim Chang 44:427–443. doi:10.1023/A:1005582929198
Plattner G-K et al (2008) Long-term climate commitments projected with climate-carbon cycle models. J Clim 21:2721–2751
Ramanathan V, Feng Y (2008) On avoiding dangerous anthropogenic interference with the climate system: formidable challenges ahead. Proc Natl Acad Sci 105:14245–14250. doi:10.1073/pnas.0803838105
Rotstayn LD, Lohmann U (2002) Tropical rainfall trends and the indirect aerosol effect. J Clim 15:2103–2116
Shindell D, Faluvegi G (2009) Climate response to regional radiative forcing during the twentieth century. Nat Geosci 2:294–300. doi:10.1038/ngeo473
Shindell D et al (2012) Simultaneously mitigating near-term climate change and improving human health and food security. Science 335:183–189
Shine KP (2009) The global warming potential: the need for an interdisciplinary retrial. Clim Chang 96:467–472. doi:10.1007/s10584-009-9647-6
Shine KP, Fuglestvedt JS, Hailemariam K, Stuber N (2005) Alternatives to the global warming potential for comparing climate impacts of emissions of greenhouse gases. Clim Chang 68:281–302. doi:10.1007/s10584-005-1146-9
Shine KP, Berntsen TK, Fuglestvedt JS, Skeie RBS, Stuber N (2007) Comparing the climate effect of emissions of short- and long-lived climate agents. Phil Trans R Soc A 365:1903–1914. doi:10.1098/rsta.2007.2050
Smith SJ, Wigley TML (2000) Global warming potentials: 1. Climatic implications of emissions reductions. Clim Chang 44:445–457. doi:10.1023/A:1005584914078
Solomon S, Kasper Plattner G, Knutti R, Friedlingstein P (2009) Irreversible climate change due to carbon dioxide emissions. Proc Natl Acad Sci 106:1704–1709
Solomon S et al (2010) Persistence of climate changes due to a range of greenhouse gases. Proc Natl Acad Sci 107:18354–18359. doi:10.1073/pnas.1006282107
UNEP (2011) Towards an action plan for near-term climate protection and clean air benefits, UNEP Science-policy Brief, 17 pp
Winton M, Takahashi K, Held IM (2010) Importance of ocean heat uptake efficacy to transient climate change. J Clim 23:23332344. doi:10.1175/2009JCLI3139.1

Building Adaptive Capacity to Climate Change in Less Developed Countries

Maria Carmen Lemos, Arun Agrawal, Hallie Eakin, Don R. Nelson, Nathan L. Engle, and Owen Johns

Abstract This paper focuses on the relevance of adaptive capacity in the context of the increasing certainty that climate change impacts will affect human populations and different social groups substantially and differentially. Developing and building adaptive capacity requires a combination of interventions that address not only climate-related risks (specific capacities) but also the structural deficits (lack of income, education, health, political power, etc.—generic capacities) that shape vulnerability. We argue that bolstering both generic and specific adaptive capacities, with careful attention to minimizing the potential tensions between these two types of capacities, can help vulnerable groups maintain their ability to address risks in the long run at the same time as they respond effectively to short term climate impacts. We examine the relationship between generic and specific capacities, taking into consideration that they are not always positively related. We then propose a conceptual model describing positive and negative feedbacks between the two.

M.C. Lemos (✉) • A. Agrawal
School of Natural Resources and Environment, University of Michigan,
440 Church St, Ann Arbor, MI 48109, USA
e-mail: lemos@umich.edu; arunagra@umich.edu

H. Eakin
School of Sustainability, Arizona State University,
PO Box 875302, Tempe, AZ 85287-5302, USA
e-mail: hallie.eakin@asu.edu

D.R. Nelson
Department of Anthropology, University of Georgia, 250A Baldwin Hall, Jackson St,
Athens, GA 30602-1619, USA
e-mail: dnelson@uga.edu

N.L. Engle • O. Johns
School of Natural Resources and Environment, University of Michigan, 440 Church St,
Ann Arbor, MI 48109, USA
e-mail: nengle@umich.edu; orjohns@umich.edu

G.R. Asrar and J.W. Hurrell (eds.), *Climate Science for Serving Society: Research, Modeling and Prediction Priorities*, DOI 10.1007/978-94-007-6692-1_16,

Keywords Adaptive capacity • Adaptive development • Generic and specific capacity • Livelihoods framework • Vulnerability • Climate variability and change

1 Introduction

Around the world, the devastation of climate-related impacts has undermined livelihoods, threatened ecosystems, and stretched the capacity of sociopolitical institutions. Droughts, storms, and floods have often caused serious agricultural losses and human suffering: images of famines in Africa, human displacement in the Caribbean, and water-covered settlements in Bangladesh illustrate just some of the disastrous impacts of climate on vulnerable populations. In recent years, the possibility of more frequent and extreme events as a result of climate change has fueled new avenues of inquiry to understand and address the vulnerability of human and social systems to these events. As adaptation becomes prominent on the social and governmental agendas of both rich and poor countries, we need to understand better the factors that increase or constrain their adaptive capacity, or the ability of different socio-ecological systems and agents to respond and recover from climate impact. Such an improved understanding is particularly important for less developed regions where these negative impacts will likely interact with and exacerbate other stressors already affecting those most vulnerable (Eakin and Lemos 2006; Wilbanks and Kates 2010).

In these regions, although climate change poses a grave and emerging threat, vulnerabilities are generally symptomatic of deep socio-economic and political inequalities that have historically characterized their societies (Blaikie et al. 1994; Adger 2006; Eakin and Luers 2006). In other words, vulnerability is as much – or more – determined by the political economy of risk than by changing climate circumstances. Under these conditions, we argue that efforts to build adaptive capacity must simultaneously and iteratively address climate threats and longstanding development needs (Lemos et al. 2007).

In practice, building adaptive capacity means designing and implementing policy that both addresses: (a) structural deficits (which we call *generic* adaptive capacity) such as universal access to education and health, income and land distribution and redistribution (e.g. cash transfers and entitlements programs, land reform), political reform (e.g. increased accountability, democratic decision-making and transparency), and institutional and administrative capacity-building (e.g. greater enforcement of regulations and norms, investment in human capital, decreasing corruption and inefficiencies); and (b) risk management (which we call *specific* adaptive capacity) such as investment in adaptation technology (e.g. public works for water storage and distribution, coastal protection, development of drought resistant crops), social innovation (e.g. disaster response, insurance, alert systems) and specific interventions that either mitigate exposure of different groups to a particular climate threat (e.g. drought-related famine prevention, creation of early warning systems for storms, and relocation of vulnerable populations in the face of recurrent and unmanageable floods). These interventions and policies will necessarily need to be carried out across different levels of government and across different sectors (Adger et al. 2005a; Wilbanks and Kates 2010) and are likely to be controversial and politically costly (Lemos 2007; Eakin and Patt 2011). However, the implications of the interaction between specific and generic capacities and the relative importance of each in affecting the overall ability to respond and recover from climate change impact have received relatively little empirical and theoretical attention (but see Adger and Vincent 2005; Lemos 2007).

In this article, we specifically discuss these interactions and theorize about different ways that generic and specific adaptive capacity intersect and shape each other in the context of building adaptive capacity in less developed regions. We hypothesize that in the best-case scenario, the combination of generic and specific adaptive capacity is synergistic, creating a virtuous cycle in which overall capacity is sustainably enhanced, fostering long-term adaptation (Lemos 2007; Lemos and Tompkins 2008). However, in less desirable scenarios, tensions in the relationship between generic and specific adaptive capacity may lead to negative feedbacks such as those that foster poverty and rigidity traps and resilient undesirable states such as those existing in clientelistic political situations. In these cases adaptation interventions can actually exacerbate inequalities or perpetuate maladaptation (Lemos 2007; Nelson and Finan 2009; Maru et al. 2012). For example, at the household level, the goal is to avoid an emphasis on interventions that focus on risk management without increasing the household's overall asset base because while these interventions may allow for short term coping, they fail to assure long-term adaptation (delNinno et al. 2003; Nelson and Finan 2009). In contrast, targeted capacity building for specific subpopulations or sectors may result either in complacency or rigidity traps in which endogenous efforts at specific risk management are thwarted (Eakin et al. 2011; Murtinho 2011).

Although there is growing consensus that adaptation policy must take into consideration structural deficits and long-term sustainability, addressing inequalities that create and sustain poverty and propagate vulnerabilities will likely require politically difficult policies that profoundly challenge the existing distribution of power and assets (Pelling 2009). At best, implementation of such structural changes has been slow and incremental in most countries, while virtually impossible in others. In this context, it is not surprising that adaptation interventions so far have mostly been technical and palliative (Lemos 2003). In some respect, linking progress on climate change adaptation to development goals can risk bogging adaptation policy down in the same politics of resource access and distribution that have impeded social development for decades (Eakin and Patt 2011). On the other hand, failing to integrate adaptation and development policy may result in distortions and inefficiencies that threaten sustainability in the long-run (Huq et al. 2003; Agrawala 2004; Bizikova et al. 2007).

To foster development that addresses climate change risk in the context of multiple stressors and enables adaptation, policy makers must decide whether it is more effective to invest in measures that will reduce vulnerability to a broad range of both climatic and non-climatic stressors, or whether it is best to focus on enhancing specific capacities to manage particular hazards. At the level of individuals and households, policy makers may wish to build capacities for autonomous risk management and adaptation as part of social contracts to disadvantaged citizens. Yet deciding which of the diversity of assets and entitlements that constitute livelihoods need to be strengthened through public investment and support is complex and uncertain. Additionally, the implementation of interventions that positively interact with household and community level capacities rather than detracting from them by stifling or constraining local level ingenuity and resources (such as the mobilization of cultural and social capitals) should also be taken into account in the design and deployment of risk management. In this sense, understanding the relationship between generic and specific adaptive capacity at different scales of governance is a critical component of informing policy-making and planning to respond to climate change impact. In the next sections, we review the literature focusing on adaptive capacity and develop a conceptual model theorizing the relationship between generic and specific capacities across scales in the context of less developed regions (Box 1).

Box 1 Governance and Adaptive Capacity in the Brazilian Water Sector

Brazil's national reform of water management in 1997 brought changes to the water resources sector that have contributed to both better governance, including deeper democratic participation, and improvements in disaster risk response (Engle and Lemos 2010; Johns 2011). Results of the reform in the drought-prone Jaguaribe basin in NE state of Ceará reveal how governance factors at the institutional scale contribute to adaptive capacity and how generic improvements in institutional capacity interact with specific risk reduction interventions. However, challenges to inclusion and equality remain that may limit the potential synergies between governance and adaptive capacity (Johns 2011).

In Jaguaribe, state policy makers sought to design a new set of institutions to manage water resources based on emerging models (Integrated Water Resources Management – IWRM), which included participatory user commissions and basin-level committees to deliberate about water allocation (Lemos and De Oliveira 2004). These new institutions have contributed to generic adaptive capacity by giving water users greater access to decision-making and voice. Increased transparency and legitimacy have begun to erode the legacy of clientelistic power arrangements that benefitted elites in the distribution of drought aid by giving preference to irrigation and local elites. The negotiated allocation of water has reduced conflict among users and increased equality, thereby reflecting the positive relationship between generic governance factors in increasing the efficacy and accountability in specific risk reduction interventions (Johns 2011).

(continued)

Box 1 (continued)

However, there have been limitations in the quality and scope of democratization in which centralized institutions maintain high levels of power, attenuating the decision-making capacity of the new participatory institutions by exercising veto power over democratic decisions that run contrary to the official position. Within user commissions and committees, non-elite and poorer users, such as rural workers and small producers, are still marginalized in part due to their lack of resources, social and political capital (Taddei 2005). Alienation and continued exclusion is also a function of the control of knowledge in the form of technical climate information, which is not equally accessible to all participants (Lemos 2007). Thus, while the reform has improved governance and adaptive capacity, there are still constraints to risk response due to skewed power relationships.

The Jaguaribe case illustrates how integration and stakeholder participation contribute to limited gains in adaptive capacity in the case of a severe drought in 2001. The multiple agencies tasked with water management worked together to craft a solution to the water shortage by compensating water-intensive rice producers for foregoing their water allocation and thereby saving perennial fruit orchards. While the coordination enabled by the reform allowed for such a response, there were limitations in using this opportunity for installing bulk water charges in the agricultural sector, mainly due to the limited nature of democratic participation, which stalled a more nuanced and locally-informed implementation of water charges (Johns 2011).

The reform in Jaguaribe has led to increases in generic and specific adaptive capacity over time by allowing water users and small agriculturalists greater access to decision-making through participatory governance, but there are tradeoffs between centralization, knowledge access and participation that complicate the maturity of institutional changes. The reform has complemented wider national anti-poverty measures, such as Zero Hunger and Family Fund (conditional cash transfer schemes), and enhanced the effectiveness and equitable benefits derived from the historical reliance on measures to target specific drought risks. Despite these advances, making further gains in democratic participation is a continuing challenge.

2 Understanding Adaptive Capacity

The concept of adaptive capacity has existed for decades (Parsons 1964; Chakravarthy 1982; Staber and Sydow 2002). Current conceptual underpinnings of adaptive capacity are most closely associated with the Intergovernmental Panel on Climate Change's (IPCC) characterization of *adaptation* as an "adjustment in

Table 1 Determinants of AC

Determinant	Encompasses
Human Capital	Knowledge (scientific, "local", technical, political), education levels, health, individual risk perception, labor
Information & Technology	Communication networks, freedom of expression, technology transfer and data exchange, innovation capacity, early warning systems, technological relevance
Material Resources & Infrastructure	Transport, water infrastructure, buildings, sanitation, energy supply and management, environmental quality
Organization & Social Capital	State-civil society relations, local coping networks, social mobilization, density of institutional relationships
Political Capital	Modes of governance, leadership legitimacy, participation, decentralization, decision and management capacity, sovereignty
Wealth & Financial Capital	Income and wealth distribution, economic marginalization, accessibility and availability of financial instruments (e.g. insurance, credit), fiscal incentives for risk management
Institutions & Entitlements	Informal and formal rules for resource conservation, risk management, regional planning, participation, information dissemination, technological innovation property rights, risk sharing mechanisms

Source: Eakin and Lemos (2006) (Based on Smit et al. 2001 and Yohe and Tol 2002)

natural or human systems in response to actual or expected climatic stimuli or their effects" (Parry et al. 2007). Successful adaptation should result in an equal or improved situation when compared with the initial condition while less successful responses (such as coping) would allow for short term recovery but continued vulnerability. But what ultimately determines the success or failure of adaptation is a system's adaptive capacity, for it describes the ability of a socio-ecological system, group, or individual to mobilize resources to prepare for and respond to current or perceived stresses. Table 1 summarizes the determinants of adaptive capacity often found in the literature.

Understanding what influences adaptive capacity is rooted in the IPCC's categorization of the determinants of adaptive capacity: economic resources, technology, information and skills, infrastructure, institutions, and equity (Smit et al. 2001). A number of scholars have expanded on and redefined this initial list of six categories, and, depending on the analytical lens of the researcher, have emphasized the importance of some elements over others. For example, some research suggests that communities are limited in their capacity to adapt by their ability to act collectively. Here, social capital, trust, and organization greatly influence this capability (Adger and Neil 2003; Pelling and High 2005). Others narrow in on institutions, governance, and management as critical influences on the system or individual's capacity adapt to climate change (Yohe and Tol 2002; Adger et al. 2005a; Eakin and Lemos 2006; Agrawal 2008; Brown et al. 2010; Engle and Lemos 2010; Gupta et al. 2010). In this emphasis, the degree to which governance is inclusive, just and participatory can have an important influence on what populations are able to effectively cope and adapt to stressors and which populations are most likely to suffer from harm (O'Brien and Leichenko 2003). Adaptive capacity is not equally distributed (Adger et al. 2007) and differential capacities among households, between different communities and even between nations can often be traced to histories of inequitable trajectories of development and differential access to power and resources (Dow et al. 2006).

Despite a long conceptual history and increasing emphasis in climate and sustainability literatures, adaptive capacity has yet to receive sustained empirical examination. In particular, analyses that move from a normative and theoretical understanding of adaptive capacity to test and unpack the theorized determinants of adaptive capacity are lacking. Moreover, it is increasingly evident that focusing on adaptive capacity can have practical and theoretical benefits. Not only is adaptive capacity an integral concept to both vulnerability and resilience studies uniquely positioned to draw from the benefits of both frameworks, but it also better resonates with practitioners and policy makers than concepts such as resilience and sensitivity (Engle 2011).

Adaptive capacity affects vulnerability by modulating exposure and sensitivity (Yohe and Tol 2002; Adger et al. 2007) and influencing both the biophysical and human elements of a socio-ecological system (Eakin and Luers 2006). Political-economy approaches to vulnerability analysis have particularly emphasized that adaptive capacity is socially and politically determined (Kelly and Adger 2000; Eakin 2005; Eakin and Bojorquez-Tapia 2008; Adger et al. 2009; Eriksen and Lind 2009). Adaptive capacity is thus both an aspect of vulnerability directly amenable to human influence and intervention, but particularly challenging to enhance because doing so may threaten existing power relations and resource distribution (Lemos 2003; Eakin and Patt 2011). In resilience studies, adaptive capacity, or adaptability, is the capacity of actors within the system to manage and influence resilience (Walker et al. 2004, 2006). Thus, the more adaptive capacity within a system, the greater the likelihood is that the system will be resilient in the face of climate stress. There is less attention in resilience studies, however, to how the capacities of individuals or groups – particularly those who are politically marginalized or disempowered – can be enhanced in order to effectively manage systemic resilience (but see Tschakert and Dietrich 2010; Brown and Westaway 2011).

These two perspectives, vulnerability and resilience, combine to suggest that there are two important temporal aspects of adaptive capacity. First, adaptive capacity is important for a system or for the actor(s) that constitute that system to cope in the short-term so as to maintain the status quo (i.e., resilience), recognizing that a return to the status quo without challenging existing power structures or resource allocation may not address underlying drivers of vulnerability (Lemos et al. 2007). Second, adaptive capacity is important to facilitate transitions and transformations – the long-term adaptation directed to more desirable states (Nelson et al. 2007). Yet high adaptive capacity does not necessarily translate into long-term adaptation. Rather than being discrete processes, resilience, transitions and transformations are part of a continuum to which most adaptation action can contribute. What differentiates between them is the quality of the outcome, with transformation leading to highly desirable political, social and rights regimes (Pelling 2009). And while 'desirability' is usually defined by those human elements within a given system (i.e., as negotiated between actors and various interests), the greater the adaptive capacity, the more likely the system or actor(s) will wind up in a 'desirable' situation in the face of a climate variability and change. However, it is important to take into consideration that different actors within a system may have competing and even conflicting interests, and that these actors may have different levels of power to pursue their interests. Depending on the scale of the system in question and the structure of governance, the voices of the most vulnerable populations may not have influence over how "desirability" is defined and achieved. Moreover, there may be tradeoffs between these two elements of adaptive capacity (short-term coping and long-term adaptation) as well as with other aspects of adaptation implementation. For example, synergy between coping and adaptation for one population may mean failure in adaptation for others or enhancing resilience at one scale may exacerbate vulnerabilities at another (Eriksen and Brown 2011). Finally, adaptive capacity is a relative concept both in terms of spatial distribution and the way it is realized in different contexts. For example, within a given country or region there may be a great diversity of levels of adaptive capacity both generic and specific and first order interventions may lead to second and third order adaptations ("adaptations to the adaptations"). In this context, policy makers and decision makers should focus efforts on aligning development initiatives and goals in a manner that can make building adaptive capacity synergistic, rather than leading to competing or incompatible outcomes. In this pursuit, it is important that we improve understanding of what builds adaptive capacity and/or functions as barriers or limits to adaptation through more systematic empirical evaluations (Adger et al. 2009; Engle 2011). Identifying what has led successful and desirable adaptations can help to build empirical evidence for the factors necessary to facilitate these adaptations.

3 Generic and Specific Adaptive Capacity

As mentioned above, generic adaptive capacity is defined as those assets and entitlements that build the ability of different systems to cope with and respond to a range of stressors. Poor households are usually vulnerable to a number of overlapping and interdependent disturbances that shape their overall vulnerability. For example, in India, agricultural households are affected not only by climate impacts but also by globalization that shapes their access to markets and incomes – that is, they are double exposed to climate impacts and globalization processes (O'Brien et al. 2004). Specific adaptive capacity refers to conditions that prepare systems to cope and recover from a particular event, in this case, a climate-related impact such as drought, flooding, or extreme weather (Sharma and Patwardhan 2008).

Based on case-study evidence, Lemos and her colleagues (Lemos 2007; Tompkins et al. 2008) have argued that building adaptive capacity is a dialectic, two-tiered process in which risk management (specific adaptive capacity) and deeper level socioeconomic and political reform (generic adaptive capacity) iterate to shape overall vulnerability. In principle, risk management approaches can create positive synergies across the state-society divide through participatory and transparent approaches (such as participatory vulnerability mapping or local disaster relief committees) that empower local households and institutions which in turn mobilize for further socio-political reform (Lemos 2007; Nelson et al. 2009). Similarly, by increasing households' overall adaptive capacity, anti-poverty programs (especially those that couple with education and health programs) may positively influence their ability to better take advantage of risk management mechanisms (e.g. access to social programs and insurance, identification of effective drought response).

Yet, empirically, the distinction between generic and specific adaptive capacity has received little attention despite widespread recognition of its critical implications for policy choice and design. These policy implications are twofold. First, policy makers in less developed regions and development scholars increasingly argue that it makes little sense to design policy to build adaptive capacity to climate stressors that ignores the multitude of other factors at the root of different systems' vulnerability. In this sense, this scholarship argues that adaptation policy needs to be *mainstreamed* into development policy to be effective (Huq et al. 2005; Jerneck and Olsson 2008; Kok et al. 2008). Second, some scholars argue that the concept of generic adaptive capacity can only take us so far. Some variables are not generalizable between different stresses and systems (Adger and Vincent 2005) and there is the suggestion that the prospect of adaptive capacity across a range of stresses is essentially a myth (Tol and Yohe 2007). In the next two sections we discuss the relationship between generic and specific adaptive capacity first at the national level, and second, at the household level. We use the concept of adaptive development to argue for a new approach to development that takes into consideration climate risk in policy-making and planning so as to enable national states to respond and recover from current and projected negative impacts of climate change. Formally integrating generic and specific capacity through an adaptive development approach at the national level could effectively balance climatic and developmental challenges. Using a livelihood approach at the household level (Scoones 1998; Ellis 2000), we theorize the relationship between generic and specific adaptive capacity and propose a simple conceptual model of potential synergies and trade-offs between the two.

4 Adaptive Development

Historically, the failure of economic growth alone to solve pressing societal problems has encouraged the emergence of new approaches to development. For example, dominant development paradigms over the past five decades have included human and sustainable development as attempts to address inequality and environmental degradation respectively (Parpart and Veltmeyer 2004). As unprecedented risks represented by climate change impacts become more palpable, the next frontier of developmental policy-making will have to take into account not only past concerns but also climate adaptation.

The effects of climate change will fall unequally and disproportionally on poor communities, and will create greater stress around issues of sustainability (Adger et al. 2005b; Parks and Roberts 2010). Impacts will also bring already stressed human and ecological systems closer to the thresholds of undesirable and irreversible changes (Rockstrom et al. 2009). Climate change also enhances uncertainty in development planning, such that intended economic and social outcomes of policy are potentially jeopardized if climate risks are not accounted for (Box 2).

Box 2 Disaster Risk Reduction in Bangladesh

Bangladesh lowland's exposure to climate-related disasters is well documented; between 1970 and 2004 around 0.7 million people have been killed and economic losses in excess of 5.5 billion dollars have been incurred as a result of cyclones and flooding (Chowdhury et al. 1993; delNinno et al. 2002). Perhaps the worst climate-related disaster was the 1970 Bhola cyclone that hit then East Pakistan (now Bangladesh), killing over half a million people. As recently as 1991, another cyclone, this one hitting at night, killed over 130,000 people and negatively affected other five million. Despite early warning (15 h ahead) and greater availability of shelters (built after the Bhola cyclone by public and private organizations), 67,000 died on impact and property worth US$ 2.4 billion was destroyed (Financial Indicators Bangladesh, 1991 cited by Chowdhury et al. 1993). Human-induced climate change is expected to exacerbate the problem; projected half-meter sea-level rise by 2050 is likely to permanently inundate about 11 % of Bangladesh territory (Khan and Rahman 2007). Bangladesh is the most densely populated country in the world with more than 1,000 people per sq. km (Khan and Rahman 2007). Agriculture, which provides about a quarter of the country's GDP, is largely nature-dependent due to heavy reliance on favorable seasonal conditions, particularly on monsoon rainfall.

Building adaptive capacity in Bangladesh has involved developing both generic and specific capacities. Over the past 30 years, Bangladesh has significantly reduced poverty. While the proportion of the population living below the poverty line was as high as 74 % in 1973–1974, between 1991 and 1992 and 2000, the incidence of national poverty declined from 50 to 40 %,

(continued)

Box 2 (continued)

indicating a reduction rate of 1 % per year (Sen 2003). However, a significant portion of the population remains vulnerable, especially in areas of low "geographic capital". In these locations, social and geographical disadvantages overlap and residents derive few benefits from the economic and social opportunities created by economic growth. Natural resources crises (including disasters) are especially threatening in these areas, being responsible for 15 % of the reason for increasing household poverty (Sen 2003). Specific AC has also been built through risk management programs, especially disaster response and anti-famine interventions. For example, since the 1970s a diverse network of shelters (including hundreds of one-story and two-stories concrete buildings, multi-purpose cyclone shelters and rehabilitating houses) has been built with the help of organizations such as the World Bank and NGOs. The government has also built 150 *killas* (artificial hills), mainly to protect household animals from flooding (Chowdhury et al. 1993). In the 1998 "flood of the century", the government was able to avoid a famine crisis like the one that killed tens of thousands of people in 1974 through a combination of trade liberalization, importation of food and aid (delNinno et al. 2003). Moreover, following the initial flood period, immediate relief was available through the Gratuitous Relief program which provided 35.7 % of severely flood-exposed households with direct relief. The overall handling of the crisis kept prices from rising despite larger losses in rice production than in 1974; indeed the government seems to have learned from successive droughts both in terms of preparedness (public stocks) and longer term planning (role of private markets) (delNinno et al. 2003).

However, vulnerability has persisted as households have remained sensitive (delNinno et al. 2003). After a successful response in 1998, long-term negative impacts included lower calorie consumption, damage to infrastructure (houses) and negative health impacts. Rather than adapting, most households coped with the shock of the flood in several major ways, including reducing expenditures, selling assets and borrowing. While immediate post-disaster relief programs facilitated coping, they were small relative to the needs of households (only one-sixth to one-eighth the size of household borrowing). Borrowing from the private sector to purchase food and to fund other expenses such as education, health, farming, business, repayment of loans, marriage and dowry, purchases and mortgage of land or agricultural equipment constituted the main coping strategy, leaving many households in debt even a year after the event. Fifteen months after the flood, household debts still averaged 146 % of 1 month's average consumption for the 64.2 % of flood-exposed households in the bottom 40 % of the expenditure distribution (delNinno et al. 2003). Although debt declined with time, it still constituted a great part of household hardship and left them vulnerable to future shocks. The Bangladesh case suggests that while focusing on risk management greatly reduces casualties and facilitates coping in the short run, it fails to foster long-term adaptation.

New approaches to help govern social and individual risks must explicitly consider the negative synergy between climate risks and structural deficits in its many forms. As mentioned above, poverty, lack of access to health and education, lack of political power, and social inequalities exacerbate vulnerability to climate impacts while recurrent impacts (drought, storms, etc.) increase vulnerability (Heltberg et al. 2009). By focusing on how risks can be reduced in the pursuit of development and vice-versa, it becomes possible to identify the essential difference between development in the face of climate change and development as growth, human development, and/or sustainable development. Yet, this distinction does not mean that we believe policy to address risk should not be to integrated and reconciled into other developmental policy; rather, we argue that adaptive development pays specific attention to how risk management intersects (positively and negatively) with policies aiming at economic growth, human and sustainable development. For example, in drought ravaged Northeast Brazil, risk management interventions such as crop insurance or emergency provision of drinking water can allow affected households to respond to short-term drought stress. However, the extent to which these interventions allow families to cope and also develop longer term adaptive capacity is likely to be predicated on the combination of specific risk management with generic anti-poverty programs such as the Zero Hunger or Family Fund initiative which provide households with fungible cash resources and long-term access to education and health. In NE Brazil, such programs may be fundamentally changing the relationship between exposure and sensitivity to drought and improving the ability of households to use monthly cash allowances for short-term survival while simultaneously engendering long-term resilience through better health and educational access.

When considered as a means to address risks faced by diverse populations, the concept of adaptive development provides a clear conceptual basis upon which to elaborate strategies aimed at improving the life chances of the poor and the long-term sustainability of ecosystems. Adaptive development strategies would work to reduce the riskiness of development choices, even as they attend to the criteria of equity and sustainability. The idea of adaptive development can help take into account the dynamic, non-incremental, synergistic and often surprising nature of climate change hazards that will need to be addressed in the future. Going back to the NE Brazil example above, it would be precisely in the positive synergy between short term risk interventions and long-term development programs that our ability as a society to prepare for both extreme events and long-term incremental change brought about by climate change lie. Adaptive development provides the social infrastructure that bridges individual actions to reduce personal vulnerability into a framework in which such actions contribute to collective capacity to manage risk. In addition, thinking about development through a risk and risk governance lens enables policy makers and scholars to draw upon a vast body of historical and emerging scholarly work that has sought to examine the nature of risks, and how risks can be and have been addressed in the past. Better understanding these responses leads us squarely to the scholarship focusing on the political economy of hazards, disaster risk and adaptation to climate-related impacts (especially climate variability) (Blaikie et al. 1994; Pelling and High 2005).

From a policy point of view, beyond conceptualizing the relationship between development and risk, there is a need to understand the dynamics of adaptive action, that is, how the practice of implementing risk management interplays with development policy negatively and positively. The adaptive nature of this implementation requires monitoring and experimentation that lead to evaluation and learning, aimed especially at increasing understanding of how positive synergies between more traditional development policies (i.e. those which aim to address structural deficits) interact and intersect with new ones designed to address climate-related risk. It also requires that we understand the direct and indirect effects of adaptation policy and make sure that the solutions pursued yield desirable outcomes for those populations who are particularly at risk and do not trade off negatively with sustainability and equity (Brown 2011; Eriksen and Brown 2011). Next, we look at specific and generic adaptive capacity at the household level and discuss their implications for mitigating vulnerability to climate change.

5 Livelihoods and Adaptation

At the household level, the combination of generic and specific adaptive capacity (or lack thereof) is associated with two kinds of actions: (1) those that enable households to maintain their level of assets even after the climate-related impact (defined as adaptations); and (2) those that allow households to respond to extreme events in the short term, but in ways that may erode their asset-base in the long-term (defined as coping). For example, when a household adapts in anticipation of drought, it might invest in water harvesting or the infrastructure for silage. When a drought hits this household it is less exposed and therefore able to 'ride the drought' relatively unscathed. In contrast, a household might otherwise sell some livestock to pay for fodder for the rest of the herd, subsequently losing part of its asset base forcing it to rebuild the herd in less than optimal circumstances (Carter et al. 2007). In this case, it copes rather than adapts because it fails to maintain or improve over its original state. In other words, while some extreme event-coping actions such as the sale of livestock or land might allow the household to recover in the short run, they will diminish its asset base in the long run, making the household more vulnerable. Broadly stated, households with enhanced adaptive capacity – and presumably more secure assets, entitlements and thus livelihood–may be more likely to engage in welfare-enhancing adaptations because they have the stock of capital from which to make these investments. Unlike asset-constrained households, they are less likely to rely on coping strategies that threaten their long-term welfare (Dercon 1998; Siegel and Alwang 1999; Carter et al. 2007). Typically there is a history to such differences in assets and entitlements: households are embedded in political structures that institutionalize resource access and distribution in ways that are often path dependent, creating poverty traps for those households who are excluded.

Livelihood analysis provides a pragmatic approach to assessing capacities and entitlements at the household level. Drawing from Sen's (1981) entitlement theory,

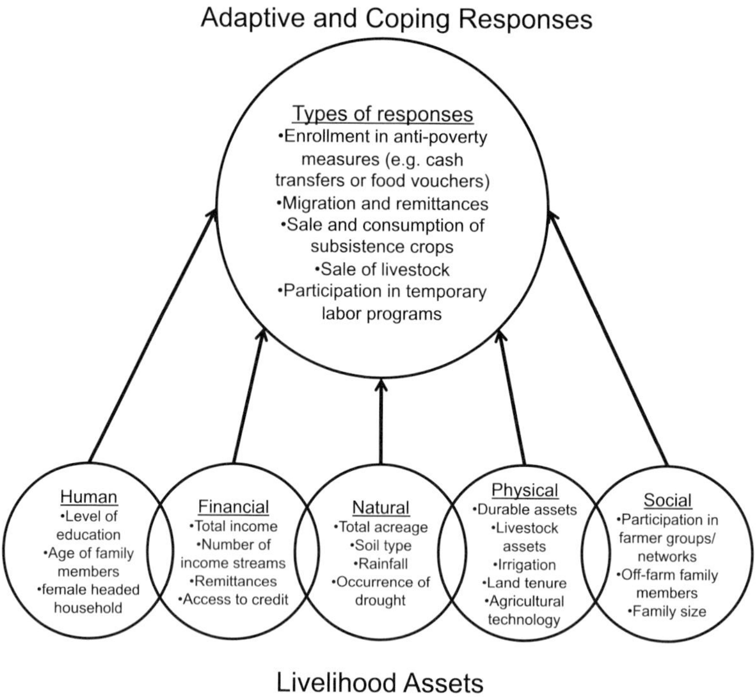

Fig. 1 Relationship between capitals, and adaptive and coping responses

sustainable livelihood research (Scoones 1998; Carney et al. 1999) addresses the relationships among a household's resource base (assets), its entitlements (the institutional context affecting rights and access to resources), and the result of these activities for aggregate household welfare (outcomes, or what we define as responses). Household capacity attributes can be categorized into five classes of livelihood capital: human capital (education, health, attitudes, belief systems); natural capital (soil quality, water endowments); physical capital (equipment, transport); social capital (connectivity in social or political networks); and financial capital (monetary savings, income composition) (Scoones 1998; Ellis 2000). Depending on the specific circumstances of the household and the political and economic structures in which the household exists, these different capitals play different functions in livelihood strategies and are differentially weighted in relation to risk management (Eakin and Bojorquez-Tapia 2008). These types of livelihood capital interact to engender coping and adaptation strategies (i.e. responses). Whether the strategies households engage in ultimately enhance (adaptation) or maintain/diminish their welfare over time (coping), such strategies typically can be classified as those that involve mobility, storage, diversification, communal pooling, and market exchange (Agrawal 2008). Figure 1 above depicts the five types of capital in relation to adaptive and coping responses.

As mentioned above, to support household adaptation in developing countries, adaptation policy makers must decide whether it is more effective to invest in measures that will reduce vulnerability to a broad range of stressors (climatic and non-climatic), or whether it is best to focus on enhancing capacities to manage specific hazards. In terms of the livelihood framework, policy makers must decide not only which types of livelihood assets and risk management should be strengthened through public investment and support but also how their design and implementation positively synergize with rather than detract from existing desirable responses (e.g. local mobilization of social capital and risk pooling) (Box 3).

Box 3 Poverty Traps and Disaster in Ethiopia

Poverty traps are "self-reinforcing feedback loops that keep social-ecological systems in persistent poverty" (Azariaidis and Stachurski 2005, Dasgupta 2007) (Maru et al. 2012). Carter et al. (2007) define poverty traps as a "minimum asset threshold" below which dynamic accumulation and livelihood growth towards greater well-being, that is – in climate parlance – adaptation, is not feasible. In the context of climate vulnerability, poverty traps define poor households' coping capacity to respond to climate-driven impacts such as drought and flooding and ultimately shape their inability to adapt. In some areas of both the developed and less developed world, poverty traps represent undesirable resilient states that critically limit the asset base of poor communities (e.g. income, access to health and educational services, social and political capital, etc.) (Lemos and Tompkins 2008; Nelson and Finan 2009; Maru et al. 2012).

The Ethiopian drought-driven famine crisis of 1998–2000 exemplifies both the progress that LDCs have made in improving disaster response and the role poverty traps can play in staving long term adaptive capacity building (Hammond and Maxwell 2002; Carter et al. 2007). The crisis itself was the result of both the relative failure of three consecutive rainy seasons and the inability of Ethiopian policy makers and the international aid system to fully prevent and respond to post-disaster impacts on poor households, especially highlands pastoralists (Hammond and Maxwell 2002). While government response markedly improved in relation to the 1983 El Niño-driven drought famine, in these households poverty traps resulted in an asset smoothing function (i.e. when households hold on to their livestock assets rather than selling them at the expense of an increase in food consumption after the shock). However, despite trying to hold on to their animals many of these households soon reached a threshold – a lower equilibrium – at which they settle down and stop growing (Carter et al. 2007) or, in other words, they cope rather than adapt and, in consequence, position themselves poorly to respond to the next set of stressors coming their way. To break out of this undesirable state beyond disaster response, it is necessary to build and diversify the asset base of these households by tackling several types of their capital shortage including income, social networks, food security, political participation, etc.

At the household level, we theorize that the relationship between specific and generic adaptive capacity is twofold. First, the ability of households to benefit from risk management may be predicated on a minimum level of generic capacity. For example, some households may be so vulnerable that they lack the minimum level of resources to benefit from or engage in specific risk management interventions. This may be the case for households lacking basic education and enough financial resources to enroll and benefit from programs such as crop insurance or rural credit. In this case, their adaptive capacity may be enhanced by specific educational and social policies such as Oportunidades in Mexico or Zero Hunger in Brazil. It can also be enhanced by their membership in rural labor unions or cooperatives through which they pool risk or share resources. Another example relates to the usability of seasonal climate forecasting (SCF) information. Empirical research has repeatedly uncovered that certain communities of groups in least developed countries are severely limited in their ability to benefit from SCF because of their lack of minimum capacity to respond to the projections. In this case, even if farmers had access to SCF, their lack of financial capital constrains their ability either to change crops (to shorter or longer grains, for example) or engage in other forms of adaptation (Finan and Nelson 2001; Ingram et al. 2002; Lemos et al. 2002). In many cases, households with constrained entitlements have not benefited from development interventions adequately, or have been marginalized in national economic trajectories (Eakin 2005). Here, if households had the socioeconomic preconditions to change their crops or participate in seed distribution programs, there would be the possibility of a synergistic relationship between generic and specific adaptive capacity as climate information could be effectively employed to mitigate climate variability risk.

In contrast, reliance on cash transfers may erode households' long-term capacities through the issue of "lock-in", that is, when welfare programs create relationships and dependencies between state and society that are difficult to uproot and may create rigidity rather than flexibility to respond to multiple stressors. Saldaña-Zorilla (2008), for example, found that despite the decline in public investment and support for the rural sector, there was a persistent expectation among farmers in Mexico that the government should be responsible for disaster risk mitigation, contributing to enhanced vulnerability and passivity. Eakin and Bojorquez-Tapia (2008) found that larger-scale private sector farmers in northern Mexico who had historically benefited from preferential access to land, financial services and commercialization support were more sensitive and ultimately more vulnerable to climatic shocks than their relatively resource-poor *ejidal* (a form of collective tenure) neighbors. As public support for farmers of almost all types declined in the 1990s in Mexico, and the government no longer guaranteed insurance or provided financial support, the larger-scale and more privileged farm class found it lacked the crop and livelihood diversity to cope effectively with extreme events. The *ejidatarios*, having never relied on public support as a means of coping with shocks, were far more autonomous and self-reliant in terms of risk management, although also less commercially engaged and productive than their counterparts. In other cases in Mexico, larger-scale commercial producers moved quickly to secure public support

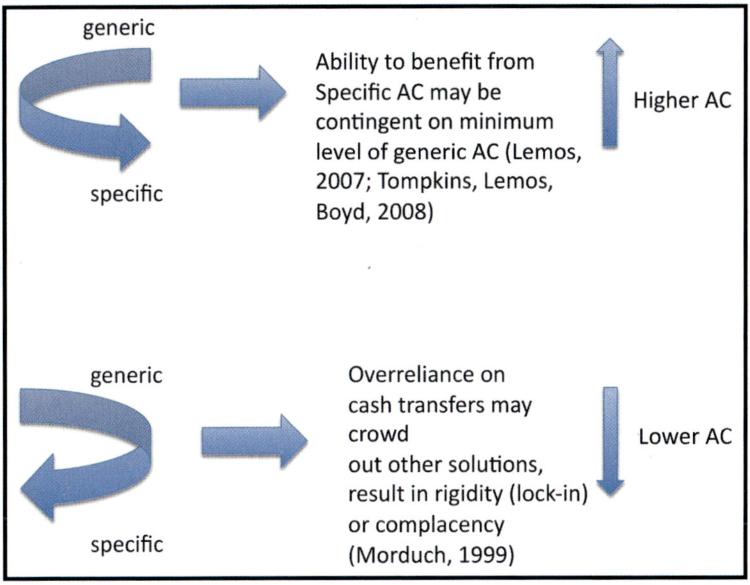

Fig. 2 Positive and negative feedbacks between generic and specific AC

following agricultural market liberalization in Mexico in the early 1990s. Their actions, designed to ensure that federal and state policy are closely aligned with their sectoral interests, resulted in a dangerous degree of complacency and neglect of risk such that famers require unprecedented federal support after their crops failed to frost in February 2011 (see Eakin et al. 2013).

Moreover, cash transfer programs may "crowd out" other initiatives (such as private investments) that may enhance adaptive capacity. For example, Murtinho (2011) found that in some rural Andean communities, autonomous adaptations to address problems of water scarcity were effectively "crowded out" by unsolicited public sector interventions. Rather than enhancing capacities to collectively manage current and future risk, the heavy-handed support of government was diminishing the probability that the community would take action. Figure 2 above shows a conceptual model of some of the relationships between generic and specific adaptive capacity.

6 Conclusions

This paper focuses on the relevance of adaptive capacity in the context of the increasing certainty that climate change impacts will affect human populations and different social groups substantially and differentially. The paper does so by arguing for greater attention to increasing climate risks in the design of development policies. The argument builds on two conceptual distinctions. The first is between specific and general adaptive capacity where specific adaptive capacity refers to the ability of agents and systems to address the risks specific to a particular climate threat and generic adaptive capacity refers to household endowments and system characteristics that enable more flexible responses to a diverse range of climate threats and other stressors. While we recognize that building both kinds of capacity may require different strategies and face diverse levels of resistance, bolstering generic and specific adaptive capacities with careful attention to minimizing the potential tensions between these two types of adaptive capacity can help vulnerable groups maintain their ability to address risks in the long run at the same time as they respond effectively to short term climate impacts.

An analogous distinction that the paper advances concerns the idea of adaptive development and development as usual. Adaptive development focuses on how to address livelihoods and welfare in increasingly risky contexts compared to earlier variants of development that focused on growth, equity, and/or sustainability. The paper highlights how future development policies and interventions are likely to require greater attention to risk reduction to secure the objective of greater welfare because more frequent, intense, and widespread climate threats may otherwise undermine development gains.

The paper also emphasizes the fact that specific and generic adaptive capacity are not always positively related, just as development interventions and growth-focused development outcomes can sometimes reduce the ability to cope with risks. Using a number of case examples, the paper identifies how to enhance the potentially synergistic relationship between specific and generic adaptive capacity or between risk reduction and growth, equity, and sustainability.

Acknowledgement This material is based upon work supported by the US National Science Foundation under Grant No. SES 1061930.

References

Adger WN (2006) Vulnerability. Glob Environ Chang-Hum Policy Dimens 16(3):268–281

Adger NW, Neil W (2003) Social capital, collective action, and adaptation to climate change. Econ Geogr 79(4):387–404

Adger WN, Vincent K (2005) Uncertainty in adaptive capacity. Compte Rendus Geosci 337(4):399–410

Adger NW, Arnell NW et al (2005a) Successful adaptation to climate change across scales. Glob Environ Chang 15:77–86

Adger WN, Brown K et al (2005b) The political economy of cross-scale networks in resource co-management. Ecol Soc 10:9

Adger WN, Agrawala S et al (2007) Assessment of adaptation practices, options, constraints and capacity. In: Parry ML, Canziani OF, Palutikof JP, van der Linden PJ, Hanson CE (eds) Climate change 2007: impacts, adaptation and vulnerability. Contribution of working group II to the fourth assessment report of the Intergovernmental Panel on Climate Change. Cambridge University Press, Cambridge, p.p717–743

Adger WN, Dessai S et al (2009) Are there social limits to adaptation to climate change? Clim Chang 93(3 – 4):335–354

Agrawal A (2008) The role of local institutions in adaptation to climate change. International Forestry Resources and Institutions Program, Working paper #W081-3, University of Michigan, Ann Arbor

Agrawala S (2004) Adaptation, development assistance and planning: challenges and opportunities. IDS Bull-Inst Dev Stud 35(3):50–54

Bizikova L, Robinson J et al (2007) Linking climate change and sustainable development at the local level. Clim Policy 7(4):271–277

Blaikie P, Cannon T et al (1994) At risk. Natural hazards, people's vulnerability and disasters. Routledge, London

Brown K (2011) Sustainable adaptation: an oxymoron? Clim Dev 3:21–31

Brown K, Westaway E (2011) Agency, capacity, and resilience to environmental change: lessons from human development, well-being, and disasters. Annu Rev Environ Resour 36:321–342

Brown H, Nkem J et al (2010) Institutional adaptive capacity and climate change response in the Congo Basin forests of Cameroon. Mitig Adapt Strateg Glob Chang 15(3):263–282

Carney J, Drinkwater M et al (1999) Livelihood approaches compared. Department for International Development, London, p 19

Carter MR, Little PD et al (2007) Poverty traps and natural disasters in Ethiopia and Honduras. World Dev 35(5):835–856

Chakravarthy BS (1982) Adaptation: a promising metaphor for strategic management. Acad Manag Rev 7(1):35–44

Chowdhury AMR, Bhuyia AU et al (1993) The Bangladesh cyclone of 1991 – why so many people died. Disasters 17(4):291–304

delNinno C, Dorosh PA et al (2002) Reducing vulnerability to natural disasters – lessons from 1998 floods in Bangladesh. IDS Bull-Inst Dev Stud 33(4):98–107

delNinno C, Dorosh PA et al (2003) Public policy, markets and household coping strategies in Bangladesh: avoiding a food security crisis following the 1998 floods. World Dev 31(7): 1221–1238

Dercon S (1998) Wealth, risk and activity choice: cattle in Western Tanzania. J Dev Econ 55:1–42

Dow K, Kasperson R et al (2006) Exploring the social justice implications of adaptation and vulnerability. In: Adger N, Paavola J, Huq S, Mace MJ (eds) Fairness in adaptation to climatic change. MIT Press, Cambridge

Eakin H (2005) Institutional change, climate risk, and rural vulnerability: cases from central Mexico. World Dev 33(11):1923–1938

Eakin H, Bojorquez-Tapia LA (2008) Insights into the composition of household vulnerability from multicriteria decision analysis. Glob Environ Chang-Hum Policy Dimens 18(1):112–127

Eakin H, Lemos MC (2006) Adaptation and the state: Latin America and the challenge of capacity-building under globalization. Glob Environ Chang-Hum Policy Dimens 16(1):7–18

Eakin H, Luers AL (2006) Assessing the vulnerability of social-environmental systems. Annu Rev Environ Resour 31:365–394

Eakin HC, Patt A (2011) Are adaptation studies effective, and what can enhance their practical impact? Wiley Interdiscip Rev: Clim Chang 2(2):141–153

Eakin H, Perales H et al (2011) Between rigidity and lock-in traps: adaptability and the future of maize in Mexico. Paper presented at Resilience 2011, Tempe

Eakin H, Bausch JC, Sweeney S (2013) Agrarian winners of neoliberal reform: the 'Maize Boom' of Sinaloa, Mexico. J Agrarian Change. doi:10.1111/joac.12005

Ellis F (2000) Rural livelihoods and diversity in developing countries. Oxford University Press, Oxford

Engle NL (2011) Adaptive capacity and its assessment. Glob Environ Chang 21(2):647–656

Engle N, Lemos MC (2010) Unpacking governance: building adaptive capacity to climate change for river basins in Brazil. Glob Environ Chang 20(1):4–13

Eriksen S, Brown K (2011) Sustainable adaptation to climate change. Clim Dev 3

Eriksen S, Lind J (2009) Adaptation as a political process: adjusting to drought and conflict in Kenya's drylands. Environ Manag 43(5):817–835

Finan TJ, Nelson DR (2001) Making rain, making roads, making do: public and private adaptations to drought in Ceará, Northeast Brazi. Clim Res 19:97–108

Gupta J, Termeer C et al (2010) The adaptive capacity wheel: a method to assess the inherent characteristics of institutions to enable the adaptive capacity of society. Environ Sci Policy 13(6):459–471

Hammond L, Maxwell D (2002) The Ethiopia crisis of 1999–2000: lessons learned, questions unanswered. Disasters 26(3):262–279

Heltberg R, Siegel PB et al (2009) Addressing human vulnerability to climate change: toward a 'no-regrets' approach. Glob Environ Chang 19(1):89–99

Huq S, Rahman A et al (2003) Mainstreaming adaptation to climate change in least developed countries (LDCs). Cambridge University Press, International Institute for Environment and Development: Climate Change Program. Intergovernmental Panel on Climate Change. Working group II, climate change 2001: impacts, adaptation and vulnerability, Chap 18

Huq S, Yamin F et al (2005) Linking climate adaptation and development: a synthesis of six case studies from Asia and Africa. IDS Bull-Inst Dev Stud 36(4):117–122

Ingram KT, Roncoli C et al (2002) Opportunities and constraints for farmers of west Africa to use seasonal precipitation forecasts with Burkina Faso as a case study. Agric Syst 74(3):331–349

Jerneck A, Olsson L (2008) Adaptation and the poor: development, resilience and transition. Clim Policy 8(2):170–182

Johns O (2011) Climate and change in Brazilian water management: a case study of governance and adaptive capacity. MSc., University of Michigan, Practicum

Kelly PM, Adger WN (2000) Theory and practice in assessing vulnerability to climate change and facilitating adaptation. Clim Chang 47:325–352

Khan MR, Rahman MA (2007) Partnership approach to disaster management in Bangladesh: a critical policy assessment. Nat Hazard 41(2):359–378

Kok M, Metz B et al (2008) Integrating development and climate policies: national and international benefits. Clim Policy 8(2):103–118

Lemos MC (2003) A tale of two policies: the politics of seasonal climate forecast use in Ceará, Brazil. Policy Sci 32(2):101–123

Lemos MC (2007) Drought, governance and adaptive capacity in North East Brazil: a case study of Ceará. Fighting climate change: human solidarity in a divided world, Human Development Report 2007/2008. H. D. R. O. O. paper, United Nations Development Program

Lemos MC, De Oliveira JLF (2004) Can water reform survive politics? Institutional change and river basin management in Cear Northeast Brazil. World Dev 32(12):2121–2137

Lemos MC, Tompkins EL (2008) Creating less disastrous disasters. IDS Bull-Inst Dev Stud 39(4):60–66

Lemos MC, Finan T et al (2002) The use of seasonal climate forecasting in policymaking: lessons from Northeast Brazil. Clim Chang 55(4):479–507

Lemos MC, Boyd E et al (2007) Developing adaptation and adapting development. Ecol Soc 12(2):26

Maru YT, Fletcher CS et al (2012) A synthesis of current approaches to traps is useful but needs rethinking for indigenous disadvantage and poverty research. Ecol Soc 17(2):7

Murtinho F (2011) Adaptation to environmental change among water user associations in the Colombian Andes. PhD, University of California, Santa Barbara

Nelson DR, Finan TJ (2009) Praying for drought: persistent vulnerability and the politics of patronage in Ceara, NE Brazil. Am Anthropol 11(3):302–316

Nelson DR, Adger WN et al (2007) Adaptation to environmental change: contributions of a resilience framework. Annu Rev Environ Resour 32:395–419

Nelson DR, Folhes MT et al (2009) Mapping the road to development: a methodology for inclusion and scaling-up of participation in policy processes. Dev Pract 19(3):386–395

O'Brien K, Leichenko RM (2003) Winners and losers in the context of global change. Ann Assoc Am Geogr 93(1):89–103

O'Brien K, Leichenko R et al (2004) Mapping vulnerability to multiple stressors: climate change and globalization in India. Glob Environ Chang 14:303–313

Parks BC, Roberts JT (2010) Climate change, social theory and justice. Theory Cult Soc 27(2–3):134–166

Parpart JL, Veltmeyer H (2004) The development project in theory and practice: a review of its shifting dynamics. Can J Dev Stud 25(1):39–59

Parry ML, Canziani OF et al (2007) Technical summary. In: Parry ML, Canziani OF, Palutikof JP, v. d. Linden PJ, Hanson CE (eds) Climate change 2007: impacts, adaptation and vulnerability. Contribution of working group II to the fourth assessment report of the Intergovernmental Panel on Climate Change. Cambridge University Press, Cambridge

Parsons T (1964) Evolutionary universals in society. Am Sociol Rev 29(3):339–357

Pelling M (2009) Adaptation to climate change. From resilience to transformation. Routledge, Abingdon

Pelling M, High C (2005) Understanding adaptation: what can social capital offer assessments of adaptive capacity? Glob Environ Chang 15:308–319

Rockstrom J, Steffen W et al (2009) A safe operating space for humanity. Nature 461(7263): 472–475

Saldaña-Zorilla SO (2008) Stakeholders' views in reducing rural vulnerability to natural disasters in Southern Mexico: hazard exposure and coping and adaptive capacity. Glob Environ Chang 18:583–597

Scoones I (1998) Sustainable rural livelihoods: a framework for analysis. IDS working paper

Sen B (2003) Drivers of escape and descent: changing household fortunes in rural Bangladesh. World Dev 31(3):513–534

Sharma U, Patwardhan A (2008) An empirical approach to assessing generic adaptive capacity to tropical cyclone risk in coastal districts of India. Mitig Adapt Strateg Glob Chang 13(8):819–831

Siegel P, Alwang J (1999) An asset-based approach to social risk management: a conceptual framework. Social protection discussion paper series. Social Protection Unit, The World Bank, Washington, DC

Smit B, Pilifosova O et al (2001) Adaptation to climate change in the context of sustainable development and equity. In: McCarthy JJ, Canziani OF, Leary NA, Dokken DJ, White KS (eds) Climate change 2001: impacts, adaptation and vulnerability. Contribution of working group II to the third assessment report of the Intergovernmental Panel on Climate Change. Cambridge University Press, Cambridge

Staber U, Sydow J (2002) Organizational adaptive capacity: a structuration perspective. J Manag Inq 11(4):408–424

Taddei R (2005) Of clouds and streams, prophets and profits: the political semiotics of climate and water in the Brazilian Northeast. PhD dissertation, Columbia University

Tol RSJ, Yohe GW (2007) The weakest link hypothesis for adaptive capacity: an empirical test. Glob Environ Chang 17(2):218–227

Tompkins EL, Lemos MC, Boyd E (2008) A less disastrous disaster: managing response to climate-driven hazards in the Cayman Islands and NE Brazil. Glob Environ Chang 18(4):736–745

Tschakert P, Dietrich KA (2010) Anticipatory learning for climate change adaptation and resilience. Ecol Soc 15(2)

Walker B, Holling CS et al (2004) Resilience, adaptability and transformability in social–ecological systems. Ecol Soc 9(2):5. [online] URL: http://www.ecologyandsociety.org/vol9/iss2/art5

Walker B, Gunderson L et al (2006) A handful of heuristics and some propositions for understanding resilience in social-ecological systems. Ecol Soc 11(1):13. [online] URL:http://www.ecologyandsociety.org/vol11/iss11/art13/

Wilbanks TJ, Kates RW (2010) Beyond adapting to climate change: embedding adaptation in responses to multiple threats and stresses. Ann Assoc Am Geogr 100(4):719–728

Yohe G, Tol RSJ (2002) Indicators for social and economic coping capacity – moving toward a working definition of adaptive capacity. Glob Environ Chang-Hum Policy Dimens 12(1):25–40

A Climate and Health Partnership to Inform the Prevention and Control of Meningoccocal Meningitis in Sub-Saharan Africa: The MERIT Initiative

Madeleine C. Thomson, E. Firth, M. Jancloes, A. Mihretie, M. Onoda, S. Nickovic, H. Broutin, S. Sow, W. Perea, E. Bertherat, and S. Hugonnet

Abstract Many human diseases are climate-sensitive: climate acting as an important driver of spatial and seasonal patterns, year-to-year variations (including epidemics), and longer-term trends. Although climate is only one of the many drivers of both infectious and non-infectious disease, public health policy makers and practitioners are increasingly concerned about the potential impact of climate change on the health of populations.

The authors Stéphane Hugonnet, Emily Firth and William Perea are (or were at the time of writing) staff member of the World Health Organization. These authors alone are responsible for views expressed in this publication and they do not necessarily represent the decisions or policies of the World Health Organization.

M.C. Thomson (✉)
International Research Institute for Climate and Society, The Earth Institute, Columbia University, 104 Monell, 61 Route 9 W, PO Box 1000, Palisades, NY 10964, USA
e-mail: mthomson@iri.columbia.edu

E. Firth • W. Perea • E. Bertherat • S. Hugonnet
Pandemic and Epidemic Diseases Department, World Health Organization, Avenue Appia 20, CH-1211 Geneva, Switzerland
e-mail: emily.firth@gmail.com; pereaw@who.int; bertherate@who.int; hugonnets@who.int

M. Jancloes
Health and Climate Foundation, Champ Courtet, Marchissy 1261, Switzerland
e-mail: michel.jancloes@gmail.com

A. Mihretie
Health, Development and Anti Malaria Association, PO Box 27279, Addis Ababa, Ethiopia
e-mail: aberemihretie@yahoo.com

M. Onoda
Group on Earth Observations, 7 bis, avenue de la Paix, Case postale 2300 Geneva 2, CH-1211, Switzerland
e-mail: secretariat@geosec.org

G.R. Asrar and J.W. Hurrell (eds.), *Climate Science for Serving Society: Research, Modeling and Prediction Priorities*, DOI 10.1007/978-94-007-6692-1_17,

The MERIT Initiative was launched in 2007 to provide a platform for enabling health specialists (public health specialists, epidemiologists, immunologists, microbiologists, demographers, etc.) and climate and environment specialists to work together to help solve a pressing health problem. The main objective of the initiative is to address meningococcal meningitis epidemics in Africa in the context of perceived environmental, biological, economic and demographic influences. The effort is designed to create new knowledge that can be used to improve the current (reactive) and future (preventive) vaccination strategies.

Preliminary results of this research to policy and practice consortium have advanced the understanding of how climate-related information can be tailored to inform and, where possible, strengthen public health decisions. Specifically, the MERIT experience to date indicates new evidence on the contribution that climate and environment make to the spatio-temporal distribution of meningococcal meningitis and demonstrates a multi-sectoral strategic approach to the creation of evidence, together with the development of a cumulative knowledge base. The MERIT Initiative is establishing an effective means for the dissemination of new knowledge and provides a platform to facilitate access to this knowledge by public health practitioners. These developments, along with an increase in the uptake of evidence in both policy and practice have the potential to impact health outcomes in vulnerable at-risk populations in Africa's Meningitis Belt.

The collaborative partnership model of MERIT provides an innovative framework to support public health preparedness and control strategies for climate sensitive diseases. Public health decision-makers have been willing to explore unfamiliar territory and opportunities for improving well-established control strategies by leveraging new knowledge and expertise from other disciplinary communities including climate and environmental researchers. Equally important have been the investments made by a multi-disciplinary research and practice community to adapt research projects in line with the evolving public health strategy across the Meningitis Belt. The lessons learned from the MERIT project offer valuable input and new ideas for improving global public health strategies for other climate and environmentally sensitive epidemic prone diseases.

S. Nickovic
World Meteorological Organization, 7bis, avenue de la Paix Case postale No. 2300 Geneva 2, CH-1211, Switzerland
e-mail: snickovic@wmo.int

H. Broutin
MIVEGEC, UMR 5290 CNRS- 224IRD- UM1- UM2 Centre de Recherche IRD, 911, Avenue Agropolis, BP 64501, 34394, Montpellier, Cedex 5, France
e-mail: helene.broutin@ird.fr

S. Sow
Center for Vaccine Development, CVD-Mali, Centre National D'Appui A La Lutte Contre La Maladie, Ministère de la Santé, Institut Marchoux, BP 251, Bamako, Mali
e-mail: ssow@medicine.umaryland.edu

Keywords Meningococcal meningitis • African meningitis belt • Climate • Environment • Dust • Epidemic early warning • Meningitis risk mapping • MERIT partnership • Knowledge translation

1 Introduction

With increasing recognition of the importance of climate as a driver of many infectious disease occurrences and the potential for climate change to exacerbate global health concerns (61st World Health Assembly, Agenda Item 11.11, Climate Change and Health WHA61.19, May 19–24, 2008) there is growing interest from the climate and environmental communities to contribute knowledge and resources towards improving climate sensitive health outcomes. The scientific literature on the impact of climate on infectious diseases transmission dynamics (including parasitic, viral and bacterial pathogens such as malaria, leishmaniasis, schistosomiasis, leptospirosis, dengue rotavirus, meningococcal meningitis, etc) indicates significant research interest in this area (Kelly-Hope and Thomson 2008). However the public health community lags behind others in the use of climate and environmental information for climate-sensitive decision-making. Recent developments in climate science and, more recently, climate services, along with new technologies for data management, analysis and sharing provide unprecedented opportunities for rapidly advancing this area. However, new developments must be responsive to the real needs of the global health decision-making community and empower their associated research and practitioner communities if this potential is to be fully realized.

2 Problem Identification – Epidemic Meningitis in Sub-Saharan Africa

Epidemic bacterial meningitis (causal agent *Neisseria meningitidis, Nm*) devastates the lives of individuals and communities across the 'Meningitis Belt' of Africa, a sub-Saharan zone extending from Senegal to Ethiopia and first described by Lapeyssonnie in 1963 (Lapeyssonnie 1963). The bacteria is transmitted through respiratory droplets throughout the year but invasive disease and associated epidemics are largely restricted to the dry season (Greenwood et al. 1985) which is marked by certain climatic factors considered favorable for epidemics, as illustrated in Fig. 1.

The mechanism by which environmental and climatic factors may influence meningitis epidemic occurrence remains unclear; the most common hypothesis for this role is that physical damage to the epithelial cells lining the nose and throat in hot, dry and dusty conditions permits the easy passage of the bacteria (found here frequently in asymptomatic form) into the blood stream causing invasive disease (Greenwood 1999). It has also been hypothesized that meningitis epidemics may be preceded by viral infections (also associated with specific climatic conditions) which may facilitate the transition from carrier to case (Mueller et al. 2008).

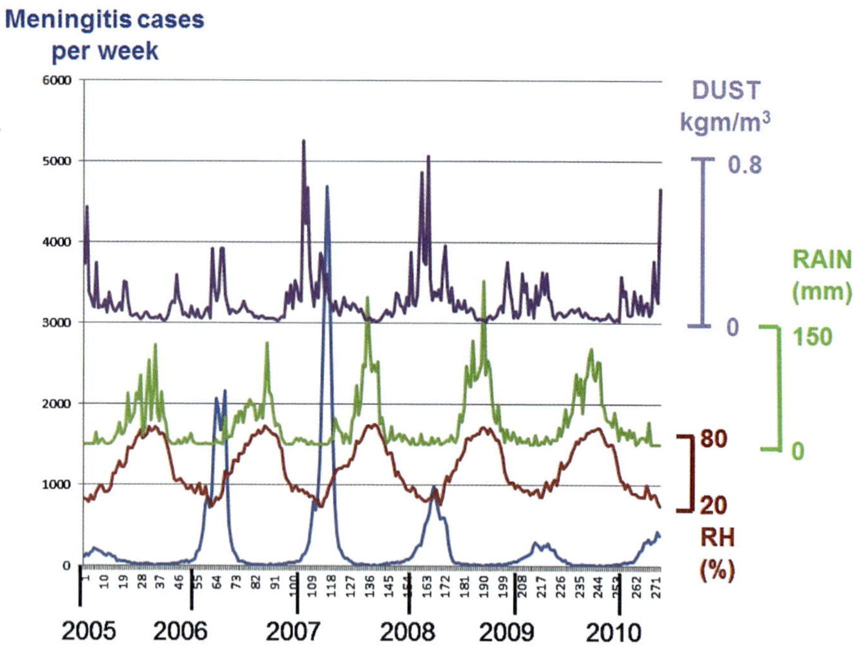

Fig. 1 Seasonality of meningitis epidemics. The number of meningitis cases per week (*blue line*) is influenced by various parameters that may include relative humidity (RH – *red line*), rainfall (*green line*) and surface dust concentration (*purple line*) (While data shown here is from 2005 to 2011 for Burkina Faso, this picture is indicative of the situation across the Meningitis Belt)

Neisseria bacteria need iron to grow and become virulent (Noinaj et al. 2012) and it has been postulated that the mineralogical properties of dust aerosol (especially the iron content) may also facilitate transition (Thomson et al. 2009). Many non-climatic factors have also been associated with epidemics including new bacterial strains, overcrowding, population movement etc.

The disease has severe social and economic consequences at the individual, household and community level and is widely feared because of its rapid onset, high mortality rate and frequent long term sequelae (Colombini 2009). The need for rapid vaccination in response to epidemics means that health staff are diverted from other important service activities, creating an additional burden to the health sector (Colombini et al. 2011).

3 Current and Emerging Control Strategies

Historically, meningitis control activities in Africa have largely relied on the early identification of epidemics followed by a rapid deployment of polysaccharide vaccines (Anon 2000). This 'reactive' strategy has meant that people at risk of being infected have only been immunized in an emergency situation once a meningitis outbreak has started. The effectiveness of the reactive strategy can determine whether an outbreak is controlled or risks spreading to neighboring districts and or countries. With the recent development of a conjugate meningitis A vaccine however, public health practitioners are now moving towards a more preventive strategic approach to minimize the risk of epidemics across the Meningitis Belt. The strengths and limitations of these two approaches are outlined in further detail below, leading into a discussion on the role and value of the MERIT Initiative in supporting the progressive public health strategies for reducing epidemic meningitis control.

3.1 *Reactive Vaccination Strategy*

Thresholds have been developed to inform the decision process as part of the reactive vaccination strategy. The epidemic threshold distinguishes a situation that is likely due to seasonal fluctuation from a situation that will develop into an epidemic. Implementing mass immunization campaign in any situation that is not an epidemic would be a misuse of scarce resources. In practice, weekly attack rates of suspect meningitis cases are monitored, and once the epidemic threshold is crossed, the decision to implement a reactive vaccination campaign is made (Anon 2000). To be effective and prevent as many cases as possible, vaccination needs to be implemented quickly. It has been estimated that 60 % of the cases could be prevented if vaccination is implemented within 4 weeks after crossing the epidemic threshold (Moore et al. 1992). This time window is extremely short, considering the eventual delay in the flow of surveillance data, the time required to deliver the vaccine and to organize and run the campaign.

Although much debate has been dedicated to the effectiveness of this reactive vaccination strategy, it is widely recognized that there is only a short lead-time for vaccination once an epidemic is underway and the impact of the vaccination response largely depends on the quality and timeliness of the surveillance system. Finally, the characteristics of the polysaccharide vaccine (short duration of the immunity, absence of herd immunity, weakly immunogenic in infants) and its limited availability precludes its use for truly preventive vaccination campaigns (WHO 2007).

3.2 *Preventive Vaccination Strategy*

As a consequence of the limitations of the reactive vaccination strategy, a preventive approach involving a conjugate A vaccine has recently been adopted by WHO and its partners. The MenAfriVac™ vaccine, which protects against infection caused by *Neisseria meningitidis* A (Sow et al. 2011); a serogroup which is responsible for the vast majority of the outbreaks, offers great potential to eliminate large meningitis outbreaks as a public health problem (Roberts 2008). Almost 10 years after the initiation of its development by the Meningitis Vaccine Project (MVP, http://www.meningvax.org/), the conjugate A vaccine was successfully introduced in Burkina

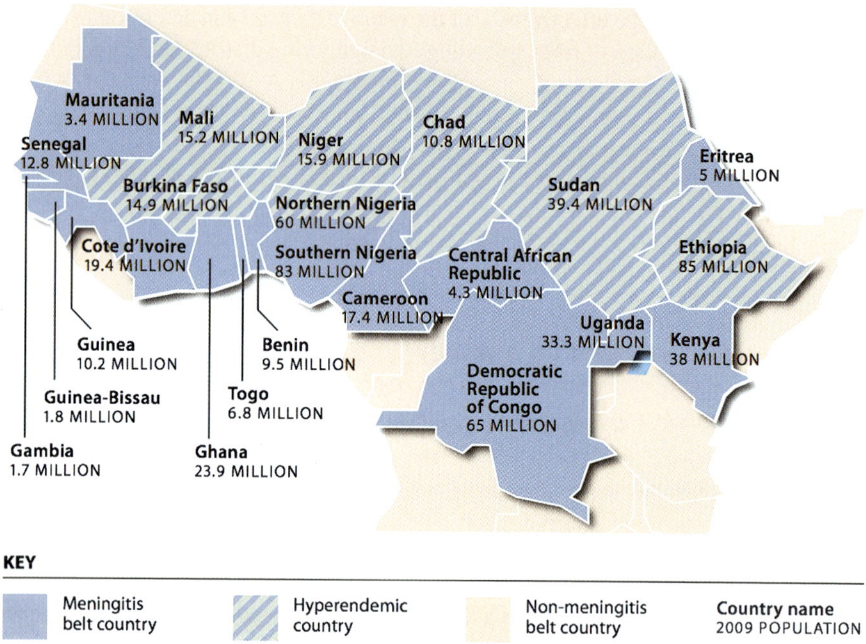

Fig. 2 Countries of the 'Meningitis Belt' and the introduction plan of the conjugate meningitis A vaccine. Approximately 450 million people are at risk of meningitis in sub-Saharan Africa. Certain areas are considered hyperendemic with populations considered to be at highest risk of developing the disease, and are targeted along with other at-risk countries for the phased roll-out of the new vaccine (Source: Meningitis Vaccine Project)

Faso, Mali and Niger in the last quarter of 2010 and in Nigeria, Chad and Cameroun in 2011 resulting in more than 55 million people vaccinated. A further 5–10 years are now required to immunize individuals at highest risk living in the 26 countries targeted by the project.

In the meantime, epidemiological and microbiological surveillance and forecasting systems which can help identify populations at risk remain a public health priority to detect and respond to meningitis outbreaks (whether due to Nm serogroup A or other that A, such as W135 or X) and to evaluate the impact of the vaccine (Cuevas et al. 2008). In order to better understand the impact of the conjugate vaccine on reducing transmission in Africa, a global research effort led by the African Meningococcal Carriage Consortium 'MenAfriCar', is performing carriage studies in line with the introduction plan of the conjugate vaccine.

While the introduction of the conjugate A vaccine promises to significantly reduce the problem of meningitis epidemics in Africa, the reactive vaccination approach remains an important part of the control strategy for populations not yet immunized with the conjugate A vaccine and in response to epidemics caused by other serogroups such as C, W135 and X. In this light, the MERIT approach remains valid, providing an opportunity to strengthen the reactive vaccination strategy and help evaluate the impact of the conjugate A vaccine following its introduction across the Meningitis Belt (Fig. 2).

4 Climate, Environment and the Risk of Epidemics

There is substantive evidence that climatic and environmental factors describe the overall spatial distribution of the disease in the Meningitis Belt of Africa (Lapeyssonnie 1963) and the indication that climate likely influences the seasonality of the disease is also widely accepted (Greenwood et al. 1985); epidemics start in the latter half of the Sahelian dry season when the weather is dry and dusty and subside at the onset of the rains. Much is still unclear about why they occur but it is likely that a combination of environmental, demographic, and behavioral factors as well as those relating to the hosts' immunity to disease will determine the occurrence of epidemic meningitis. In particular, the fact that the disease incidence tends to fall off rapidly once the moist pre-monsoon air arrives is seen as indicative of an important climatic control to disease occurrence. For a number of years, research focused on the relationship of environmental and climatic variables to meningitis incidence produced tantalizing results suggesting a significant interaction (Besancenot et al. 1997; Cheesbrough et al. 1995; Molesworth et al. 2003; Sultan 2005; Sultan et al. 2005; Yaka et al. 2008). However limitations in the data sets and modeling frameworks used precluded a definitive answer (Thomson et al. 2006).

In order to make progress in understanding the relative importance of climate as a driver of meningococcal meningitis epidemics, it was deemed necessary to first identify how such knowledge might inform operational decisions under field conditions. After discussions with key policy makers the priority concerns and research questions were identified in relation to (1) improving the reactive strategy with the polysaccharide vaccines in emergency situations in response to epidemics, and (2) supporting the longer-term preventive strategy with the introduction of the new conjugate A vaccine to 450 million people at risk across the Meningitis Belt.

Taking these concerns into consideration, a key identified need was for better data collection and research that could contribute robust estimates of the environmental contribution (alongside other factors such as carriage and immunity) to spatial and seasonal risk, year to year variation and longer term trends in meningitis epidemic occurrence and intensity. The identified research needs according to specific spatial scales and time horizons are elaborated further in the following section and in the table below.

5 The MERIT Initiative

Converging interests from the health and climate communities around the problem of epidemic meningitis were explored at a meeting hosted by the Group on Earth Observations (GEO) Secretariat in Geneva in 2007. Here a multidisciplinary group of participants (practitioners and researchers from both public health and climate communities) led by the World Health Organization and including the World Meteorological Organization (WMO), the Group of Earth Observations (GEO), the International Federation of the Red Cross (IFRC), the International Research

Institute for Climate and Society (IRI) the Health and Climate Foundation (HCF) and the Agencia Estatal de Meteorología (AEMET) agreed to the creation of the Meningitis Environmental Risk Information Technologies (MERIT, http://merit.hc-foundation.org) project. A Steering Committee for the initiative was formed the following year in June 2008. In 2011 the Steering Committee was joined by two additional members, one from the National Center of Research Institute/Institute of Research for Development Research Unit in Montpellier, France and the other from the Center for Vaccine Development, CVD-Mali, CNAM, Ministère de la Santé, in Bamako Mali. All current Steering Committee members are authors on this chapter.

6 MERIT Objectives

The original objectives of MERIT were clearly expressed in the purpose statement for the 1st MERIT meeting which took place at the John Knox Centre in Geneva on 26–27 September, 2007 in relation to advancing (a) partnerships (b) new knowledge and (c) improved decision-making. This purpose stemmed from expressed interests of the public health community to:

1. find a common platform between relevant communities to address meningococcal meningitis epidemics in Africa in the context of perceived environmental, biological, economic and demographic influences;
2. gain a greater understanding of the current knowledge and active research surrounding the epidemic risk indicators; and
3. communicate the information needs of the public health community to the research community to enhance epidemic meningitis control strategies in Africa.

The past 5 years has shown that the MERIT Initiative has to varying degrees of success responded to each of its original objectives. First and foremost, MERIT has encouraged and facilitated greater cross-disciplinary interactions established on a platform of well-defined needs and opportunities. MERIT has connected research more directly to the evolving needs of public health practitioners by providing a unique model for building and sustaining effective health-climate partnerships within the framework of improving health outcomes. While new knowledge has been generated, it is yet to be integrated into practical decision-making processes. To advance the transition from research to operations, an exercise was held during the 2012 meningitis season to provide public health practitioners in four countries in West Africa the chance to evaluate in near-real-time the output of predictive models with the disease dynamics at the district level, the results of which are currently being determined.

The specific objective of the Steering Committee was that of charting the course of the MERIT consortium throughout its expected lifetime of a decade. The consortium aims to extend current capabilities to more effectively combine environmental information with knowledge of epidemic meningococcal meningitis through analysis of the spatial and temporal distribution of cases, populations, environmental and climatic conditions, vaccination status and strain characteristics. Ultimately, this research seeks to inform three operational areas:

1. the reactive vaccination strategy (improve the impact of the reactive mass vaccination campaigns), prepare for the following epidemic season, and refine the response strategy for outbreaks due to serogroups other than A;
2. the preventive vaccination campaigns with the conjugate A vaccine (guide the introduction of the conjugate A vaccine and estimate the impact of the conjugate A vaccine); and
3. 5–10 years time-horizon forecasting to gather information on the possible vaccine needs in the medium and long term.

Despite much progress in surveillance and biological research in recent years, no explanation exists to date for the epidemic pattern of meningitis in the African Meningitis Belt (Mueller and Gessner 2010). Hence MERIT has tried to stimulate and support modeling efforts that might better explain the epidemic pattern of meningitis across the Belt as well as identify opportunities for prediction.

Key to the MERIT concept was consensus among partners that research needs would be demand-led, i.e. identified by those that were responsible for solving the health problem. Taking this approach, MERIT seeks to serve WHO, Ministries of Health, the Meningitis Vaccine Project and other relevant research initiatives such as MenAfriCar in the prevention and control of meningitis epidemics in Africa. This approach has been adopted by others engaged in strengthening prevention and control strategies for climate-sensitive diseases, as has been seen in the establishment of the Global Leptospirosis Environmental Action Network 'GLEAN' initiative.

While the development and refinement of research questions are guided by the public health needs, the momentum of the MERIT Initiative is in large part sustained by long-term research grants which support specific research questions, often over a period of several years. The challenge that has arisen is to determine how to maintain a degree of flexibility in the research arena in such a way that enables the research projects to stay in line with a changing public health strategy.

Summary table of analysis scale and research needs identified by WHO

Research need	Spatial scale	Time scale
To improve the impact of the reactive mass vaccination campaigns	District	Forecast the weekly attack rate several weeks ahead of time
To prepare for the following epidemic season	Region, country	Forecast the magnitude of an epidemic (yearly cumulative rate) 1 year ahead of time
To refine the response strategy for outbreaks due to serogroups other than A (NmW135, NmX)	District	Forecast the weekly attack rate several weeks ahead of time
To assess the risk of NmA outbreak in an area previously vaccinated with the conjugate A vaccine	District	Forecast the weekly attack rate several weeks ahead of time

(continued)

(continued)

Research need	Spatial scale	Time scale
To guide the introduction of the conjugate A vaccine	Region, country, district	Seasonal risk and historical trends
To estimate the impact of the conjugate A vaccine	Region, country, district	Predict the number and magnitude of epidemics one/several year(s) ahead of time
To gather information on the possible vaccine needs in the medium and long term	Region, country	Predict changes in the meningitis belt 5–10 years ahead of time

7 MERIT Research Networking Capabilities

In its initial phase, the MERIT consortium has focused on core countries of the Meningitis Belt (including Niger, Burkina Faso, Ethiopia, Ghana, and Nigeria). Six international technical meetings have been held in Geneva (2007, 2011), Ethiopia (2008 and 2010), Niger (2009) and Ghana (2012). Participants come from the public health and environmental sectors, governments, regional and international organizations as well as research institutions. Over 100 research papers have been presented in oral or poster format. Smaller 'Mini MERIT' meetings have been held in New York (USA), Montpellier (France), Lancaster (UK) and Boulder (USA) and have focused on specific research modeling questions. Presentations and MERIT meeting reports have been made available on the web via http://merit.hc-foundation.org. MERIT members are in the process of publishing their results in peer review journals.

Research groups from Europe, the USA and Africa have attended each of the technical meetings. Despite these meetings occurring in Ethiopia and Niger, participation by African research groups were initially limited; at least in part because funding for MERIT has been largely based on 'contributions in kind' to MERIT participants and the bringing together of ongoing research efforts. However, with the initiation of the Outbreak Prediction Tool exercise in 2012" focussed on Togo, Benin, Chad and Nigeria (see below) there has been a substantive increase in engagement by regional researchers, policy-makers and practitioners.

8 Summary of Research Projects Performed Under the MERIT Umbrella or Relevant for MERIT

As a research consortium without core funds, MERITs achievements are in reality the achievements of MERIT members some of which have developed and implemented specific MERIT related research projects, others of which have contributed research efforts developed under another umbrella to the overall MERIT Initiative. Research outputs are ongoing and results to date were formally reviewed at an

International Technical meeting in Geneva in November 2011. Information on the partner institutions and engagement in MERIT related activities at the international, regional and country levels can be found on the MERIT website (http://merit.hc.foundation.org). Details on key institutional contributions are indicated below.

8.1 *International Research and Initiatives*

World Health Organization (WHO): Central to the work of MERIT and guiding the research projects are the operational activities of WHO country, regional and international teams.

WHO provides technical assistance and support to Ministries of Health to help improve the prevention and control of meningitis epidemics in the region. The WHO office in Geneva responsible for meningitis control is within the Pandemic and Epidemic Diseases (PED) Department and works closely with the WHO African Regional Office Intercountry Support Team based in Ouagadougou, Burkina Faso to strengthen data management and surveillance of meningitis across the Meningitis Belt countries. As seen in recent years, the unpredictable nature of the extent and intensity of meningitis epidemics and the changing influence of meningococcal strains from 1 year to the next, emphasize the importance of robust meningitis surveillance across the Meningitis Belt. Furthermore, in its role as Secretariat of the International Coordinating Group (ICG) for Meningitis Vaccine Provision, WHO is responsible for the management of the global emergency vaccine stockpile funded by the GAVI Alliance and works closely with other ICG partners (UNICEF, Médecins Sans Frontières and the International Federation of the Red Cross), vaccine manufacturers and Member States to ensure a rapid response to meningitis outbreaks.

Within the MERIT framework, WHO has helped guide the design of research projects with a view to integrating the increasing understanding of environmental influences on meningitis epidemics into the meningitis control strategy. Ultimately, WHO would like to synthesize research outcomes from various MERIT projects in such a way as to support its activities in the field and strengthen the decision algorithm which determines the timing and distribution of vaccines to countries during an epidemic season.

The engagement of WHO with MERIT partners has initiated several new areas of activity, including:

1. Chairing the MERIT Steering Committee and facilitating the interactions between MERIT partners through annual technical meetings. Since 2007, these international technical meetings have been held in Geneva, Addis Ababa, Niamey and Accra. Smaller 'mini-MERIT' meetings have been held on an *ad hoc* basis in order to engage public health specialists with research groups to present outcomes and review the direction of specific research projects.
2. Leading the recent development of the district prioritization tool (DPT) to support the roll-out of the new conjugate A vaccine in countries and districts across the Meningitis Belt. This tool is at the heart of the current public health vaccination

strategy and integrates factors such as vaccine availability, current epidemiology, country capacities for implementation and surveillance, and political situations. While the DPT tool is not a result of the MERIT project, there is potential for results from MERIT research activities to feed into the tool and inform the strategy for the introduction of the conjugate A vaccine.
3. Participating with research scientists, developers of predictive models and disease focal points in several countries of the Meningitis Belt (see below) in cross-sectoral monitoring exercises of the 2010 and 2012 meningitis epidemic seasons.

The **CHICAS** (Combining Health Information, Computation and Statistics) **research group** at Lancaster University has invested in the development of several spatio-temporal models designed to support national- and district-level short-term forecasting of meningitis epidemics. The project has advanced through collaboration between Lancaster University and Columbia University's International Research Institute for Climate and Society (IRI). Combining weekly epidemiological data from Niger (1986–2007) with gridded reanalysis climate data for the region aggregated to the national level (including humidity, temperature dust and wind) the preliminary results suggest that climate data is of limited value for short-term (subseasonal) forecasting but would add value to longer-term forecasting of the meningitis season ahead. While plans to field-test one of the models in Niger in early 2011 were abandoned due to the political situation at the time, the group is engaged in an exercise with authorities in Togo, Benin, Chad and Nigeria to assess on a weekly basis the predictive output of the models during the 2012 meningitis season. At the end of the season, a formal analysis was held to assess the performance of three different types of models (Markov, Dynamic Linear Model and Dynamic Poisson-log-linear) in line with the actual epidemic activity during 2012.

Integrated Geophysical Modeling for Regional Climate Studies – the South East European Virtual Climate Change Center (SEEVCCC) hosted by the Republic Hydrometeorological Service of Serbia is developing a regional Earth modeling system with a dust component integrated to perform subseasonal/seasonal/climate studies. With the global database on mineral in arid soils, SEEVCCC should contribute to MERIT by assessing environmental conditions (dusty weather; mineral composition of dust) on time scales longer than current 2–3 days.

International Research Institute for Climate and Society (IRI), the Earth Institute, Columbia University: The IRI has been active in the MERIT initiative since its inception through support to the Steering Committee, the International Meetings and as host to New York based 'mini-MERIT' meetings contributing its scientific and technical capacity in the area of climate information for public health. Along with partners at Columbia University (the Center for International Earth Science Information Network (CIESIN), Goddard Institute of Space Studies (GISS) and Mailman School of Public Health) it has led several research projects under the MERIT framework to advance the understanding of the environmental factors (climate and aerosols) and population dynamics as determinants of meningitis epidemics in the Meningitis Belt. These include projects focused on the development of ground observations, remote sensing products and model outputs (including seasonal climate predictands) of direct relevance to modeling the climate and environmental drivers of meningococcal meningitis. In collaboration with other research and operational groups within MERIT the IRI engaged in projects with funding from NOAA, NIEHS, NASA and Google.org to help improve the application of available information and knowledge on the factors which influence meningitis epidemics, with a view to improving decision-making in meningitis control. The IRI has also contributed to the MERIT Initiative through its innovative Summer Institute 'Climate Information for Public Health' and has enabled the public distribution of relevant data sets through the IRI Data Library meningitis map room (Del Corral et al. 2012).

MACC – Monitoring Atmospheric Composition and Climate – is the current pre-operational atmospheric service of the European GMES program (http://www.gmes-atmosphere.eu/). MACC provides data records on atmospheric composition for recent years, data for monitoring present conditions and forecasts of the distribution of key constituents for a few days ahead. MACC WP 3.1 (funded by the EU and NSF Spain) "Meningitis linked to mineral dust transport in the Sahel", a collaborative project between the Meteorological State Agency of Spain (AEMET), The Earth Institute at Columbia University (IRI and NASA-GISS), the Barcelona Supercomputing Center (BSC-CNS) and the Spanish National Research Council (IDAEA-CSIC) has provided detailed validation of the dust model in the Sahel and elsewhere (Cuevas et al. 2011).

MAMEMA – Multidisciplinary Approach for Meningitis Epidemiology and Modeling in Africa: a consortium of MERIT partners was formed in 2010 to help increase the sharing of information between research groups on projects related to meningitis transmission dynamics and modeling in the Meningitis Belt. The initial areas of focus of the group under the MERIT framework concern the identification and the estimations of key parameters that should be included in epidemiological models based on the current knowledge and data availability. Another aspect is to define the future projects that could provide crucial parameters estimations for the models under development. The group gathers researchers from different complementary disciplines including epidemiology, medicine, public health, immunology, epidemiological modeling, climatology, anthropology and biostatistics. A first meeting occurred in Montpellier in 2011 and several kinds of models were presented to improve the reactive (see DTP presented above) and the preventive vaccination strategies (Irving et al. 2011). A second meeting was held in April 2012 in Montpellier. Presentations focused on the identification and estimation of environmental, climatic, epidemiological and societal parameters relevant for epidemiological modeling, to inform the long term vaccination strategies at different spatial scales (health center, districts, national) (Bharti et al. 2011; Irving et al. 2011; Paireau et al. 2011; Agier et al. 2013).

The MAMEMA consortium aims understand the drivers of localized epidemics and includes the National Center of Scientific Research (CNRS) Institute of Research for Development (IRD) Research Unit, in Montpellier, France, University of Bourgogne, Princeton University, Penn State University, University of Bristol, Washington state University, Agence de Médecine Préventive (AMP), Lancaster

University, and Ecole des Hautes Etudes en Santé Publique (EHESP), Paris and is connected with MenAfriCar.

UCAR Project: The University Corporation for Atmospheric Research (UCAR) in Boulder, Colorado USA was funded by Google.org to research methodologies for short-term forecasting during the meningitis season, with initial emphasis on the end of the meningitis season. Building on the historical work of the Navrongo Health Research Centre in Ghana and working with North Carolina State University, Regional Maritime University in Ghana and the IRI, the project aims to help increase the understanding of environmental variables which may influence the epidemic status of a district, and use that information to better respond to epidemics already in progress or about to start. Using a differential-equation based model of disease transmission, physical insight into meningitis transmission, 10 years of regional data and 2 years of data from across the Belt, the team was able to show that humidity, NE winds and heat all show positive correlations with future cases, as compared to historical persistence. An additional test with a generalized additive model confirmed that the temperature, humidity, and carbon monoxide concentration (as a proxy for burning) were the variables most persistently related to meningitis (Dukic et al. 2012). These two models, along with current epidemiological data and weather forecasts, provide the basis for weekly predictions of meningitis cases that can help guide vaccination. Other areas of research include determining the economic benefits of the forecast, surveying to document the knowledge, attitudes and practices of area residents, identifying the impacts of these predictions and other public health interventions and, most importantly, developing an information system for surveillance, data collection and disease management.

WMO Sand and Dust Storm Warning System: Of particular relevance to the MERIT project has been the WMO Sand and Dust Storm Warning Advisory and Assessment System (SDS-WAS) which was developed to study, among other questions, the impacts of dust aerosol on health, including the possible link of dust with meningitis and other diseases. The initial premise being that the prediction of dust events can help to better understand the hypothesized role of mineral dust and dry hot air in outbreaks of meningitis across the Meningitis Belt in the Sahel region. Under the SDS-WAS umbrella there are several activities and studies of relevance for MERIT; some of which have been specifically undertaken to support the MERIT Initiative. For instance, in 2010 a SDS-WAS regional centre for Northern Africa, Middle East and Europe established a "one-stop-shop" portal (http://sds-was.aemet.es/) which delivers near-real-time observations and short-medium range forecasts of dust-related parameters as well as forecasts of associated meteorological conditions.

SDS-WAS observations and products include:

(1) Ground observations;
(2) Satellite products;
(3) Dust forecast products for periods of several days ahead combining information available from several organizations which run dust modeling systems.

The parameters of potential importance for MERIT include:

(a) Columnar dust amount;
(b) Surface concentration;
(c) Wet and dry deposition.

Of notable significance is the fact that, since the dust concentration component in numerical models is driven by the atmospheric model, meteorological values such as temperature, air moisture, wind and precipitation are available simultaneously with the dust-related parameters (Nickovic et al. 2011). All data mentioned above are available on a daily basis for the Meningitis Belt and beyond and can now be accessed via the IRI Data Library (www.iridl.ldeo.columbia.edu/) (Del Corral et al. 2012) as well as from their original source.

8.2 Country-Led MERIT Initiatives

MERIT partners in disease endemic countries have formed local consortia with an aim to apply MERIT research outputs into operational decision-making at the country level as outlined below. While the importance of supporting country-led activities is well recognized by the MERIT community, it is critical that additional resources are allocated to ensure the development and application of research projects targeted to country-specific needs while helping build capacity in countries to sustain the activities.

Burkina Faso. In Burkina Faso the National Meteorological Services has run an "Environment and Bioclimatology Desk" for the last 10 years in order to (a) promote research on the relationship between climate – environment and diseases distribution – emergence, (b) to use observations forecasts and information on environment and climate to predict diseases outbreak and spatial distribution and (c) to contribute to elaboration of an integrated early warning system for prediction of climate sensitive diseases and promotion of suitable public health policies.

Among others activities, at the beginning of the meningitis epidemic season every year, a joint bulletin from National Meteorological and Health Services on prediction of meningococcal meningitis yearly incidence trend in Burkina Faso and Niger is produced by integrating climate, environment and health data and information. At the end of the year, a common evaluation is done. This information is jointly transmitted to national, regional and international organization and research centers working in medical and climate sectors as a contribution to help mitigate the diverse consequences of meningococcal meningitis epidemics. Using reanalysis data from the National Centers for Environmental Prediction (NCEP) and the National Center for Atmospheric Research (NCAR) reanalysis project (with an up-to-date version of the Medium Range Forecast model) statistical analysis of annual incidence of meningococcal meningitis and climatic variables for Niger indicated that 25 % of the year-to-year disease variance in this country can be explained by the winter climate; a similar analysis failed to accurately represent the disease dynamics in Burkina Faso (Yaka et al. 2008).

Ethiopia. Following the 2nd MERIT Technical Meeting in Addis Ababa in 2008 (MERIT 2008), the 'MERIT Ethiopia' project was established to help apply MERIT research across four key operational areas: (1) socio-economic impact of the disease, (2) determinants and risk factors of meningitis outbreaks, (3) education and training, and (4) disease surveillance. The MERIT-Ethiopia project was expected to benefit from and leverage the high level political momentum which had been generated by the formalization of a collaborative framework between the health and climate sectors under the title 'Climate and Health Working Group'. Despite this initial optimism however, the engagement of donors to support the country-led activities has not been actively pursued at the country level or by the MERIT Steering Committee, and as such the potential results have not yet been realized.

Niger. This is historically one of the most active countries in the Meningitis Belt in terms of epidemic activity and has one of the strongest epidemiological databases in the region over the past 20 years. In November 2009, the 3rd International MERIT Technical meeting (MERIT 2009) was hosted by the African Centre for Meteorological Applications for Development (ACMAD) and the Centre de Recherche Médicale et Sanitaire (CERMES) in Niamey, Niger. The meeting was followed by a 1-day national training workshop which provided an opportunity for national meteorological staff, health practitioners and medical students to engage with local MERIT partners and assess the benefits and applications of MERIT-related activities in Niger.

In relation to the MERIT objectives, the Centre de Recherche Médicale et Sanitaire (CERMES – a national medical research laboratory in Niger that is part of the Institute Pasteur Network) is leading a project entitled "Spatial epidemiology of acute bacterial meningitis in Niger. Role of climatic, environmental, health and socio-demographic factors on the spatio-temporal dynamic of the epidemics" in collaboration with Institut Pasteur in Paris (Emerging Diseases Epidemiology' Research Unit). The 3-year project is financed by the Coopération Monégasque for the period October 2010 to September 2013. The project consists of three components: (1) a descriptive study which aims to detect clusters of meningitis cases during the meningitis epidemic seasons, generally occurring from February to April each year; (2) an ecological and geographical study of the role of the climatic, environmental, health and socio-demographic factors on the occurrence of meningitis epidemics and to build a risk map for meningitis epidemics; and (3) to build an early warning system to help decision-making in relation to the implementation of reactive vaccination campaigns.

Other activities led by CERMES which consider the link between the climatic factors and the occurrence of meningitis are based on the use of time series studies and incorporate other risk factors such as respiratory infections. A first study has been conducted in the Niamey area whereby a generalized additive model was used to relate the daily change in bacterial meningitis cases and climatic factors.

First results were presented at the Hong Kong International Tropical Medicine Forum in January 2012.

Nigeria. In Nigeria, a state level MERIT committee was established in Katsina State bordering Niger in northern Nigeria. With an initial level of financial support provided by the State Focal Office on MDGs in the Governor's Office, the 'MERIT Katsina' project aims to: (1) fully realize current meningitis research and development initiatives, (2) build capacity within health and climate communities to improve health outcomes in the state, and (3) identify gaps and accelerate new warning and intervention strategies for meningitis. The MERIT Katsina group has also initiated a strong collaboration with the MAMEMA consortium.

8.3 *Regional MERIT Initiatives Across the Meningitis Belt*

ACMAD – African Centre of Meteorological Application for Development. Based in Niamey, Niger, ACMAD was instrumental in the organization of the 3rd MERIT meeting held in November 2009. The Centre produces a weekly 'Special Climate-Health Outlook Bulletin' to help translate meteorological observations and forecasts into a more meaningful language for the meningitis public health community. The bulletin is distributed on a weekly basis throughout the meningitis season and highlights: (1) the observed climate situation across the region for the previous 2 weeks, (2) the epidemiological situation based on weekly updates from the WHO Inter-country Support Team (IST) West Africa office in Ouagadougou, Burkina Faso, and (3) the climate outlook for the next week based on NCEP/NCAR, NOAA/NCEP/CPC and BSC-DREAM8b models. According to the observed and forecast climatic parameters (relative humidity, temperature and dust events), potential meningitis risk zones are identified with a view to informing the decisions of public health practitioners in relation to meningitis outbreaks in the region.

Following the Season: 2010. Following the 3rd MERIT meeting held in Niamey, Niger in November 2009, a cross-sectoral monitoring of the 2010 meningitis epidemic season was initiated. Throughout the 2010 meningitis epidemic season, WHO participated in an exploratory exercise with the IRI and the African Center of Meteorological Applications for Development (ACMAD) to follow the meteorological and epidemiological developments of a meningitis season. Between December 2009 and April 2010, WHO (Geneva, Switzerland), ACMAD (Niamey, Niger) and the IRI (New York, USA) met via teleconference on a 10-daily basis to share information on (1) the developments of epidemic outbreaks in districts and areas of alert across the Meningitis Belt, and (2) the climatic, meteorological and air quality observations and forecasts at the global and regional level. Climate information discussed global Sea Surface Temperature anomalies (known to influence atmospheric circulation on seasonal time-scales), large scale atmospheric circulation indices such as NAO (past and projected conditions) relevant for dust bearing circulation patterns, information on recent wind and humidity conditions including the location of the Inter-tropical Discontinuity (ITD-a region of convergence between the dry and dusty air from the Sahara and moist and dust-free air from the Atlantic) as well as on dust (past conditions and 8-day forecasts).

In total, 12 teleconference calls were held over the 4-month period, and while anecdotal feedback has been recorded, a formal assessment of the exercise is yet to be finalized. Throughout the season observed, climate showed conditions not favoring meningitis outbreaks with frequent higher than usual humidity, lower wind and dustiness conditions, consistent with a relatively quiet meningitis season until mid-March.

Retrospective analyses of the 2009 and 2010 seasons showed that, while epidemic behavior differed substantially between those years, climatic conditions averaged over the entire season did not exhibit such substantial differences. However, when analyzed at the sub-seasonal scale, climatic characteristics were found to differ between years: in 2010 higher variations in low level wind and humidity patterns brought moist and dust-free air to the region more often than in 2009, except in Burkina Faso, where the higher sub-seasonal variability in wind patterns translated in higher occurrence of dry and dusty episodes in 2010 as compared to 2009 (Trzaska et al. 2010). This points to the potential importance of sub-seasonal characteristics of the season in meningitis outbreaks but needs to be further documented.

Outbreak Prediction Tool: Tested in Real Time in Season: 2012. Building on the experience of the 2010 exercise outlined above and following the recommendations from the MERIT Strategic Review in 2011, WHO led an exercise in 2012 to prospectively assess the output of several statistical models developed to provide sub-seasonal, district-level predictions of epidemic activity (Agier et al. 2012; Stanton 2012) alongside the observed incidence of cases and epidemics during the 2012 meningitis season (January–May 2012). With a focus on Togo, Benin, Chad and Nigeria, disease experts and focal points from WHO (Geneva, African Regional Office and Country Offices) engaged on a weekly basis with the developers of the models from the CHICAS group at Lancaster University and climate science researchers from UCAR and NASA GISS (Columbia University).

The purpose of the exercise was to assess the performance of statistical models tailored for predicting meningitis epidemics with a lead-time of 1–4 weeks, alongside the observed epidemic behavior in the Meningitis Belt. The observed and forecast relative humidity and dust events in the region were integrated into the discussions in order to consider the environmental conditions favorable for epidemics. A formal analysis of the exercise will be conducted at the end of 2012 to determine the performance of the models and their potential to support the ongoing public health strategy and preparation of vaccination campaigns in the region.

Of particular interest in 2012 as compared to the previous 2 years, is the relatively high level of epidemic activity due to serogroup W135 and low activity due to serogroup A. This could be an area of further investigation, incorporating the results of the two seasonal exercises from 2010 to 2012 with microbiological, immunological, epidemiological, environmental and vaccination data from the past few years.

9 Bringing It All Together

Through the MERIT Initiative a much broader understanding by natural scientists of the problem of epidemic meningitis in the Sahel has been achieved. GEO, AEMET and WMO have played a significant role in ensuring that policy makers in the environmental community are aware of MERIT and have sought their support through its network of high-level partnerships.

Disparate data sets have been enhanced and brought together for analysis. These include epidemiological data sets from national surveillance systems facilitated through WHO, population data from the Global Rural-Urban Mapping Project (GRUMP) from CIESIN (at Columbia University) and a wide range of environmental and climatic variables now made available via open access portals such as the IRI (Del Corral et al. 2012). The challenge of integrating these disparate datasets should not be underestimated. Combined with practical knowledge of the data sources and their constraints, a raft of modeling exercises have been undertaken by the different research teams. In countries such as Niger where meaningful information can be extracted from large historical databases, statistical analyses such as those based on a Bayesian network approach (Beresniak et al. 2012) or geo-spatial analysis (Stanton et al. 2011) have demonstrated that innovative techniques can be developed to help understand and predict the risk of meningitis outbreaks at the district and sub-district level. This modeling effort has been combined with detailed field studies in some countries, such as those at Navrongo, Northern Ghana.

More recently models which can incorporate both extrinsic (e.g. climate/environmental) and intrinsic (e.g. immunity) drivers have been explored.

9.1 Possible Role of Dust Mineral Composition in Meningitis Epidemics

The mechanism by which dust may cause meningitis epidemics remains unclear. A common explanation is that physical damage to the nose and throat epithelial cells by dust particles permits invasion of bacteria into the blood stream. It is hypothesized that the activation of the meningococcal bacteria is fostered with high iron content in Fe-rich minerals in dust (Thomson et al. 2009). Current dust models are not capable of simulating/predicting in details the mineral fractions and trace metals such as Fe in dust concentration. However, recently developed global high-resolution (1-km) datasets on soil mineralogical composition (Nickovic et al. 2012) if used as input in dust models could help better understand the possible links between meningitis and iron fraction in dust.

9.2 *Limitations of Current Modeling Studies and Environmental Products*

Most of the products relevant for MERIT are publicly available. However, short-term forecasts have limited value for MERIT because, in the case of operational dust modeling, forecasts are valid for 3–5 days in advance which is too short a period to be of practical use for planning vaccination actions in the field. On the other hand, observational data (including remote sensing products) and re-analyses made for multi-decadal periods can help understand if and how meningitis outbreaks depend on environmental conditions and their seasonality.

Medium range sub-seasonal forecasts may prove relevant for MERIT with promising opportunities for designing such concepts. Several operational centers (including the European Centre for Medium-Range Weather Forecasts (ECMWF), UK Met Office, Japan Meteorological Agency (JMA), US National Centers for Environmental Prediction (NCEP), Environment Canada and the Bureau of Meteorology in Australia) are already providing global experimental sub-seasonal/seasonal forecasts. Several other organizations such as UCAR are also working on predictions for the same time scales while using regional modeling facilities. Following interest for such forecasts over extended periods, WMO is currently launching a new project "Sub-seasonal to Seasonal Prediction". With additional dust components in such systems, it will be possible to explore the value of longer-term dust forecasts for the needs of the MERIT health community.

10 Contributions from the MERIT Climate Community to the Assessment of the Links Between Weather/Climate and Meningitis Occurrence

A number of significant modeling results are emerging from different research teams as a result of the creation of new sources of weather, climate and environmental data tailored to MERIT needs. For example a series of analyses have resulted through the development of the dust-modeling activities of the SDS-WAS.

In a study based on a 30-year simulation model (1979–2010) recently developed with a $0.5° \times 0.5°$ resolution (NASA-GISS, IRI, BSC) (Pérez et al. 2011) the relation between climate indices and simulated dust aerosol concentrations over the Meningitis Belt is being examined (Pérez et al. 2009). Studies suggest that three is a certain level of correlation between dust and climate variability parameters, such as the NAO index. This result forms a basis to use climate indices as first-approximation indicators on favorable conditions for meningitis outbreaks.

Using the same simulation model (NASA-GISS, IRI, BSC, CHICAS) the relation between meningitis outbreaks in sub-Saharan Africa and simulated dust aerosol concentrations over the Meningitis Belt is being examined. The outputs of the simulations have been validated at the daily, seasonal, annual, inter-annual and trend scales using in situ and satellite dust data (Pérez et al. 2011; Haustein et al. 2012). They are also being used to perform a seasonal and weekly analysis of meningitis epidemic outbreaks at national and district levels in Niger (Pérez et al. 2009; Stanton et al. 2011).

Preliminary results show that climate parameters (including wind and dust) prior to January and early season meningitis cases explain about one quarter of the epidemic year-to-year variability at national and district levels. At the district level, both national-level covariates and district-level covariates of climate and dust variables and early cases together with population density and latitude represent the spatio-temporal variability of the diseases moderately well. Although the study outlines the need for other sources of data to better represent between-district variability (susceptibility, viral infections, new strains or previous vaccinations), the study shows the potential of climate information in the early season to explain part of the variability of the disease at the seasonal and district scale (Stanton 2012). Another study under progress in the framework of the MAMEMA group shows that aerosols represent a relevant climate and environmental parameter (together with wind, relative humidity and temperature) to explain the seasonal pattern of meningitis at the district level in Niger (Agier et al. 2013).

The non-linear interaction of different co-factors, many of them not known, partly hampers the assessment of the impact of climate and dust upon epidemics. This problem is even more critical at the weekly scale. In a recent study using the regional dust and climate database (Stanton et al. 2011), at the national scale, zonal wind and dust concentration made modest improvements in meningitis incidence forecasting ability, but the majority of temporal variation could be explained using a seasonal trend and previous incidence. At the district level, the inclusion of climate variables made no real difference to the forecasting performance of the models and the majority of temporal variation could be explained using a seasonal cycle and previous incidence (Stanton et al. 2011).

11 Data Policy Challenges and Opportunities

The MERIT Steering Committee has sought to create a data policy designed to achieve maximum research opportunities and outcomes while promoting open use by health practitioners and decision makers for the public benefit. However no single solution has been found to the diverse types of data involved and the diverse needs of the MERIT community since different data and products carry different access and dissemination conditions (e.g. restrictions due to privacy) and intellectual property rights. As a consequence the MERIT Steering Committee is actively discussing data sharing through the WHO information platform OpenHealth, the IRI Data Library and the GEO WebPortal (http://www.earthobservations.org). The IRI Data Library is making MERIT related environmental and climate data available to the global research community.

The 4th MERIT Technical Meeting in Addis Ababa, Ethiopia (MERIT 2010) extensively discussed the issue of data sharing, interoperability and developments of tools, including a proposal of a central MERIT information system. While so far the MERIT information systems are a combination of independently developed information systems, the community is continuing to explore technical options in order to realize the MERIT goals.

11.1 Meningitis Map Room

The Meningitis Map Room, situated in the IRI Data Library, makes meningitis-related environmental, demographic, and epidemiological data available for visualization, integration, analysis, and download (http://iridl.ldeo.columbia.edu/maproom/.Health/.Regional/.Africa/.Meningitis/). The data are web-accessible through a set of easy-to-use map pages and links to datasets. The spatial extent of the area of interest is controlled by click and drag cursor controls. First and second order administrative boundaries are used to calculate time varying spatial averages of gridded environmental data. These environmental time series can be correlated with time series of epidemiological data or downloaded to the user's desktop. New datasets are added when the spatial resolution or quality of the data is an improvement upon existing map room datasets (Del Corral et al. 2012).

The environmental data include rainfall, temperature, relative humidity, visibility, and wind speed. There are environmental quantities from a regional dust model run that spans 1979–2008. The demographic data is the CIESIN Gridded Population of the World Version 3 (GPW3).

12 MERIT Today and Future Direction

The 5th MERIT Technical Meeting and strategic review took place in Geneva, Switzerland in November 2011. At this meeting the achievements of MERIT to date were evaluated by an independent group of experts and a new chart for MERIT activities was developed based on the changing needs of the meningitis control community. Key outcomes from this meeting and highlights of the preliminary research results are outlined as follows.

1. New evidence has been produced on the contribution that climate and environment make to the spatio-temporal distribution of meningococcal meningitis.
   (a) Meningitis – climate/environment linkages have been further elaborated using robust statistical techniques taking into account the natural history of the disease, non-climatic factors and verified and relevant climate and environmental information. However the lack of understanding of the mechanisms and the interaction of infection, disease and immunity remains a challenge to the interpretation of these results.

2. Transitioning research into policy and practice is strengthened by a strategic approach for the creation of evidence, together with the development of a cumulative knowledge base.
   (a) The multi-sectoral Steering Committee, led by WHO, enables new communities to be brought together that are focused on policy-relevant problem-focused research.
   (b) The creation of mini-MERIT meetings has enabled highly focused scientific discussions to be developed and shared.
   (c) The increased profile of meningitis as a climate sensitive disease has created funding opportunities for researchers.
3. An effective means for the dissemination of new knowledge together with development of means to broadly access this knowledge has been undertaken.
   (a) MERIT international meetings provide a platform for sharing new data, research innovation and scientific knowledge.
   (b) Peer review publications and presentations at scientific and policy conferences/workshops have been used to disseminate research findings.
   (c) Engagement with national research and health decision-making partners has enabled learning and dialogue between south-south-north MERIT partners.
   (d) The development of a MERIT website (http://www.merit.hc-foundation.org) has enabled widespread sharing of MERIT information.
4. Initiatives to increase the uptake of evidence in both policy and practice are in development.
   (a) As data and learning accumulate, the opportunity to engage new partners with a move to a multi-disease approach is opening up new operational research opportunities.
   (b) MERIT responsiveness to changes in policy environment – i.e. the move from a reactive to preventive vaccine strategy has resulted in prioritization of longer term changes in the Belt.
   (c) Creation of training opportunities such as the IRI Summer Institute 'Climate Information for Public Health Action' has increased understanding of researchers and decision-makers in use of climate information.

The strategic review and technical meeting succeeded in large part by engaging both MERIT partners who have been actively involved in the Initiative to date and a small group of independent, external advisers representing the areas of meningitis control, environmental information and policy makers. Technical partners provided updates on latest research activities and preliminary outputs were discussed extensively in light of the current public health priority areas. The advisory group acknowledged the high commitment and quality of the multi-disciplinary teams involved in MERIT, as well as the willingness to adapt and streamline their research in line with the changing public health situation. Recognizing the importance and relevance of preliminary research results, the advisory group highlighted the need to translate new knowledge into operational activities in order to lead to tangible public health impacts and improved decisions.

As a result of the strategic review, consensus was reached on the importance for MERIT to continue with clarification of priority research needs in light of a changing epidemiological landscape and renewed focus on translating new knowledge into decisions and operations. Future developments of MERIT may extend beyond the limitations of current research in terms of providing an early warning system, and add considerable value to the development of an impact assessment of the conjugate vaccine and carriage studies. The MERIT Steering Committee is reviewing the structural and financial needs of the Initiative to ensure its future sustainability and to support MERIT activities in countries of the Meningitis Belt.

The collaborative partnership model that the MERIT Initiative has demonstrated over the past 5 years provides a promising, innovative framework to support public health preparedness and control strategies. MERIT's strength is that the research conducted has been driven in large part by clearly articulated public health questions. Public health decision-makers have been willing to explore unfamiliar territory and opportunities for improving well-established control strategies by leveraging new knowledge and expertise from the climate, environmental and research sectors. Equally important have been the investments made by the scientific and practice communities across various disciplines to adapt research projects in line with the evolving public health strategy across the Meningitis Belt. Not limited to the problem of meningitis epidemics in Africa, the lessons learned from the MERIT Initiative offer valuable input and new ideas for improving global public health strategies for other climate/environmentally sensitive epidemic prone diseases of international concern.

Acknowledgements The MERIT Steering Committee thanks the broad stakeholder community represented by the names of those included in the many presentations associated with MERIT Technical Meetings for their comments on this report. The individuals and research groups draw on a wide range of funding sources to support their work. The MERIT Steering Committee is grateful to all MERIT partners who have shared their experiences and insights into the MERIT Initiative to date, and to Dr. David Rogers of the Health and Climate Foundation for facilitating the partnership network and communications activities. The discussions with partners and views expressed have been a valuable part of the MERIT Situational Analysis initially presented at the World Climate Research Programme (WCRP) Open Science Conference in October 2011 and as described in this chapter. The full list of MERIT partners as well as MERIT meeting reports and related publications are available on the MERIT website at http://merit.hc-foundation.org/.

References

Agier L, Stanton M, Diggle PJ (2012) A multi-state spatiotemporal Markov model for categorized incidence of meningitis in Sub-Saharan Africa. Epidemiology and Infection 21:1–8

Agier L, Deroubaix A, Martiny N, Yaka P, Djibo A, Broutin H (2013) Seasonality of meningitis in Africa and climate forcing: aerosols stand out. J R Soc Interface 10(79):20120814

Anon (2000) The use of polysaccharide trivalent ACW vaccine for the control of epidemic meningococcal meningitis disease outbreaks in countries of the African meningitis belt. Wkly Epidemiol Rec 75:306–309

Beresniak A, Bertherat E, Perea W, Soga G, Souley R, Duponte D, Hugonnet S (2012) A Bayesian network approach to the study of historical epidemiological databases: modelling meningitis outbreaks in the Niger. Bulletin of the World Health Organization, 1/90(6):412-417A. doi:10.2471/BLT.11.086009

Besancenot JP, Boko M, Oke PC (1997) Weather conditions and cerebrospinal meningitis in Benin (Gulf of Guinea, West Africa). Eur J Epidemiol 13:807–815

Bharti N, Broutin H, Grais RF, Ferrari MJ, Djibo A, Tatem AJ, Grenfell BT (2011) Spatial dynamics of meningococcal meningitis in Niger: observed patterns in comparison with measles'. Epidemiol Infect 140(8):1356–1365, Epidemiology and Infection, 1–10

Cheesbrough JS, Morse AP, Green SDR (1995) Meningococcal meningitis and carriage in Western Zaire – a hypoendemic zone related to climate. Epidemiol Infect 114:75–92

Colombini A (2009) Cost for household and community perception of meningitis epidemics in Burkina Faso. Clin Infect Dis 49:1520–1525

Colombini A, Badolo O, Gessner BD, Jaillard P, Seini E, Da Silva A (2011) Costs and impact of meningitis epidemics on the public health system i Burkina Faso. Vaccine 29:5474–5480

Cuevas L, Hart CA, Thomson MC (2008) Climate and environmental information and the control of epidemic meningitis in Africa. In: Thomson MC, Herrera RG, Beniston M (eds) Seasonal forecasts, climatic change and human health. Springer, Dordrecht, pp 85–100

Cuevas E, Pérez C, Baldasano JM, Camino C, Alonso-Pérez S, Basart S (2011) MACC O-INT WP3.1 "Meningitis linked to mineral dust transport in the Sahel". http://sds-was.aemet.es/projects-research/MACCOINTWP3.1SeconddeliveryReport_Final.pdf

Del Corral J, Blumenthal MB, Mantilla G, Ceccato P, Connor SJ, Thomson MC (2012) Climate information for public health: the role of the IRI climate data library in an integrated knowledge system. Geosp Health 6(3):S15–S24

Dukic V, Hayden M, Adams Forgor A, Hopson T, Akweongo P, Hodgson A, Wiedinmyer C, Yoksas T, Thomson M, Trzaska S, Pandya R (2012) The role of weather in meningitis outbreaks in Navrongo, Ghana: a generalized additive modeling approach. J Agric Biol Environ Stat 17:442–460. doi:10.1007/s13253-012-0095-9

Greenwood B (1999) Meningococcal meningitis in Africa. Trans R Soc Trop Med Hyg 93:341–353

Greenwood BM, Bradley AK, Wall RA (1985) Meningococcal disease and season in sub-Saharan Africa. Lancet 2:829–830

Haustein K, Coauthors (2012) Atmospheric dust modeling from meso to global scales with the online NMMB/BSC-Dust model – part 2: experimental campaigns in Northern Africa. Atmos Chem Phys 12:2933–2958

Irving TJ, Blyuss K, Colijn C, Trotter CL (2011) Modelling meningococcal meningitis in the African meningitis belt. Epidemiol Infect, 1–9

Kelly-Hope L, Thomson MC (2008) Climate and infectious diseases. In: Thomson MC, Herrera RG, Beniston M (eds) Seasonal forecasts, climatic change and human health. Springer, Dordrecht, pp 31–70

Lapeyssonnie L (1963) La méningite cérébro-spinale en Afrique. Bull World Health Org 28:53–114

MERIT (2008) Report of the 2nd MERIT technical meeting and Ethiopian workshop, 45 pp

MERIT (2009) Report of the third MERIT technical meeting and national workshop Niger, 55 pp

MERIT (2010) Report of the 4th MERIT technical meeting, 59 pp

Molesworth A, Cuevas LE, Connor SJ, Morse AP, Thomson MC (2003) Environmental risk and meningitis epidemics in Africa. Emerg Infect Dis 9:1287–1293

Moore PS, Coauthors (1992) Detection of meningitis epidemics in Africa: a population-based analysis. Int J Epidemiol 21:155–162

Mueller J, Gessner BD (2010) A hypothetical explanatory model for meningococcal meningitis in the African meningitis belt. Int J Infect Dis 14:e553–e559

Mueller JE, Coauthors (2008) Association of respiratory tract infection symptoms and air humidity with meningococcal carriage in Burkina Faso. Trop Med Int Health 13:1543–1552

Nickovic S, Vukovic A, Vujadinovic M, Djurdjevic V, Pejanovic G (2011) Technical note: minerals in dust productive soils – impacts and global distribution. Atmos Chem Phys: Discuss 11:26009–26034

Nickovic S, Vukovic A, Vujadinovic M, Djurdjevic V, Pejanovic G (2012) Technical note: high-resolution mineralogical database of dust-productive soils for atmospheric dust modeling. Atmos Chem Phys 12:845–855

Noinaj N, Coauthors (2012) Structural basis for iron piracy by pathogenic Neisseria. Nature 483:53–58

Paireau J, Girond F, Collard J-M, Maïnassara HB, Jusot J-F (2011) Analysing spatio-temporal clustering of meningococcal meningitis outbreaks in Niger reveals opportunities for improved disease control. PLoS Negl Trop Dis 6:e1577

Pérez C, Baldasano JM, Jiménez-Guerrero P, Jorba O, Haustein K, Cuevas E, Basart S, Nickovic S (2009) Dust modelling and forecasting in the Barcelona supercomputing center: activities and developments. WMO/GEO expert meeting on an international sand and dust storm warning system. IOP Publishing. IOP Conf Ser: Earth Environ Sci 7:012013. doi:10.1088/1755-1307/7/1/012013

Pérez C, Haustein K, Janjic Z, Jorba O, Huneeus N, Baldasano JM, Black T, Basart, Nickovic S, Miller RL, Perlwitz JP, M. Schulz M, Thomson M (2011) Atmospheric dust modeling from meso to global scales with the online NMMB/BSC-Dust model: 1. Model description, annual simulations and evaluation. Atmos Chem Phys Discuss 11:13001–13027

Roberts L (2008) An ill wind, bringing meningitis. Science 320:1710–1715

Sow SO, Coauthors (2011) Immunogenicity and safety of a meningococcal a conjugate vaccine in Africans. N Engl J Med 364:2293–2304

Stanton M (2012) Chapter 5 exploring the relationship between meningitis incidence and climatic factors in Niger, West Africa'. In: Spatial and spatio-temporal statistical methods and their applications in epidemiology. School of Medicine, University of Lancaster, Lancashire, UK

Stanton M, Diggle P, Perez C, Trzaska S, Thomson M (2011) Exploring the relationship between meningitis incidence and climatic factors is Niger, West Africa. In: 17th European young statisticians meeting, Lisbon, Portugal, 5–9 Sept 2011

Sultan B (2005) Influence of climate upon the meningitis onset in West Africa. M S-Med Sci 21:470–471

Sultan B, Labadi K, Guegan JF, Janicot S (2005) Climate drives the meningitis epidemics onset in West Africa. Plos Med 2:43–49

Thomson MC, Molesworth AM, Djingarey MH, Yameogo KR, Belanger F, Cuevas LE (2006) Potential of environmental models to predict meningitis epidemics in Africa. Trop Med Int Health 11:781–788

Thomson MC, Jeanne I, Djingarey M (2009) Dust and epidemic meningitis in the Sahel: a public health and operational research perspective. IOP Conf Ser: Earth Environ Sci 7:4

Trzaska S, Perez C, Thomson M (2010) The role of intraseasonal atmospheric variability in meningitis outbreaks in the Sahel – 2010 vs. 2009 case study. In: Progress report to NOAA NA05OAR4311004 1 Nov 2009–31 Oct 2010, pp 57–58

WHO (2007) Risk of meningitis epidemic in Africa: a cause for concern. Wkly Epidemiol Rec 82:79–87

Yaka P, Sultan B, Broutin H, Janicot S, Philippon S, Fourquet N (2008) Relationships between climate and year-to-year variability in meningitis outbreaks: a case study in Burkina Faso and Niger. Int J Heal Geogr 7:13